HOMEOSTASIS AND TOXICOLOGY OF NON-ESSENTIAL METALS

This is Volume 31B in the

FISH PHYSIOLOGY series
Edited by Chris M. Wood, Anthony P. Farrell and Colin J. Brauner
Honorary Editors: William S. Hoar and David J. Randall

A complete list of books in this series appears at the end of the volume

HOMEOSTASIS AND TOXICOLOGY OF NON-ESSENTIAL METALS

Edited by

CHRIS M. WOOD
Department of Biology
McMaster University
Hamilton, Ontario
Canada

ANTHONY P. FARRELL
Department of Zoology and Faculty of Land and Food Systems
The University of British Columbia
Vancouver, British Columbia
Canada

COLIN J. BRAUNER
Department of Zoology
The University of British Columbia
Vancouver, British Columbia
Canada

ELSEVIER

AMSTERDAM • BOSTON • HEIDELBERG • LONDON • OXFORD
NEW YORK • PARIS • SAN DIEGO • SAN FRANCISCO
SINGAPORE • SYDNEY • TOKYO

Academic Press is an imprint of Elsevier

Academic Press is an imprint of Elsevier
32 Jamestown Road, London NW1 7BY, UK
225 Wyman Street, Waltham, MA 02451, USA
525 B Street, Suite 1800, San Diego, CA 92101-4495, USA

First edition 2012

Cover Image
Cover image from figure 1 in Paquin, P.R., Gorsuch, J.W., Apte, S., Batley, G.E., Bowles, K.C., Campbell, P.G.C., Delos, C.G., Di Toro, D.M., Dwyer, R.L., Galvez, F., Gensemer, R.W., Goss, G.G., Hogstrand, C., Janssen, C.R., McGeer, J.C., Naddy, R.B., Playle, R.C., Santore, R.C., Schneider, U., Stubblefield, W.A., Wood, C.M., and Wu, K.B. (2002a). The biotic ligand model: a historical overview. *Comp. Biochem. Physiol.* 133C, 3-35. Copyright Elsevier 2002.

Notice
No responsibility is assumed by the publisher for any injury and/or damage to persons or property as a matter of products liability, negligence or otherwise, or from any use or operation of any methods, products, instructions or ideas contained in the material herein. Because of rapid advances in the medical sciences, in particular, independent verification of diagnoses and drug dosages should be made

British Library Cataloguing-in-Publication Data
A catalogue record for this book is available from the British Library

Library of Congress Cataloging-in-Publication Data
A catalog record for this book is available from the Library of Congress

ISBN: 978-0-12-378634-0
ISSN: 1546-5098

For information on all Academic Press publications
visit our website at www.elsevierdirect.com

Typeset by MPS Limited, a Macmillan Company, Chennai, India
www.macmillansolutions.com

Printed and bound in United States of America
Transferred to Digital Printing, 2011

CONTENTS

1. Silver
   Chris M. Wood

2. Aluminum
   Rod W. Wilson

5. Mercury
Karen Kidd and Katharina Batchelar

6. Arsenic
Dennis O. McIntyre and Tyler K. Linton

CONTENTS OF
HOMEOSTASIS AND TOXICOLOGY OF
ESSENTIAL METALS, VOLUME 31A

3.   Zinc
Christer Hogstrand

4.   Iron
Nicolas R. Bury, David Boyle and Christopher A. Cooper

8.   Molybdenum and Chromium
Scott D. Reid

9.   Field Studies on Metal Accumulation and Effects in Fish
Patrice Couture and Greg Pyle

CONTRIBUTORS

The numbers in parentheses indicate the pages on which the authors' contributions begin.

KATHARINA BATCHELAR *(237), University of New Brunswick, Saint John, New Brunswick, Canada*

RONNY BLUST *(351), University of Antwerp, Antwerp, Belgium*

M. JASIM CHOWDHURY *(351), International Lead Zinc Research Organization, Durham, NC, USA*

CLAUDE FORTIN *(391), INRS-Eau, Terre et Environnement, Université du Québec, Québec, QC, Canada*

RICHARD R. GOULET *(391), Environmental Risk Assessment Division, Canadian Nuclear Safety Commission, Ottawa, Ontario, Canada*

KAREN KIDD *(237), University of New Brunswick, Saint John, New Brunswick, Canada*

TYLER K. LINTON *(297), Great Lakes Environmental Center, Columbus, OH, USA*

EDWARD M. MAGER *(185), University of Miami, Miami, FL, USA*

JAMES C. MCGEER *(125), Wilfrid Laurier University, Waterloo, Ontario, Canada*

DENNIS O. MCINTYRE *(297), Great Lakes Environmental Center, Columbus, OH, USA*

SOM NIYOGI *(125), University of Saskatchewan, Saskatoon, Canada*

PAUL PAQUIN *(429), HydroQual, Inc., Mahwah, NJ, USA*

AARON REDMAN *(429), HydroQual, Inc., Providence, UT, USA*

ADAM RYAN *(429), HydroQual, Inc., East Syracuse, NY, USA*

ROBERT SANTORE *(429)*, *HydroQual, Inc., East Syracuse, NY, USA*

D. SCOTT SMITH *(125)*, *Wilfrid Laurier University, Waterloo, Ontario, Canada*

DOUGLAS J. SPRY *(391)*, *National Guidelines and Standards Office, Environment Canada, Gatineau QC, Canada*

ROD W. WILSON *(67)*, *University of Exeter, Exeter, UK*

CHRIS M. WOOD *(1)*, *McMaster University, Hamilton, Ontario, Canada, and University of Miami, Miami, FL, USA*

PREFACE

We are pleased to present this two-volume book on the homeostasis and toxicology of metals to the *Fish Physiology* series, the brainchild of Bill Hoar and Dave Randall, which has become the bible of our field since its inception more than 40 years ago. Physiology and toxicology are particularly closely linked in the aquatic sciences, and all three editors are practitioners of both fields. Indeed, we prefer to work at the interface of the two fields where physiological understanding of mechanisms explains toxic response, and toxicological phenomena illuminate physiological theory. We believe the book captures this interface. We trust it will appeal to the regular readers of the *Fish Physiology* series, as well as to a much broader audience including nutritional physiologists, toxicologists, and environmental regulators.

The motivation for this two-volume book has two origins:

Firstly, there has been an explosion of new information on the molecular, cellular, and organismal handling of metals in fish in the past 15 years. While most of the research to date has focused on waterborne metals, there is a growing realization of the importance of diet-borne metals. These elements are no longer viewed by fish physiologists as evil "heavy metals" (an outdated and chemically meaningless term) that kill fish by suffocation. Rather, they are now viewed as interesting moieties that enter and leave fish by specific pathways, and which are subject to physiological regulation. These regulatory pathways may be ones dedicated for essential metal uptake (e.g. copper-specific, iron-specific, zinc-specific transporters) or ones at which metals masquerade as nutrient ions ("ionic mimicry" e.g. copper and silver mimic sodium; cobalt, zinc, lead, strontium, and cadmium mimic calcium; nickel mimics magnesium). Internally, homeostatic mechanisms include regulated storage and detoxification (e.g. metallothioneins, glutathione, granule formation) and protein vehicles for transporting metals around the body in the circulation (e.g. ceruloplasmin, transferrin).

Molecular and genomic techniques have allowed precise characterization of these pathways, and how they respond to environmental challenges such as metal loading and deficiency. Bioaccumulation of metals is now widely studied in both the laboratory and the field, but interpretation of the data remains controversial. New techniques such as subcellular fractionation and modeling of metal-sensitive and metal-insensitive pools are providing clarification and new pathways for further research.

Secondly, this same period has seen a progressively increasing concern about the potential toxicity of metals in the aquatic environment. At present, the European Union, the United States, Canada, Australia/New Zealand, China, several Latin American countries, and diverse other jurisdictions around the world are all in the process of revising their ambient water quality criteria for metals. Coupled to this has been a sharp growth in toxicological research on metal effects on fish. Much of this research has focused on the physiological mechanisms of uptake, storage, and toxicity, and from this various modeling approaches have evolved which have proven very useful in the regulatory arena. For example, tissue residue models, to relate internal metal burdens to toxic effects, and biotic ligand models (BLMs), to relate gill metal burdens in different water qualities to toxic effects, are two physiological models that are now being considered by regulatory authorities in setting environmental criteria for metals (e.g. residue models for selenium and mercury regulations; BLMs for copper, zinc, silver, cadmium, and nickel criteria).

This work was conceived as a single book to cover all the metals for which a sizeable database exists. Its division into two published volumes (Vol. 31A dealing with essential metals, Vol. 31B dealing with non-essential metals) was solely for practical reasons of size, stemming from each metal being dealt with in a uniform and comprehensive manner. Regardless, the two volumes are fully integrated by cross-referencing between the various chapters, and they share a common index.

Three chapters in particular tie the package together with real-world scenarios and applications: Chapter 1 of Vol. 31A on *Basic Principles* serves as an Introduction to the whole book, while Chapter 9 of Vol. 31A on *Field Studies on Metal Accumulation and Effects in Fish* and Chapter 9 of Vol. 31B on *Modeling the Physiology and Toxicology of Metals* serve as integrative summaries dealing with both essential and non-essential metals.

The other 15 chapters each deal with specific metals, and authors were strongly urged to adopt a unified format which is explained in Chapter 1 of Vol. 31A. This format includes consideration of the following topics:

1. Chemical Speciation in Freshwater and Seawater
2. Sources of Metals and Economic Importance

3. Environmental Situations of Concern
4. Acute and Chronic Ambient Water Quality Criteria
5. Mechanisms of Acute and Chronic Toxicity
6. Evidence of Essentiality or Non-Essentiality of Metals
7. Potential for Bioconcentration and/or Biomagnification of Metals
8. Characterization of Uptake Routes
9. Characterization of Internal Handling
10. Characterization of Excretion Routes
11. Behavioral Effects of Metals
12. Molecular Characterization of Metal Transporters, Storage Proteins, and Chaperones
13. Genomic and Proteomic Studies
14. Interactions with Other Metals
15. Knowledge Gaps and Future Directions.

As a result, the book should serve as a one-stop source for a synthesis of current knowledge on both the physiology and toxicology of a specific metal, and selective readers should be able to quickly find the specific information they require. Furthermore, the chapters should help guide future research by pointing out significant data gaps for particular metals.

This book would not have been possible without a vast contribution of time and effort from many people. First and foremost, our gratitude to the authors of the chapters, who represent some of the leading experts in the world in metals physiology and toxicology. Not only did these researchers sacrifice nights, weekends, and holidays to craft their chapters, they also constructively reviewed many of the other chapters. In addition, more than 20 anonymous external peer-reviewers contributed greatly to the quality of the chapters. Pat Gonzalez, Kristi Gomez, Caroline Jones, and Charlotte Pover at Elsevier provided invaluable guidance and kept the project on track. Finally, special thanks are due to Sunita Nadella at McMaster University, who proofread and corrected every chapter before submission to Elsevier.

This book is dedicated to the memory of Rick Playle, a good friend and a superb scientist who pioneered physiological understanding and modeling of the effects of metals on fish.

Chris M. Wood
Anthony P. Farrell
Colin J. Brauner
April 2011

1

SILVER

CHRIS M. WOOD

Homeostasis and Toxicology of Non-Essential Metals: Volume 31B
FISH PHYSIOLOGY

In pristine natural waters, silver (Ag) occurs at low ng L^{-1} levels, rising to 1000-fold at highly contaminated sites. The free ion Ag^+ appears to be the sole cause of acute toxicity in freshwater, being among the most toxic of the metals in this regard, but probably does not occur to any significant extent in natural waters, where speciation is dominated by complexation to sulfide, dissolved organic carbon, chloride, and particles. Nevertheless, Ag^+ is the form used in regulatory tests, and on which environmental water quality criteria are based. Such criteria are often related to water hardness, but this probably reflects a misinterpretation of original test data, since other water quality parameters are far more protective than calcium and magnesium. The biotic ligand model approach holds promise for improving water quality criteria for Ag. In freshwater fish, waterborne Ag^+ poisons two key enzymes of ion transport in the gills (Na^+/K^+-ATPase and carbonic anhydrase), and death results from ionoregulatory failure. Dietborne toxicity is negligible. Silver appears to be taken up by sodium and copper transport pathways in the gills, as well as by diffusion of neutral complexes. Chronic toxicity occurs at much lower Ag concentrations, and may again involve ionoregulatory disturbance as well as other mechanisms. Acclimation may occur. In saltwater, Ag speciation is dominated by salinity-dependent chloride complexation, and Ag is far less toxic on an acute basis than in freshwater. Since seawater fish drink the medium, both gills and gut are targets of acute toxicity and potential routes of Ag uptake, but mechanisms remain unclear. Marine elasmobranchs are far more sensitive, and take up much more Ag, than marine teleosts. Bioconcentration of Ag clearly occurs in both freshwater and saltwater fish, but the bioconcentration factor approach is not a useful regulatory tool for Ag. Silver accumulates preferentially in the liver, which serves as a scavenging organ. Silver is a powerful inducer of metallothionein synthesis for detoxification. Biological half-lives are long and excretory mechanisms remain poorly characterized. Trophic transfer efficiency is low and biomagnification of Ag does not occur.

1. INTRODUCTION

The physiology and toxicology of silver (Ag) in aquatic animals were only sparsely studied before about 1990. However, over the past two decades, there has been an immense amount of research activity, much of it sponsored by the photographic industry working in cooperation with government agencies. This occurred because photoprocessing effluent disposal, an important source of Ag in aquatic environments, came under increasing regulatory pressures in the 1990s. Various aspects of some of this work have been reviewed (Eisler, 1996; Hogstrand and Wood, 1998; Wood et al., 1999; Ratte, 1999). Particularly useful sources are the book edited by Andren and Bober (2002), which provides an overview of the conclusions of the *Argentum International Conference* series (1993–2000), and a compendium edited by Gorsuch et al. (2003), republishing all relevant papers appearing on Ag from 1982 to 2003 in the journal *Environmental Toxicology and Chemistry*. However, important additional progress has been made in recent years, making the present review timely.

2. SOURCES OF SILVER AND OCCURRENCE IN NATURAL WATERS

Silver (atomic number 47, atomic weight 107.86) is a naturally occurring rare element (67th most abundant in the Earth's crust) and is a class B soft metal. Natural concentrations average about 100 μg kg^{-1} (~1 μmol kg^{-1}) in typical rocks and soil. Peru, Mexico, China, Australia, and Chile are the major producers of Ag by mining, and the USA, Japan, European Union (EU) countries, and India are the major consumers. About 900 million troy ounces (28 million kg) of Ag is "manufactured" each year, which is about a 30% excess over the amount that is mined as there is significant recycling of this precious metal. At current world prices, this translates to a market value of about US $20 billion annually. Formerly, Ag halides in photographic film were the major use and disposition, but with the advent of digital photography, this has dropped steadily in the past decade to about 13% of the market. Industrial applications have been steadily increasing owing to Ag's excellent conductive properties, and now account for about 54% of the market, whereas jewelry (18%), coinage (8%), and tableware (7%) are the other important uses (Silver Institute, 2010). The discharge of Ag to the environment is largely as solid waste (66%) which, owing to its low solubility, has minimal environmental impact (Purcell and Peters, 1998). The 34% discharged in liquid form is of much greater concern; China, India, Indonesia, the USA, and the EU countries are the largest sources of waterborne discharge (Eckelman and Graedel, 2007).

Silver is only sparingly soluble, such that concentrations in pristine natural waters are 0.1–5 ng L^{-1} (1–50 pmol L^{-1}) and not detectable with the technology used in most routine analytical laboratories (Adams and Kramer, 1999a, b; Kramer et al., 2002). The reader should beware of reports of naturally occurring Ag levels in the high ng L^{-1} to low μg L^{-1} range that were obtained before the widespread adoption of clean sampling technology towards the end of the twentieth century (see Wood, Chapter 1, Vol. 31A;

Benoit, 1994). For example, statements in water quality criteria documents (Environment Canada, 1980; US EPA, 1980; CCME, 2007) that natural background levels of Ag in surface waters are in the 0.1–$0.5\,\mu g\,L^{-1}$ (1–$5\,nmol\,L^{-1}$) range are undoubtedly too high by several orders of magnitude. Kramer et al. (2002) estimated that Ag concentrations even in contaminated natural waters rarely exceed a few hundred $ng\,L^{-1}$ (i.e. 1–$3\,nmol\,L^{-1}$). However, occasionally, Ag concentrations may be elevated above this value (indeed occasionally into the $\mu g\,L^{-1}$) range on a site-specific basis by inputs from sources detailed in Section 5. If highly concentrated industrial and domestic effluents are processed through sewage treatment plants, Ag concentrations are generally greatly reduced (e.g. by 75–94%) (Lytle, 1984; Shafer et al., 1998) to no more than $10\,\mu g\,L^{-1}$ ($100\,nmol\,L^{-1}$) in the discharged water (Kramer et al., 2002), and then diluted about 100-fold in the mixing zone. More importantly, they are changed in speciation such that relatively inert Ag sulfide complexes predominate (Purcell and Peters, 1998). The recent extensive use of Ag in nanoparticles may also result in locally high concentrations of Ag (see Wood, Chapter 1, Vol. 31A), but nanoparticles will not be considered in the present review.

Research in Ag toxicology and physiology has been greatly facilitated by a commercially available radioisotope, [110m]Ag, which is a strong beta and gamma emitter (peak at 658 keV). Hansen et al. (2002) discovered that there are important methodological precautions that must be applied in the use of this radioisotope to avoid artifacts in interpretation. [110m]Ag is made by neutron activation of natural Ag, which is 52% [107]Ag and 48% [109]Ag; transmutation events during this irradiation result in trace amounts of [109]Cd (e.g. 0.16% of total beta radioactivity in the stock solution studied by Hansen et al., 2002). [109]Cd is a strong beta and weak gamma emitter (peak at 88 keV). This may be very important because of the affinity of many biological ligands and processes for Cd. For example, Hansen et al. (2002) were able to detect the problem in a study of [110m]Ag uptake kinetics, because trout gills took up substantial [109]Cd radioactivity from this contaminant whereas eel gills did not. If this problem had not been detected by comparative beta versus gamma counting, the uptake and depuration patterns of Ag would have been completely misinterpreted. While beta counting is more efficient than gamma counting for [110m]Ag, it is problematic as it is virtually impossible to separate out the potential [109]Cd component by this technique alone. The lesson here is that beta counting of [110m]Ag in biological samples should always be associated with gamma counting to check for [109]Cd contamination, using an energy window well above 88 keV. In practice, gamma counting alone using a high-energy window is an easier and more reliable solution now adopted by most workers in this area.

3. SPECIATION IN FRESHWATER

In pristine natural freshwaters, dissolved Ag concentrations range from 0.5 to 5 ng L^{-1} (5–50 pmol L^{-1}), rising up to 1000-fold at highly contaminated sites (Section 2). While these are "dissolved" concentrations by operational definition (i.e. not retained by a 0.45 μm filter; Wood, Chapter 1, Vol. 31A), more precise characterization indicates that much of this is in the fraction considered as "colloidal" (i.e. > 10 kDa) by geochemists (Adams and Kramer, 1999a, b). Contrary to popular belief, modern analytical techniques have shown that metastable reduced sulfur is present at nanomolar levels in most natural freshwaters even under oxygenated conditions (reviewed by Kramer et al., 2002). Given the extremely high affinity of reduced sulfur for Ag, most Ag likely occurs as Ag sulfide complexes in nature. Although these complexes are thought to be generally smaller than 0.45 μm, they are retained by adsorption on 0.45 μm filter membranes, thereby further complicating the operational definition of the "dissolved fraction" (Bowles et al., 2002; Bianchini and Bowles, 2002).

In contrast to many other metals, the speciation of Ag in freshwater is not greatly dependent on pH. In the absence of dissolved organic carbon (DOC), chloride, sulfide, thiosulfate, or particles, chemical speciation programs indicate that in simplified freshwaters, most Ag, if added in the free ion form (Ag$^+$), will stay in that form. This is of major concern to both regulatory agencies and regulated organizations, because this is the form (generally the AgNO$_3$ salt) usually mandated for toxicity tests, along with the requirement that simplified or synthetic "laboratory waters" be used for the exposures. Such freshwaters have usually been prepared from base water that has been chlorinated/dechlorinated, distilled, and/or treated by reverse osmosis. As a result, these waters are extremely low in particles, DOC, and sulfides, and generally quite low in chloride, moities that normally complex Ag$^+$. The simplified water chemistry of these laboratory-based test media results in an artificial condition that overemphasizes the toxicity of Ag, because of this speciation behavior of Ag$^+$. It is not surprising that Ag is highly toxic under these conditions, because current theory suggests that only Ag$^+$ causes acute toxicity (see Sections 7.1, 7.2). The result is often misplaced concern about Ag's toxicity in natural waters.

In natural freshwaters, Ag$^+$ readily complexes with relatively high affinity to the appreciable quantities of sulfide, DOC, chloride, and particles that are present, thereby rendering it largely unavailable to bind to "toxic sites" on the gills of aquatic organisms. The same will occur with thiosulfate, a predominant anion in photoprocessing effluent. Approximate log K values are 13.5 for sulfide, 9.0 for DOC, 8.8–14.2 for thiosulfate (various complexes), and 3.3–5.5 for chloride, but 10.5 for the precipitated form cerargyrite (Hogstrand and Wood, 1998), while the partition coefficient of dissolved Ag onto particles can exceed 100,000 (Wen et al., 1997). This means that when Ag is added to natural freshwaters in laboratory tests, it will first bind to the high log K anions and particles, and then the "left-over" Ag will form the neutral $AgCl_0$ complex, some of which may precipitate as cerargyrite, especially at higher Cl^- levels. Cerargyrite formation is unlikely under field conditions, because Ag concentrations are rarely high enough. Thus, even when added in a highly dissociatable form as $AgNO_3$, Ag is much less toxic in most natural freshwater than in "laboratory waters" because the concentration of the free Ag^+ ion is extremely low. The water effect ratio is a regulatory approach designed to take this into account when applying regulations on a site-specific basis (Paquin et al., 2000); the biotic ligand model (BLM) may serve a similar purpose (see Wood, Chapter 1, Vol. 31A, Paquin et al., Chapter 9, and Section 7.4).

It is worth pointing out that sorption occurs not only onto particles but also onto surfaces such as glass, plastic, ceramics, and food added to test chambers This explains why when Ag salts are added in toxicity tests, especially to natural waters, much of the added Ag may "disappear" from solution. Nominal dissolved Ag concentrations reported in many papers should not be trusted; they are undoubtedly overestimates. Furthermore, the loss is not immediate, but rather progressive over time (Bowles et al., 2002), and the toxicity-reducing effects of ameliorating agents such as food, DOC, and metastable sulfide develop over the first few hours after addition (Bianchini et al., 2002a; Glover et al., 2005; Kolts et al., 2006, 2008). Equilibration of exposure solutions for 24 h prior to tests is advisable (Wood et al., 2002b).

4. SPECIATION IN SEAWATER

Dissolved Ag concentrations in seawater tend to be even lower than those in freshwater, ranging from 0.1–2.0 ng L^{-1} (1–20 pmol L^{-1}) in pristine offshore waters up to 30 ng L^{-1} (300 pmol L^{-1}) in estuaries close to point-source discharges (Kramer et al., 2002). However, Ag levels in the low $\mu g\ L^{-1}$ range have been reported in bays and in intertidal areas close to sewage outfalls and industrial sites (Eisler, 1996).

Complex changes in Ag distribution occur in brackish and estuarine waters. As freshwater mixes with seawater, some DOC molecules lose solubility and precipitate, and particles tend to settle out, which may remove some Ag from the water column. The fraction in the colloidal phase (i.e. > 10 kDa, $< 0.45\ \mu m$) tends to decrease also (Wen et al., 1997), but a substantial fraction remains. Sulfide complexes likely persist in solution but, in general, con-centrations reported in saltwater appear to be lower than those in freshwater (Bianchini and Bowles, 2002); the importance of metastable reduced sulfur in Ag speciation at different salinities has been little studied. Similarly, the role of DOC in Ag speciation in seawater is poorly understood and sometimes dismissed (Miller and Bruland, 1995; Ward and Kramer, 2002), though recent geochemical (Buck et al., 2007) and toxicological results (Arnold, 2005) with Cu, a somewhat similar metal (Grosell, Chapter 2, Vol. 31A), suggest that DOC complexation may be very important in the estuarine and marine environment.

Most focus has been on Ag complexation with chloride, which changes progressively with increasing salinity, as illustrated in Fig. 1.1. This figure, adapted from Wood et al. (2004), uses the speciation scheme recommended by Ward and Kramer (2002). Free Ag^+ ion concentrations are vanishingly low, being replaced by chloride complexes above about 0.1–0.2% seawater (which is really freshwater). $AgCl_0$, the principal chloride complex in freshwater, is then replaced by $AgCl_2^-$, which starts to predominate above about 2% seawater, and then by $AgCl_3^{2-}$, which starts to predominate above about 75% seawater (Fig. 1.1). Other geochemical programs include another species ($AgCl_4^{3-}$), but according to Ward and Kramer (2002) these schemes

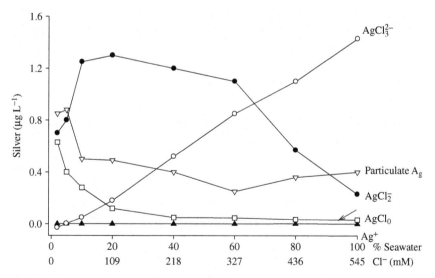

Fig. 1.1. Speciation of 2.18 µg L^{-1} of total Ag (added as $AgNO_3$) as a function of salinity and Cl^- concentration, calculated using the mean measured particulate fraction of Ag (0.45 µm filtration) and partitioning the dissolved component into its various species. Data from Wood et al. (2004) and the computational framework of Ward and Kramer (2002).

are in error because this species does not occur at normal salinities, and furthermore because they do not incorporate the proper corrections for activity coefficients. Direct measurements with an Ag^+ ion-specific electrode during toxicity testing in 60% seawater revealed that even at 410 μg L^{-1} (4 μmol L^{-1}) dissolved Ag, a concentration that would never occur in nature, ionic Ag^+ was only 25 ng L^{-1} (0.25 nmol L^{-1}) (Ward and Kramer, 2002). The slow precipitation of cerargyrite is a common complication during toxicity testing in seawater, often occurring at dissolved Ag concentrations greater than 500 μg L^{-1} (Wood et al., 1999).

5. ENVIRONMENTAL SITUATIONS OF CONCERN

The author is aware of no documented situation where a "fish kill" or a loss of natural fish populations has been attributed specifically to Ag. Historically, most concern has focused on photoprocessing effluent, which can contain Ag in the mg L^{-1} range, mainly complexed with thiosulfate (Cooley et al., 1988). As noted in Section 2, such concentrations are greatly reduced and altered in speciation to Ag sulfide after processing through sewage treatment plants, but there is concern about rural areas and underdeveloped countries where effluents may escape treatment. Untreated discharges from jewelry, coinage, battery, mirror, and tableware manufacturers, even household effluents (e.g. "Ag ion" dishwashers) are of similar concern. Concerns that cloud-seeding ("rain-making") using Ag iodide could cause ecological damage in surface waters have been largely discounted (Cooper and Jolly, 1970). Acidic drainage from abandoned mines into streams is a situation that occasionally raises dissolved Ag levels into the μg L^{-1} range (surveyed by Kramer et al., 1999), but such scenarios are almost invariably confounded by low pH and elevated concentrations of other metals, making it very difficult to factor out effects due to Ag alone. Human drinking water standards for Ag are typically in the range of 10–200 μg L^{-1} (0.1–2 μmol L^{-1}), and if such levels were achieved by addition of $AgNO_3$, as is sometimes done for disinfection in emergency situations, the discharge of this untreated water could be hazardous to aquatic life.

6. ACUTE AND CHRONIC AMBIENT WATER QUALITY CRITERIA IN FRESHWATER AND SEAWATER

In general, ambient water quality criteria (AWQC) for Ag are old, derived in the 1980s (Purcell and Peters, 1999). Although there have been many efforts to revise them, particularly with the BLM (Paquin et al., 1999, 2000; McGeer et al., 2000; Niyogi and Wood, 2004; Section 7.4), these have not moved into practice, and the original numbers persist (Table 1.1). Some

Table 1.1
Acute and chronic water quality criteria for silver, for the protection of aquatic life, in various jurisdictions

FRESH WATER	Reference	Acute (μg L^{-1})	Chronic (μg L^{-1})	Notes
U.S.	U.S. EPA (1980, 1986)	0.22	–	hardness 20 mg L^{-1}, dissolved
		1.05	–	hardness 50 mg L^{-1}, dissolved
		11.37	–	hardness 200 mg L^{-1}, dissolved
	U.S. EPA (1980, 1986)	–	0.12	Proposed, not implemented, total
Canada	CCME (2007)	–	0.10	total
British Columbia, Canada	BC (1995)	0.1	0.05	Hardness < 100 mg L^{-1}, total
		3.0	1.50	Hardness > 100 mg L^{-1}, total
Australia-NZ	ANZECC/ARMCANZ (2000)		0.02–0.20	tiered approach, total
			0.05	trigger value
Brazil	CONAMA (2005)	0–10	–	tiered approach, total
The Netherlands	RIVM (1999)	–	0.082	total above background
SALT WATERS				
U.S.	U.S. EPA (1980)	1.92	–	dissolved
Canada	CCME (2007)	–	0.10	total
British Columbia, Canada	BC (1995)	3.00	1.50	total
Australia-NZ	ANZECC/ARMCANZ (2000)	–	0.80–2.60	tiered approach, total
			1.47	trigger value
Brazil	CONAMA (2005)	0–5.0	–	tiered approach, total
The Netherlands	RIVM (1999)	–	1.20	total, above background
U.K.	European Commission (1979) Shellfish Water Directive 79/923/EEC	10.0	–	maximum allowable total

of the AWQC were clearly derived by the application of safety factors, because the technology was not generally available in regulatory laboratories to measure Ag at levels lower than $1\ \mu g\ L^{-1}$ at the time they were derived. Some maximum allowable levels are below values that were believed to be natural background concentrations before the adoption of clean technology (e.g. CCME, 2007).

Virtually all jurisdictions have criteria for allowable levels of Ag in drinking water (generally in the $10\text{--}200\ \mu g\ L^{-1}$ range), but these are so high as to be meaningless for the protection of aquatic life. Notably, some prominent jurisdictions (e.g. India, South Africa, Japan, China, the EU) appear to have no AWQC for Ag for aquatic life protection.

The United States Environmental Protection Agency regulations (USEPA, 1980) for freshwater have only an acute value; a chronic level was suggested but not implemented (Table 1.1). The allowable acute value is based on a hardness equation:

$$\text{Maximum allowable total Ag } (\mu g\ L^{-1}) = e^{(1.72[\ln\text{hardness}] - 6.52)} \qquad (1)$$

The outcome is then multiplied by 0.85 to yield values of allowable dissolved (rather than total) Ag (USEPA, 1992). As outlined in Section 7.1.1, the rationale behind the use of hardness in this instance is questionable; at high hardness, the equation yields allowable levels greater than $10\ \mu g\ L^{-1}$. In most other jurisdictions, the allowable levels are lower, with a tendency for lower permitted concentrations in freshwater than in saltwater. Some jurisdictions (e.g. Australia/New Zealand, Brazil) use a tiered approach, with the level of protection adjusted for the intended site-specific use, thereby yielding the range of values reported in Table 1.1. There appear to be no tissue residue criteria for Ag.

7. WATERBORNE SILVER TOXICITY IN FRESHWATER

7.1. Acute Toxicity

There is now a comprehensive literature on fish deaths occurring within a few days of exposure to waterborne Ag, as for example during a 96 h or 168 h toxicity test. These data indicate that water chemistry, and therefore Ag speciation, are critically important.

7.1.1. INFLUENCE OF WATER CHEMISTRY ON ACUTE TOXICITY

In freshwater, only the free ion Ag^+ appears to exert acute toxicity and indeed in this form, Ag appears to be the most toxic of the metals when

$AgNO_3$ is tested in simplified "laboratory waters" which lack significant amounts of complexing and competing agents (Jones, 1939). The 96 h LC50 values are typically in the range of 2–30 μg L^{-1} (20–300 nmol L^{-1}) for many fish species, and certain invertebrates such as cladocerans (e.g. daphnia) are about 10-fold more sensitive than fish (see summary tables in Ratte, 1999, and Wood et al., 2002b). Strong complexing agents with high log K values such as sulfide, thiosulfate, and DOC offer almost complete protection when present in excess of the total Ag concentration (Terhaar et al., 1972; LeBlanc et al., 1984; Hogstrand et al., 1996; Erickson et al., 1998; Bury et al., 1999a; Bianchini et al., 2002a; VanGenderen et al., 2003), because there is virtually no free Ag$^+$ ion present in solution (see Section 3). Different sources of DOC differ in their protective abilities, but VanGenderen et al. (2003) were not able to relate this variability to differences in the physicochemical or spectroscopic properties of different DOC types. Na$^+$ appears to exert almost no protective influence, while higher alkalinity and higher pH have a modest protective influence against acute toxicity (Erickson et al., 1998).

Hardness (Ca^{2+} + Mg^{2+}) has only a weak protective effect (Davies et al., 1978; Erickson et al., 1998; Bury et al., 1999a), which appears to be mainly due to the Ca^{2+} rather than the Mg^{2+} component (Schwartz and Playle, 2001). This is surprising because hardness is used as the basis for environmental regulations in some jurisdictions, most notably the US EPA (1980) [see Eq. (1) in Section 6]. Hogstrand et al. (1996) reanalyzed some of the key input data used in deriving this relationship (Lemke, 1981) and concluded that the discrepancy was the result of mistaken identity – the real protective agent in the original toxicity tests was probably Cl$^-$ rather than Ca^{2+} (Fig. 1.2), as well as other factors such as pH and HCO$_3^-$ which are likely to have been associated with increased hardness in natural waters. Subsequent tests confirmed that Cl$^-$ was far more protective than Ca^{2+} against acute toxicity to rainbow trout (Hogstrand et al., 1996; Galvez and Wood, 1997; Bury et al., 1999a). Indeed, whereas the 7 day LC50 for AgNO$_3$ was only 9.1 μg L^{-1}, the comparable value for AgCl$_0$ in the same test water was greater than 100,000 μg L^{-1}, comparable to that (137,000 μg L^{-1}) when Ag was completely bound up as Ag(S$_2$O$_3$)$_n^-$ by a strong complexing ligand (thiosulfate) (Hogstrand et al., 1996). LeBlanc et al. (1984) reported very similar protection for fathead minnows: 96 h LC50 values greater than 4600 μg L^{-1} for AgCl$_0$, and greater than 280,000 μg L^{-1} for Ag(S$_2$O$_3$)$_n$.

Curiously, the protective effect of Cl$^-$ against Ag$^+$ toxicity appears to vary greatly among species, from highly protective in salmonids to almost completely ineffective in some studies on fathead minnows, eels, zebrafish, and killifish (Bury et al., 1999a; Karen et al., 1999; Grosell et al., 2000; Bielmyer et al., 2008). This is particularly puzzling, because: (1) LeBlanc et al. (1984) reported that Cl$^-$ was highly protective in fathead minnows, but

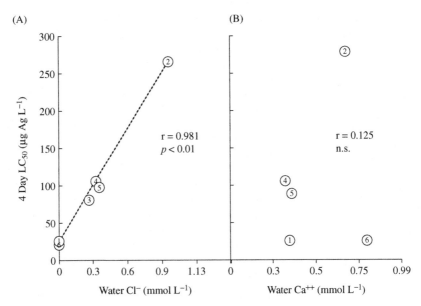

Fig. 1.2. Reanalysis of the data of Lemke (1981) on the acute toxicity (96 h LC50 values) of Ag (as $AgNO_3$) to juvenile rainbow trout in various freshwaters showing (A) a strong correlation with water $[Cl^-]$ and (B) a lack of correlation with water $[Ca^{2+}]$ in the same tests. Numbers refer to coded laboratories in the original report. The data have been recalculated from the analysis of Hogstrand et al. (1996), and suggest that water $[Cl^-]$ rather than hardness was the real protective agent in these tests.

set up the exposure in such a way that substantial cerargyrite precipitation may have occurred (Section 3); (2) Cl^- was also highly protective against chronic toxicity in early life stages of fathead minnows (Naddy et al., 2007); and (3) the presence of excess Cl^- does not inhibit gill binding or branchial uptake of Ag into the body, even in salmonids (McGeer and Wood, 1998; Bury et al., 1999b; Wood et al., 2002a; Hogstrand et al., 2003). One can only speculate that the discrepancy is related to differences in ion transport mechanisms, notably that eel and killifish lack active Cl^- uptake mechanisms at the gills in freshwater, whereas Cl^- uptake in adult fathead minnow appears to be insensitive to Ag (Bielmyer et al., 2008), and/or to differences in the affinity (log K) values of target sites. Given the relatively weak log K values (3.3–5.5) for silver chloride complexes in freshwater, it is possible that the affinity of specific target sites (e.g. sites on Na^+/K^+-ATPase and carbonic anhydrase in mitochondria-rich cells, as elaborated subsequently) may be strong enough to repartition Ag^+ from $AgCl_0$ in most species, but not in salmonids. Another idea is that the replacement of Ag^+ with $AgCl_0$ means that most Ag enters through a different route, thereby bypassing the target sites in the mitochondria-rich cells (Bury and Hogstrand, 2002). More work on this interesting difference is required, because it has regulatory significance in addition to mechanistic importance (McGeer et al., 2000; Paquin et al., 1999, 2002a).

7.1.2. MECHANISMS OF ACUTE TOXICITY

Initially, Ag was thought to be a respiratory toxicant, killing fish by suffocation due to branchial mucification, edema, and generalized gill damage (Jones, 1939; Cooper and Jolly, 1970; Coleman and Cearley, 1974). However, this is now known to be mainly restricted to "industrial" concentrations of no environmental relevance. Even after several days' exposure to concentrations close to the LC50, there is negligible disturbance of blood gases or elevation of internal lactate levels (Wood et al., 1996a). However, Janes and Playle (2000) did notice subtle evidence of respiratory disturbance in addition to marked ionoregulatory disturbance in rainbow trout exposed to Ag (11 μg L^{-1} as AgNO$_3$) in very soft water. The hyper-ventilation that has been observed in these studies is probably mainly due to metabolic acidosis, the cause of which will be discussed subsequently.

Freshwater fish are hypertonic to the dilute medium, and produce a copious urine to excrete the water which is continually entering by osmosis (Evans et al., 2005). They must continually absorb Na$^+$, Cl$^-$, and other ions by active transport at the gills to replace diffusive losses across the branchial and body surfaces, as well as ions lost in the urine. In most freshwater situations, Ag$^+$ is a highly potent ionoregulatory toxicant to the gill with resulting internal effects in the fish reminiscent of those seen during exposure to low pH (Milligan and Wood, 1982; Wood, 2001) and Cu (Wilson and Taylor, 1993; see Grosell, Chapter 2, Vol. 31A), i.e. progressive decline in blood Na$^+$ and Cl$^-$ levels, fluid volume disturbance, and circulatory failure (Wood et al., 1996a; Webb and Wood, 1998). Because ions are lost from the extracellular compartment more rapidly than from the intracellular compartment, there is a compensatory fluid shift into tissues and blood volume decreases. The red blood cells do not swell. Hematocrit, hemoglobin, and plasma protein concentrations rise. Blood viscosity and blood pressure both increase, compounded by a mobilization of stress hormones (catecholamines, cortisol) which accelerate heart rate, cause discharge of erythrocytes from the spleen into the circulation, and promote systemic vasoconstriction. Blood glucose rises markedly, indicative of a generalized stress response. Eventually, the fish dies of hypovolemic, hypertensive cardiovascular failure, but the proximate cause of the syndrome is the disturbance of Na$^+$ and Cl$^-$ regulation. Death occurs once plasma [Na$^+$] and [Cl$^-$] fall by about 30% (Wood et al., 1996a).

In general, under control conditions, the smaller the organisms, the greater the mass-specific surface areas of the gills for ion leakage, and therefore the higher the mass-specific Na^+ and Cl^- uptake rates that are needed to balance efflux rates (Bianchini et al., 2002b). Inhibition of active influx therefore has more serious acute consequences in small organisms as body pools of Na^+ and Cl^- run down more rapidly, explaining the general inverse relationship between acute toxicity and body mass both within species (Klaine et al., 1996; Bielmyer et al., 2007) and among different species (Bianchini et al., 2002b). Paquin et al. (2002b) have developed a physiologically based model for survival time in organisms exposed to ionoregulatory toxicants, and have calibrated it for Na^+ losses and exposure durations in organisms exposed to Ag^+ [see ion balance model (IBM) in Paquin et al., Chapter 9, and Section 7.4].

The mechanism of action of Ag^+ appears to be far more specific than that of low pH (Wood, 2001) or Cu (Grosell, Chapter 2, Vol. 31A): an extremely potent inhibition of the active uptake of both Na^+ and Cl^-, with only minor stimulation of the diffusive efflux components (Wood et al., 1996a; Morgan et al., 1997; Webb and Wood, 1998) (Fig. 1.3). At least for Na^+ uptake of trout in vivo, measured after 48 h exposure to a sublethal concentration (2 µg L^{-1}) of $AgNO_3$, the inhibition is non-competitive as demonstrated by a reduction in maximum transport rate (J_{max}) with unchanged affinity (K_m) (Fig. 1.4A) (Morgan et al., 1997). Efflux rates generally decline as the internal gradients run down (Fig. 1.3A) (Morgan et al., 1997; Webb and Wood, 1998). The latter is a clear difference from the actions of both low pH and Cu which greatly stimulate Na^+ and Cl^- effluxes, in addition to inhibiting influxes. However, this is in accord with the very limited protective effect of water Ca^{2+} and the rather weak affinity of the Ag^+-binding ligands on the gills for Ca^{2+} (Section 7.1.1). These observations all suggest that Ag^+ does not attack the Ca^{2+}-binding sites that stabilize branchial paracellular permeability. Notably, active Ca^{2+} uptake at the gills is also not affected (Wood et al., 1996a).

The blockade of active Na^+ and Cl^- uptake was originally explained by a potent inhibition of basolateral membrane Na^+/K^+-ATPase activity in the ion-transporting cells (Morgan et al., 1997; McGeer and Wood, 1998; Bury et al., 1999b; McGeer et al., 2000), related to the strong affinity of Ag^+ for sulfhydryl groups (Nechay and Saunders, 1984; Hussain et al., 1994, 1995). When assessed in vitro (Fig. 1.4B), this inhibition appears to be non-competitive with respect to Na^+, in accord with in vivo observations (Morgan et al., 1997) (Fig. 1.4A), but competitive with respect to Mg^{2+}, suggesting that the mechanism of action is blockade of access of Mg^{2+} to the Mg^{2+} activation site on the intracellular domain of the α-subunit (Ferguson et al., 1996). Each Na^+/K^+-ATPase molecule appears to bind two atoms of

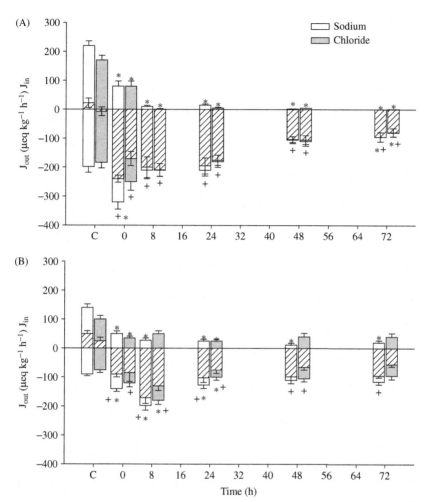

Fig. 1.3. Effects of waterborne exposure to (A) 10 μg L^{-1} Ag or (B) 2 μg L^{-1} Ag (as AgNO$_3$) on the unidirectional fluxes of Na$^+$ (open bars) and Cl$^-$ (shaded bars) in adult rainbow trout in freshwater. Means ± 1 SEM. Positive values represent movement into the fish (J_{in}), negative values represent movement out of the fish (J_{out}), and hatched bars represent the net movement of the ion (J_{net}). *Significant difference ($p < 0.05$) from the pre-exposure control (C) for J_{in} or J_{out}; +significant difference from the pre-exposure control for J_{net}. Data from Morgan et al. (1997).

Ag$^+$ (Hussain et al., 1994). Na$^+$/K$^+$-ATPase is generally thought to be the key enzyme energizing ion transport in the ionocytes of freshwater fish (Evans et al., 2005). However, in some freshwater ion transport models, apical membrane V-type H$^+$-ATPase also plays an important role, by

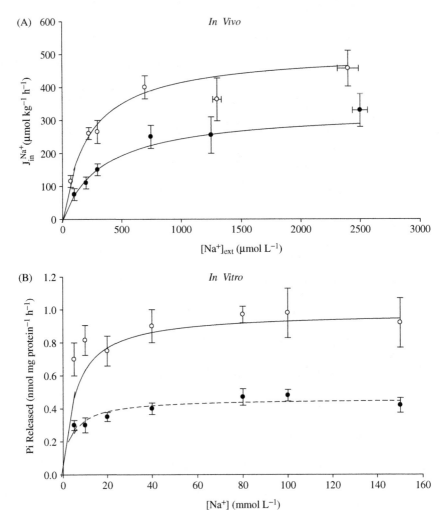

Fig. 1.4. (A) Effects of 48 h waterborne exposure to $2 \, \mu g \, L^{-1}$ Ag (as $AgNO_3$) on the concentration kinetics of unidirectional Na^+ influx ($J_{in}^{Na^+}$) in adult rainbow trout in freshwater, measured *in vivo*. Means $\pm$ 1 SEM. Open circles represent fish under control conditions, closed circles represent Ag-exposed fish. Data from Morgan et al. (1997). (B) Effects of addition of $42 \, nmol \, L^{-1}$ Ag (as $AgNO_3$) on the concentration kinetics of Na^+/K^+-ATPase activity from dog kidney, measured *in vitro*. Means $\pm$ 1 SEM. Open circles represent control conditions, closed circles represent the presence of Ag. Data from Ferguson et al. (1996). In both cases note the decrease in maximum transport capacity (A) or activity (B) without a shift in the position of the curves, i.e. no change in affinity (K_m) of the transport system (A) or enzyme (B) for Na^+, indicative of non-competitive inhibition by Ag.

energizing Na^+ entry through an as yet unidentified apical Na^+ channel, whereas in others, an apical Na^+/H^+ exchanger provides the initial step in Na^+ uptake. There has been speculation (Morgan et al., 2004a, b; Schnizler et al., 2007) but as yet no concrete information on whether the V-type H^+-ATPase and/or the Na^+ channel and/or the Na^+/H^+ exchanger are inhibited by Ag^+. However, very recently Goss et al. (2011) showed that $AgNO_3$ at 10–1000 $\mu g\ L^{-1}$ will almost instantaneously (within 2 min) inhibit acid-stimulated Na^+ uptake *in vitro* by one subtype of mitochondria-rich cell ("PNA-negative cell") from the gills of rainbow trout, suggesting an apical site of action. Further research on this important topic is needed.

While it is very clear that inhibition of Na^+/K^+-ATPase plays a key role in the blockade of active Na^+ uptake by Ag^+, a second documented site of action both *in vitro* (Christensen and Tucker, 1976) and *in vivo* (Morgan et al., 1997, 2004a) is carbonic anhydrase in the branchial ionocytes. The molecular mechanism of inhibition is as yet unknown, but Soyut et al. (2008) reported that the Ag^+ inhibition of carbonic anhydrase from trout brain appeared to be competitive, though Ag^+ did not appear to be displacing Zn from the active site on the enzyme. Conversely, Ag^+ inhibition of carbonic anhydrase from trout liver appeared to be uncompetitive (Soyut and Beydemir, 2008). Carbonic anhydrase is the enzyme that catalyzes the hydration of CO_2 to produce the acidic (H^+) and basic (HCO_3^-) counterions against which Na^+ and Cl^- uptake are exchanged at the apical surface (Evans et al., 2005). Carbonic anhydrase also occurs in erythrocytes, but in trout exposed to sublethal Ag^+ *in vivo*, the observed inhibition of activity occurs in the gill cells, rather than in the blood trapped in the gills (Morgan et al., 2004a).

A detailed time-course analysis (Morgan et al., 2004a) in rainbow trout exposed to 4.3 $\mu g\ L^{-1}$ $AgNO_3$ has demonstrated that there is actually a two-step process to the inhibition of Na^+ uptake (see Fig. 1.4). The first step, a 30% inhibition, occurs very quickly, within 1–2 h of exposure to waterborne Ag^+, long before the inhibition of branchial Na^+/K^+-ATPase activity becomes significant (Fig. 1.5A). This first step temporally correlates with a rapid 30% inhibition of carbonic anhydrase activity in the gills (Fig. 1.5D) and an almost complete blockade of Cl^- uptake (Fig. 1.5B). The second step of Na^+ uptake inhibition develops more slowly and progressively, with close to 100% inhibition by 8–24 h (Fig. 1.5A). This second step temporally correlates with a progressive inhibition of Na^+/K^+-ATPase activity, which reaches about 40% blockade by 24 h (Fig. 1.5C). This time course is similar to that reported by Morgan et al. (1997), who interpreted the second step as the time it took for Ag^+ to fully penetrate to the basolateral membrane of the ionocytes. Morgan et al. (1997) measured Na^+/K^+-ATPase and carbonic anhydrase activities only after 48 h, and reported that 10 $\mu g\ L^{-1}$ $AgNO_3$

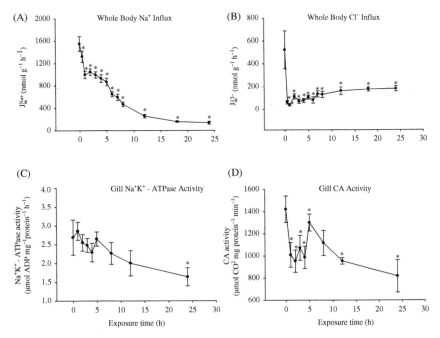

Fig. 1.5. Detailed time-course analysis in juvenile rainbow trout of the effects of waterborne exposure to $4.3\,\mu g\,L^{-1}$ Ag (as $AgNO_3$) in freshwater. (A) Whole-body unidirectional Na^+ influx ($J_{in}^{Na^+}$). (B) Whole-body unidirectional Cl^- influx ($J_{in}^{Cl^-}$). (C) Gill Na^+/K^+-ATPase activity. (D) Gill carbonic anhydrase (CA) activity. Means $\pm$ 1 SEM. *Significant difference ($p < 0.05$) from the 0 h control value. Note the two-stage inhibition of $J_{in}^{Na^+}$, with the first rapid stage corresponding to carbonic anhydrase inhibition, and the second slower phase corresponding to Na^+/K^+-ATPase inhibition. Note also the more rapid inhibition of $J_{in}^{Cl^-}$, corresponding to the inhibition of carbonic anhydrase activity. Data from Morgan et al. (2004a).

caused 85% and 30% inhibition, respectively, with almost complete blockade of both Na^+ and Cl^- uptake (Fig. 1.3A). In contrast, $2\,\mu g\,L^{-1}$ $AgNO_3$ caused only a partial (~70%) inhibition of Na^+ uptake with no significant effect on Cl^- uptake at 48 h (Fig. 1.3B). These latter responses were associated with only a 50% inhibition of Na^+/K^+-ATPase and no effect on carbonic anhydrase activity. Thus, the relative contribution of the two target sites to Na^+ uptake blockade may vary with the dose. Furthermore, the inhibition of Cl^- uptake appears to be more closely associated with reductions in carbonic anhydrase activity.

A surprising feature of the blockade of Na^+ uptake is its very rapid (within 2 h) and complete reversibility, once sublethal waterborne Ag^+ exposure is suspended (Morgan et al., 1997). Thus, the target enzymes are not irreversibly damaged. Hussain et al. (1994, 1995) demonstrated almost instantaneous reversal of Na^+/K^+-ATPase inhibition by Ag^+ *in vitro* when thiol reagents with strong sulfhydryl groups (including metallothionein) were added to the incubation media. In this regard it is interesting that the log K for Ag^+ binding by trout gills is approximately 10.0 (Janes and Playle, 1995) whereas the log K for Ag^+ inhibition of Na^+/K^+-ATPase, as well as that for acute toxicity in intact trout *in vivo*, is only about 7.3–8.0 (McGeer et al., 2000; Paquin et al., 1999; Morgan and Wood, 2004). One may speculate that there are endogenous Ag^+-binding sites in the ionocytes with a stronger affinity for Ag^+ than the target enzymes, so that when Ag^+ stops flooding in, the enzyme-bound Ag^+ is quickly repartitioned to these non-toxic binding sites.

Another prominent feature of the toxic syndrome which develops during acute Ag^+ exposure is an internal metabolic acidosis in both the extracellular and intracellular compartments, associated with an uptake of acidic equivalents across the gills, and a marked decline in plasma HCO_3^- (Wood et al., 1996a; Webb and Wood, 1998). The cause of this is unclear, but since net Na^+ and Cl^- losses are approximately equimolar and therefore would not constrain an acid uptake by the strong ion difference theory (Stewart, 1981), some other strong ion is likely involved. One likely possibility is that the acid uptake is related to the large loss of K^+ across the gills which occurs in the face of unchanged plasma $[K^+]$ (Webb and Wood, 1998). Exchange of H^+ for K^+ is the classic mechanism by which metabolic acid enters muscle, accompanied by a net excretion of K^+ to prevent hyperkalemia from developing in the blood plasma. Inhibition of branchial carbonic anhydrase by Ag^+, as discussed earlier, may also play a role in systemic acid–base disturbance. There is no inhibition of ammonia excretion, but internal ammonia levels and ammonia excretion rates rise during acute Ag^+ exposure owing to an increased metabolic production rate. This is probably caused by mobilization of the stress hormone cortisol, which drives proteolysis (Webb and Wood, 1998). The rise in internal ammonia levels may have a beneficial effect in partially counteracting metabolic acidosis.

When Ag^+ is complexed by an excess of thiosulfate (log $K = 8.8$–14.2), essentially all physiological symptoms of acute toxicity are eliminated (Wood et al., 1996b; Janes and Playle, 2000). Indeed, when trout were exposed to 30,000 $\mu g\ L^{-1}$ as $Ag(S_2O_3)_n^-$ (as opposed to 10 $\mu g\ L^{-1}$ as $AgNO_3$ used in Ag^+ studies), physiological disturbances still remained negligible (Wood et al., 1996b). Another strong complexing agent, DOC (log $K = 9.0$) at 35 mg C L^{-1}, offered complete protection against physiological disruption caused by exposure to 11 $\mu g\ L^{-1}$ as $AgNO_3$ (Janes and Playle, 2000). Even the much weaker complexing agent Cl^- (log $K = 3.3$–5.5) offered concentration-dependent protection against inhibition of branchial Na^+/K^+-ATPase (Fig. 1.6) and Na^+ influx (McGeer and Wood, 1998; Bury et al., 1999b).

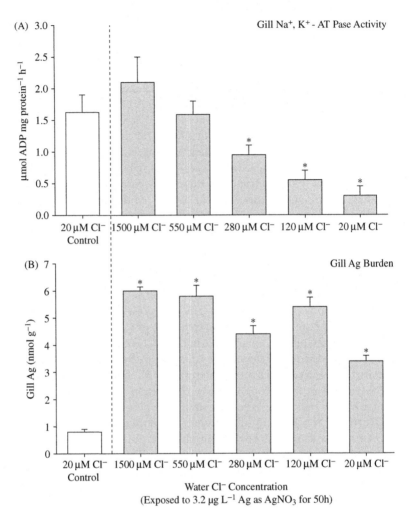

Fig. 1.6. Analysis in adult rainbow trout of the influence of water $[Cl^-]$ on the responses of (A) gill Na^+, K^+-ATPase activity and (B) gill total Ag burden to 50 h waterborne exposure to 3.2 μg L^{-1} Ag (as $AgNO_3$) in freshwater. Control values were taken from fish held in comparable soft water conditions with water $[Cl^-] = 20$ μmol L^{-1} and the absence of $AgNO_3$. Means + 1 SEM. *Significant difference ($p < 0.05$) from the control value. Note the concentration-dependent protective effect of water $[Cl^-]$ against the inhibition of Na^+/K^+-ATPase activity, and its complete lack of protective effect against gill Ag accumulation. Data from McGeer and Wood (2008).

In contrast, competitive agents such as Na^+ and Ca^{2+} had only very slight protective effects against the same physiological endpoints (Bury et al., 1999c). All these observations are in accord with acute lethality data (Section 7.1.1) but certainly do not correlate with measurements of gill or internal Ag burdens. Indeed, note that gill total Ag burdens tended to be greater when waterborne Cl^- was increased, yet Na^+/K^+-ATPase activity was completely protected (Fig. 1.6B).

7.2. Chronic Toxicity

The amount of available information on chronic toxicity (i.e. that occurring over extended durations with endpoints of decreased survival, growth, or reproduction) is only a fraction of that for acute toxicity, and more work is clearly needed. In contrast to the large database for acute toxicity, Wood et al. (2002b) were able to tabulate chronic toxicity data from only three species for $AgNO_3$ tests in simplified waters: rainbow trout (Davies et al., 1978; Nebeker et al., 1983), brown trout (Davies et al., 1998), and fathead minnow (Nebeker et al., 1983; Holcombe et al., 1983). Notably, all chronic threshold values were less than 1.0 μg L^{-1}. More recent studies on early life stages of two of these species (fathead minnows, Naddy et al., 2007; rainbow trout, Brauner and Wood, 2002a, b; Morgan et al., 2005a, b; Dethloff et al., 2007) have confirmed these low effect thresholds. The effects seen included reduced growth rate, premature hatching, and elevated mortality. There appear to be no studies using reproduction as an endpoint. In general, these values were at least 10-fold lower than acute LC50 values for the same species, suggesting that the true acute-to-chronic ratio is rather high for Ag^+. In this regard, the trend is opposite that seen for most freshwater invertebrates, where chronic threshold concentrations are often higher than acute LC50 concentrations. The latter is probably an artifact of the test conditions, because the need to intensively feed invertebrates in long-term tests results in complexation of Ag^+ by food particles and DOC, and in consequence, marked reduction in toxicity (Nebeker et al., 1983; Campbell et al., 2002; Kolts et al., 2006, 2008). In other words, the fish results provide the correct answer; Ag^+ really is much more toxic during chronic exposures than during acute exposures.

7.2.1. INFLUENCE OF WATER CHEMISTRY ON CHRONIC TOXICITY

Information on the effects of different water chemistries is similarly sparse. There appeared to be no damaging effects when fathead minnows (and other ecosystem components) were exposed to 5000 μg L^{-1} silver thiosulfate [Ag $(S_2O_3)_n$] in a multitrophic level mesocosm for 70 days, but speciation was not reported (Terhaar et al., 1977). LeBlanc et al. (1984) carried out 30 day embryo–larval exposures of fathead minnow to suspensions of silver sulfide at concentrations up to 11,000 μg L^{-1}, and found no deleterious effects, in accord with acute tests on the same species. Coleman and Cearley (1974) reported that bluegills and small-mouth bass survived 70 day exposures to 70 and 7 μg L^{-1} AgNO$_3$, respectively, without deleterious effects, but the exposures were static, thus likely resulting in very high DOC concentrations, and water Cl$^-$ levels were also high (5.4 mmol L^{-1}). The lowest effect threshold (0.17 μg L^{-1}) for AgNO$_3$ came from a rainbow trout early life stage study (Davies et al., 1978) performed in very soft, ion-poor water (hardness = 26 mg L^{-1} as CaCO$_3$). In later studies, this same group confirmed this very low threshold (0.20 μg L^{-1}) for both rainbow and brown trout in similar water chemistry (hardness = 25 mg L^{-1}, [Cl$^-$] = 0.08 mmol L^{-1}) and found that it increased to 0.50 μg L^{-1} (at hardness = 200 mg L^{-1}, [Cl$^-$] = 0.37 mmol L^{-1}) and to 1.08 μg L^{-1} (at hardness = 450 mg L^{-1}, [Cl$^-$] = 0.70 mmol L^{-1}) when more ion-rich water was mixed into the exposure (Davies et al., 1998). However, as all ions and DOC were likely changed by this mixing approach, it is not possible to factor out the true protective agent.

A series of chronic exposures with rainbow trout early life stages has subsequently been carried out to address this issue. Brauner et al. (2003) manipulated [Cl$^-$] alone by adding KCl (0.03, 0.3, and 3 mmol L^{-1}) to otherwise ion-poor water (hardness = 20 mg L^{-1}; DOC = 0.3 mg L^{-1}) and showed only very slight protection, far less than during acute toxicity tests (see Section 7.1.1). This conclusion was confirmed by Dethloff et al. (2007), who added NaCl (up to 0.85 mmol L^{-1}) rather than KCl. Morgan et al. (2005a, b) adopted a similar approach by manipulating hardness alone (2, 150, and 400 mg L^{-1}) by adding CaSO$_4$:MgSO$_4$ (as 3:1 ratio on a molar basis) to similar ion-poor water and observed somewhat greater protection, comparable to that seen during acute tests, whereas Brauner and Wood (2002a) found that DOC (as Aldrich humic acid) at 12 mg L^{-1} had only a very slight protective effect. The overall conclusion is that all water chemistry components manipulated in the Davies et al. (1998) study likely contributed some protection, but that two of them which are very protective in acute toxicity tests (Cl$^-$ and DOC) are much less protective in chronic tests. Curiously, in chronic exposures of early life stages of fathead minnow (Naddy et al., 2007), a species in which Cl$^-$ may not protect against acute toxicity (Section 7.1.1), elevation of both [Na$^+$] and [Cl$^-$] to about 1.7 mmol L^{-1} from background levels of 0.05–0.15 mmol L^{-1} raised the chronic threshold about three-fold, from 0.30 to 0.95 μg L^{-1}. However, here it is unclear whether Na or Cl$^-$ (or both) was the true protective agent(s).

7.2.2. MECHANISMS OF CHRONIC TOXICITY

For exposure conditions where some free Ag$^+$ remains available, one mechanism of chronic toxicity appears to be similar to that of acute toxicity, i.e. interference with Na$^+$ and Cl$^-$ regulation, such that whole-body levels of these ions gradually run down (Galvez et al., 1998; Guadagnolo et al., 2001; Brauner and Wood, 2002a, b; Brauner et al., 2003; Naddy et al., 2007). While it is easy to see how this could cause increased mortality, it is not clear how this affects hatching time and growth. In developing early life stages, the chorion plays a large protective role, trapping the majority of the Ag before it reaches the embryo (Guadagnolo et al., 2000). Sensitivity appears to be greatest just before hatch and thereafter (Guadagnolo et al., 2000; Brauner et al., 2003; Morgan et al., 2005a, b), as the organism relies more and more on active ion uptake from the water (Brauner and Wood, 2002b; Brauner et al. 2003). Accompanying physiological disturbances are similar to those seen during acute toxicity, and include reductions in whole-body Na^+ uptake (Fig. 1.7A), Na^+/K^+-ATPase activity (Fig. 1.7B), Na^+ content (Fig. 1.7C), and Cl^- content (Fig. 1.7D), and increases in whole-body cortisol and ammonia levels (Brauner and Wood, 2002b; Brauner et al., 2003).

Silver also readily penetrates freshwater fish at all life stages, and at least in juvenile and older life stages (Hogstrand et al., 1996; Galvez et al., 1998; Wood et al., 1996b), potently induces the detoxifying protein metallothionein, particularly in the liver (Section 12.5). The synthesis and maintenance of elevated metallothionein levels may well be metabolically costly processes that negatively affect growth, hatching, and ultimately survival. Galvez et al. (1998) reported that chronic exposure of juvenile trout to $0.5\,\mu g\,L^{-1}$ (as $AgNO_3$) increased voluntary food consumption but did not affect growth rate, indicative of increased costs. Higher exposure concentrations reduced both growth rates and feeding rates, as well as food conversion efficiency and aerobic swimming performance, indicating that sublethal Ag was acting as a loading stressor (Galvez et al., 1998; Galvez and Wood, 2002). Silver accumulation may also disturb the homeostasis of other metals by displacing them from metallothionein (Coleman and Cearley, 1974; Galvez and Wood, 1999), and in higher organisms such as mammals, Ag is known to cause oxidative damage (Ercal et al., 2001). All of these effects may also contribute to chronic toxicity. Finally, in several invertebrates, Ag is known to be a potent inhibitor of reproduction through disruption of vitellogenin synthesis (e.g. Hook and Fisher, 2001). This is clearly a topic that should be investigated in fish.

7.3. Acclimation

Physiological acclimation refers to the recovery of normal physiological homeostasis during the continued presence of the stressor (McDonald and Wood, 1993). Several studies (Galvez et al., 1998; Galvez and Wood, 2002; Brauner and Wood, 2002b; Bury, 2005) have demonstrated that physiological acclimation may occur during chronic sublethal exposure of freshwater

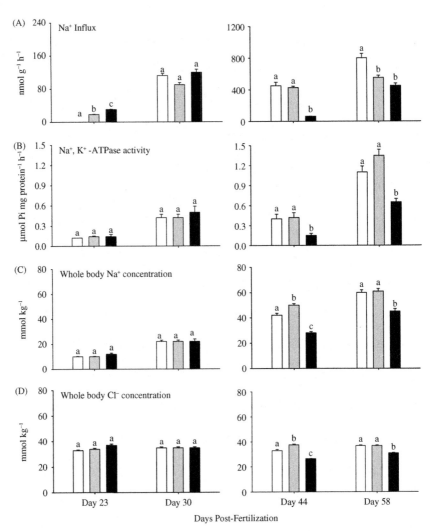

Fig. 1.7. Developmental analysis, in early life stages of rainbow trout, of the influence of chronic waterborne exposure to Ag (as $AgNO_3$) at either 0.1 (shaded bars) or 1.0 µg L^{-1} (black bars) in soft water. Control fish (open bars) were held in the same water quality in the absence of $AgNO_3$. (A) Whole-body unidirectional Na^+ influx ($J_{in}^{Na^+}$). Note the difference in scale of the two panels. (B) Whole-body Na^+/K^+-ATPase activity. (C) Whole-body Na^+ concentration. (D) Whole-body Cl^- concentration. Means + 1 SEM. Within a sampling period, means sharing the same letter were not significantly different ($p > 0.05$). Fifty percent hatching occurred by day 29, and 50% swim-up by day 46. Note the general increase in most values as development proceeds. Note also the lack of inhibitory effects of Ag exposure before hatch, and the greater inhibitory effects after hatch. Data from Brauner et al. (2003).

trout to waterborne $AgNO_3$ in simplified waters; other forms of Ag have not been tested. The phenomena observed include recovery of whole-body and/or plasma Na^+ and Cl^- concentrations, Na^+ influx rates, and Na^+/K^+-ATPase activities, as well as reductions in Ag uptake rates and increased basolateral export capacity at the gills over a time course of days to weeks. Although there have been no measurements of key indicators such as mRNA and protein expression levels, or general protein turnover rates, this physiological acclimation appears to be fairly typical of that seen with other gill-toxicant metals (e.g. Cu, Zn, Cd). As such, it fits into the "damage–repair" paradigm proposed by McDonald and Wood (1993), whereby an initial "shock phase" (e.g. ion loss, non-functional transport proteins), resulting from damage to the gills, triggers specific repair processes. If the initial physiological disturbance is great enough, this damage repair may result in increased tolerance of higher waterborne metal concentrations (i.e. toxicological acclimation) as assessed by acute toxicity tests.

Only a single study (Galvez and Wood, 2002) has examined whether fish exhibit toxicological acclimation (i.e. gain increased resistance to a more severe challenge) during chronic waterborne Ag exposure. The answer was very definitely positive, with significantly increased 7 day LC50 values in juvenile trout chronically exposed to 3 or $5 \mu g L^{-1}$ (as $AgNO_3$). The phenomenon did not occur at lower concentrations (0.1 and $1.0 \mu g L^{-1}$) even though there was evidence of small physiological disturbance and recovery at these concentrations, and it did not occur until day 15 of exposure. By this time, plasma Na^+ and Cl^- concentrations, branchial Na^+ uptake rate, and branchial Na^+/K^+-ATPase activities had been restored back to or above control values. The threshold concentration for toxicological acclimation represented 13% of the acute LC50; both this and the timing were very typical of toxicological acclimation to other metals (8–18%, 7–20 days). The authors speculated that damage repair had in some way changed the gill surface by decreasing the access of Ag^+ to key toxic sites, and/or by decreasing the diffusive permeability of the branchial epithelium to ion losses, points that need to be confirmed by future research.

7.4. Biotic Ligand Models for Prediction of Toxicity

Paquin et al. (1999, 2000, 2002a,b), McGeer et al. (2000), Niyogi and Wood (2004), Wood (Chapter 1, Vol. 31A), and Paquin et al. (Chapter 9) provide the theoretical and computational background for the BLM. Building on the original gill Ag-binding study of Janes and Playle (1995) in rainbow trout, as augmented by Schwartz and Playle (2001), several BLMs have been developed to predict the acute toxicity of Ag to freshwater organisms. Figure 1.8 presents a general overview of the key reactions and

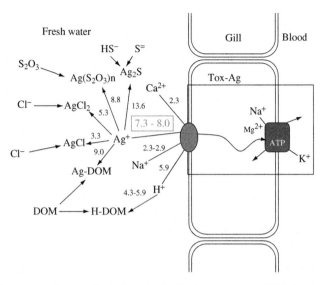

Fig. 1.8. Conceptual diagram illustrating the biotic ligand model (BLM) for Ag. The toxic mechanism (Tox-Ag) is shown as Ag^+ blocking the Mg^{2+} activation site on a basolateral Na^+/K^+-ATPase molecule (purple) in a gill ionocyte, with Ag^+ first targeting a "biotic ligand" (red) on the apical surface of the ionocyte. The numbers on the arrows represent the log K values for the various complexation and competition reactions, with the values outlined in red representing the range of log K values for Ag^+ binding to the biotic ligand used in various BLMs. Data from Janes and Playle (1995), Paquin et al. (1999), Wood et al. (1999), McGeer et al. (2000), Morgan and Wood (2004), Mann et al. (2004), and Niyogi and Wood (2004). **SEE COLOR PLATE SECTION.**

log K values incorporated into these models. At present, there is no BLM for the prediction of chronic toxicity.

The first BLM, developed by the Hydroqual Company (Paquin et al., 1999), was synthesized by integrating acute toxicity data for rainbow trout and fathead minnow with the 2 h gill binding data of Janes and Playle (1995), and was adapted for daphnids through downward adjustments of the LA50 value. The second (McGeer et al., 2000) is a physiologically based BLM calibrated to the inhibition of gill Na^+/K^+-ATPase in rainbow trout, while the third (Bury et al., 2002) is directly based on acute toxicity data for daphnids. All three versions of the model are conceptually quite similar to the initial gill Ag-binding model. However, the original affinity constants for gill–cation complexes derived by Janes and Playle (1995) were found to be too high and thus overpredictive of acute Ag toxicity in freshwater fish. An explanation was provided by Wood et al. (1999), who calculated that only a small fraction of the measured accumulation of Ag in the gills was likely associated with binding to physiologically active sites of toxicity (e.g. Na^+/K^+-ATPase, carbonic anhydrase), especially when ligands such as thiosulfate and Cl^- were used by Janes and Playle (1995). As a result, a lower set of log K values for gill–cation complexes was adopted in all three versions of the acute Ag BLM, derived primarily through the calibration of the models with available Ag toxicity datasets. A key distinction is that each of the three BLMs deals with the important DOC–Ag binding prediction (which is often the most protective parameter in natural waters) in a fundamentally different manner, though the details are beyond the scope of this chapter. Only the two BLMs that deal with fish (Paquin et al., 1999; McGeer et al., 2000) will be considered further.

The Hydroqual BLM (Paquin et al., 1999) was calibrated and validated using the same toxicity datasets from fathead minnow (Erickson et al., 1998) and rainbow trout (Bury et al., 1999a, b) as those used in its original development. Not surprisingly, in all cases, the model satisfactorily predicted (within an acceptable error range of $\pm \times 2.0$ or "factor 2") the variations in 96 h LC50 values due to different Cl^-, Ca^{2+}, and DOC levels in the water. However, a significant shortcoming was that this BLM predicted 96 h LC50 values in fish based on an assumed critical gill Ag burden (LA50), yet at the time, no measured relationship between short-term gill Ag accumulation and 96 h percent mortality existed. Morgan and Wood (2004) subsequently demonstrated that short-term gill Ag accumulation indeed correlated with percent mortality in trout at 96 h under flow-through exposure in low ionic strength soft water. However, the measured LA50 (1–2 nmol g^{-1} gill wet weight) was only a fraction of the previously assumed LA50 (17 nmol g^{-1} gill wet weight). The lower value seems more reasonable as it is close to the estimated number of Ag^+ binding sites on Na^+/K^+-ATPase molecules in the gills (1–4 nmol g^{-1} gill wet weight), the presumed binding sites of toxicity (Wood et al., 1999). Furthermore, the estimated log K for Ag^+ at the biotic ligand was 8.0, five-fold higher than the original value of 7.3 in the Hydroqual BLM. However, these changes were not incorporated into subsequent versions of the Hydroqual BLM, which may explain why this BLM (version 2.1.2) tended to underpredict toxicity of $AgNO_3$ to fathead minnows in soft water in a later validation study (Bielmyer et al., 2007).

In contrast, the second BLM (McGeer et al., 2000) predicts acute toxicity in fish not from the total gill Ag burden but rather by quantifying the percent inhibition of branchial Na^+/K^+-ATPase, which is the assumed key toxic mechanism (Morgan et al., 1997; McGeer and Wood, 2008; Section 7.1.2). The log K for Ag^+ at the biotic ligand was estimated at 7.8 based on the concentration of Ag^+ associated with 85% inhibition of branchial Na^+/K^+-ATPase, which in turn was the degree of inhibition at the 96 h

LC50 (Wood et al., 1999). This value was then adjusted downward to 7.6 based on fitting to eight toxicity datasets (Bury et al., 1999a) for rainbow trout. This physiological BLM was then tested over a wide range of water chemistries (Cl^-, Ca^{2+}, Na^+, pH, DOC) using 23 independent measurements of $AgNO_3$ toxicity in trout from 10 different published studies. All but one point agreed with measured values within "factor 2.0".

The role of Cl^- has been particularly troubling in the development of acute Ag BLMs, because this moiety clearly protects trout, but not fathead minnows against acute Ag toxicity (see Section 7.1.1), yet does not prevent Ag accumulation on the gills in either species (McGeer and Wood, 1998; Bury et al., 1999a, b). Apparently, the neutral $AgCl_0$ complex, like $Ag(S_2O_3)_n^-$ complexes, can penetrate the epithelial cells without inhibiting the Na^+/K^+-ATPase (Hogstrand et al., 1996; McGeer and Wood, 2008; Bury et al., 1999b). Therefore, the Hydroqual BLM (Paquin et al., 1999) includes a conditional log K value (6.7) for $AgCl_0$ at the biotic ligand associated with toxicity in fathead minnow only, whereas this is not included in the trout version, or in the two other BLMs (McGeer et al., 2000; Bury et al., 2002).

In summary, despite some uncertainties, the available BLMs offer a far more reliable option for generating site-specific AWQC for Ag than the traditional approach of hardness-dependent adjustments (see Section 6, Table 1.1). They may also effectively and economically substitute for water effect ratio tests to adjust AWQC to site-specific conditions (Paquin et al., 2000). An important development is the model developed by Paquin et al. (2002b) integrating the Hydroqual BLM (Paquin et al., 1999) with a physiologically based model of Na^+ homeostasis. By relating the formation of the Ag^+–biotic ligand complex on the gills to the inhibition of active Na^+ uptake, an IBM was derived that predicted the time required for plasma Na^+ stores to run down by 30%, the generally accepted threshold for fish death (Wood et al., 1996a). The IBM successfully predicted survival time in a number of literature datasets (see Paquin et al., Chapter 9). This approach holds great promise for the extension of the BLM to predict acute toxicity in new species depending on their Na^+ balance parameters and to predict the outcome of pulse exposures. Another interesting development is the *in vitro* BLM test system developed by Zhou et al. (2004) using reconstructed gill epithelia grown in primary culture from dispersed gill cells of freshwater rainbow trout (Fletcher et al., 2000). Natural freshwaters may be tested on the apical surface, and Ag-binding and Na^+/K^+-ATPase inhibition quantified in the cultured epithelium. This method has subsequently been broadened to include transcriptomic profiling of metal-responsive genes in the cultured epithelium (Walker et al., 2007, 2008). This *in vitro* approach may provide a simple, rapid, and cost-effective way for evaluating the protective effects of site-specific waters.

8. WATERBORNE SILVER TOXICITY IN SALTWATER

8.1. Acute Toxicity

Overall, the data available on the acute toxicity of Ag in saltwater are far less numerous than for freshwater, and most are restricted to full-strength seawater. Hogstrand and Wood (1998) and Wood et al. (2002b) tabulated data from 10 marine teleost species, many of which were from the non-peer-reviewed literature: acute 96 h LC50 values for $AgNO_3$ ranged from 110 μg L^{-1} (1 μmol L^{-1}) in silversides to > 1000 μg L^{-1} (10 μmol L^{-1}, the approximate solubility limit) in several other species. As in freshwater, teleost fish appear to be far less sensitive than many invertebrates. Overall, these data are sufficient to conclude that Ag exerts far less acute toxicity to fish in saltwater than in most freshwaters, where 96 h LC50 values for $AgNO_3$ are typically at least an order of magnitude lower (see Section 7.1). Silver concentrations ≥110 μg L^{-1} likely never occur in even the most contaminated saltwater in nature (see Sections 2 and 4).

8.1.1. INFLUENCE OF WATER CHEMISTRY ON ACUTE TOXICITY

The only variable that has been studied in any detail is salinity. It should be appreciated that in most situations, when salinity varies, many components that may play a role in Ag toxicity (e.g. Na^+, Cl^-, Ca^{2+}, Mg^{2+}) will vary proportionally, though effects are commonly attributed to the dominant role of Cl^- complexation in changing Ag speciation, as illustrated in Fig. 1.1.

Based partly on chloride complexation geochemistry and partly on physiological understanding, Hogstrand and Wood (1998) and Wood et al. (1999) presented similar models (averaged together as a single model in Fig. 1.9) predicting that acute $AgNO_3$ toxicity should decline precipitously as salinity increased from freshwater and Ag^+ disappeared, being progressively replaced by various chloride complexes (first $AgCl_0$, then $AgCl_2^-$, then $AgCl_3^{2-}$; see Section 4; Fig. 1.1). Toxicity was predicted to reach a minimum at about 30–40% seawater. Here, Ag solubility would be at its lowest because of cerargyrite precipitation, and the fish would be close to its isosmotic point, so interference by Ag with ion transport processes would have little effect. As salinity increased above this point, acute toxicity was predicted to increase moderately as ionoregulatory gradients and costs increased (Fig. 1.9).

In general, the few data available support this general model. Early studies which raised freshwater Cl^- levels up to the 1–25% seawater range in killifish (Dorfmann, 1977), fathead minnow (LeBlanc et al., 1984), and rainbow trout (Hogstrand et al., 1996; Galvez and Wood, 1997) noted a marked reduction or abolition of $AgNO_3$ toxicity. Using a constant $AgNO_3$

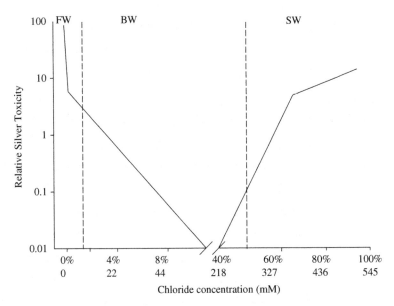

Fig. 1.9. Conceptual model illustrating the relative acute toxicity of waterborne Ag (added as AgNO$_3$) to fish as a function of salinity and Cl$^-$ concentration. The model represents a blend of those presented by Hogstrand and Wood (2008) and Wood et al. (2009). Note that toxicity was predicted to reaching a minimum at about 30–40% seawater. Here, Ag solubility would be at its lowest because of cerargyrite precipitation, and the fish would be close to its isosmotic point. FW: freshwater; BW: brackish water; SW: seawater.

concentration (401 μg L^{-1}) in rainbow trout tests, Ferguson and Hogstrand (1998) reported that acute mortality changed from 0% in 45–60% seawater to 50% in 75% seawater, and 67% in full-strength seawater. Shaw et al. (1998) found that AgNO$_3$ toxicity was two to four times greater in tidepool sculpins in 100% seawater than in 75% seawater. These patterns also correlate with short-term (24 h) [110m]Ag bioavailability studies in gulf toadfish, where whole-body accumulation progressively declined as salinity was raised from 2.5% to 40% seawater, thereafter increasing only moderately at higher salinities (Wood et al., 2004; see Section 11.2.1).

8.1.2. MECHANISMS OF ACUTE TOXICITY

As in freshwater fish, the key mechanism of acute toxicity again appears to involve osmoregulatory failure. However, in contrast to freshwater fish where this can be ascribed to a single organ (gills) and function (branchial ionoregulation; see Section 7.1.2), there are at least two possible target sites (gills and gut) and associated target functions (branchial ionoregulation, gastrointestinal ionoregulation) in marine teleosts. This is because marine fish drink the medium extensively, in contrast to freshwater fish, thereby bringing waterborne Ag in direct contact with the gastrointestinal surface. Continuous obligatory drinking is part of an overall osmoregulatory strategy which keeps internal body fluids substantially hypotonic to the external seawater (Evans et al., 2005; Grosell, 2006). Imbibed water must be continually absorbed to replace water which is being lost across the gills and body surface by osmosis. Since water itself cannot be actively transported, Na^+ and Cl^- are actively absorbed across the gut, and water follows passively by osmosis. Seawater is also rich in potentially toxic divalent ions (Mg^{2+}, Ca^{2+}, and SO_4^{2-}) but these are largely excluded from absorption. HCO_3^- secretion makes an important contribution to Cl^- uptake and the precipitation of Ca^{2+} and Mg^{2+} as insoluble carbonates, thereby keeping osmolality low in the intestinal fluids and favoring water absorption. The Na^+ and Cl^- which are absorbed at the gut, together with additional Na^+ and Cl^- entering across the body surface, must be actively excreted across the gills.

Thus, in marine teleosts, the acute toxicity of Ag to ionoregulatory/ osmoregulatory processes manifests as an increase in plasma Na^+ and Cl^- concentrations, as first demonstrated by Hogstrand et al. (1999) in the starry flounder exposed to $AgNO_3$ in full-strength seawater. This effect was minimal at 250 μg L^{-1} but pronounced at 1000 μg L^{-1}, with 50% increases in plasma $[Na^+]$ and $[Cl^-]$ recorded in fish close to death. Drinking rate was inhibited by more than 50%, even at 250 μg L^{-1}, pointing to the gut as a target. Other key features of the toxic syndrome included marked increases in plasma glucose and cortisol levels, indicative of a generalized stress response as in freshwater fish (Wood et al., 1996a; Webb and Wood, 1998). Disturbances in blood gases and acid–base status were minimal, indicating a lack of respiratory toxicity. A marked increase in plasma $[Mg^{2+}]$ was transient. Ammonia excretion rate exhibited a sustained increase, perhaps reflecting the proteolytic influence of cortisol. However, an initial rise in plasma ammonia concentration disappeared before death occurred, suggesting that this was not part of the key toxic mechanism, even though elevated waterborne ammonia was shown to elevate Ag toxicity in tidepool sculpins (Shaw et al., 1998). Regardless, ammonia excretion rates did not change in tidepool sculpins exposed to 250 μg L^{-1} as $AgNO_3$ (Webb et al., 2001).

Additional evidence that the gut is an important target was provided by Grosell et al. (1999), who confirmed that drinking rate was reduced in another flatfish, the lemon sole, after exposure to 660–1000 μg L^{-1} of waterborne $AgNO_3$. Furthermore, when the gut alone was exposed to similar levels of Ag by *in situ* perfusion of the intestinal tracts of otherwise intact animals, there were marked reductions in net Na^+ and Cl^- absorption and HCO_3^- secretion. While the expected accompanying reduction in net water absorption could not be detected in these *in situ* perfusions, this reduction was seen using *in vitro* gut sac incubations in the same study.

In the studies of Hogstrand et al. (1999) and Grosell et al. (1999) there was no direct evidence that waterborne Ag affected the gills. Indeed, in both, branchial Na^+/K^+-ATPase activity actually increased after 4–6 days' exposure to waterborne Ag, suggesting a compensatory response. However, a follow-up study in the same species (Grosell and Wood, 2001) demonstrated that waterborne exposure to 448 μg L^{-1} $AgNO_3$ caused a progressive decline in unidirectional Na^+ efflux rate at the gills. Furthermore, *in situ* perfusion of the intestine with clean saline, so as to protect the tract during waterborne Ag exposure (448 μg L^{-1}), did not attenuate this decline, so the effect was directly at the gills. However, this *in situ* perfusion did attenuate increases in plasma $[Na^+]$, $[Cl^-]$, and osmolality by about 50%, so there were clearly two sites of action, gills and gut. The mechanisms of these effects remain unknown at the cellular level. However, it is interesting that effects at the gut were not associated with inhibition of intestinal Na^+/K^+-ATPase activity (Grosell et al., 1999; Grosell and Wood, 2001), and appeared to be irreversible over 24 h, both very different from the situation at the gills in freshwater fish (Morgan et al., 1997; Section 7.1.2). On the other hand, the inhibition of unidirectional Na^+ efflux rate at the gills (the active transport component) did appear to correlate with a reduction in branchial Na^+/K^+-ATPase activity, as in freshwater fish.

8.2. Chronic Toxicity

Again, the available data are sparse. Voyer et al. (1982) and Klein-MacPhee et al. (1984) carried out short, early life-stage assays with $AgNO_3$ in winter flounder to assess chronic toxicity. Based on effects on hatching, growth, and mortality, the chronic thresholds were in the range of 70–115 μg L^{-1}. Several unpublished studies, summarized in Wood et al. (2002b), found chronic thresholds for $AgNO_3$ tests in the range of 400–600 μg L^{-1} for early life-stage tests with summer flounder and sheepshead minnow, while Ward et al. (2006) measured similar values (EC20 = 170–620 μg L^{-1}) in 7–28 day tests on developing silversides in 90–100% seawater. Relative to freshwater studies (Section 7.1.2), all these chronic thresholds are extremely high. This suggests that the differences seen between seawater and freshwater for acute toxicity are even more pronounced for chronic toxicity. However, curiously, this may not be true at lower salinities (Section 8.2.1), or for tests with adult fish (Section 8.2.2). Exposure of adult rainbow trout to only 1.5, 14.5, or 50 μg L^{-1} $AgNO_3$ for 21 days in 90% seawater increased mortality from 15% (control) to 57%, 70%, and 85%, respectively (Webb et al., 2001).

8.2.1. INFLUENCE OF WATER CHEMISTRY ON CHRONIC TOXICITY

In contrast to the acute toxicity model and data (see Section 8.1.1; Fig. 1.9), where increases in salinity above 30–40% seawater tend to increase toxicity, exactly the opposite was seen in chronic tests with developing silversides (Ward et al., 2006). Here, the EC20 increased from a fairly sensitive value of 37 µg L^{-1} in 30% seawater to 100 µg L^{-1} in 60% and 170 µg L^{-1} in 90% seawater. The reason for this difference was unclear, but the authors noted that the vanishingly small free Ag$^+$ ion concentration would have been about the same (0.008–0.016 µg L^{-1}) at these three thresholds. Notably, survival was a more sensitive endpoint than growth, but no measurements of physiological indices were taken. This study also examined the influence of DOC in 7 day tests, but only in 90% seawater. Increasing [DOC] from background levels of 1 mg L^{-1} to anywhere in the range of 2–6 mg L^{-1} by the addition of Aldrich humic acid was moderately protective, causing an approximately 2.5-fold increase in the threshold concentration. This suggests that DOC may alter Ag speciation in seawater, in contrast to some geochemical reports (Miller and Bruland, 1995; Ward and Kramer, 2002; Section 4).

8.2.2. MECHANISMS OF CHRONIC TOXICITY

Most chronic toxicity studies have focused on early life stages, showing generally high threshold concentrations, whereas the only mechanistic study (Webb et al., 2001) has focused on adults, showing much lower effect concentrations for physiological disturbance. Thus, chronic exposures (7–21 days) of adult tidepool sculpins to AgNO$_3$ in 90% seawater caused reductions in oxygen consumption rate, ammonia excretion rate, drinking rate, gill Na$^+$/K$^+$-ATPase activity, and intestinal Na$^+$/K$^+$-ATPase activity with thresholds for significant effects at either 1.5 or 14.5 µg L^{-1} AgNO$_3$ in all cases (Webb et al., 2001). In this same study, adults of two other species, rainbow trout and midshipman, also exhibited some of these chronic responses, and elevated mortality occurred in the trout at both 1.5 and 14.5 µg L^{-1}. At least in sculpins, disturbances in branchial and intestinal Na$^+$/K$^+$-ATPase activities were somewhat less when the same exposures were conducted in 55% seawater, in accord with the acute toxicity–salinity model of Hogstrand and Wood (1998) and Wood et al. (1999) (see Fig. 1.9). Overall, these results are descriptive rather than mechanistic. Nevertheless, they suggest that emphasis on early life-stage tests to assess chronic Ag toxicity in saltwater may be misguided. As physiological disturbances occur in adult teleosts at levels in the range of regulatory guidelines (see Table 1.1), there is a need for a renewed focus on chronic tests with adult fish.

8.3. Acclimation

To date, there appear to be no studies examining whether marine fish exhibit toxicological acclimation during prolonged exposure, i.e. gain increased resistance to a more severe challenge (McDonald and Wood, 1993). However, there is some evidence of physiological acclimation. Branchial and intestinal Na^+/K^+-ATPase activities increased back to or above control levels in some sublethal exposures (Hogstrand et al., 1999; Grosell et al., 1999; Webb et al., 2001). Gulf toadfish exposed to about 200 μg L^{-1} $AgNO_3$ for 23 days in full-strength seawater exhibited an increased ability to clear a tracer dose of ^{110m}Ag injected into the bloodstream (Wood et al., 2010). They were also able to minimize its entry into red blood cells relative to naïve fish, thereby potentially protecting hemoglobin and carbonic anhydrase function in the erythrocytes. Increased clearance was associated with greater appearance of ^{110m}Ag radioactivity in both bile and urine, suggesting that both liver and kidney excretory abilities were upregulated.

8.4. Marine Elasmobranchs: A Special Case?

Marine elasmobranchs have a fundamentally different osmoregulatory strategy than marine teleosts, maintaining internal osmolality slightly above that of the external seawater by the retention of nitrogenous wastes, primarily urea (see Hazon et al., 2003, for review). Therefore, they largely avoid the need to drink, removing the intestine as a potential target tissue. Na^+ and Cl^- entry rates into the whole body are therefore much lower, and the rectal gland, an organ unique to elasmobranchs which is buried in the colon and not exposed to external seawater, may replace the gills as the primary route of active NaCl excretion. Gill transport functions appear to be more similar to those of freshwater teleosts than marine teleosts. All these differences might be anticipated to make marine elasmobranchs far less sensitive to waterborne Ag than marine teleosts, but in fact, exactly the opposite appears to be true (De Boeck et al., 2001, 2010). Silver uptake rates are many times higher than in marine teleosts (Section 11.2.1) and acute sensitivity is considerably greater. In the Pacific spiny dogfish, the 96 h LC50 in $AgNO_3$ tests was about 100 μg L^{-1}, several times lower than most teleosts (De Boeck et al., 2001).

Physiological analysis of dogfish dying over 24–72 h at 200 and 685 μg L^{-1} revealed dramatic respiratory disturbance (increases in arterial PCO_2, decreases in arterial PO_2, lactacidosis), increases in all plasma ions, especially $[Mg^{2+}]$ which rose five-fold, marked falls in plasma [urea], and massive losses of urea to the external water (De Boeck et al., 2001). Na^+/K^+-ATPase activity was inhibited in the rectal gland as well as the gills, probably associated with Ag accumulation in both tissues. The buildup of

Ag in the rectal gland reflected generally high penetration of Ag to most tissues, presumably entering via the gill and being distributed by the circulatory system. There was frank gill damage, evidenced by epithelial swelling and lamellar fusion. However, there was no change in the very low drinking rate. Even at 30 µg L^{-1}, there was subtle evidence of urea loss and ionoregulatory impairment. The reason for this high sensitivity remains unknown. It may be related to the similarity of gill function to that in freshwater fish, as earlier speculated by Wood et al. (1999). De Boeck et al. (2001) postulated that it reflected an inhibitory and/or entry action of Ag on the Na^+-coupled active backtransport mechanism for urea on the basolateral membranes of the gill cells (Pärt et al., 1998; Fines et al., 2001). This urea transporter is critical in minimizing urea losses at the branchial epithelium, and therefore maintaining ionoregulatory and osmoregulatory homeostasis in internal fluids. Finally, a recent report (De Boeck et al., 2010) suggests that elasmobranchs may not synthesize the detoxifying protein metallothionein in response to Ag challenge.

9. ESSENTIALITY OR NON-ESSENTIALITY OF SILVER

As in other vertebrate classes, there is no evidence whatsoever that Ag is essential in fish, but then again, there have been no experimental tests. Silver is not known to be a cofactor or component of any biological molecule. However, it is clear that Ag is avidly taken up at the gills and intestine (Section 11), is well regulated in certain tissues such as blood, gills, and white muscle by homeostatic mechanisms (Section 12.6), and is accumulated in internal organs such as liver and kidney. Even animals collected from so-called pristine marine environments, where background Ag levels should be very low, have substantial levels of Ag in their tissues (e.g. Webb and Wood, 2000; De Boeck et al., 2001). This may occur because Ag is taken up "accidentally" by Na and/or Cu transport pathways, and sequestered by metallothionein (Section 12.5), but it raises the question of whether it could be a micronutrient. Experimental tests to critically evaluate the potential essentiality of Ag would make an interesting contribution.

10. POTENTIAL FOR BIOCONCENTRATION AND/OR BIOMAGNIFICATION OF SILVER

This area has been extensively reviewed (Eisler, 1996; Hogstrand and Wood, 1998; Ratte, 1999; Wood et al., 2002b; Luoma et al., 2002; McGeer et al., 2003). The conclusions of these syntheses are very clear. Waterborne Ag bioaccumulates in the tissues of fish, especially the liver, to concentrations (on a per wet weight basis) that at equilibrium are orders of magnitude greater than those in the environment (on the same weight of water basis). The same is true of other aquatic organisms. Thus, bioconcentration clearly occurs, and in this regard Ag is similar to virtually all other metals. However, because the trophic transfer efficiency of Ag is extremely low (Section 11.1.2), biomagnification up the food chain does not occur, i.e. at equilibrium, concentrations in the predator will not exceed those in the prey from foodborne exposure alone.

The bioconcentration factor (BCF) is simply the mean concentration of metal in the whole organism (at equilibrium) divided by that in the water. Depending on the jurisdiction, BCFs greater than 500 or 5000 are often used to classify substances as hazardous; however, the use of BCFs as a regulatory tool for metals has been controversial, and so this was critically evaluated by McGeer et al. (2003). Based on available literature data from all aquatic organisms, screened for equilibrium conditions, these authors showed that tissue Ag concentrations increased with exposure Ag concentrations, but not proportionally. Note the low slope and displaced intercept in the log–log plot of Fig. 1.10A. Thus, when BCFs were calculated (Fig. 1.10B), the highest BCFs (approximately $10^4 = 10,000$) occurred under the lowest, close to pristine exposure concentrations ($< 10^{-5}$ mg L$^{-1} = < 10$ ng L^{-1}), and the lowest BCFs (approximately $10^2 = 100$) occurred under the highest exposure concentrations (approximately 10^{-1} mg L$^{-1} = 100$ μg L^{-1}) where chronic toxicity would undoubtedly occur. This same pattern was seen for a wide range of other metals, both essential and non-essential. Thus, a fundamental tenet of the BCF application as a regulatory tool, that BCF should be independent of exposure concentration, was violated. It is illogical that an index of hazard should be highest at low exposure concentrations, and lowest at high exposure concentrations. As for most other metals, the calculation and use of BCFs for Ag is not useful for environmental regulation or hazard classification, and indeed, may well be misleading (McGeer et al., 2003; DeForest et al., 2007).

11. CHARACTERIZATION OF UPTAKE ROUTES

11.1. Freshwater

To the author's knowledge, there is no information on the potential uptake of Ag through the skin or olfactory epithelium, and essentially all waterborne uptake is assumed to occur via the gills, because drinking rates

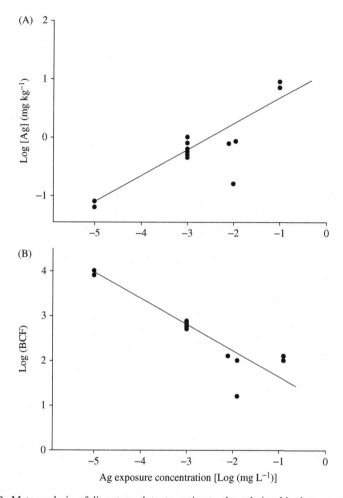

Fig. 1.10. Meta-analysis of literature data to estimate the relationship between total Ag concentration in aquatic organisms and the concentration of Ag in the water to which they were chronically exposed for a period sufficient to achieve equilibrium conditions. Data have been plotted on a log–log basis and the best fit regression lines are shown. (A) Raw data; (B) relationship when the data were converted to bioconcentration factors (BCFs, see text for calculation details). Note that the highest BCFs occurred under the lowest exposure concentrations, close to pristine conditions, and the lowest BCFs occurred under the highest exposure concentrations, where chronic toxicity would undoubtedly occur. Data from McGeer et al. (2003).

are very low in freshwater fish. Uptake through the gastrointestinal tract is thought to be restricted to Ag in the food. However, at least for one electrolyte (Ca^{2+}), significant uptake from water does occur via the skin, presumably facilitated by the secondary circulation close to the surface

(Perry and Wood, 1985). Other metals are known to be taken up by the olfactory route (Tjälve and Henriksson, 1999). These potential pathways for Ag should be examined in the future.

11.1.1. FRESHWATER GILLS

It is now clear that Ag may be taken up at the gills from freshwater in at least four forms: as the free ion Ag^+, as the neutral complex $AgCl_0$, as silver thiosulfate complexes $[Ag(S_2O_3)_n]$, and as silver sulfide complexes (probably Ag_2S). Appearance in the gills may reflect adsorption to the external surface or actual absorption, so internal accumulation is a better indicator of true uptake. Hogstrand and Wood (1998) calculated concentration-specific accumulation rates (CSARs) for the first three of these species based on the buildup of Ag in the livers of freshwater trout over the course of a few days, and concluded that the rates scaled in the order $Ag^+ > AgCl_0 >>> Ag(S_2O_3)_n$. The low CSARs of $Ag(S_2O_3)_n^-$ were recorded from very high concentration exposures in which toxicity was completely prevented by the thiosulfate complexation. It is therefore likely that this $Ag(S_2O_3)_n^-$ uptake is due to simple diffusive entry and/or anion transporter-mediated uptake of these negatively charged complexes. Surprisingly, in the only quantitative examination of silver sulfide complexes, Mann et al. (2004) reported liver Ag burdens from trout that would put the CSAR of Ag_2S comparable to or slightly greater than that of Ag^+. Therefore, in the experiments of Mann et al. (2004), the addition of sulfide to $AgNO_3$ exposures either did not change or moderately accelerated the rate of Ag uptake by the fish, yet it prevented acute $AgNO_3$ toxicity. The mechanism(s) responsible for this effect is unknown, and an important topic for future investigation. In this regard Ag_2S may be comparable to $AgCl_0$. Most research to date has focused on the pathways and differences between Ag^+ and $AgCl_0$ uptake.

In order to characterize the apical entry step of $^{110m}Ag^+$ at the gills of juvenile rainbow trout, Bury and Wood (1999) used low-level, simplified water exposures with radiolabeled $AgNO_3$ in which the free ion $[Ag^+]$ would predominate. Uptake was saturable, exhibited Michaelis–Menten kinetics with a very high affinity ($K_M = 55$ nmol L^{-1} or 6 µg L^{-1}), was competitively inhibited by external $[Na^+]$, and was unaffected by large excesses (up to 10 mmol L^{-1}) of external $[Ca^{2+}]$ or $[K^+]$. Chronic low-level waterborne exposure (0.9 µg L^{-1}) resulted in downregulation of apical entry (Bury, 2005). Furthermore, $^{110m}Ag^+$ uptake was substantially reduced by treatment with bafilomycin A_1, a selective blocker of V-type H^+-ATPases, and by phenamil, a selective blocker of Na^+ channels (Bury and Wood, 1999). These agents reduced the uptake rates of both Ag^+ and Na^+ by about 50%. The clear conclusion is that Ag^+ enters, at least in part, in competition with

Na^+ via apical Na^+ channels situated on branchial epithelial ionocytes, driven by the electrochemical gradient created via the H^+-ATPase which is thought to be coupled to the Na^+ channel (Evans et al., 2005). Recent observations by Goss et al. (2011) have shown that the site of these Na^+/Ag^+ channels in trout may be one subtype of mitochondria-rich cell ("PNA-negative cell"), though their molecular nature remains unknown. In social hierarchies, subordinate trout have much higher branchial Na^+ uptake rates than dominant trout, to compensate for higher stress-induced Na^+ loss rates; this coupling of Ag^+ uptake to the Na^+ uptake pathway may explain why subordinate trout also have much higher Ag^+ uptake rates (Sloman et al., 2003).

Inasmuch as only about 50% of Ag^+ uptake was blocked by bafilomycin A_1, phenamil, or Na^+ competition (Bury and Wood, 1999), there may be a second apical uptake pathway separate from the Na^+ pathway. A likely candidate here is the high-affinity Cu transporter Ctr1, which has been implicated in a portion of Cu uptake in freshwater fish gills (Grosell and Wood, 2002; Bury et al., 2003). Ag^+ has been shown to serve as a substrate for this transporter in other systems, and a 100-fold excess of Cu^{2+} inhibited Ag^+ uptake in juvenile rainbow trout (Bury and Hogstrand, 2002). However, as Cu^{2+} also inhibits Na^+ uptake (see Grosell, Chapter 2, Vol. 31A), this is not necessarily diagnostic of Ctr1 involvement.

With respect to $AgCl_0$ uptake, Bury and Hogstrand (2002) experimentally increased water $[Cl^-]$ and confirmed earlier observations (McGeer and Wood, 1998; Bury et al., 1999a) that partial replacement of Ag^+ with $AgCl_0$ does not significantly reduce the uptake of Ag until the point (about $[Cl^-] =$ 1 mmol L^{-1}) where $AgCl_0$ becomes more than 50% of the total. Well above this point ($[Cl^-] = 5$ mmol L^{-1}), where only $AgCl_0$ is present, Ag uptake is somewhat reduced and exhibits linear kinetics, indicating that it is probably taken up by simple diffusion. These authors speculated that this entry occurs across all gill cells, rather than across the relatively rare mitochondria-rich cells which are thought to be the sites of Ag^+ uptake. The higher octanol/water partition coefficient (K_{ow}) value of $AgCl_0$ (0.09 versus 0.03 for Ag^+) (Reinfelder and Chang, 1999) may be responsible for higher diffusive permeability through lipoprotein membranes. This diffuse pattern would dilute the Ag dose to the mitochondria-rich cells, thereby potentially explaining the protective influence of $[Cl^-]$ in salmonids. In both trout and eel, depuration of ^{110m}Ag from the gills after a pulse exposure was much faster when it was loaded mainly as $^{110m}AgCl_0$ rather than as $^{110m}Ag^+$, providing further evidence of the diffusive lability of the neutral $AgCl_0$ complex (Wood et al., 2002a; Hogstrand et al., 2003).

In contrast to branchial Na^+ and Cl^- transport, where apical entry is the rate-limiting step, basolateral export appears to be rate limiting for Ag transport across the gills (Bury and Wood, 1999; Morgan et al., 2004a, b).

This efflux step has been studied *in vitro* using basolateral membrane vesicles (BLMVs) from trout gill epithelial cells (Bury et al., 1999c, 2003; Bury, 2005). Basolateral transport of Ag was upregulated during chronic low-level waterborne exposure (0.9 μg L^{-1}). Transport was clearly active and carrier mediated, exhibiting Michaelis–Menten saturation kinetics and ATP dependence, though the affinity was three orders of magnitude lower than for the apical entry of Ag$^+$ ($K_M = 63$ μmol L^{-1} versus 55 nmol L^{-1}), perhaps partly a function of protein binding in the assay buffer. Transport exhibited sensitivity to orthovanadate, diagnostic of a P-type ATPase. Ag$^+$ was capable of activating the enzyme in the presence of ATP, as shown by the formation of acyl phosphate intermediates. Inasmuch as the highest Ag transport rates occurred in the absence of K$^+$, and Ag is known to poison Na$^+$/K$^+$-ATPase (Section 7.1.2), it seems unlikely that this is the responsible P-type ATPase. Instead, the authors proposed the involvement of a P-type Cu-ATPase because many members of this enzyme class can transport Ag, and Ag uptake into the BLMVs was strongly inhibited by a 10-fold excess of Cu, but not by other metals (Bury et al., 2003).

At present it is unclear whether the basolateral exit mechanism for Ag is the same when the apical entry pathway is as Ag$^+$ or as AgCl$_0$. Given that the [Cl$^-$] in the whole cell water of gill cells is above 10 mmol L^{-1} (Perry, 1997), one might expect all Ag$^+$ that enters to become AgCl$_0$ in the intracellular milieu, or alternately that Ag from either source would bind to stronger intracellular ligands such as sulfhydryl groups, i.e. that speciation of the entering Ag would not affect the efflux mechanism. However, this conclusion is challenged by the fact that depuration rates from the gills and subsequent internal handling in the body clearly differ for Ag that has entered via the two different pathways, in both trout and eel (Wood et al., 2002a; Hogstrand et al., 2003). Either different cellular sites of entry (i.e. mitochondria-rich cells only for Ag$^+$ versus all gill cells for AgCl$_0$) and/or targeted subcellular trafficking may be more important than simple geochemical equilibria might indicate, and may in turn influence the basolateral exit pathway.

11.1.2. FRESHWATER GUT

At least in laboratory trials, Ag uptake through the gastrointestinal tract in freshwater fish is extremely low and non-toxic. The highest trophic transfer efficiency was about 10%, reported by Garnier and Baudin (1990) who fed brown trout with pieces of carp that had been contaminated with radiolabeled silver cyanide ([110m]AgCN) by low-level waterborne exposure. However, when rainbow trout were fed silver sulfide (Ag$_2$S) for 58 days, there was negligible internal bioaccumulation, and there were no effects on growth even up to dietary concentrations of 3000 mg kg^{-1}, which is 100-fold above levels recorded in even the most contaminated sediments or invertebrate prey (Galvez and Wood, 1999). While Ag_2S is expected to predominate in sediments, other forms of Ag (e.g. protein-bound) might be expected to predominate in prey organisms. Therefore, Galvez et al. (2001) formulated such a diet by exposing trout for 7 days to a high concentration of waterborne silver thiosulfate $[Ag(S_2O_3)_n^-]$, harvesting their tissues, then pelleting them together with other constituents to make a nutritionally complete diet containing 3.1 mg kg^{-1} of biologically incorporated Ag. This is an environmentally realistic concentration for prey from highly contaminated sites. Again, there were no effects on growth or on a wide variety of physiological parameters over 128 days of satiation feeding, but Ag accumulated in internal tissues. Levels were highest in liver, stabilizing at about 12-fold over background after 36 days. Calculation of the CSAR (Section 11.1.1) yielded a very low value comparable to that seen during waterborne $Ag(S_2O_3)_n^-$ exposures, and about 20,000-fold lower than that seen during waterborne Ag^+ exposures (Hogstrand and Wood, 1998). Nevertheless, this was 4.6 orders of magnitude greater than reported for the dietary Ag_2S trials of Galvez and Wood (1999).

In vitro tests with trout gut sac preparations where 50 μmol L^{-1} $^{110m}AgNO_3$ was incubated in isotonic saline (i.e. most Ag existing as $AgCl_2^-$; see Fig. 1.1B), revealed that the anterior intestine, including the pyloric ceca, was by far the major site, accounting for 80% of uptake (Ojo and Wood, 2007). This reflected both its larger surface area and four-fold greater area-specific uptake rates in comparison to stomach, mid, and posterior intestine. At present, nothing is known about the nature of Ag transport pathways across the gut in freshwater fish. By analogy to the gill, future studies should target possible Na-specific and Cu-specific pathways, as well as simple diffusion and "piggyback" carriage on absorbed amino acids.

11.2. Saltwater

Again, nothing is known about potential uptake through the olfactory epithelium or skin. However, marine teleosts drink the medium at an appreciable rate for osmoregulatory purposes. It is now clear that both gut and gills participate in the uptake of waterborne Ag, in contrast to the freshwater situation. However, the gut is probably not involved in marine elasmobranchs, which drink very little (Webb and Wood, 2000; De Boeck et al., 2001), though it may be very important in dietborne uptake in these animals (Pentreath, 1977). Salinity may have a major impact on both the rates of Ag uptake and the relative contributions of gut versus gill, because drinking rate, intestinal osmoregulatory functions, and branchial ionoregulatory physiology, as well as speciation (see Fig. 1.1), all change with salinity. For example, in the gulf toadfish exposed to 2.2 µg L^{-1} Ag (added as ^{110m}AgNO$_3$) for 24 h, whole-body uptake varied nine-fold across salinities, being greatest in 2.5% seawater, lowest in 40% seawater (close to the isosmotic point), and then increasing only 1.6-fold up to 100% seawater (Fig. 1.11A) (Wood et al., 2004). However, in chronic 21 day exposures to 1.5, 14.5, or 50 µg L^{-1}, tidepool sculpins took up six times more Ag in 55% seawater than in 90% seawater (Webb and Wood, 2000). This difference was not seen in a 48 h study at a higher concentration (250 µg L^{-1}) (Webb and Wood, 2000), so the influence of salinity may vary with the length of the exposure and/or with the dose. Notably, using ^{110m}Ag radioactivity in the gills versus that in the esophagus–stomach as an indicator of the importance of the two routes in the 24 h study on the gulf toadfish, the contribution of the gills dropped dramatically as salinity increased, with the two routes becoming of similar magnitude at salinities of 40% or higher (Fig. 1.11B) (Wood et al., 2004).

11.2.1. SALTWATER GILLS

At the low end of the salinity range (below the isosmotic point of 30–40% seawater), fish will still be taking up Na$^+$ by active transport, so it is possible that some of the Ag influx occurs as Ag$^+$ through the Na$^+$ pathway discussed in Section 11.1.1. However, as Ag$^+$ levels will be vanishing because of the increasing [Cl$^-$], it is more likely that the diffusive uptake of AgCl$_0$ predominates; note that as salinity increased in this range, the gill burden of ^{110m}Ag decreased (Fig. 1.11B) (Wood et al., 2004) in parallel to the decline in [AgCl$_0$] (Fig. 1.1). Above the isosmotic point, the gills are actively excreting Na$^+$ and Cl$^-$, and AgCl$_0$ is largely replaced by AgCl$_2^-$ and then AgCl$_3^{2-}$ (Fig. 1.1), so active Na$^+$ and/or Cl$^-$ uptake pathways cannot contribute to branchial Ag uptake, and AgCl$_0$ diffusion should become increasingly less important. However, it is worth noting that apical Na$^+$/H$^+$ exchangers are still present in the gills for acid–base regulation at higher salinities (Evans et al., 2005), so it remains possible that these could serve as a pathway of Ag uptake. Unfortunately, the mechanism of branchial Ag uptake at these higher salinities remains completely unknown, and is an important topic for future investigation. Yet, at the same time, there is evidence that the gills remain the predominant pathway of entry even at these higher salinities. For example, in the toadfish exposed to 2.2 µg L^{-1}, Wood et al. (2004) calculated that drinking could account for only about 20% of whole-body ^{110m}Ag uptake in 100% seawater. In the lemon sole in 95% seawater, Grosell and Wood (2001) reported that protecting the gut (via surgically short-circuiting drinking and perfusing the intestine with Ag-free saline *in situ*) during waterborne Ag exposure (448 µg L^{-1}) did not significantly reduce the internal tissue-specific accumulation of Ag.

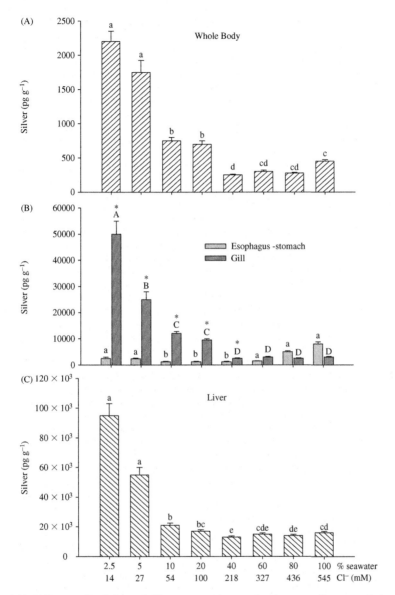

Fig. 1.11. Influence of salinity and Cl⁻ concentration on the tissue-specific accumulation of radiolabeled Ag in marine gulf toadfish exposed to 2.18 µg L⁻¹ of total waterborne Ag (added as AgNO₃) for 24 h. (A) Whole-body concentrations. (B) Esophagus–stomach (a single organ in the toadfish) and gill concentrations. (C) Liver concentrations. Means + 1 SEM. Means sharing the same letter were not significantly different ($p > 0.05$). *Significant differences ($p < 0.05$) between esophagus–stomach and gill concentrations at the same salinity. Data from Wood et al. (2004).

In elasmobranchs, the ionoregulatory mechanisms are completely different (Section 8.4), and the absence of drinking means that gills must account for virtually all waterborne Ag entry. Uptake rates are much higher than in teleosts (Pentreath, 1977; Webb and Wood, 2000; De Boeck et al., 2001, 2010). In some way this may relate to the very different physiology of the gills, which appears to be more similar to those of freshwater teleosts (Wood et al., 1999). Apical Na^+/H^+ exchange mechanisms are present in the gills of marine elasmobranchs (Evans et al., 2005). Notably, there is also a Na^+-coupled active backtransport mechanism for urea on the basolateral membranes of the gill cells (see Section 8.4; Pärt et al., 1998; Fines et al., 2001). Potentially, this could provide a pathway for entry, given the general affinity of Ag for Na^+ transport pathways.

11.2.2. SALTWATER GUT

It is generally believed from modeling studies that trophic uptake via the gut is the dominant source for the bioaccumulation of Ag (Rouleau et al., 2000) and other metals (Mathews and Fisher, 2009) in marine fish under natural circumstances, but there have been very few experimental feeding studies to back this up. In American plaice force-fed a single meal which had been artificially contaminated with $^{110m}AgNO_3$, the trophic transfer efficiency was 4–16% (Rouleau et al., 2000). Similar values (about 5%) were reported for European plaice fed a single meal of worms which had been biologically contaminated by exposing them to waterborne $^{110m}AgNO_3$ for 3 days (Pentreath, 1977). However, in this same study, it was very high (49%) for elasmobranch rays fed a similar meal. Thus, the very high Ag uptake capacity seen in the gills of elasmobranchs (see Section 11.2.1) also extends to the gut.

Furthermore, the situation is complex in teleosts inasmuch as uptake may occur from both imbibed water (i.e. waterborne Ag) and ingested food (i.e. trophic uptake), and the two routes may interact. For example, in a 22 day exposure of gulf toadfish to waterborne Ag ($200 \, \mu g \, L^{-1}$) in full-strength seawater, daily satiation feeding with non-contaminated food attenuated whole-body Ag accumulation by more than 50% relative to that in fasted animals (Wood et al., 2010). Feeding completely prevented Ag accumulation in the esophagus–stomach. Presumably this occurred because the binding of imbibed Ag to food and chyme in the gastrointestinal tract would have reduced its bioavailability for uptake. This result also argues that uptake via the gastrointestinal tract may be more important during chronic exposures than during short-term tests (Grosell and Wood, 2001; Wood et al., 2004). In both toadfish studies (Wood et al., 2004, 2010), the esophagus–stomach (a single organ in this species) appeared to be a quantitatively more important site than the intestine for Ag uptake. The possibility that the esophagus or stomach could serve as a significant site of Ag absorption was overlooked in the *in situ* gut perfusion studies of Grosell and Wood (2001). Therefore, their conclusion that the gut was unimportant in waterborne Ag uptake in the lemon sole should be re-evaluated, especially in light of other indications.

For example, in earlier studies on the lemon sole (Grosell et al., 1999) and the starry flounder (Hogstrand et al., 1999) and a more recent investigation of the gulf toadfish (Wood et al., 2004), Ag concentrations were reduced as imbibed seawater moved through the tract to the rectum *in vivo*, despite a several-fold reduction in volume due to water absorption. This suggests that the gut takes up Ag. In addition, there were substantial levels of Ag in the intestinal tissues of lemon sole after exposure to waterborne Ag without the protection of intestinal perfusion (Grosell and Wood, 2001). Furthermore, Ag did accumulate to high levels in both the stomach and intestine of plaice exposed to waterborne ^{110m}Ag or fed ^{110m}Ag-labeled worms (Pentreath, 1977). Several other studies have reported Ag accumulation in intestinal tissues during both short-term and long-term exposures (Webb and Wood, 2000; Hogstrand et al., 2002; Wood et al., 2004; Nichols et al., 2006), though no clear pattern as to regional heterogeneity has emerged.

Hogstrand et al. (2002) carried out a thorough *in vitro* characterization of intestinal Ag transport kinetics using ^{110m}AgNO$_3$ in everted intestinal sac preparations from European flounder which had been acclimated to 64% seawater. As in the freshwater gut sac study of Ojo and Wood (2007), most of the Ag in the mucosal solution would have been present as ^{110m}AgCl$_2^-$ (Fig. 1.1). However, a much lower concentration range was used (11 nmol L^{-1} to 1 μmol L^{-1} versus 50 μmol L^{-1}) than by Ojo and Wood (2007). Varying the mucosal [Cl$^-$] seemed to have little effect on the influx rate, but solvent drag appeared to be important, as Ag absorption tended to track water absorption. There was a large mucus-binding component, but the trans-flux to the serosal saline was saturable, exhibiting Michaelis–Menten kinetics with a K_M of 180 nmol L^{-1} (19 μg L^{-1}). Curiously, neither cyanide nor low temperature treatments had any influence on influx rates, so it remains problematic whether active transport was involved. There is a clear need for additional mechanistic studies to characterize the pathways of Ag transport in the gastrointestinal tracts of marine fish.

12. CHARACTERIZATION OF INTERNAL HANDLING

12.1. Biotransformation

Given what is known about speciation in natural waters (Sections 3, 4), it seems safe to assume that there is essentially no free Ag$^+$ inside the cells or body fluids of fish (Bell et al., 2002; Luoma et al., 2002). All Ag, regardless of the form in which it enters, will be bound up by chloride and sulfide, as well as by biomolecules such as glutathione, proteins, and amino acids, especially those bearing sulfhydryl groups (e.g. cysteine residues) such as metallothionein (see Section 12.5).

12.2. Transport through the Bloodstream

Appreciable levels of Ag in the range of 0.1–2 μmol L^{-1} (10–200 μg L^{-1}) are found in the blood plasma of naïve fish (Wood et al., 1996b, 2010; Hogstrand et al., 1999; Galvez et al., 2001; Bertram and Playle, 2002; Nichols and Playle, 2004; Nichols et al., 2006; De Boeck et al., 2010). In general, Ag concentrations in plasma appear to be higher in freshwater fish than in seawater fish, perhaps reflecting higher background levels in the environment (see Sections 1–3). In both, plasma levels may increase many-fold during waterborne Ag exposures. Nothing is known about transport mechanisms in the plasma in fish, but in mammals, Ag binds mainly to α_1-macroglobulin (which is rich in cysteine residues) and to a small extent to ceruloplasmin (Hanson et al., 2001). Ceruloplasmin normally serves for Cu transport (see Grosell, Chapter 2, Vol. 31A). Silver also occurs in red blood cells at background concentrations similar to or higher than those in plasma. However, during acute exposures, the erythrocytes appear to be somewhat protected, because Ag concentrations increase to a lesser extent in the red blood cells than in the plasma (Hogstrand et al., 2003; Wood et al., 2010). This may be important in protecting carbonic anhydrase function inside the red blood cells.

12.3. Accumulation in Specific Organs

Virtually every paper ever published on the physiology of Ag in fish contains some measurements of Ag burdens in tissues. Several general conclusions may be drawn. Firstly, as for blood plasma, appreciable levels of Ag in the range of 0.1–2 μmol kg^{-1} (10–200 μg kg^{-1}) are found in most tissues of naïve fish, reinforcing the view that this element is bioaccumulated (see Section 10), even from pristine environments. Secondly, background concentrations tend to be highest in the liver, but since this is generally a small fraction of total body mass in most teleosts ($<2\%$), it makes up only a small fraction of the total body burden. This appears to be true even in elasmobranchs (Webb and Wood, 2000; De Boeck et al., 2001, 2010), where the liver may exceed 10% of body mass. Thirdly, background concentrations tend to be lowest in the white muscle, skin, and bone, which are often lumped together as the carcass. However, since the carcass comprises the largest component of the fish by mass ($>60\%$), it accounts for the largest fraction of the background Ag burden. Finally, during experimental challenges with higher environmental levels, carcass tissues accumulate negligible amounts of additional Ag, whereas the liver may accumulate massive amounts (e.g. Fig. 1.11C). Gills, gut, and kidney may also accumulate under these circumstances, but to a much lesser extent than the liver. This is important from a human health perspective, as the carcass represents the portion of the fish that is usually considered edible.

Several studies have utilized [110m]Ag to look at organ-specific handling of Ag, as summarized in Table 1.2. From this, the generally dominant role of the liver in accounting for a large or largest portion of the body burden is readily apparent, and the gill, though it may be the principal site of uptake, rarely accounts for more than a small fraction. Several other interesting trends have emerged.

Firstly, time is important (Table 1.2). As seen following short-term (24–48 h) waterborne exposures with trout and eel in freshwater (Galvez et al., 2002; Hogstrand et al., 2003), accumulation in the liver became increasingly dominant. As [110m]Ag was depurated from other tissues, the liver appeared to act as a scavenger, accounting for more and more of the body burden as depuration time increased. This occurred more quickly in trout than in eel. This progressive movement of [110m]Ag to the liver was less apparent when exposures were prolonged, as in the 57 day waterborne exposure of brown trout (Garnier et al., 1990); the liver burden may have already reached equilibrium by this time, such that 28 day depuration had little influence on the pattern.

Secondly, the original speciation in the exposure water, which likely influences the pathway of [110m]Ag entry, appears to make a difference (Table 1.2). Both depuration from the gill and movement to the liver occurred more quickly when waterborne [110m]Ag was presented to trout mainly as [110m]AgCl$_0$ rather than as [110m]Ag$^+$ (Wood et al., 2002a; Hogstrand et al., 2003). In the gulf toadfish, not only was absolute uptake much greater during 24 h waterborne exposure to [110m]Ag in 2.5% seawater versus 100% seawater (see Fig. 1.11A), but a much greater percentage of the body burden went initially to the liver (Table 1.2). These two salinities represent conditions in which both the chloride speciation of Ag (see Fig. 1.1) and the relative contributions of gills versus gut to uptake (see Sections 11.2.1, 11.2.2) are vastly different.

Thirdly, after the liver, the gut is often the second most important site of [110m]Ag accumulation (Table 1.2). While this is to be expected following dietary exposures (e.g. Pentreath, 1977; Garnier and Baudin, 1990), it is surprising to see the same pattern after waterborne exposures of plaice to [110m]Ag in seawater (Pentreath, 1977) and rainbow trout (Galvez et al., 2002; Hogstrand et al., 2003) and brown trout (Garnier et al., 1990) to [110m]Ag in

Table 1.2

Organ-specific distribution of ^{110m}Ag taken up from the environment under different waterborne and foodborne exposure conditions in fish in freshwater and saltwater

Species	Reference	Liver	Gill	Kidney	GI tract	Carcass	Treatment
FRESHWATER							
Rainbow trout	Hogstrand et al. (2003)	48.6%	6.9%	6.6%	5.7%	32.2%	1 day after 24 h waterborne exposure to ^{110m}AgNO$_3$ (1.3 µg L^{-1}) mainly as Ag$^+$
Rainbow trout	Hogstrand et al. (2003)	73.2%	5.7%	6.9%	6.7%	7.5%	1 day after 24 h waterborne exposure to ^{110m}AgNO$_3$ (1.3 µg L^{-1}) mainly as AgCl$_0$
Rainbow trout	Hogstrand et al. (2003)	92.6%	1.4%	2.7%	0.4%	2.9%	67 day after 24 h waterborne exposure to ^{110m}AgNO$_3$ (1.3 µg L^{-1}) mainly as Ag$^+$
Rainbow trout	Hogstrand et al. (2003)	97.3%	0.5%	0.6%	0.3%	1.3%	67 days after 24 h waterborne exposure to ^{110m}AgNO$_3$ (1.3 µg L^{-1}) mainly as AgCl$_0$
European eel	Hogstrand et al. (2003)	26.9%	52.4%	2.1%	2.1%	16.5%	1 day after 24 h waterborne exposure to ^{110m}AgNO$_3$ (1.3 µg L^{-1}) mainly as Ag$^+$
European eel	Hogstrand et al. (2003)	25.0%	55.9%	1.7%	2.0%	15.4%	1 day after 24 h waterborne exposure to ^{110m}AgNO$_3$ (1.3 µg L^{-1}) mainly as AgCl$_0$
European eel	Hogstrand et al. (2003)	71.8%	6.1%	4.5%	2.6%	15.0%	67 days after 24 h waterborne exposure to ^{110m}AgNO$_3$ (1.3 µg L^{-1}) mainly as Ag$^+$
European eel	Hogstrand et al. (2003)	77.4%	6.6%	1.7%	1.4%	12.9%	67 days after 24 h waterborne exposure to ^{110m}AgNO$_3$ (1.3 µg L^{-1}) mainly as AgCl$_0$
Rainbow trout	Galvez et al. (2002)	24.0%	6.0%	2.0%	35.0%	33.0%	1 day after 48 h waterborne exposure to ^{110m}AgNO$_3$ (11.9 µg L^{-1}), followed by 3.8 µg L^{-1} cold AgNO$_3$
Rainbow trout	Galvez et al. (2002)	58.0%	4.5%	1.0%	1.5%	35.0%	19 days after 24 h waterborne exposure to ^{110m}AgNO$_3$ (11.9 µg L^{-1}), followed by 3.8 µg L^{-1} cold AgNO$_3$
Brown trout	Garnier et al. (1990)	70.3%	1.9%	1.7%	9.2%	16.9%	Immediately after 57 day waterborne exposure to ^{110m}AgCN (0.7 µg L^{-1})
Brown trout	Garnier et al. (1990)	61.6%	3.1%	1.6%	11.6%	22.1%	28 days after 57 day waterborne exposure to ^{110m}AgCN (0.7 µg L^{-1})

Species	Reference	Liver	Gill	Kidney	GI tract	Carcass	Treatment
Brown trout	Garnier and Baudin (1990)	62.8%	1.1%	2.2%	13.6%	20.3%	Immediately after 34 day dietary exposure to biologically incorporated ^{110m}AgCN
Brown trout	Garnier and Baudin (1990)	79.2%	0.6%	3.5%	6.1%	10.6%	27 days after 34 day dietary exposure to biologically incorporated ^{110m}AgCN
SALTWATER							
Gulf toadfish	Wood et al. (2004)	41.4%	26.6%	0.3%	3.0%	28.7%	Immediately after 24 h exposure to waterborne ^{110m}AgNO$_3$ in 2.5% seawater
Gulf toadfish	Wood et al. (2004)	24.5%	9.2%	0.5%	22.4%	43.4%	Immediately after 24 h exposure to waterborne ^{110m}AgNO$_3$ in 100% seawater
European plaice	Pentreath (1977)	9.4%	5.4%	2.9%	17.7%	64.6%	Immediately after 63 day exposure to waterborne ^{110m}AgNO$_3$ (0.04 μg L^{-1}) in 100% seawater
European plaice	Pentreath (1977)	17.4%	5.4%	2.8%	19.3%	55.1%	41 days after 63 day exposure to waterborne ^{110m}AgNO$_3$ (0.04 μg L^{-1}) in 100% seawater
European plaice	Pentreath (1977)	41.0%	3.0%	3.2%	14.2%	38.6%	5 days after dietary exposure (a single meal) to biologically incorporated ^{110m}AgNO$_3$ in 100% seawater
European plaice	Pentreath (1977)	31.9%	3.8%	1.7%	11.4%	51.2%	46 days after dietary exposure (a single meal) to biologically incorporated ^{110m}AgNO$_3$ in 100% seawater
Thornback ray	Pentreath (1977)	96.0%	0.5%	0.1%	1.2%	2.2%	Immediately after 63 day exposure to waterborne ^{110m}AgNO$_3$ (0.04 μg L^{-1}) in 100% seawater
Thornback ray	Pentreath (1977)	96.8%	1.5%	0.2%	0.5%	1.0%	35 days after 63 day exposure to waterborne ^{110m}AgNO$_3$ (0.04 μg L^{-1}) in 100% seawater
Thornback ray	Pentreath (1977)	74.4%	0.1%	0.3%	21.1%	4.1%	5 days after dietary exposure (a single meal) to biologically incorporated ^{110m}AgNO$_3$ in 100% seawater
Thornback ray	Pentreath (1977)	99.2%	0.0%	0.0%	0.2%	0.6%	71 days after dietary exposure (a single meal) to biologically incorporated ^{110m}AgNO$_3$ in 100% seawater

Data are presented as percentage of estimated whole-body burden.

freshwater. Indeed, the gastrointestinal burdens were far greater than could be explained by drinking. In part, this could be because biliary excretion recycles ^{110m}Ag from the liver to the intestinal tissue (Section 13). However, Galvez et al. (2002) pointed out an additional possibility, that intestinal tissues have a high concentration of basement membrane material which is rich in cysteine residues; their sulfhydryl groups have a great avidity for Ag.

12.4. Subcellular Partitioning

Two studies using homogenization plus differential centrifugation have concurred that most (60–95%) of the radiolabeled Ag in the gills of freshwater trout exposed to waterborne ^{110m}AgNO$_3$ localizes to the so-called nuclear fraction (Galvez et al., 2002; Wood et al., 2002a). This does not necessarily mean that Ag accumulates in the nucleus, because the fraction also includes mucus and cellular debris. The next largest fraction was cytosolic (5–20%), with only small amounts (<15%) in the mitochondrial and lysosomal, and in the microsomal fractions. This was true regardless of whether the ^{110m}Ag was presented to trout mainly as ^{110m}AgCl$_0$ or mainly as ^{110m}Ag$^+$. Therefore, the differential toxicity (see Section 8.1.1) and uptake/depuration kinetics (see Section 11.1.1) of the two Ag species cannot be explained by differential subcellular partitioning, at least at the level of resolution achieved. As depuration proceeded over 67 days, the percentage in the nuclear fraction of gill tissue progressively increased while that in the cytosolic fraction decreased proportionately (Wood et al., 2002a), suggesting that the latter was more labile. The pattern of ^{110m}Ag distribution was entirely different in the liver, where the cytosolic fraction generally dominated (40–65%), with each of the other three accounting for 5–30% (Galvez et al., 2002). As depuration proceeded, the percentage in the cytosolic fraction of the liver increased, at the expense of the other three.

When the cytosolic fraction was subjected to size-exclusion chromatography, most of the ^{110m}Ag eluted at a molecular weight (MW) characteristic of metallothionein (apparent MW ~10 kDa), with two smaller peaks at high MW (~220 kDa) and low MW (22 kDa) (Galvez et al., 2002). The amount in the metallothionein fraction increased over time. In contrast, in the gills, the elution pattern was diffuse, with only about 15% in the metallothionein fraction. However, about 60% was in the same high MW fraction (~220 kDa) as in the liver, and 15% in an even lower MW fraction (~1.2 kDa). Thus, the cytosol of the liver appeared to be well protected by metallothionein (Section 12.5), whereas in the gills, large proteins may have suffered as a result of Ag binding.

12.5. Detoxification and Storage Mechanisms

It is well established that Ag is a powerful inducer of metallothionein in teleost fish, particularly in liver, kidney, and gills (Cosson, 1994a, b; Hogstrand et al., 1996; Wood et al., 1996b; Mayer et al., 2003; Walker et al., 2007, 2008), though perhaps not in elasmobranchs (De Boeck et al., 2010). At least in fish cell lines, Ag appears to induce metallothionein transcription directly, i.e. without displacing Zn from pre-existing Zn-metallothionein (Mayer et al., 2003). Each metallothionein molecule contains sufficient sulfhydryl groups to complex and effectively detoxify 12 atoms of Ag^+ owing to its very high log $K > 17$. As noted in Section 12.4, about 70% of the ^{110m}Ag in the liver cytosol of rainbow trout pre-exposed to waterborne $^{110m}AgNO_3$ eluted at a molecular weight appropriate for metallothionein, whereas the corresponding figure for gill cytosol was only about 15% (Galvez et al., 2002). The high and low molecular weight fractions where most of the remaining ^{110m}Ag eluted were not identified; potentially, these could represent either additional detoxification mechanisms or pathological targets. Glutathione often plays a role in metal detoxification, but the low molecular weight fractions appeared to be somewhat larger than this small molecule. Some invertebrate organisms also form Ag_2S granules for detoxification, but the author is aware of no concrete evidence that this occurs in fish.

12.6. Homeostatic Controls

While Ag concentrations in plasma of trout quickly increase upon waterborne exposure, they also stabilize after a few days during continued exposure (Wood et al., 1996b). When the exposure stops, most of the plasma Ag burden is cleared within a few days, and depuration from the gills occurs at a similar rate (Wood et al., 2002a; Hogstrand et al., 2003). However, the latter processes represent internal distribution to storage sites such as the liver, as excretion from the whole body appears to be very slow. Very little Ag accumulates in the white muscle. These observations, together with clear evidence of physiological acclimation during prolonged exposure (Sections 7.3, 8.3), suggest that Ag is under homeostatic control, at least in the extracellular compartment, but the possible sensing and feedback mechanisms involved remain completely unknown. Possibly the phenomenon may reflect clearance of binding proteins such as metallothionein. Alternately or additionally, it is possible that this apparent homeostasis is an accidental or passive process, owing to the similarities between Ag and other elements (see Section 15).

13. CHARACTERIZATION OF EXCRETION ROUTES

Curiously, while there has been a substantial amount of research on the pathways of Ag uptake, the pathways of excretion remain relatively unexplored. One practical reason for this deficit is that excretion appears to be an extremely slow process, and therefore difficult to study. For example, in two studies on rainbow trout exposed to waterborne $^{110m}AgNO_3$ for 24–48 h, there was no detectable decrease in the whole-body ^{110m}Ag burden over the following 67 days in clean water (Hogstrand et al., 2003), or 19 days in water contaminated with a sublethal level of non-radioactive Ag (Galvez et al., 2002). Similar results were reported with freshwater brown trout and marine thornback rays contaminated via either waterborne (Pentreath, 1977; Garnier et al., 1990) or dietary routes (Pentreath, 1977; Garnier and Baudin, 1990). However, excretion was somewhat faster in freshwater eel (Hogstrand et al., 2003) and seawater plaice (Pentreath, 1977), with half-times of elimination of 67 and 12 days, respectively.

The possible excretion routes include gills, gut, kidney (i.e. urinary excretion), and liver (i.e. biliary excretion). There are no reports on Ag excretion by the first two pathways. However, neither large differences in temperature (3°C versus 16°C) (Nichols and Playle, 2004) nor feeding versus fasting (Bertram and Playle, 2002) had any effect on the very slow Ag depuration kinetics seen in trout previously exposed to waterborne $AgNO_3$. These observations could be taken as indirect evidence that neither gills nor gut are important excretion routes. There are also no reports on urinary Ag excretion in freshwater fish. However, two studies on marine gulf toadfish exposed to waterborne $AgNO_3$ have reported the appearance of Ag in the urine. Concentrations were lower than in blood plasma after a 24 h exposure to 2.2 µg L^{-1} (Wood et al., 2004) but several-fold higher than plasma after a 22 day exposure to 200 µg L^{-1} (Wood et al., 2010). Urinary excretion rates were not measured. Nevertheless, given the relatively low Ag concentrations and low urine flow rates of marine fish, this is unlikely to be a major route of excretion.

Biliary excretion is probably a far more important pathway, in light of the Ag scavenging function of the liver (Section 12.3). There are several reports of Ag concentrations in bile samples taken from the gall bladders of both freshwater (Bertram and Playle, 2002; Hogstrand et al., 2003; Nichols and Playle, 2004) and marine teleosts (Wood et al., 2004, 2010) challenged with sublethal waterborne $AgNO_3$. All agree that biliary Ag levels are many-fold higher than plasma levels, and generally similar to the concentrations in the liver tissue which produces the bile, varying in parallel to the latter over time and treatment. The mechanism of Ag entry into the bile and the actual rates of biliary excretion are important topics for future investigation. For example, gall bladder discharge occurs more frequently when fish are fed, but the chemistry of the bile changes, so feeding may (Wood et al., 2010) or may not (Bertram and Playle, 2002) increase the rate of biliary Ag excretion.

14. BEHAVIORAL EFFECTS OF SILVER

Behavior has been shown to influence the response to waterborne $AgNO_3$, with subordinate freshwater trout taking up more Ag through the gills at low concentrations and exhibiting greater inhibition of branchial Na^+ uptake at high concentrations relative to dominant individuals (Sloman et al., 2003). These effects likely arise because subordinate fish have higher gill Na^+ uptake rates to compensate for greater loss rates, and Ag^+ can enter via the Na^+ transport pathway (see Section 11.1.1), eventually blocking it (see Section 7.1.2). However, as yet there appears to be no information on the reciprocal issue, i.e. does Ag exposure influence behavior? Waterborne metals have long been known as powerful chemosensory inhibitors that can distort natural behavioral patterns in aquatic animals (e.g. Pyle and Wood, 2007). Copper in particular has been well studied in this regard, exerting deleterious effects at concentrations close to environmental guidelines (see Grosell, Chapter 2, Vol. 31A), but the author is aware of no comparable studies on Ag. However, relatively high levels of both waterborne Cu and waterborne Ag inhibited the olfactory bulb electrical responses of rainbow trout to model odorants (Brown et al., 1982). Given the immense ecological consequences of disturbed chemosensory function (e.g. failure to detect predators or to find food, mates, and native streams), this is an important area for future study.

15. MOLECULAR CHARACTERIZATION OF SILVER TRANSPORTERS, STORAGE PROTEINS, AND CHAPERONES

Sections 11 and 12 have dealt with all the various aspects of this topic, so only three key points will be briefly reiterated here. Ag^+ tends to mimic Na^+, so may be transported by so-called Na^+-specific pathways, yet at the same time may block those pathways by competition (e.g. at Na^+ channels) or enzyme inhibition (e.g. at Na^+/K^+-ATPase). Silver also tends to mimic Cu (perhaps Ag^+ mimicking the reduced form Cu^+), so may be transported by so-called Cu-specific pathways such as P-type Cu-ATPases and/or Cu transporter Ctr1 (Bury et al., 2003). Silver is a powerful inducer of metallothionein which can immobilize 12 atoms of Ag^+; this is likely the most important intracellular "storage protein", serving as an important agent of detoxification.

16. GENOMIC AND PROTEOMIC STUDIES

There have been no such studies to date, but the time is ripe for such approaches, as illustrated by recent advances in this regard in understanding the impacts of Cu (Grosell, Chapter 2, Vol. 31A) and Zn (Hogstrand, Chapter 3, Vol. 31A) on transcriptomic responses in fish.

17. INTERACTIONS WITH OTHER METALS

A priori, one may predict that Ag should interact most strongly with the two other strong sulfhydryl-seeking metals Hg and Cu, especially the latter, because of the well-documented affinity of both Cu and Ag for common transport mechanisms (i.e. Na transporters, Cu transporters; see Section 11.1.1). However, there is little evidence that this is true, though available data are sparse. In a heroic multifactorial experiment, Ribeyre et al. (1995) measured metal bioaccumulation in zebrafish exposed for 12 days to various combinations of waterborne Ag, Cu, Se, Zn, and Hg. The most striking result was a 30% inhibition of Hg uptake by elevated Ag; however, as Hg was presented as methyl mercury rather than inorganic Hg, and there was no reciprocal antagonism, there is no clear interpretation of these results. Elevated Cu actually slightly stimulated Ag bioaccumulation by 10%, but both Se and Zn also had the same effect, so again the interpretation is unclear. When rainbow trout were first acclimated in the field to control water or water from a site polluted with Cd and Zn, and then exposed to trace levels of various radiolabeled metals in the same water, the uptake of ^{110m}Ag was dramatically reduced by 62–90% in the fish held in the polluted water (Ausseil et al., 2002). However, the uptakes of radiolabeled Cs and Co were also reduced by 20–68%, so the response may have reflected a general acclimation effect rather than a specific interaction. There is a pressing need for more work on this topic, as multimetal exposure is the normal situation in polluted waters in the field. Playle (2004) pointed out that the BLM (see Section 7.4, Wood, Chapter 1, Vol. 31A, and Paquin et al., Chapter 9) is an ideal tool for making testable predictions on how metals may interact with respect to both uptake and toxicity under such conditions.

18. KNOWLEDGE GAPS AND FUTURE DIRECTIONS

As with most other metals, research on the physiological and toxicological effects of Ag has been largely funded by industries concerned about the environmental impacts of their discharges. To date, the photographic industry has borne the brunt of these costs, but with the replacement of film-based silver halide photography by digital photography, this source of support has ended. Nevertheless, the industrial use of Ag, and therefore its dispersal into the environment, both continue to increase, especially in the manufacture of electronics, nanoparticles, and jewelry (Silver Institute, 2010). It is hoped that these industries will exhibit the same responsibility in supporting environmental research as did the photographic industry. The following are key knowledge gaps and suggestions for future research.

- The clear protective role of water Cl^- against Ag^+ toxicity to salmonids, but not to several other species, remains perplexing and of regulatory concern. Related to this issue, it is also unclear to what extent the apical entry and basolateral exit mechanisms of $AgCl_0$ and Ag^+ at the gills of freshwater fish differ from one another, and differ among various species.
- While two key intracellular proteins of ion transport (Na^+/K^+-ATPase, carbonic anhydrase) in gill ionocytes are now recognized as targets for acute Ag^+ toxicity in freshwater fish, there is a lack of information on other potential target or transport sites, such as apical Na^+ channels, Na^+/H^+ antiporters, V-type H^+-ATPase, P-type ATPases, and Ctr1. It is also unclear whether any of these proteins are important targets at other sites such as the gastrointestinal tract in either freshwater or marine teleosts, as well as gill ionocytes in the latter.
- In general, mechanisms of chronic Ag toxicity remain poorly understood in both freshwater and saltwater fish. It will be particularly important to investigate the possibility of endocrine disruption by interference with vitellogenin metabolism, as seen in some invertebrates, as well as possible relationships between tissue-specific residues or whole-body burden of Ag and chronic toxic responses. The latter information will be critical to the development of a chronic BLM for Ag.
- The limited literature available suggests that increasing salinity above the isosmotic point increases acute Ag toxicity but ameliorates chronic toxicity in euryhaline marine fish. Furthermore, DOC is very effective in protecting against acute Ag toxicity in freshwater fish, but less so in marine fish or during chronic exposures. The explanations for these two apparent dichotomies deserve further investigation.
- Relative to marine teleosts, marine elasmobranchs are far more sensitive to waterborne Ag toxicity, and take up far more Ag at both gills and gut. The mechanistic explanations for these differences require further investigation.
- In saltwater, waterborne Ag is known to exert physiological toxicity at both gills and gut, but the relative contribution of the two target surfaces to Ag uptake remains unclear.
- Silver is generally considered to be a non-essential element. However, this has never been properly investigated in fish. Experimental studies of potential essentiality, homeostatic sensing, and regulatory mechanisms for Ag would be very welcome. In this regard, almost nothing is known about excretory mechanisms, an important subject for future investigation.
- While some other waterborne metals are taken up through the olfactory tract and skin of fish, these potentially important uptake sites have never been investigated for Ag.
- Copper and Ag are similar in many respects, and the former is now known to be a potent disruptor of olfactory mechanisms, and therefore of behavior in fish. Comparable studies on the chemosensory effects of waterborne Ag are needed.
- In real-world situations, Ag contamination rarely occurs in isolation from other metals; studies are urgently needed on possible interactive effects with other metals.

ACKNOWLEDGMENTS

Work in the author's laboratory on silver has been supported by Eastman Kodak, Kodak Canada, and NSERC. Sunita Nadella is thanked for excellent bibliographic assistance and for drawing the figures. This chapter is dedicated to the memory of Steve Munger, who was instrumental in the author's early investigations into silver effects on fish.

REFERENCES

Adams, N. W. H., and Kramer, J. R. (1999a). Silver speciation in wastewater effluent, surface waters, and pore waters. *Environ. Toxicol. Chem.* **18**, 2667–2673.

Adams, N. W. H., and Kramer, J. R. (1999b). Determination of silver speciation in wastewater and receiving waters by competitive ligand equilibration/solvent extraction. *Environ. Toxicol. Chem.* **18**, 2674–2680.

Andren, A.W. and Bober, T.W. (eds) (2002). *Silver in the Environment: Transport, Fate, and Effects.* Society of Environmental Toxicology and Chemistry (SETAC), Pensacola, FL.

ANZECC/ARMCANZ (2000). *Australian and New Zealand Guidelines for Fresh and Marine Water Quality.* CSIRO, Sydney.

Arnold, W. R. (2005). Effects of dissolved organic carbon on copper toxicity: implications for saltwater copper criteria. *Integr. Environ. Assess. Manag.* **1**, 34–39.

Ausseil, O., Adam, C., Garnier-Laplace, J., Baudin, J.-P., Casellas, C., and Porcher, J.-M. (2002). Influence of metal (Cd and Zn) waterborne exposure on radionuclide (^{134}Cs, ^{110m}Ag and ^{57}Co) bioaccumulation by rainbow trout (*Oncorhynchus mykiss*): a field and laboratory study. *Environ. Toxicol. Chem.* **21**, 619–625.

B.C. (British Columbia) (1995). *Ambient Water Quality Criteria for Silver.* Compiled by Warrington, P.D., Ministry of Environment, Lands and Parks, Province of British Columbia. Water Quality Branch, Environmental Protection Department, Victoria, Canada. 169 pp.

Bell, R. A., Ogden, N., and Kramer, J. R. (2002). The biotic ligand model and a cellular approach to class B metal aquatic toxicity. *Comp. Biochem. Physiol. C* **133**, 175–188.

Benoit, G. (1994). Clean technique measurement of Pb, Ag, and Cd in freshwater: a redefinition of metal pollution. *Environ. Sci. Technol.* **28**, 1987–1991.

Bertram, B. O. B., and Playle, R. C. (2002). Effects of feeding on waterborne silver uptake and depuration in rainbow trout (*Oncorhynchus mykiss*). *Can. J. Fish. Aquat. Sci.* **59**, 350–360.

Bianchini, A., and Bowles, K. C. (2002). Metal sulfides in oxygenated aquatic systems. Implications for the biotic ligand model. *Comp. Biochem. Physiol. C* **133**, 51–64.

Bianchini, A., Bowles, K. C., Brauner, C. J., Gorsuch, J. W, Kramer, J. R., and Wood, C. M. (2002a). Evaluation of the effect of reactive sulfide on the acute toxicity of silver (I) to *Daphnia magna.* Part 2. Toxicity results. *Environ. Toxicol. Chem.* **21**, 1294–1300.

Bianchini, A., Grosell, M., Gregory, S. M., and Wood, C. M. (2002b). Acute silver toxicity in aquatic animals is a function of sodium uptake rate. *Environ. Sci. Technol.* **36**, 1763–1766.

Bielmyer, G. K., Grosell, M., Paquin, P. R., Mathews, R., Wu, K. B., Santore, R. C., and Brix, K. V. (2007). Validation study of the acute biotic ligand model for silver. *Environ. Toxicol. Chem.* **26**, 2241–2246.

Bielmyer, G., Brix, K., and Grosell, M. (2008). Is Cl$^-$ protection against silver toxicity due to chemical speciation? *Aquat. Toxicol.* **87**, 81–87.

Bowles, K. C., Bianchini, A., Brauner, C. J., Kramer, J. R., and Wood, C. M. (2002). Evaluation of the effect of reactive sulfide on the acute toxicity of silver (I) to *Daphnia magna*. Part 1. Description of the chemical system. *Environ. Toxicol. Chem.* **21**, 1286–1293.

Brauner, C. J., and Wood, C. M. (2002a). Effect of long-term silver exposure on survival and ionoregulatory development in rainbow trout (*Oncorhynchus mykiss*) embryos and larvae, in the presence and absence of added dissolved organic matter. *Comp. Biochem. Physiol. C* **133**, 161–173.

Brauner, C. J., and Wood, C. M. (2002b). Ionoregulatory development and the effect of chronic silver exposure on growth, survival, and sublethal indicators of toxicity in early life stages of rainbow trout (*Oncorhynchus mykiss*). *J. Comp. Physiol. B* **172**, 153–162.

Brauner, C., Wilson, J., Kamunde, C., and Wood, C. M. (2003). Water chloride provides partial protection during chronic exposure to waterborne silver in rainbow trout (*Oncorhynchus mykiss*) embryos and larvae. *Physiol. Biochem. Zool.* **76**, 803–815.

Brown, S. B., Evans, R. E., Thompson, B. E., and Hara, T. J. (1982). Chemoreception and aquatic pollutants. In: *Chemoreception in Fishes* (T.J. Hara, ed.), pp. 363–393. Elsevier, New York.

Buck, K. N., Ross, J., Flegal, A. R., and Bruland, K. W. (2007). A review of total dissolved copper and its chemical speciation in San Francisco Bay, CA. *Environ. Res.* **105**, 5–19.

Bury, N. R. (2005). The changes to apical silver membrane uptake, and basolateral membrane silver export in the gills of rainbow trout (*Oncorhynchus mykiss*) on exposure to sublethal silver concentrations. *Aquat. Toxicol.* **72**, 135–145.

Bury, N. R., and Hogstrand, C. (2002). Influence of chloride and metals on silver bioavailability to Atlantic salmon (*Salmo salar*) and rainbow trout (*Oncorhynchus mykiss*) yolk-sac fry. *Environ. Sci. Technol.* **36**, 2884–2888.

Bury, N. R., and Wood, C. M. (1999). Mechanism of branchial apical silver uptake by rainbow trout is via the proton-coupled Na^+ channel. *Am. J. Physiol.* **277**, R1385–R1391.

Bury, N. R., Galvez, F., and Wood, C. M. (1999a). The effects of chloride, calcium, and dissolved organic carbon on silver toxicity: comparison between rainbow trout and fathead minnows. *Environ. Toxicol. Chem.* **18**, 56–62.

Bury, N. R., McGeer, J. C., and Wood, C. M. (1999b). Effects of altering freshwater chemistry on the physiological responses of rainbow trout to silver exposure. *Environ. Toxicol. Chem.* **18**, 49–55.

Bury, N. R., Grosell, M., Grover, A. K., and Wood, C. M. (1999c). ATP-dependent silver transport across the basolateral membrane of rainbow trout gills. *Toxicol. Appl. Pharmacol.* **159**, 1–8.

Bury, N., Shaw, J., Glover, C., and Hogstrand, C. (2002). Derivation of a toxicity-based model to predict how water chemistry influences silver toxicity to invertebrates. *Comp. Biochem. Physiol. C* **133**, 259–270.

Bury, N. R., Walker, P. A., and Glover, C. N. (2003). Nutritive metal uptake in teleost fish. *J. Exp. Biol.* **206**, 11–23.

Campbell, P. G. C., Paquin, P. R., Adams, W. J., Brix, K. V., Juberg, D. R., Playle, R. C., Ruffing, C. J., and Wentsel, R. S. (2002). Risk assessment. In: *Silver in the Environment: Transport, Fate, and Effects* (A.W. Andren and T.W. Bober, eds), pp. 97–139. Society of Environmental Toxicology and Chemistry (SETAC), Pensacola, FL.

CCME (2007). *Canadian Water Quality Guidelines for the Protection of Aquatic Life.* Canadian Council of Ministers of the Environment, Government of Canada, Winnipeg.

Christensen, G. M., and Tucker, J. H. (1976). Effects of selected water toxicants on the *in vitro* activity of fish carbonic anhydrase. *Chem. Biol. Interact.* **13**, 181–192.

Coleman, R. L., and Cearley, J. E. (1974). Silver toxicity and accumulation in largemouth bass and bluegill. *Bull. Environ. Contam. Toxicol.* **12**, 53–61.

CONAMA (2005). *Ministério do Meio Ambiente. Resolução no. 357, de 17 de Março de 2005.* Conselho Nacional do Meio Ambiente, Brasília.

Cooley, A. C., Dagon, T. J., Jenkins, P. W., and Robillard, K. A. (1988). Silver and the environment. *J. Imag. Technol.* **14**, 183–189.

Cooper, C. F., and Jolly, W. C. (1970). Ecological effects of silver iodide and other weather modification agents. A review. *Water Resour. Res.* **6**, 88–98.

Cosson, R. P. (1994a). Heavy metal intracellular balance and relationship with metallothionein induction in the gills of carp after contamination by silver, cadmium, and mercury following or not pretreatment by zinc. *Biol. Trace Elem. Res.* **46**, 229–245.

Cosson, R. P. (1994b). Heavy metal intracellular balance and relationship with metallothionein induction in the liver of carp after contamination by silver, cadmium, and mercury following or not pretreatment by zinc. *Biol. Metals* **7**, 9–19.

Davies, P. H., Goettl, J. P., Jr., and Sinley, J. R. (1978). Toxicity of silver to rainbow trout (*Salmo gairdneri*). *Water Res.* **12**, 113–117.

Davies, P. H., Brinkman, S., and McIntyre, M. (1998). Acute and chronic toxicity of silver to aquatic life at different water hardness, and effects of mountain and plains sediments on the bioavailability and toxicity of silver. In: *Water Pollution Studies*, pp. 17–48. Federal Aid Project No. F-243R-5. Fort Collins, CO: Colorado Division of Wildlife.

De Boeck, G., Grosell, M., and Wood, C. M. (2001). Sensitivity of the spiny dogfish (*Squalus acanthias*) to waterborne silver exposure. *Aquat. Toxicol.* **54**, 261–275.

De Boeck, G., Eyckmans, M., Lardon, I., Bobbaers, R., Sinha, A. K., and Blust, R. (2010). Metal accumulation and metallothionein induction in the spotted dogfish *Scyliorhinus canicula*. *Comp. Biochem. Physiol. A* **155**, 503–508.

DeForest, D. K., Brix, K. V., and Adams, W. J. (2007). Assessing metal bioaccumulation in aquatic environments: the inverse relationship between bioaccumulation factors, trophic transfer factors and exposure concentration. *Aquat. Toxicol.* **84**, 236–246.

Dethloff, G. M., Naddy, R. B., and Gorsuch, J. W. (2007). Effects of sodium chloride on chronic silver toxicity to early life stages of rainbow trout (*Oncorhynchus mykiss*). *Environ. Toxicol. Chem.* **26**, 1717–1725.

Dorfmann, D. (1977). Tolerance of *Fundulus heteroclitus* to different metals in saltwater. *Bull. N. J. Acad. Sci.* **22**, 21–33.

Eckelman, M. J., and Graedel, T. E. (2007). Silver emissions and their environmental impacts: a multilevel assessment. *Environ. Sci. Technol.* **41**, 6283–6289.

Eisler, R. (1996). *Silver Hazards to Fish, Wildlife and Invertebrates: A Synoptic Review.* National Biological Service Biological Report 32. US Department of the Interior.

Environment Canada (1980). *Guidelines for Surface Water Quality*, Vol. 1. *Inorganic Chemical Substances.* Inland Waters Directorate, Water Quality Branch, Ottawa.

Ercal, N., Gurer-Orhan, H., and Aykin-Burns, N. (2001). Toxic metals and oxidative stress. Part I: Mechanisms involved in metal-induced oxidative damage. *Curr. Top. Med. Chem.* **1**, 529–539.

Erickson, R. J., Brooke, L. T., Kahl, M. D., Venter, F. V., Harting, S. L., Marker, T. P., and Spehar, R. L. (1998). Effects of laboratory test conditions on the toxicity of silver to aquatic organisms. *Environ. Toxicol. Chem.* **17**, 572–578.

European Commission (1979). EC Shellfish Water Directive 79/923/EEC. *Official Journal* **L 281**,10.11.1979

Evans, D. H., Piermarini, P. M., and Choe, K. P. (2005). The multifunctional fish gill: dominant site of gas exchange, osmoregulation, acid–base regulation, and excretion of nitrogenous waste. *Physiol. Rev.* **85**, 97–177.

Ferguson, E. A., and Hogstrand, C. (1998). Acute silver toxicity to seawater-acclimated rainbow trout: influence of salinity on toxicity and silver speciation. *Environ. Toxicol. Chem.* **17**, 589–593.

Ferguson, E. A., Leach, D. A., and Hogstrand, C. (1996). Metallothionein protects against silver blockage of the Na^+/K^+-ATPase. In: *4th International Conference on Transport, Fate, and Effects of Silver in the Environment* (A.W. Andren and T.W. Bober, eds), pp. 191–196. University of Wisconsin Sea Grant Institute, Madison, WI.

Fines, G. A., Ballantyne, J. S., and Wright, P. A. (2001). Active urea transport and an unusual basolateral membrane composition in the gills of a marine elasmobranch. *Am. J. Physiol.* **280**, R16–R24.

Fletcher, M., Kelly, S., Pärt, P., O'Donnell, M. J., and Wood, C. M. (2000). Transport properties of cultured branchial epithelia from freshwater rainbow trout: a novel preparation with mitochondria-rich cells. *J. Exp. Biol.* **203**, 1523–1537.

Galvez, F., and Wood, C. M. (1997). The relative importance of water hardness (Ca) and chloride levels in modifying the acute toxicity of silver to rainbow trout (*Oncorhynchus mykiss*). *Environ. Toxicol. Chem.* **16**, 2363–2368.

Galvez, F., and Wood, C. M. (1999). The physiological effects of dietary silver sulphide exposure in rainbow trout. *Environ. Toxicol. Chem.* **18**, 84–88.

Galvez, F., and Wood, C. M. (2002). The mechanisms and costs of physiological and toxicological acclimation to waterborne silver in juvenile rainbow trout (*Oncorhynchus mykiss*). *J. Comp. Physiol. B* **172**, 587–597.

Galvez, F., Hogstrand, C., and Wood, C. M. (1998). Physiological responses of juvenile rainbow trout to chronic low level exposures to waterborne silver. *Comp. Biochem. Physiol. C* **119**, 131–137.

Galvez, F., Hogstrand, C., McGeer, J. C., and Wood, C. M. (2001). The physiological effects of a biologically incorporated silver diet in rainbow trout (*Oncorhynchus mykiss*). *Aquat. Toxicol.* **55**, 95–112.

Galvez, F. H., Mayer, G. D., Wood, C. M., and Hogstrand, H. (2002). The distribution kinetics of waterborne silver-110m in juvenile rainbow trout. *Comp. Biochem. Physiol. C* **131**, 367–378.

Garnier, J., and Baudin, J. P. (1990). Retention of ingested [110m]Ag by a freshwater fish, *Salmo trutta* L. *Water Air Soil Pollut.* **50**, 409–421.

Garnier, J., Baudin, J. P., and Foulquier, L. (1990). Accumulation from water and depuration of [110m]Ag by a freshwater fish, *Salmo trutta* L. *Water Res.* **24**, 1407–1414.

Glover, C. N., Playle, R. C., and Wood, C. M. (2005). Heterogeneity in natural organic matter (NOM) amelioration of silver toxicity to *Daphnia magna*: effect of NOM source and silver–NOM equilibration time. *Environ. Toxicol. Chem.* **24**, 2934–2940.

Gorsuch, J.W., Kramer, J.R. and La Point, T.W. (eds) (2003). *Papers from Environmental Toxicology and Chemistry, 1983–2002.* Society of Environmental Toxicology and Chemistry (SETAC), Pensacola, FL.

Goss, G., Gilmour, K., Hawkins, G., Brumbach, J., Huynh, M., and Galvez, F. (2011). Mechanism of sodium uptake in PNA negative MR cells from rainbow trout, *Oncorhynchus mykiss* as revealed by silver and copper inhibition. *Comp. Biochem. Physiol. A* **159**, 234–241.

Grosell, M. (2006). Intestinal anion exchange in marine fish osmoregulation. *J. Exp. Biol.* **209**, 2813–2827.

Grosell, M., and Wood, C. M. (2001). Branchial versus intestinal silver toxicity and uptake in the marine teleost (*Parophrys vetulus*). *J. Comp. Physiol. B* **171**, 585–594.

Grosell, M., and Wood, C. M. (2002). Copper uptake across rainbow trout gills: mechanisms of apical entry. *J. Exp. Biol.* **205**, 1179–1188.

Grosell, M., De Boeck, G., Johannsson, O., and Wood, C. M. (1999). The effects of silver on intestinal ion and acid–base regulation in the marine teleost fish. *Parophrys vetulus. Comp. Biochem. Physiol. C* **124**, 259–270.

Grosell, M., Hogstrand, C., Wood, C. M., and Hansen, J. M. (2000). A nose-to-nose comparison of the physiological effects of exposure to ionic silver versus silver chloride in the European eel (*Anguilla anguilla*) and the rainbow trout (*Oncorhynchus mykiss*). *Aquat. Toxicol.* **48**, 327–343.

Guadagnolo, C. M., Brauner, C. J., and Wood, C. M. (2000). Effects of an acute silver challenge on survival, silver distribution and ionoregulation within developing rainbow trout eggs (*Oncorhynchus mykiss*). *Aquat. Toxicol.* **51**, 195–211.

Guadagnolo, C. M., Brauner, C. J., and Wood, C. M. (2001). Chronic effects of silver exposure on ion levels, survival, and silver distribution, within developing rainbow trout embryos (*Oncorhynchus mykiss*). *Environ. Toxicol. Chem.* **20**, 553–560.

Hansen, H., Grosell, M., Jacobsen, U., Jorgensen, J. C., Hogstrand, C., and Wood, C. M. (2002). Precautions in the use of 110mAg, containing transmutation produced 109Cd, as a tracer of silver metabolism in ecotoxicology. *Environ. Toxicol. Chem.* **21**, 1004–1008.

Hanson, S. R., Donley, S. A., and Linder, M. C. (2001). Transport of silver in virgin and lactating rats and relation to copper. *J. Trace Elem. Med. Biol.* **15**, 243–253.

Hazon, N., Wells, A., Pillans, R. D., Good, J. P., Anderson, W. G., and Franklin, C. E. (2003). Urea based osmoregulation and endocrine control in elasmobranch fish with special reference to euryhalinity. *Comp. Biochem. Physiol. B* **136**, 685–700.

Hogstrand, C., and Wood, C. M. (1998). Towards a better understanding of the bioavailability, physiology, and toxicology of silver in fish: Implications for water quality criteria. *Environ. Toxicol. Chem.* **17**, 547–561.

Hogstrand, C., Galvez, F., and Wood, C. M. (1996). Toxicity, silver accumulation and metallothionein induction in freshwater rainbow trout during exposure to different silver salts. *Environ. Toxicol. Chem.* **15**, 1102–1108.

Hogstrand, C., Ferguson, E. A., Galvez, F., Shaw, J. R, Webb, N. A., and Wood, C. M. (1999). Physiology of acute silver toxicity in the starry flounder (*Platichthys stellatus*) in seawater. *J. Comp. Physiol. B* **169**, 461–473.

Hogstrand, C., Wood, C. M., Bury, N. R., Wilson, R. W., Rankin, J. C., and Grosell, M. (2002). Binding and movement of silver in the intestinal epithelium of a marine teleost fish, the European flounder (*Platichthys flesus*). *Comp. Biochem. Physiol. C* **133**, 125–135.

Hogstrand, C., Grosell, M., Wood, C. M., and Hansen, H. (2003). Internal redistribution of radiolabelled silver among tissues of rainbow trout (*Oncorhynchus mykiss*) and European eel (*Anguilla anguilla*): the influence of silver speciation. *Aquat. Toxicol.* **63**, 139–157.

Holcombe, G. W., Phipps, G. L., and Fiandt, J. T. (1983). Toxicity of selected priority pollutants to various aquatic organisms. *Ecotoxicol. Environ. Saf.* **7**, 400–409.

Hook, S. E., and Fisher, N. S. (2001). Sublethal effects of silver in zooplankton; importance of exposure pathways and implications for toxicity testing. *Environ. Toxicol. Chem.* **20**, 568–574.

Hussain, S., Meneghini, E., Moosmayer, M., LaCotte, D., and Anner, B. M. (1994). Potent and reversible interaction of silver with pure Na,K-ATPase and Na,K-ATPase liposomes. *Biochim. Biophys. Acta* **1190**, 402–408.

Hussain, S., Anner, R. M., and Anner, B. M. (1995). Metallothionein protects purified Na, K-ATPase from metal toxicity *in vitro*. *In Vitro Toxicol.* **8**, 25–30.

Janes, N., and Playle, R. C. (1995). Modeling silver binding to gills of rainbow trout (*Oncorhynchus mykiss*). *Environ. Toxicol. Chem.* **14**, 1847–1858.

Janes, N., and Playle, R. C. (2000). Protection by two complexing agents, thiosulphate and dissolved organic matter, against the physiological effects of silver nitrate to rainbow trout (*Oncorhynchus mykiss*) in ion-poor water. *Aquat. Toxicol.* **51**, 1–18.

Jones, J. R. E. (1939). The relation between the electrolytic solution pressures of the metals and their toxicity to the stickleback (*Gasterosteus aculeatus* L.). *J. Exp. Biol* **16**, 425–437.

Karen, D. J., Ownby, D. R., Forsythe, B. L., Bills, T. P., La Point, T. W., Cobb, G. B., and Klaine, S. J. (1999). Influence of water quality on silver toxicity to rainbow trout (*Oncorhynchus mykiss*), fathead minnows (*Pimephales promelas*), and water fleas (*Daphnia magna*). *Environ. Toxicol. Chem.* **18**, 63–70.

Klaine, S. J., Bills, T. L., Wenholz, M. D., La Point, T. W., Cobb, G. P., and Forsythe, B. L., II (1996). Influence of age sensitivity on the acute toxicity of silver to fathead minnows at various water quality parameters. In: *4th International Conference on Transport, Fate, and Effects of Silver in the Environment* (A.W. Andren and T.W. Bober, eds), pp. 125–129. University of Wisconsin Sea Grant Institute, Madison, WI.

Klein-MacPhee, G., Cardin, J. A., and Berry, W. J. (1984). Effects of silver on eggs and larvae of the winter flounder. *Trans. Am. Fish. Soc.* **113**, 247–251.

Kolts, J. M., Boese, C. J., and Meyer, J. S. (2006). Acute toxicity of copper and silver to *Ceriodaphnia dubia* in the presence of food. *Environ. Toxicol. Chem.* **25**, 1831–1835.

Kolts, J. M., Brooks, M. L., Cantrell, B. D., Boese, C. J., Bell, R. A., and Meyer, J. S. (2008). Dissolved fraction of standard laboratory cladoceran food alters toxicity of waterborne silver to *Ceriodaphnia dubia*. *Environ. Toxicol. Chem.* **27**, 1426–1434.

Kramer, J. R., Adams, N. W. H., Manolopoulos, H., and Collins, P. V. (1999). Silver at an old mining camp, Cobalt, Ontario, Canada. *Environ. Toxicol. Chem.* **18**, 23–29.

Kramer, J. R., Benoit, G., Bowles, K. C., DiToro, D. M., Herrin, R. T., Luther, G. W., III, Manolopoulos, H., Robillard, K. A., Shafer, M. M., and Shaw, J. R. (2002). Environmental Chemistry of Silver. In: *Silver in the Environment: Transport, Fate, and Effects* (A.W. Andren and T.W. Bober, eds), pp. 1–25. Society of Environmental Toxicology and Chemistry (SETAC), Pensacola, FL.

LeBlanc, G. A., Mastone, J. D., Paradice, A. P., and Wilson, B. F. (1984). The influence of speciation on the toxicity of silver to fathead minnow (*Pimephales promelas*). *Environ. Toxicol. Chem.* **3**, 37–46.

Lemke, A. E. (1981). Interlaboratory Comparison: Acute Testing Set. *EPA 600/3-81-005*. Office of Pesticides and Toxic Substances, US Environmental Protection Agency, Washington, DC.

Luoma, S. N., Hogstrand, C., Bell, R. A., Bielmyer, G. K., Galvez, F., LeBlanc, G. A., Lee, B.-G., Purcell, T. W., Santore, R. C., Santschi, P. F., and Shawa, J. R. (2002). Biological processes. In: *Silver in the Environment: Transport, Fate, and Effects* (A.W. Andren and T.W. Bober, eds), pp. 65–95. Society of Environmental Toxicology and Chemistry (SETAC), Pensacola, FL.

Lytle, P. E. (1984). Fate and speciation of silver in publicly owned treatment plants. *Environ. Toxicol. Chem.* **3**, 21–30.

Mann, R. M., Ernste, M. J., Bell, R. A., Kramer, J. R., and Wood, C. M. (2004). Evaluation of the protective effect of reactive sulfide on the acute toxicity of silver to rainbow trout (*Oncorhynchus mykiss*). *Environ. Toxicol. Chem.* **23**, 1204–1210.

Mathews, T., and Fisher, N. S. (2009). Dominance of dietary intake of metals in marine elasmobranch and teleost fish. *Sci. Total Environ.* **407**, 5156–5161.

Mayer, G. D., Leach, A., Kling, P., Olsson, P.-E., and Hogstrand, C. (2003). Activation of the rainbow trout metallothionein-A promoter by silver and zinc. *Comp. Biochem. Physiol. B* **134**, 181–188.

McDonald, D. G., and Wood, C. M. (1993). Branchial acclimation to metals. In: *Fish Ecophysiology* (J.C. Rankin and F.B. Jensen, eds), pp. 297–321. London: Chapman.

McGeer, J. C., and Wood, C. M. (1998). Protective effects of water Cl⁻ on physiological responses to waterborne silver in rainbow trout. *Can. J. Fish. Aquat. Sci.* **55**, 2447–2454.

McGeer, J. C., Playle, R. C., Wood, C. M., and Galvez, F. (2000). A physiologically based biotic ligand model for predicting the acute toxicity of waterborne silver to rainbow trout in fresh waters. *Environ. Sci. Technol.* **34**, 4199–4207.

McGeer, J. C., Brix, K. V., Skeaff, J. M., Deforest, D. K., Brigham, S. I., Adams, W. J., and Green, A. (2003). Inverse relationship between bioconcentration factor and exposure concentration for metals: implications for hazard assessment of metals in the aquatic environment. *Environ. Toxicol. Chem.* **22**, 1017–1037.

Miller, L. A., and Bruland, K. W. (1995). Organic speciation of silver in marine waters. *Environ. Sci. Technol.* **29**, 2616–2621.

Milligan, C. L., and Wood, C. M. (1982). Disturbances in hematology, fluid volume distribution, and circulatory function associated with low environmental pH in the rainbow trout, *Salmo gairdneri*. *J. Exp. Biol.* **99**, 397–415.

Morgan, T. P., and Wood, C. M. (2004). A relationship between gill silver accumulation and acute silver toxicity in the freshwater rainbow trout: support for the acute silver biotic ligand model. *Environ. Toxicol. Chem.* **23**, 1261–1267.

Morgan, I. J., Henry, R. P., and Wood, C. M. (1997). The mechanism of acute silver nitrate toxicity in freshwater rainbow trout (*Oncorhynchus mykiss*) is inhibition of gill Na^+ and Cl^- transport. *Aquat. Toxicol.* **38**, 145–163.

Morgan, T. P., Grosell, M., Gilmour, K. M., Playle, R. C., and Wood, C. M. (2004a). Time course analysis of the mechanism by which silver inhibits active Na^+ and Cl^- uptake in the gills of rainbow trout. *Am. J. Physiol.* **287**, R234–R242.

Morgan, T. P., Grosell, M. G., Playle, R. C., and Wood, C. M. (2004b). The time course of silver accumulation in rainbow trout during static exposure to silver nitrate: physiological regulation or an artifact of the exposure conditions? *Aquat. Toxicol.* **66**, 55–72.

Morgan, T. P., Guadagnolo, C. M., Grosell, M., and Wood, C. M. (2005a). Effects of water hardness on the physiological responses to chronic waterborne silver exposure in early life stages of rainbow trout (*Oncorhynchus mykiss*). *Aquat. Toxicol.* **74**, 333–350.

Morgan, T. P., Guadagnolo, C. M., Grosell, M., and Wood, C. M. (2005b). Effects of water hardness on toxicological responses to chronic waterborne silver exposure in early life stages of rainbow trout (*Oncorhynchus mykiss*). *Environ. Toxicol. Chem.* **24**, 1642–1647.

Naddy, R. B., Rehner, A. B., McNerney, G. R., Gorsuch, J. W., Kramer, J. R., Wood, C. M., Paquin, P. R., and Stubblefield, W. A. (2007). Comparison of short-term chronic and chronic silver toxicity to fathead minnows in unamended and NaCl-amended waters. *Environ. Toxicol. Chem.* **26**, 1922–1930.

Nebeker, A. V., McAuliffe, C. K., Mshar, R., and Stevens, D. G. (1983). Toxicity of silver to steelhead and rainbow trout, fathead minnows and *Daphnia magna*. *Environ. Toxicol. Chem.* **2**, 95–104.

Nechay, B. R., and Saunders, J. P. (1984). Inhibition of adenine triphosphatases *in vitro* by silver nitrate and silver sulfadiazine. *J. Am. Coll. Toxicol.* **3**, 37–42.

Nichols, J. W., and Playle, R. C. (2004). Influence of temperature on silver accumulation and depuration in rainbow trout. *J. Fish Biol.* **64**, 1638–1654.

Nichols, J. W., Brown, S., Wood, C. M., Walsh, P. J., and Playle, R. C. (2006). Influence of salinity and organic matter on silver accumulation in gulf toadfish (*Opsanus beta*). *Aquat. Toxicol.* **78**, 253–261.

Niyogi, S., and Wood, C. M. (2004). Biotic ligand model, a flexible tool for developing site-specific water quality guidelines for metals. *Environ. Sci. Technol.* **38**, 6177–6192.

Ojo, A. A., and Wood, C. M. (2007). *In vitro* analysis of the bioavailability of six metals via the gastro-intestinal tract of the rainbow trout (*Oncorhynchus mykiss*). *Aquat. Toxicol.* **83**, 10–23.

Paquin, P., Di Toro, D., Santore, R., Trivedi, D., and Wu, B. (1999). A Biotic Ligand Model of the Acute Toxicity of Metals III. Application to Fish and Daphnia Exposure to Silver. *EPA 822-E-00-001*. US Government Printing Office, Washington, DC.

Paquin, P. R., Santore, R. C., Wu, K. B., Kavvadas, C. D., and Di Toro, D. M. (2000). The biotic ligand model: a model of the acute toxicity of metals to aquatic life. *Environ. Sci. Pollut.* **3**, 175–182.

Paquin, P. R., Gorsuch, J. W., Apte, S., Batley, G. E., Bowles, K. C., Campbell, P. G. C., Delos, C. G., Di Toro, D. M., Dwyer, R. L., Galvez, F., Gensemer, R. W., Goss, G. G., Hogstrand, C., Janssen, C. R., McGeer, J. M., Naddy, R. B., Playle, R. C., Santore, R. C., Schneider, U., Stubblefield, W. A., Wood, C. M., and Wu, K. B. (2002a). The biotic ligand model: a historical overview. *Comp. Biochem. Physiol. C* **133**, 3–35.

Paquin, P. R., Zoltay, V., Winfield, R. P., Wu, K. B., Mathew, R., Santore, R. C., and Di Toro, D. M. (2002b). Extension of the biotic ligand model of acute toxicity to a physiologically-based model of the survival time of rainbow trout (*Oncorhynchus mykiss*) exposed to silver. *Comp. Biochem. Physiol. C* **133**, 305–343.

Pärt, P., Wright, P. A., and Wood, C. M. (1998). Urea and water permeability in dogfish (*Squalus acanthias*) gills. *Comp. Biochem. Physiol. A* **199**, 117–123.

Pentreath, R. J. (1977). The accumulation of [110m]Ag by the plaice, *Pleuronectes platessa* L. and the thornback ray, *Raja clavata* L. *J. Exp. Mar. Biol. Ecol.* **29**, 315–325.

Perry, S. F. (1997). The chloride cell: Structure and function in the gills of freshwater fishes. *Annu. Rev. Physiol.* **59**, 325–347.

Perry, S. F., and Wood, C. M. (1985). Kinetics of branchial calcium uptake in the rainbow trout: effects of acclimation to various external calcium levels. *J. Exp. Biol.* **116**, 411–433.

Playle, R. C. (2004). Using multiple metal–gill binding models and the toxic unit concept to help reconcile multiple-metal toxicity results. *Aquat. Toxicol.* **67**, 359–370.

Purcell, T. W., and Peters, J. J. (1998). Sources of silver in the environment. *Environ. Toxicol. Chem.* **17**, 539–546.

Purcell, T. W., and Peters, J. J. (1999). Historical impacts of environmental regulation of silver. *Environ. Toxicol. Chem.* **18**, 3–8.

Pyle, G., and Wood, C. M. (2007). Predicting "non-scents": rationale for a chemosensory-based biotic ligand model. *Austral. J. Ecotoxicol.* **13**, 47–51.

Ratte, H. T. (1999). Bioaccumulation and toxicity of silver compounds: a review. *Environ. Toxicol. Chem.* **18**, 89–108.

Reinfelder, J. R., and Chang, S. I. (1999). Speciation and microalgal bioavailability of inorganic silver. *Environ. Sci. Technol.* **33**, 1860–1863.

Ribeyre, F., Amiard-Triquet, C., Boudou, A., and Amiard, J.-C. (1995). Experimental study of interactions between five trace elements – Cu, Ag, Se, Zn and Hg – toward their bioaccumulation by fish (*Brachydanio rerio*) from the direct route. *Ecotoxicol. Environ. Saf.* **32**, 1–11.

RIVM (1999). *Risk Limits for Boron, Silver, Titanium, Tellurium, Uranium and Organosilicon Compounds in the Framework of EU Directive 76/464/EEC. "Setting Integrated Environmental Quality Standards"* (compiled by E. Van de Plassche, M. Van De Hoop, R. Posthumus and T. Crommentuijn). RIVM REPORT 601501 005. The Hague: Rijksinstuut voor Volksgezondheid en Milieu.

Rouleau, C., Gobeil, C., and Tjalve, H. (2000). Accumulation of silver from the diet in two marine benthic predators: the snow crab (*Chionoecetes opilio*) and American plaice (*Hippoglossoides platessoides*). *Environ. Toxicol. Chem* **19**, 631–637.

Schnizler, M. K., Bogdan, R., Bennert, A., Bury, N. R., Fronius, M., and Clauss, W. (2007). Short-term exposure to waterborne free silver has acute effects on membrane current of *Xenopus* oocytes. *Biochim. Biophys. Acta* **1768**, 317–323.

Schwartz, M. L., and Playle, R. C. (2001). Adding magnesium to the silver–gill binding model for rainbow trout (*Oncorhynchus mykiss*). *Environ. Toxicol. Chem.* **20**, 467–472.

Shafer, M. M., Overdier, J. T., and Armstong, D. E. (1998). Removal, partitioning and fate of silver and other metals in wastewater treatment plants and effluent-receiving streams. *Environ. Toxicol. Chem.* **17**, 630–641.

Shaw, J. R., Wood, C. M., Birge, W. J., and Hogstrand, C. (1998). Toxicity of silver to a marine teleost, the tidepool sculpin (*Oligocottus maculosus*): effects of salinity and ammonia. *Environ. Toxicol. Chem.* **17**, 594–600.

Silver Institute (2010). *Silver – The Indispensable Metal.* http://www.silverinstitute.org/supply_demand.php

Sloman, K. A., Morgan, T. P., McDonald, D. G., and Wood, C. M. (2003). Socially-induced changes in sodium regulation affect the uptake of waterborne copper and silver in the rainbow trout. *Oncorhynchus mykiss. Comp. Biochem. Physiol. A.* **135**, 393–403.

Soyut, H., and Beydemir, S. (2008). Purification and some kinetic properties of carbonic anhydrase from rainbow trout (*Oncorhynchus mykiss*) liver and metal inhibition. *Prot. Pept. Lett.* **15**, 528–535.

Soyut, H., Beydemir, S., and Hisar, O. (2008). Effects of some metals on carbonic anhydrase from brains of rainbow trout. *Biol. Trace Elem. Res.* **123**, 179–190.

Stewart, P. A. (1981). *How to Understand Acid–Base, A Quantitative Acid–Base Primer for Biology and Medicine.* Elsevier, New York.

Terhaar, C. J., Ewell, W. S., Dziuba, S. P., and Fassett, D. W. (1972). Toxicity of photographic processing chemicals to fish. *Photograph. Sci. Eng.* **16**, 370–377.

Terhaar, C. J., Ewell, W. S., Dziuba, S. P., White, W. W., and Murphy, P. J. (1977). A laboratory model for evaluating the behavior of heavy metals in an aquatic environment. *Water Res.* **11**, 101–110.

Tjälve, H., and Henriksson, J. (1999). Uptake of metals in the brain via olfactory pathways. *Neurotoxicology* **20**, 181–196.

USEPA (1980). Ambient Water Quality Criteria for Silver. *EPA-440/5-80-071.* Office of Water, Regulations and Standards, Criteria and Standards Division, US Environmental Protection Agency, Washington, DC.

USEPA (1986). Quality Criteria for Water. *EPA 440/5-86-001.* Office of Water, Regulations and Standards, Criteria and Standards Division, US Environmental Protection Agency, Washington, DC.

USEPA (1992). Draft Ambient Aquatic Life Water Quality Criteria for Silver. *EPA-440:5-87-011.* NTIS, US Environmental Protection Agency, Springfield, VA.

VanGenderen, E. J., Ryan, A. C., Tomasso, J. R., and Klaine, S. J. (2003). Influence of dissolved organic matter source on silver toxicity to *Pimephales promelas. Environ. Toxicol. Chem.* **22**, 2746–2751.

Voyer, R. A., Cardin, I. A., Heltshe, J. F., and Hoffman, G. L. (1982). Viability of embryos of the winter flounder *Pseudopleuronectes americanus* exposed to mixtures of cadmium and silver in combination with selected fixed salinities. *Aquat. Toxicol.* **2**, 223–233.

Walker, P. A., Bury, N. R., and Hogstrand, C. (2007). Influence of culture conditions on metal-induced responses in a cultured rainbow trout gill epithelium. *Environ. Sci. Technol.* **41**, 6505–6513.

Walker, P. A., Kille, P., Hurley, A., Bury, N. R., and Hogstrand, C. (2008). An *in vitro* method to assess toxicity of waterborne metals to fish. *Toxicol. Appl. Pharmacol.* **230**, 67–77.

Ward, T. J., and Kramer, J. R. (2002). Silver speciation during chronic toxicity tests. *Comp. Biochem. Physiol. C* **133**, 75–86.

Ward, T. J., Boeri, R. L., Hogstrand, C., Kramer, J. R., Lussier, S. M., Stubblefield, W. A., Wyskiel, D. C., and Gorsuch, J. W. (2006). Influence of salinity and organic carbon on the chronic toxicity of silver to mysids (*Americamysis bahia*) and silversides (*Menidia beryllina*). *Environ. Toxicol. Chem.* **25**, 1809–1816.

Webb, N. A., and Wood, C. M. (1998). Physiological analysis of the stress response associated with acute silver nitrate exposure in freshwater rainbow trout. *Environ. Toxicol. Chem.* **17**, 579–588.

Webb, N. A., and Wood, C. M. (2000). Bioaccumulation and distribution of silver in four marine teleosts and two marine elasmobranchs: influence of exposure duration, concentration, and salinity. *Aquat. Toxicol.* **49**, 111–129.

Webb, N. A., Shaw, J. R., Morgan, I. J., Hogstrand, C., and Wood, C. M. (2001). Acute and chronic physiological effects of silver exposure in three marine teleosts. *Aquat. Toxicol.* **54**, 161–178.

Wen, L. S., Santschi, P. H., Gill, G., and Paternostro, C. (1997). Colloidal and particulate silver in river and estuarine waters of Texas. *Environ. Sci. Technol.* **31**, 723–731.

Wilson, R. W., and Taylor, E. W. (1993). The physiological responses of freshwater rainbow trout, *Oncorhynchus mykiss*, during acutely lethal copper exposure. *J. Comp. Physiol. B* **163**, 38–47.

Wood, C. M. (2001). Toxic responses of the gill. In: *Target Organ Toxicity in Marine and Freshwater Teleosts, Vol. 1. Organs.* (D.W Schlenk and W.H Benson, eds), pp. 1–89. Taylor and Francis, Washington, DC.

Wood, C. M., Hogstrand, C., Galvez, F., and Munger, R. S. (1996a). The physiology of waterborne silver toxicity in freshwater rainbow trout (*Oncorhynchus mykiss*): 1. The effects of ionic Ag⁺. *Aquat. Toxicol.* **35**, 93–109.

Wood, C. M., Hogstrand, C., Galvez, F., and Munger, R. S. (1996b). The physiology of waterborne silver toxicity in freshwater rainbow trout (*Oncorhynchus mykiss*): 2. The effects of silver thiosulphate. *Aquat. Toxicol.* **35**, 111–125.

Wood, C. M., Playle, R. C., and Hogstrand, C. (1999). Physiology and modelling of the mechanisms of silver uptake and toxicity in fish. *Environ. Toxicol. Chem.* **18**, 71–83.

Wood, C. M., Grosell, M., Hogstrand, C., and Hansen, H. (2002a). Kinetics of radiolabeled silver uptake and depuration in the gills of rainbow trout (*Oncorhynchus mykiss*) and European eel (*Anguilla anguilla*): the influence of silver speciation. *Aquat. Toxicol.* **56**, 197–213.

Wood, C. M., La Point, T. W., Armstrong, D. E., Birge, W. J., Brauner, C. J., Brix, K. V., Call, D. J., Crecelius, E. A., Davies, P. H., Gorsuch, J. W., Hogstrand, C., Mahony, J. D., McGeer, J. C., and O'Connor, T. P. (2002b). Biological effects of silver. In: *Silver in the Environment: Transport, Fate, and Effects* (A.W. Andren and T.W. Bober, eds), pp. 27–63. Society of Environmental Toxicology and Chemistry (SETAC), Pensacola, FL.

Wood, C. M., McDonald, M. D., Walker, P., Grosell, M., Barimo, J. F., Playle, R. C., and Walsh, P. J. (2004). Bioavailability of silver and its relationship to ionoregulation and silver speciation across a range of salinities in the gulf toadfish (*Opsanus beta*). *Aquat. Toxicol.* **70**, 137–157.

Wood, C. M., Grosell, M., McDonald, D. M., Playle, R. C., and Walsh, P. J. (2010). Effects of waterborne silver in a marine teleost, the gulf toadfish (*Opsanus beta*): effects of feeding and chronic exposure on bioaccumulation and physiological responses. *Aquat. Toxicol.* **99**, 138–148.

Zhou, B., Nichols, J., Playle, R. C., and Wood, C. M. (2004). An *in vitro* biotic ligand model (BLM) for silver binding to cultured gill epithelia of freshwater rainbow trout (*Oncorhynchus mykiss*). *Toxicol. Appl. Pharmacol.* **202**, 25–37.

2

ALUMINUM

ROD W. WILSON

Homeostasis and Toxicology of Non-Essential Metals: Volume 31B
FISH PHYSIOLOGY

Aluminum (Al) has no established biological function but can be extremely toxic to fish when solubilized under acidic (pH < 6) or alkaline conditions (pH > 8), and is particularly important in explaining fish population crashes associated with freshwater acidification. The gill is the target organ for waterborne Al toxicity, with accumulation in internal organs being extremely slow and no internal toxic effects in fish yet documented. Dietary toxicity is negligible and bioaccumulation via the food chain does not occur. Toxic mechanisms include impairment of gill ionoregulation (acceleration of passive ion losses and inhibition of their active uptake) by cationic Al species (especially Al^{3+}), and/or respiratory dysfunction due to precipitation of $Al(OH)_3$ or polymerization of aluminum hydroxides on the gill surface during alkalinization of water passing over the gills. The latter occurs particularly in the pH range 5–6 and causes gill inflammation, histopathologies, and excessive mucus accumulation. Secondary symptoms of these toxic mechanisms include impairment of cardiovascular and aerobic swimming performance, and reduced appetite, growth, spontaneous activity, and reproductive success. Episodic exposure to extremes of [Al] and low pH may be the limiting factor for many fish species, although chronic low-level exposure can provide increased resistance (i.e. acclimation). Of additional concern is the subsequent impairment of seawater tolerance in migratory salmonids (e.g. salmon smolts) following even short-term (2 days) exposure to mildly elevated Al in freshwater. Aluminum can also be a problem under alkaline conditions (pH > 8) when Al salts are used to clarify turbid water and remove phosphate nutrients that cause eutrophication, but the toxicity of the prevailing aluminate anion, $Al(OH)_4^-$, is low and the toxic mechanism is not known.

1. INTRODUCTION

Several comprehensive reviews have been published on aluminum (Al) chemistry and its biological effects in the aquatic environment (Havas, 1986a, b; Havas and Jaworski, 1986; Driscoll and Schecher, 1988; Sigel and Sigel, 1988; Lewis, 1989; Sposito, 1989, 1996; Gostomski, 1990; Howells et al., 1990; Rosseland et al., 1990; Scheuhammer, 1991; Sparling and Lowe, 1996;

Yokel and Golub, 1997; Gensemer and Playle, 1999; Lydersen and Löfgren, 2002). The reader is referred to all of these both for broader aspects of Al in the environment and for more detailed coverage of its chemistry and organismal effects that are peripheral to fish toxicology and physiology.

Aluminum has some intriguing physical and chemical properties that influence its lack of biological essentiality, economic importance, and industrial uses (see Section 3; Wood, 1984, 1985; Eichenberger, 1986; Williams, 1999; Exley, 2003). It has the atomic number 13, an atomic mass of 26.98, a melting point of 660.4°C and boiling point of 2467°C. Aluminum is a very light metal with a density of 2.70 g ml^{-3}, and it has only one naturally occurring radioisotope, ^{26}Al (Yokel, 2004). Although this isotope has been used in a number of biomedical studies, its rarity and very high cost have greatly restricted its use in environmental research projects involving fish (e.g. Playle, 1987b; Bjørnstad et al., 1992; Oughton et al., 1992). Another trivalent cation, gadolinium, Gd^{3+}, was tested as potential tracer for Al^{3+} but did not prove useful in studies on freshwater snails (Walton et al., 2010). Under natural environmental conditions Al has only one valence state, which is +3. Binding of Al^{3+} is primarily electrostatic rather than covalent and with an ionic radius of 53.5 pm, which is similar to Fe^{3+} and Mg^{2+}, it has an extremely high charge density (i.e. charge to radius ratio) (Lukiw, 2010). The speciation of Al is crucial to understanding its environmental fate and toxicology in fish, and so the following section will cover this issue in some detail.

2. CHEMICAL SPECIATION IN FRESHWATER AND SEAWATER

2.1. Freshwater

The issue of Al toxicity is of far more environmental importance in freshwater than in marine environments. This relates partly to the anthropogenic causes of elevated levels of dissolved Al (see Section 4) and partly to the very different physicochemical behavior of Al in the two aqueous media (i.e. solubility and speciation). For thorough reviews of Al chemistry and speciation the reader is referred to excellent coverage in Nordstrom and May (1996) and Gensemer and Playle (1999). For the present chapter, which focuses on Al within the context of fish physiology and toxicology, only key points that are most relevant to this topic will be summarized.

In aqueous solution, the measured concentrations of total Al can include monomeric (both inorganic and organic) and polymeric forms, as well as colloidal, particulate, and clay Al (Neville and Campbell, 1988) (Fig. 2.1).

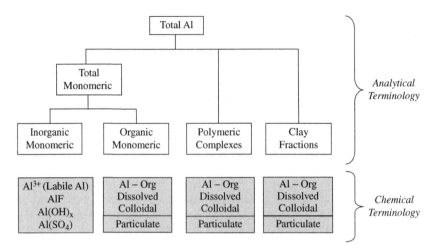

Fig. 2.1. Terminology used to describe the analytical and chemical aspects of Al in aqueous systems. After Neville and Campbell (1988); redrawn from Allin (1999).

Indeed, the complexity of Al chemistry in freshwater has been a major problem in gaining a useful understanding of the toxicity of this metal in laboratory and field studies. The greatest factor influencing the availability of Al species is pH, and Al is largely insoluble at neutral pH (the equilibrium concentration of Al^{3+} at pH 7 is estimated to be only 10^{-12} M) (Williams, 1999). As a result, significant dissolved concentrations of Al, and therefore its toxicity, are largely limited to pH values either less than 6 or greater than 8 (Driscoll and Postek, 1996; Gensemer and Playle, 1999; Winter et al., 2005). Although Al is virtually ubiquitous in the rocks and soils in contact with most freshwater ecosystems (see Section 2), its insolubility at relatively neutral pH values means that it is not a toxicological problem in the majority of freshwater environments (i.e. within a pH range of 6–8). However, the exponential increase in the solubility of Al as pH drops below 6 means that potential toxicity is a major issue within this lower pH range. This has become an increasing problem for freshwater biota as a result of increasing anthropogenic sources of freshwater acidification. Aluminum rose to infamy as an environmental problem for freshwater biota in the late 1970s and early 1980s. This was when it was first recognized that the freshwater ecosystems were being impacted by "acid rain", principally as a result of burning fossil fuels that release sulfates and nitrates into the atmosphere, which produce acidic precipitation in the form of dilute sulfuric and nitric acids (Schindler, 1988). However, the acid-affected areas suffered from not only a reduction in water pH but also a subsequently accelerated mobilization of trace metals from minerals in the surrounding soil and rocks. Indeed, it was soon realized that Al rather than pH per se may be the major factor in the decline of aquatic life observed in such systems (Cronan and Schofield, 1979; Drablos and Tollan, 1980; Muniz and Lievestad, 1980).

The solubility of Al is also greater at lower temperatures and in the presence of a range of complexing ligands (both inorganic and organic; see below), with a 15°C drop in temperature having a similar effect to an acidification of 1 pH unit (Lydersen, 1990). Both of these have important influences on the environmental effects of Al in fish. In particular, episodic acidic pulses in affected streams often occur during the winter snowmelt, thus combining low temperature and low pH, which will maximize the solubilization of Al. Furthermore, dissolved organic carbon (DOC) can vary greatly between acidic rivers, further complicating our understanding and prediction of the toxic effects of mobilized Al (McCartney et al., 2003; Laudon et al., 2005).

There is a very wide range of Al concentrations (both total and inorganic) in natural freshwaters, with the highest values recorded in the greatest pH extremes (acidic and alkaline), as expected from the influence of pH on solubilization of Al (see above). Surveys of US lakes in the Adirondacks and in Florida revealed that pH values ranged from 3.8 to 9.4 (with an average pH of 6.3 in both regions). The corresponding mean Al concentrations for the Adirondack and Florida lakes were 89 and 138 $\mu g\,L^{-1}$ for total Al, and 22 and 43 $\mu g\,L^{-1}$ for monomeric Al, respectively. However, extreme values across these two areas ranged from 8 to 1350 $\mu g\,L^{-1}$ for total Al and from 0 to 432 $\mu g\,L^{-1}$ for monomeric Al (Linthurst et al., 1986; Driscoll and Postek, 1996; Gensemer and Playle, 1999).

2.1.1. INORGANIC ALUMINUM SPECIES

Inorganic monomeric Al species are the most toxic (Driscoll, 1984; Gensemer and Playle, 1999) and are comprised of cationic Al species [Al^{3+}, $AlOH^{2+}$, and $Al(OH)_2^+$] in acidic pH conditions, $Al(OH)_3$ at intermediate pH values, and anionic $Al(OH)_4^-$ under alkaline conditions (Fig. 2.2). Aqueous Al also forms inorganic complexes with fluoride (F^-), sulfate (SO_4^{2-}), phosphate (PO_4^{3-}), and silicic acid [$Si(OH)_4$], dependent on the ligand concentrations, pH, temperature, and total ionic strength. The formation of all these complexes can affect the bioavailability and hence toxicity of Al (Birchall et al., 1989; Wilkinson et al., 1990; Gensemer and Playle, 1999; Teien et al., 2006a). However, the net effect of these complexes on fish may be difficult to predict. For example, although the presence of fluoride ions has been shown to reduce overall toxicity in juvenile Atlantic salmon, this was not explained simply by the reduction in Al^{3+} concentrations. Indeed, the presence of fluoride actually increased toxicity when based

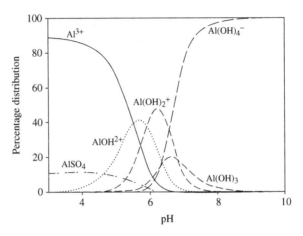

Fig. 2.2. Speciation modeling for Al in synthetic freshwater carried out using Visual MinTeq Version 2.60 over a range of pH values from 3 to 10, at 14°C (synthetic freshwater composition as used by White et al., 2008: 222 mg L^{-1} $CaCl_2$, 9.6 mg L^{-1} $MgSO_4$, 4 mg L^{-1} $KHCO_3$, 5.1 mg L^{-1} KNO_3, 58 mg L^{-1} $NaHCO_3$). Redrawn from Walton et al. (2010).

on the calculated Al^{3+} concentrations (Wilkinson et al., 1990). Other studies have shown that while low levels of fluoride (< 100 μg L^{-1}) may reduce gill damage induced by Al exposure in acid waters, higher levels of fluoride may offer no protection (Hamilton and Haines, 1995). Similarly, while silicic acid can eliminate toxicity in trout exposed to pH 5 and elevated Al (Birchall et al., 1989), the protection requires a large excess of Si over Al, which is questionable in most natural waters (LaZerte et al., 1997; Gensemer and Playle, 1999). More specifically, aluminosilicate complexes may only play a quantitatively important role in ameliorating toxicity in acidic streams which can have relatively high dissolved Si concentrations, rather than acidic lakes which typically have much lower Si concentrations (Hutchinson, 1957; Driscoll and Newton, 1985; Findlay and Kasian, 1986; Willén, 1991; Gensemer and Playle, 1999). However, there does appear to be some potential for using artificial addition of sodium silicate to ameliorate Al toxicity, for example as an alternative to liming in severely acidified streams (see Section 16 and Teien et al., 2006a).

2.1.2. Organic Aluminum Species

Of great relevance to natural freshwaters, Al can additionally form complexes (both monomeric and polymeric) with various organic ligands (proteins, lipids, sugars, and especially humic and fulvic acids), that are often collectively expressed as DOC, dissolved organic matter (DOM), or natural organic matter (NOM). Under experimental situations, citrate has also been used as an organic complexing agent for Al (e.g. Lacroix et al., 1993). Such organic acids are negatively charged at pH greater than 4, allowing them to bind Al, thus forming organic complexes. These maintain a greater total Al concentration in solution but they usually also reduce its bioavailability and toxicity, at least under acute exposures (LaCroix et al., 1993; McCartney et al., 2003; Peuranen et al., 2003; Laudon et al., 2005; Winter et al., 2005).

2.1.3. RATES OF CHEMICAL REACTION

Of particular relevance to the mechanism of acute toxicity in fish is the fact that changes in the inorganic Al species are effectively instantaneous. The speed of these chemical reactions is important when explaining the toxicity due to rapid changes in Al speciation that occur within the transit time of inhaled water along the respiratory surface of fish gills (Playle and Wood, 1990; see Section 6). In contrast, the equilibrium times for Al complexation with organic ligand (e.g. with DOC compounds) can be minutes or even hours (Gensemer and Playle, 1999), which obviously influences the design and interpretation of experimental manipulations involving these organic components. Detailed information on the stability constants for such organic complexes of Al (e.g. Nordstrom and May, 1996) has been more difficult to obtain than for inorganic complexes, largely owing to the heterogeneity of the major components of DOC (i.e. humic and fulvic acids). However, recent advances are now making modeling possible that can help to predict acute toxicity based on both inorganic and organic aspects of Al chemistry (see Section 5 on the biotic ligand model).

2.2. Seawater

Aluminum toxicity is not so relevant in seawater, and concentrations tend to be much lower than in freshwaters (at least compared to acidic and alkaline pH extremes in freshwater) (Gerensemer and Playle, 1999); oceanic values for Al concentration are $1\ \mu g\ L^{-1}$ or less (Hydes and Liss, 1977; Tria et al., 2007; Brown et al., 2010). The speciation of inorganic Al in oceanic seawater is dominated by $Al(OH)_4^-$ at normal surface pH values (e.g. 8.0–8.3), with more minor contributions from $Al(OH)_2^+$ and $Al(OH)_3^0$ (Brown et al., 2010). The major input of dissolved Al into the marine environment is from freshwater rivers, but the majority of Al entering estuaries is rapidly sedimented owing to its adsorption onto the surface of clay particles (Hydes and Liss, 1977). However, toxicity to fish has been observed as a result of the rapid changes in Al chemistry upon conversion of non-reactive Al species in freshwater inflows, during high river discharge, into reactive Al on mixing with high-salinity waters (Bjerkenes et al., 2003;

Teien et al., 2006b; see Section 4). The chemistry underlying this toxicity is poorly understood relative to freshwater toxicity. However, the Al complexed by DOC (e.g. colloidal humic and fulvic compounds), which has relatively little biological importance, can be converted to reactive Al species upon mixing with the high pH and high ionic strength coastal waters (Rosseland et al., 1998; Bjerkenes et al., 2003).

3. SOURCES (NATURAL AND ANTHROPOGENIC) OF ALUMINUM AND ECONOMIC IMPORTANCE

Aluminum is the 11th most abundant element in the universe (Frausto da Silva and Williams, 1996; Williams, 1999) and in the Earth's crust it is the most abundant metal, and the third most abundant element overall (after oxygen and silicon), comprising 8.1% by mass. There are many minerals based on Al (e.g. kaolinite, vermiculite, montmorillonite), but the mineral form that is most commonly cited when describing aqueous Al chemistry is gibbsite, which is an octahedral crystalline form of aluminum hydroxide, $Al(OH)_3$. However, the availability of Al to biota is very limited owing to its low solubility in water under most common circumstances (i.e. water pH values between 6 and 8). There are thus two major anthropogenic sources of Al toxicity in the aquatic environment:

- the indirect solubilization and release of Al from rocks and soils as freshwater becomes progressively or episodically acidified by acidic atmospheric deposition
- direct anthropogenic addition of Al salts to freshwater, e.g. to reduce phosphate concentrations to control algal blooms, or water clarification by precipitation of particulates.

In terms of its economic importance Al is the most widely used non-ferrous metal and is highly valued for its low density, resistance to corrosion, and non-magnetic properties. The main source of Al is bauxite ore $[AlO_x(OH)_{3-2x}]$, which consists mainly of the minerals gibbsite, boehmite, and diaspore. Aluminum (mostly as alloys) is vital to the transport industry in general, and aerospace in particular, because of its use in making lightweight structures. Aluminum alloys are also used in packaging (cans, foil), construction (windows, doors), cooking utensils, and housing of electronic equipment. Aluminum compounds are used as a catalyst in chemical mixtures (e.g. zeolites, Freidel–Craft reaction), including ammonium nitrate explosives (Williams, 1999). Aluminum salts are also used to clarify water (e.g. in sewage treatment) and counteract eutrophication in lakes by precipitation and removal of nutrients and colloids (see Section 4).

4. ENVIRONMENTAL SITUATIONS OF CONCERN

In order of importance, the following scenarios represent the major situations of concern to the aquatic environment with respect to Al.

4.1. Chronic Acidification of Freshwater Ecosystems

This occurs in catchments that are chronically impacted by acid input from either anthropogenic or natural sources. The best known source is a direct result of burning fossil fuels, which generates oxides of sulfur and nitrogen that dissolve in atmospheric moisture to form sulfuric and nitric acids (Haines, 1981; Driscoll et al., 2001; USEPA, 2003). These are then deposited either as dry particulate matter or as wet precipitation in the form of acid rain or snow. Natural causes of freshwater acidification include organic acid input (e.g. DOC in the form of humic and fulvic acids) and dilution of base cations, both of which are often associated with springtime (Kahl et al., 1989, 1992). An excess of windborne, oceanic Na^+ and Mg^{2+} ions can displace H^+ ions in soils, which also ultimately acidify freshwater systems (Wiklander, 1975; Kahl et al., 1992; Heath et al., 1992). In particular, these various inputs of acid become a major issue when they occur in low-calcium waters (i.e. low hardness or soft water) that are usually also poorly buffered (i.e. low alkalinity) and so have little capacity to maintain pH after acidic inputs. In these situations high levels of aqueous Al can be mobilized from natural geological sources as a result of the chronic depression of pH (below 6), which enhances Al solubility.

4.2. Acute Episodic Acidification of Freshwater Ecosystems

This is essentially the same cause as chronic acidification described above, but much higher levels of Al can be achieved acutely through the impact of episodic heavy acidification of freshwater due to sudden release of acids (snowmelt or heavy rainfall after prolonged dry deposition onto a susceptible catchment) (Fig. 2.3). Such short-term, episodic events, sometimes lasting for several days, are thought to have a greater ecological effect than chronic exposures (Cleveland et al., 1991; Kroglund et al., 2008). They may be particularly damaging to salmonids in the spring, when rapid mortality of sensitive yolk-sac fry and smolt stages has been observed (Cleveland et al., 1986; Ingersoll et al., 1990a; Handy, 1994; Kroglund et al., 2007), and also because such pulses have been shown to impair the subsequent ability of smolts to tolerate seawater following their downstream

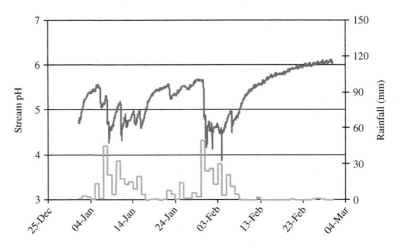

Fig. 2.3. Data from Crockern Tor (West Dart, Devon, UK) on streamwater pH (in dark grey; left-hand *y*-axis) measured continuously between January 1 and February 29, 2004 (pH readings were logged every 15 min). The right-hand *y*-axis shows the daily rainfall (mm; in light grey) for the same period. Episodic extremes of pH occur during particularly heavy rainfall, with pH values well below 5 being achieved for several days at a time, and pH below 4 on one occasion in early February 2004. Data were obtained courtesy of Trevor Cronin of the Environment Agency, Exminster, Devon, UK.

migration (Magee et al., 2001, 2003; National Academy of Science, 2004; Monette et al., 2008, 2010).

4.3. Aluminum Addition to Freshwater Lakes and Reservoirs

The direct anthropogenic addition of Al [as either acidic solutions of $AlCl_3$ or $Al_2(SO_4)_3$, or a basic solution of $NaAl(OH)_4$] to freshwater lakes and reservoirs has been successfully used as a remedial treatment for eutrophication as it reduces the primary nutrient cause (i.e. dissolved phosphorus) and thus helps to control algal blooms. The added Al removes dissolved phosphorus by forming an inorganic $AlPO_4$ complex which polymerizes, adsorbing further dissolved and particulate phosphorus, ultimately resulting in its flocculation and removal by sedimentation (Playle, 1987a; Cooke et al., 2005). A similar strategy has sometimes been used in the past for clarification of water for human use, where the addition of Al sulfate again causes flocculation (e.g. water treatment plants). Both of these processes can be used to effectively control specific environmental problems. However, eutrophication is commonly associated with alkaline pH, and the solubility of Al increases above pH 8. There is therefore potential for excess dissolved Al to be added under such circumstances, which can lead to a degree of subsequent Al toxicity (Wauer and Teien, 2010).

4.4. Acidic River Discharge into Coastal Waters

The sudden high discharge of acidic, Al-rich freshwater rivers into coastal waters can create problems for fish in coastal saline waters. For example, salmonids held in aquaculture cages in saline waters have been shown to suffer increased mortality following snowmelt and winter floods, which raise the discharge of acidic rivers into these coastal water. The effects are largely related to the insolubility and flocculation of Al once the acidic and Al-rich freshwater mixes with the new conditions in the saline coastal waters (e.g. fjords in Norway) (Bjerknes et al., 2003; Teien et al., 2006b).

5. AMBIENT WATER QUALITY CRITERIA IN FRESHWATER

Gensemer and Playle (1999) have pointed out that there are significant problems with ambient water quality criteria (AWQC) for Al. The United States Environmental Protection Agency (EPA) AWQC are based on "acid-soluble" Al (acidified to pH < 2 then 0.45 μm filtered, which is not ideal for true environmental protection) rather than dissolved or monomeric Al. Also, almost all toxicity data available are based on low hardness water with pH below the normal range for US EPA criteria (6.5–9.0). The limited data on toxicity within this range create significant errors in generating useful AWQC values (Gensemer and Playle, 1999). Furthermore, for pH values less than 6.5 the toxicity data for Al are highly variable, both between species (including closely related ones) and even within a single species. Some of this variability can undoubtedly be explained by the dependence of Al speciation upon the physicochemical factors discussed previously (e.g. pH, temperature, hardness, and the complexing actions of DOC, fluoride, and silicates). However, water quality criteria are yet to consider properly the protective actions of all these components. A further source of variability in Al toxicity data, and therefore a complication of our ability to set reliable criteria, has recently been highlighted based upon differences in the aging of Al stock solutions before their dilution and delivery to fish exposure tanks in experimental situations (R. Santore, A. Ryan, P. Paquin, R. Gensemer and W. Adams, personal communication). It would appear that transient and highly toxic effects can occur within the first seconds after a dosing solution is prepared, which can disappear following a suitable aging period that can be as short as a few minutes.

Additional complications to setting water quality criteria arise from the fact that different life stages and body sizes can have very different sensitivity to Al (e.g. Sayer et al., 1993; Wilson et al., 1994a), and therefore would require different water quality criteria for Al. Furthermore, anadromous salmonids (e.g. Atlantic salmon and brown trout) are key species affected by Al in acid waters, and for Atlantic salmon at least, levels of Al that have little or no impact on life stages in freshwater can have substantial delayed effects that only become apparent once migrating smolts reach seawater, ultimately with potential for major population-level effects (Staurnes et al., 1996; Kroglund et al., 2007).

The US EPA classifies Al as a non-priority pollutant, and in their review Gensemer and Playle (1999) cited the following AWQC published by the EPA in 1998 for waters of pH 6.5–9.0:

- acute: 750 μg L^{-1} (28 μM)
- chronic: 87 μg L^{-1} (3.2 μM).

Despite the above reservations, and in particular those raised by Gensemer and Playle in 1999, the US EPA's version of the National Recommended Water Quality Criteria for Al published a decade later (USEPA, 2009) remained the same.

Models that describe the metal–gill interactions [e.g. biotic ligand models (BLMs)] have been published and used very effectively to help predict acute toxicity for several waterborne metals, including Cu, Cd, Ag, Co, and Pb (Paquin et al., 2002; Wood, Chapter 1, Vol. 31A; Paquin et al., Chapter 9). These models are based on the gill being the main target organ (or at least site of metal entry) for acute waterborne toxicity, and rely on the assumption that the quantity of metal binding to the gill will be proportional to whole organism toxicity (Playle, 2004). Comparable data for developing a similar approach for Al toxicity have been slower to appear, but a BLM for Al has now been developed and is likely to be published in the near future (R. Santore, A. Ryan and W. Adams, personal communication).

The author is not aware of any tissue residue criteria for Al. However, gill accumulation of Al has often been shown to correlate better with toxic and sublethal impacts than measured water levels of Al (Kroglund et al., 2007; McCormick et al., 2009a). On the other hand, accumulated gill Al can depurate to control levels rapidly following an acute exposure, while toxic impacts (e.g. on behavior and physiology) and delayed mortality can persist for at least several more days (see Sections 11 and 16.3; Allin and Wilson, 2000). Accumulation in tissues other than gills may not bare any relation to toxicity, which is certainly the case for Al accumulation in soft tissues of freshwater snails (Walton et al., 2009).

6. MECHANISMS OF TOXICITY

6.1. Acute Toxicity in Freshwater Fish

Aluminum is largely insoluble at neutral pH, so toxicity is primarily limited to pH extremes, most commonly in acidic soft water (pH < 6), but toxicity in alkaline waters can also occur at pH greater than 8, although much less is known regarding mechanisms compared to Al toxicity under acidic conditions (Gensemer and Playle, 1999; Wauer and Teien, 2010). For both pH extremes the gills are the primary site of toxic action. Dietary Al has no clear acute toxic effects at realistic dietary levels (see Section 9.2; Poston, 1991; Handy, 1993). The vast majority of studies on the mechanisms of Al toxicity have been carried out under acidic soft-water conditions, which are the most relevant for the major environmental situation of concern (see Section 4). These studies therefore recognize the importance of understanding the combined effects of low pH, low Ca^{2+}, and elevated Al (Howells et al., 1983; Gensemer and Playle, 1999). The precise mechanisms of gill toxicity depend strongly upon these prevailing water chemistry conditions, but the toxic action of Al can either be ionoregulatory, respiratory, or both, depending primarily on the pH and its influence on Al speciation (Muniz and Leivestad, 1980; Neville, 1985; Gensemer and Playle, 1999).

6.1.1. Ionoregulatory Toxicity

Ionoregulatory toxicity during acute Al exposure occurs under both moderately and severely acidic water conditions, and is thought to be caused by the cationic species that predominate [especially Al^{3+} but also $Al(OH)_x^{x+}$] that will encourage interaction with the gill cell surfaces that carry a net negative charge (Gensemer and Playle, 1999). The ionoregulatory toxicity is manifested as net losses of Na^+ and Cl^- which result from both accelerated ionic effluxes and retarded ion uptake rates at the gills (Neville, 1985; Battram, 1988; Booth et al., 1988; McDonald and Milligan, 1988; Wood et al., 1988a; Witters et al., 1992).

A reduction of active ion uptake during acute exposure to Al under acid conditions probably results either from physical damage to ion-transporting mitochondria-rich cells in the gill (Jagoe and Haines, 1997) or from a direct inhibition of branchial enzymes involved in ion/acid–base transport such as the Na^+/K^+-ATPase and carbonic anhydrase (Staurnes et al., 1984, 1993, 1996; Kroglund and Staurnes, 1999; Magee et al., 2001, 2003). However, this is not always a consistent observation, as a decline in plasma ions has been observed in response to short-term acid/Al exposure in the absence of any measurable effect on gill Na^+/K^+-ATPase activity in Atlantic salmon (Monette and McCormick, 2008).

The increased ion efflux caused by Al is associated with the disruption of tight junctional permeability (i.e. paracellular diffusion pathway), leading to elevated diffusive losses of Na^+ and Cl^- from the blood to the very dilute external medium. Cationic Al (especially Al^{3+}) is thought to bring this about by displacing Ca^{2+} from anionic binding sites within the intercellular junctions where it normally plays an important role in controlling the integrity and permeability of the branchial epithelium (Potts and Fleming, 1971; Cuthbert and Maetz, 1972; McDonald and Rogano, 1986; Booth et al., 1988; Freda et al., 1991). Although it is well established that acidity alone has a similarly negative effect on gill regulation and plasma levels of Na^+ and Cl^-, it is clear that the additional presence of Al exacerbates these problems (Witters, 1986; Booth et al., 1988; Wilson and Wood, 1992; Wilson et al., 1994a). Although the majority of studies have focused on the impacts of Al upon Na^+ and Cl^- regulation, it is worth noting that Al has also been shown to impair the regulation of Ca^{2+} in brown trout fry, primarily via inhibition of active uptake (Sayer et al., 1991). As testament to the importance of understanding the summative effects of Al, Ca, and pH, it is interesting to note that Al can actually offer some protection against extremes of H^+ toxicity (e.g. pH 4) (Neville, 1985), under which conditions the predominant Al^{3+} species may be acting as a Ca^{2+} mimic, competing with H^+ for binding sites at the gill (Neville and Campbell, 1988).

6.1.2. RESPIRATORY TOXICITY

Respiratory symptoms of Al toxicity are mostly associated with Al in moderately acid waters (pH 5–6). The mechanism of respiratory toxicity centers around the concept that the chemistry within the gill microclimate changes as the inhaled water moves across the gill surface, and this change is rapid enough to accelerate superficial Al accumulation on the gill. In particular, when inspired water pH is below 6, it will become more alkaline during its passage over the gill surface owing to the continuous excretion of basic ammonia gas (NH_3), the main nitrogenous waste product of teleost fish (Playle and Wood, 1989; Lin and Randall, 1990; Playle et al., 1992; Winter et al., 2005) (Fig. 2.4B). Various authors have postulated that this rising pH will promote a change in speciation towards more cationic aluminum hydroxides, a reduction in solubility and physical precipitation of $Al(OH)_3$, and polymerization of Al, that all ultimately lead to its accumulation on the gill surface, thereby giving rise to symptoms of respiratory toxicity (Witters et al., 1987; Booth et al., 1988; Playle and Wood, 1989, 1990; Exley et al., 1991; Poleo, 1995). The rising pH within the inspired water undoubtedly causes Al solubility to decline, and so physical precipitation of $Al(OH)_3$ is plausible if the inspired [Al] and the rise in pH of inspired water over the gills

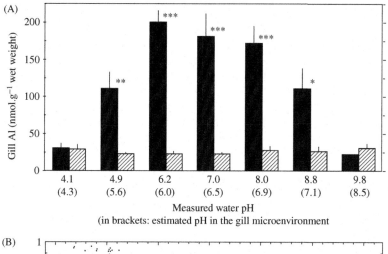

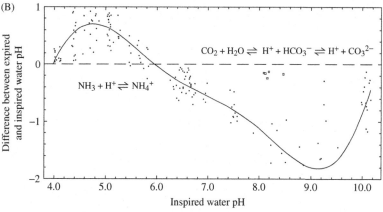

Fig. 2.4. (A) Gill Al concentrations for rainbow trout (*Oncorhynchus mykiss*) exposed for 3 h to 3 µM of Al (81 µg L^{-1}) alone (solid bars) and to Al plus Suwannee River natural organic matter (NOM) at 5 mg C L^{-1} (hatched bars) over a water pH range of pH 4–10. Bars represent means ± 1 SE ($n = 5$). Asterisks represent significant differences between gill Al values with and without NOM at a given pH (*$p < 0.05$, **$p < 0.01$, ***$p < 0.001$). Water pH values in the gill surface microenvironment are given in parentheses and were estimated from Fig. 2.4(B). (B) Differences between expired and inspired water pH as poorly buffered, ion-poor water passes over the gills of rainbow trout, modified from Playle and Wood (1989). NH$_3$ gas excreted across the gills makes acidic water more basic during the gill transit time (i.e. a positive pH difference), whereas basic water is made more acidic by excretion of CO$_2$ gas (i.e. a negative pH difference). From Winter et al. (2005).

are sufficiently high to exceed the solubility limit. Playle and Wood (1989) showed this to be the case for rainbow trout acutely exposed to 93 µg Al L^{-1} when inspired water pH was between 4.8 and 5.2. An alternative hypothesis that does not rely on physical precipitation of insoluble Al(OH)$_3$ has also been suggested, whereby the rising pH within the gill transit time will generate greater proportions of positively charged Al hydroxide species (Playle and Wood, 1989) that then bind to the negatively charged sites on the gill surface (Wilkinson and Campbell, 1993), which act as nuclei for polymerization, which occurs instantaneously as these Al hydroxides are formed (Baker and Schofield, 1982; Poleo, 1995). Higher temperatures reduce Al solubility and also promote the degree of ongoing Al polymerization and subsequently therefore the amount of Al accumulation on fish gills (Poléo and Muniz, 1993). Elevated temperature also speeds up metabolism and therefore branchial ammonia excretion in fish, which in turn exaggerates the pH-driven transformation of Al species during the transit of ventilatory water over the gills. This further supports the respiratory mechanism of Al toxicity, which is also enhanced with increasing temperature, in Atlantic salmon (Poleo et al., 1991; Poléo and Muniz, 1993; Poleo, 1995).

Regardless of the precise chemical processes occurring at the gill, accumulation of Al on the gill surface is clearly accelerated within the pH range associated with respiratory toxicity (Neville, 1985; Witters et al., 1987; Booth et al., 1988; Playle and Wood, 1989; Poleo, 1995). Such physical symptoms associated with Al accumulation on the gills include excessive mucus production and "clogging" of the interlamellar spaces, inflammation of the gill epithelium, and thickening and shortening lamellae (sometimes to the point of lamellar fusion), which all impair respiratory gas exchanges because of limitations of either diffusion or surface area (Youson and Neville, 1987; Evans et al., 1988; Tietge et al., 1988; Mueller et al., 1991; Audet and Wood 1993; Peuranen et al., 2003). Diffusion is inversely proportional to the square of distance, so even small increases in gill diffusion distance can dramatically impair gas exchange. Wilson et al. (1994b) showed that gill surface area is reduced and diffusion distance increased during acute exposure (5 days), and that although blood–water diffusion distances may recover with prolonged exposure (34 days) the area available for gas exchange exhibited no recovery even after this chronic period (see Fig. 2.7). As a secondary effect of these gill respiratory limitations, blood acidosis can also occur, related to the accumulation of carbon dioxide (CO_2; respiratory acidosis), and/or buildup of lactic acid (metabolic acidosis) in extreme cases where severely low blood oxygen levels (hypoxemia) occur (e.g. Neville, 1985; Wood et al., 1988a, c; Walker et al., 1988; Playle et al., 1989; Witters et al., 1991).

6.1.3. SECONDARY CARDIOVASCULAR EFFECTS

Secondary cardiovascular effects can ultimately result from the combination of ionoregulatory and respiratory toxicity caused by Al.

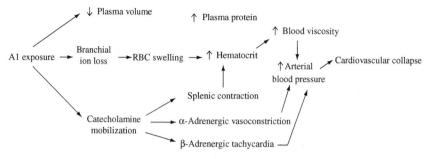

Fig. 2.5. Scenario for describing the physiological events leading to the symptoms of acute toxicity due to Al exposure in acidic water. From Dussault et al. (2001), which is an adaptation based on the model described by Milligan and Wood (1982) for the toxicity of acidity alone.

A cardiovascular collapse scenario was elegantly described for acute acid toxicity in freshwater fish by Milligan and Wood (1982), which has since also been applied to acute Al toxicity (Dussault et al., 2004). In this model, if ionic imbalance is sufficiently rapid, it can cause fluid shifts between extracellular and intracellular compartments, hemoconcentration, increased in blood viscosity, and eventually circulatory collapse (Fig. 2.5). Any respiratory symptoms of reduced blood oxygen transport (as caused by Al at moderately acidic water pH) would simply accelerate this cardiovascular collapse, owing to the added demand on cardiac output and the adrenergic responses that will exacerbate the hemoconcentration and blood pressure problems (e.g. Witters et al., 1990; Dussault et al., 2001, 2004).

6.1.4. Generalized Stress Responses to Acid and Aluminum

Acute exposures to acid and Al have repeatedly shown the induction of primary and secondary stress responses which are often associated with either ion losses or respiratory impairment. For example, elevations in circulating levels of the catecholamines epinephrine (adrenaline) and norepinephrine (noradrenaline) and/or cortisol have been shown during short-term exposures in a range of salmonid species (Goss and Wood, 1988; Whitehead and Brown, 1989; Witters et al., 1991; Brown and Whitehead, 1995; Waring and Brown, 1995; Waring et al., 1996). These hormone systems are well known for their role in the primary stress responses of fish, and are thought to serve in compensating for respiratory and/or ionregulatory impacts of such toxic exposures. In the short term, catecholamines are important in protecting oxygen transport processes against the respiratory component of Al toxicity (see Section 6.1.2), for example by increasing cardiac output and lamellar blood flow, but also protecting the intracellular pH of erythrocytes to help maintain the oxygen-carrying potential of their hemoglobin. Cortisol is known to stimulate cellular proliferation and differentiation of gill ionoregulatory cells, and specifically to increase the activity of Na^+/K^+-ATPase in the gills, which would be a clear advantage in the face of ion losses induced by Al.

Cortisol and catecholamines are also known to induce hyperglycemia, the elevation of circulating glucose levels, which is part of the secondary generalized stress response in fish. Such increases in plasma glucose concentration seem to be a common feature in salmonids acutely exposed to acid and Al (Goss and Wood, 1988; Whitehead and Brown, 1989; Brown and Whitehead, 1995; Laitinen and Valtonen, 1995; Waring and Brown, 1997; Allin and Wilson, 1999; Brodeur et al., 2001; Monette et al., 2008), which is again likely to serve a positive compensatory role by enhancing the availability of this fuel for rapid utilization during a time of enhanced energetic costs. However, as mentioned in Section 12.1, hyperglycemia can act as a satiation signal resulting in suppression of appetite and feeding, which is also a frequently observed response to acid and Al exposure (Lacroix and Townsend, 1987; Wilson et al., 1994a, 1996; Allin and Wilson, 1999, 2000; Brodeur et al., 2001). If cortisol is chronically elevated it can also have further negative whole-animal effects such as immunosuppression and increased susceptibility to disease (Mazeaud and Mazeaud, 1981; Pickering and Pottinger, 1989), and reduced growth and reproductive capacity due to the diversion of energetic resources (Pottinger, 1999).

6.1.5. LIMING AND TOXIC MIXING ZONES

Liming, the addition of ground-up $CaCO_3$-rich rocks such as limestone to acidified waters, has been used in numerous countries in Europe, the USA and Canada for over 20 years to improve water quality in acidified catchments (Wright, 1983; Schindler, 1988; Howells and Dalziel, 1990; Andersen, 2006). The dissolution of added $CaCO_3$ improves the bulk water quality for fish by increasing the water hardness and pH/alkalinity. However, although the long-term effect is to ameliorate Al toxicity, the rapid rise (seconds to minutes) in pH occurring immediately downstream of such liming practices results in an acutely toxic mixing zone for fish (Rosseland et al., 1992; Poleo et al., 1994; Exley et al., 1996; Witters et al., 1996; Teien et al., 2004, 2006c). Acute mortality of fish can occur in such mixing zones, and is thought to be exclusively due to respiratory toxicity caused by the rapid polymerization of monomeric Al (low molecular mass) to high molecular mass forms of Al, and the accumulation of precipitated or polymerized Al on the gills under these transient alkalinizing conditions (Oughton et al., 1992; Rosseland et al., 1992; Witters et al., 1996; Teien et al., 2004, 2006c). However, the downstream zones of high toxicity are limited in time and space owing to the rapid nature of the chemical reactions, and ultimately the transformation of the high molecular mass forms of Al to less bioavailable colloidal species (Teien et al., 2004). Nevertheless, significant acute mortality can occur if fish encounter this zone (e.g. during upstream spawning migrations) and fish that move back downstream as an avoidance response (e.g. Atland and Barlaup, 1995) have been shown to recover more slowly in this region than in neutral waters containing low Al (Verbost et al., 1995).

6.2. Chronic Toxicity in Freshwater Fish

Entry of Al into the blood via the gills is slow relative to other metals and internal accumulation cannot be detected over hours (Handy and Eddy, 1989) or even 2–3 days (Dussault et al., 2001). Although small, dose-dependent elevations in plasma [Al] were detected over a 4 day exposure in brown trout at pH values of 5.2–5.6, the internal accumulation of Al was not thought to be related to any toxic effect (Dietrich et al., 1989). The slow accumulation of Al in internal organs over longer exposure periods has been documented (Karlsson-Norrgren et al., 1986a; Lee and Harvey, 1986; Booth et al., 1988; Witters et al., 1988; Exley, 1996; Dussault et al., 2004). In one case, exposure of farmed rainbow trout for 1 year to pH around 6 plus 225 μg Al L^{-1} resulted in 10-fold higher Al levels in brain, liver, gonad, and heart compared to control fish (Exley, 1996). However, even in this chronic case of Al exposure, the internal accumulation was not associated with any toxic effects (Exley, 1996). Aluminum is therefore unlike most other toxic metals in this respect, and the effects on the gill (both ionoregulatory and respiratory) remain the primary target for both chronic and acute toxicity. The acute mechanisms described above therefore hold true for chronic exposures, but studies have investigated the longer term repercussions of gill toxicity (e.g. for swimming ability and behavior), and the potential for both recovery from and acclimation to Al during such prolonged exposures.

6.2.1. CHRONIC IMPAIRMENT OF AEROBIC SCOPE AND SWIMMING PERFORMANCE

As described above, impairment of gill respiratory function by Al is associated with gill histopathologies such as thickening of the lamellar epithelium and hyperplasia of the filamental epithelium which can affect the blood–water diffusion distance and surface area available for gas exchange (Youson and Neville, 1987; Evans et al., 1988; Tietge et al., 1988; Mueller et al., 1991; Audet and Wood, 1993). However, more subtle morphometric changes can be observed at sublethal levels of Al which are insufficient to impair resting aerobic metabolism (Wilson et al., 1994b). Typically, gill damage peaks during the first week of exposure and recovers thereafter

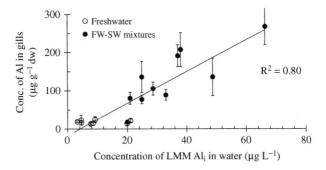

Fig. 2.6. Linear correlation ($p < 0.001$) between the low molecular mass (LMM) inorganic Al in acidic river water, before mixing with coastal seawater, and the accumulation of Al on the gills of seawater Atlantic salmon exposed to the river–seawater mixture. Open symbols represent freshwater before mixing, and closed symbols represent various river–seawater mixtures. From Teien et al. (2006b).

(Mueller et al., 1991). However, even when the initial gill damage is relatively mild, not all the gill morphometrics recover to their original state (Wilson et al., 1994b). Figure 2.6 is a schematic diagram of the gill lamellae which represents the changes in some gill morphometric parameters relevant to branchial gas exchange in rainbow trout after 5 and 34 days' exposure to sublethal Al ($30 \, \mu g \, L^{-1}$) at pH 5.2. Although the blood–water diffusion distance returned to the control level after 34 days, the lamellar height was still reduced and the filamental epithelium thicker. From the average lamellar dimensions the apparent lamellar surface area was estimated to be reduced by 22% even after 34 days' exposure. This was a time when ionoregulatory status had fully recovered (Wilson et al., 1994a, b). These longer lasting changes in gill morphology have been associated with branchial repair (McDonald and Wood, 1993) and may therefore be an unavoidable consequence of chronic exposure to Al in acid waters. Although the resting aerobic metabolism may be unaffected, or at least restored with time (Wood et al., 1988b; Walker et al., 1991), the persistence of these structural abnormalities at the gills suggests that the capacity of the respiratory gas exchange system may remain compromised during prolonged exposure to Al. Alterations to capacity will only become apparent during increased demand on aerobic metabolism such as sustained swimming activity, or recovery from exhaustive/anaerobic exercise.

Acidic pH alone (without Al) is known to reduce swimming performance (Hargis, 1976; Waiwood and Beamish, 1978; Graham and Wood, 1981; Ye and Randall, 1991; Butler et al., 1992; Dussault et al., 2004). However, exposure of rainbow trout to acidic pH with sublethal Al ($30 \, \mu g \, L^{-1}$)

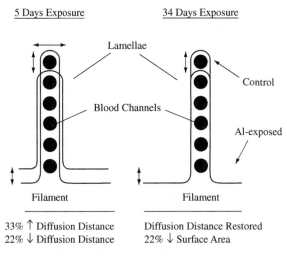

Fig. 2.7. Schematic diagram representing some of the gill morphometric changes measured in juvenile rainbow trout exposed to acidified soft water (pH 5.2) plus sublethal Al (30 µg L^{-1}) for 5 and 34 days. Gills of fish exposed to acid alone (pH 5.2) were not significantly different from those of the control fish. The solid outline represents the outer dimensions of a single lamella and the adjoining filament tissue. The dotted outline represents the relative change in the lamellar height and thickness, and the filamental thickness in fish exposed to acid plus Al for 5 days (acute damage phase) and 34 days (chronic recovery phase). Below the diagram are summaries of the calculated changes in blood-to-water diffusion distance, and the total lamellar surface area that equates to these dimensional changes. Note that although the diffusion distance completely recovered after 34 days, no recovery was observed in the total lamellar surface area, which parallels the chronic reduction in maximum oxygen uptake and aerobic swimming performance shown in Figs. 2.8 and 2.9. Redrawn from Wilson et al. (1994b).

reduced the maximum aerobic swimming speed (U_{crit}) (Brett, 1964) 1.5–2 times more than in trout exposed to acid alone, and in both media U_{crit} did not recover with time (Fig. 2.7) (Wilson and Wood, 1992; Wilson et al., 1994b; Dussault et al., 2004). Aerobic swimming performance can be affected by changes at any level of the oxygen transport system and/or muscle contraction process. Randall and Brauner (1991) demonstrated that reduced U_{crit} can occur in response to disturbances in the ion and water balance in muscle of coho salmon (*Oncorhynchus kisutch*) and suggested that this occurred via decreased muscle contractility. Severe ion losses encountered during the acute toxicity of acid/Al exposures could therefore exert a similar effect on muscle performance and swimming. However, since recovery of ionoregulation is complete within 25 days (Wilson et al., 1994a) it seems unlikely that alterations at the level of the muscle contraction are important over these longer periods.

In terms of oxygen transport and aerobic metabolism, Brett (1958) divided environmental stresses into two categories: "limiting" stresses that reduce the maximum aerobic capacity (MO_2 $_{max}$), and "loading" stresses that increase the cost of routine maintenance (MO_2 $_{basal}$). The 1.5–2-fold greater reduction in U_{crit} in trout exposed to acid plus Al (compared to acid alone) is most likely due to limiting factors related to the chronic gill morphological changes mentioned above. In support of this, oxygen uptake capacity was reduced by 26%, similar to the 22% reduction in apparent lamellar surface area, in rainbow trout after 34 days' exposure to acid plus Al (see above) (Fig. 2.8). Although the blood–water diffusion distance had returned to control values by this time, a continuation of increased gill mucus production could act as an additional diffusion barrier to oxygen uptake (Ultsch and Gros, 1979). In contrast, in the same study (Wilson et al., 1994b) trout exposed to acid alone exhibited no significant changes in either gill morphometrics or oxygen uptake capacity compared to control fish (held at pH 6.5). The chronic reduction in U_{crit} caused solely by acidity (pH 5.2) was therefore not caused by limitations in branchial oxygen uptake capacity. Instead, it can be seen from Fig. 2.7 that chronic impairment of U_{crit} after 34 days' exposure to pH 5.2 alone was the result of loading factors (i.e. an elevation of basal metabolic rate and a demand for more oxygen at lower than maximal swimming speed). Thus, acid-exposed trout reach the

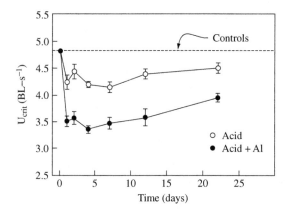

Fig. 2.8. Maximum aerobic swimming speed (U_{crit}) expressed as body lengths per second in juvenile rainbow trout at various intervals during 22 day exposure to either control soft water (pH 6.5; dashed line), acid soft water (pH 5.2; open circles) or acid soft water plus sublethal Al (30 μg L^{-1}; solid circles). The dashed line represents the mean value for control fish swum at pH 6.5 on days 0 and 24 (U_{crit} = 4.82 ± 0.11 BL s^{-1}, n = 30). Both other groups were swum in pH 5.2 water (with or without 30 μg L^{-1} added). Mean values ± SEM (n = 8) are shown. Redrawn from Wilson and Wood (1992).

same maximum oxygen uptake capacity as control fish but at a lower speed. It has been speculated that some of the increased ionoregulatory costs associated with acid (and Al) exposure would contribute to this loading stress and lead to an increase in the metabolic rate required for homeostasis (Wilson et al., 1994b). Such a loading stress would thereby impair U_{crit} by reducing the proportion of oxygen available for aerobic muscle use. Comparable loading factors must be present in trout exposed to acid plus Al, as the regression relationship between oxygen uptake and swimming speed is similarly elevated in these fish (Fig. 2.9) (Wilson et al., 1994b). However, as stated before, the greater reduction in U_{crit} in trout exposed to acid plus Al was the result of additional morphological limitations to the branchial gas exchange capacity. Thus, reduced swimming performance in trout chronically exposed to acid plus Al is due to a combination of both loading and limiting factors. Dussault et al. (2001, 2004) also suggested that

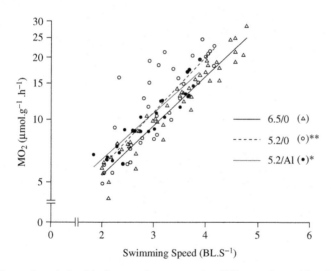

Fig. 2.9. Regression relationships between O_2 consumption (MO_2; note log scale) and swimming speed (in body lengths per second) after exposing juvenile trout for 34 days to control soft water (pH 6.5; triangles and solid line), acid only (pH 5.2; circles and dotted line), or acid plus sublethal Al (30 μg L^{-1}; solid circles and dashed line). Group regression equations, where $y = \log MO_2$, and $x = $ swimming speed, where: (i) $y = 0.239x + 0.256$ ($r = 0.962$) in the 6.5/0 group (ii) $y = 0.267x + 0.236$ ($r = 0.836$) in the 5.2/0 group, and (iii) $y = 0.225x + 0.358$ ($r = 0.874$) in the 5.2/Al group. The three slopes were not significantly different, but regression lines for both the acid and acid plus Al groups were significantly elevated above the control group ($p < 0.005$ and $p < 0.05$, respectively) indicating increased maintenance costs at all subcritical swimming speeds which represents loading factors on metabolism. Limiting factors are indicated in the group exposed to acid and Al by having a lower maximum MO_2 and lower maximum swimming speed than either the control group or fish exposed to acid alone. From Wilson et al. (1994b).

the hemoconcentration and elevated blood viscosity caused by exposure to acid plus Al (Fig. 2.5) may cause preload on the heart, thus reducing the options for increasing stroke volume (rather than just heart rate) as a means to achieving high cardiac outputs to support sustained aerobic exercise.

6.2.2. RECOVERY FOLLOWING RETURN TO LOW ALUMINUM WATER

Whereas U_{crit} recovered immediately when acid-exposed trout were transferred back to circumneutral pH water (pH 6.5), fish exposed for the same time (36 days) to acid plus Al showed no recovery of either maximum oxygen uptake rate or maximum aerobic swimming speed (Wilson et al., 1994b). This suggests that the reduced swimming performance of acid-exposed fish is entirely due to the reduced water pH during the swim test itself rather than any accumulated physiological damage caused by the prior exposure to low pH. By contrast, the lack of recovery of Al-exposed fish suggests that for fish in the wild a rapid restoration of more neutral pH (e.g. by liming) will not immediately restore swimming performance which probably requires considerably more time for normal blood viscosity, cardiac performance, branchial morphology, and oxygen uptake capacity to be achieved.

6.2.3. METABOLIC COSTS OF CHRONIC ALUMINUM EXPOSURE IN ACIDIC FRESHWATER

The causes of limiting factors for metabolism are evident (i.e. impaired gill morphology and cardiovascular performance) but the loading factors are less certain, and can be important chronic costs of exposure to Al in acidic waters. Increased ionoregulatory costs have been discussed above and repair of damaged gill tissues is another likely source of increased energy consumption. Wilson et al. (1996) attempted to measure this cost by examining protein turnover in various tissues including the gills of rainbow trout chronically exposed to sublethal Al ($30 \, \mu g \, L^{-1}$) in acidic water (pH 5.2). Interpretation of the protein turnover responses of individual tissues was complicated by the fact that this sublethal exposure to Al also caused a major disturbance to appetite, feeding and growth (see below and Wilson et al., 1994a, 1996) and so protein turnover in individual tissues had to be standardized relative to the whole body response to allow for this. Once this was taken into account ,the costs of protein synthesis and degradation were shown to be doubled in gill tissue during the early phase when observed damage to gill morphology was maximal (1 week) and protein synthesis was still elevated even after 32 days (Wilson et al., 1996). This is consistent with the concept of a metabolic cost associated with repairing tissue damage caused by a toxicant (during the early phases) and continuing additional costs during chronic exposure associated with the recovery of gill functions (e.g. turnover of ion transport proteins) or maintaining acclimation to Al (e.g. increased mucus production).

6.2.4. IMPACTS OF ALUMINUM ON EARLY LIFE STAGES AND REPRODUCTION

For many toxicants the early life stages of fish are the most sensitive, and this is certainly true for Al in acidic waters, for which enhanced mortality of embryos and fry have been cited as a major contribution to the decline of total numbers and shifts in the age structure of populations towards older fish (Mount et al., 1988a; Sayer et al., 1993; Vuorinen et al., 2003). However, migratory Atlantic salmon may be the exception to this rule, as their smolts have been shown to be the most sensitive developmental stage, suffering mortalities at levels of acid and Al that are sublethal to fry (McCormick et al., 2011; see also Section 6.5). Freshwater acidification and Al may also directly affect reproduction itself, as first reported by Beamish et al. (1975) after witnessing an inhibition of spawning. Mechanisms of reproductive toxicity may include abnormal Ca^{2+} metabolism during vitellogenesis in maturing females (Beamish et al., 1975; Rask and Tuunaine, 1990; Vuorinen et al., 1992), i.e. an indirect effect of the ionoregulatory symptoms of Al toxicity. Delayed maturation of oocytes, and delayed spawning or completely inhibited ovulation have also been observed (Mount et al., 1988b; Vuorinen et al., 1990, 2003; Vuorinen and Vuorinen, 1991).

6.2.5. THE "OLIGOTROPHICATION HYPOTHESIS" OR FOOD CHAIN EFFECTS OF CHRONIC ALUMINUM EXPOSURE

Theoretically, elevated Al in acidic freshwater could have long-term impacts on the whole ecosystem via its effect on the food web by limiting dissolved concentrations of P, the key limiting nutrient for photosynthetic organisms, especially phytoplankton, and ultimately therefore limiting primary productivity. In this way Al would be acting in much the same way as it is used for controlling eutrophication (see Section 4.3), and result in the gradual oligotrophication of acidified lakes (Gensemer and Playle, 1999). However, the various studies that have investigated this possibility suggest that there is little evidence to support this as a valid explanation of Al-induced effects on fish, i.e. through lack of available food compared to similar lakes with low Al (Schindler, 1980; Schindler et al., 1985, 1991; Persson and Broberg, 1985; Findlay and Kasian, 1986; Jansson et al., 1986; Shearer et al., 1987; Gensemer and Playle, 1999). The oligotrophication hypothesis, based on Al-induced limitation of P, has now been more or less rejected.

6.3. Aluminum Toxicity and Physiological Effects at Alkaline pH in
Freshwater

Aluminum becomes increasingly soluble above pH 8, and so alkaline
waters represent another freshwater environment in which Al from natural
sources may become an ecotoxicological issue. In addition, Al is sometimes
artificially added to eutrophic waters (that can often become alkaline as a
result of algal growth and high rates of photosynthesis) to help remove some of
the nutrient causes of eutrophication (e.g. dissolved phosphorus) (Cooke
et al., 2005). Although studies of Al effects on fish under such alkaline
conditions are relatively few, toxicity has been observed (Freeman and
Everhart, 1971; Gundersen et al., 1994; Exley et al., 1996; Poléo and Hytterød,
2003). Under steady state conditions, the Al speciation above pH 8 is
somewhat less complex than under acid conditions, with only the negatively
charged aluminate anion $[Al(OH)_4^-]$ predominating, which appears to be less
toxic to fish than the various cationic species present below pH 6 (see Fig. 2.3).
In fact, Al exposure at pH 8.7 can act to ameliorate the toxic effects of high pH,
enhancing survival compared to the same pH in the absence of Al (Heming and
Blumhagen, 1988). The much reduced toxicity of Al under alkaline conditions
has been attributed to the lack of binding of the aluminate anion to the
negatively charged gill surface. Indeed, there was no detectable additional
total Al bound to the gills of fish when exposed at pH 10 (Winter et al., 2005)
and the Al (3 μM or 81 μg L^{-1}) did not appear to have any toxic or
ameliorative effect (Winter et al., 2005). Non-steady-state conditions will occur
as inhaled water becomes acidified by excretion of CO_2 along the gill surface (see
Fig. 2.4B). In turn, this reduction in pH during the gill transit time of ventilatory
water will reduce Al solubility when starting at high pH (>8), in the same way
that rising pH in the gill boundary layer does under acidic conditions. However,
in contrast to the latter situation, this dynamic change in pH along the gill
surface appears to cause no enhancement of toxicity under alkaline conditions
(starting at pH >8) (Poléo and Hytterød, 2003; Wauer and Teien, 2010).

6.4. Mechanisms of Toxicity in Seawater

There are very few experimental studies on the effects of Al on fish
during exposure in seawater or estuarine conditions. However, acute
mortalities of salmonids have been documented as a result of sudden
storms and high Al loads delivered from acidic freshwater rivers into
seawater fjords (Bjerknes et al., 2003; Teien et al., 2006b). Under these
conditions, high concentrations of Al were found in the acidic river water,
most of which was associated with organic colloids (>8 kDa) and was
therefore non-reactive or not bioavailable. Upon mixing with the high

salinity coastal water, this colloidal Al rapidly became mobilized to gill-reactive, low molecular mass inorganic Al which then accumulated on the gills of Atlantic salmon five to 10 times more rapidly than from the native acidic river water (see Fig. 2.6) (Teien et al., 2006b). In a previous study (Bjerknes et al., 2003), such river flooding episodes were observed to cause significant increases in gill Al concentration together with toxicity and mortality to Atlantic salmon in fish farms within a high-salinity Norwegian fjord. It is therefore suggested that this phenomenon may be an additional environmental problem caused by Al for natural as well as farmed populations of fish in estuarine conditions (Teien et al., 2006b).

6.5. Impacts of Aluminum on Smolting and Seawater Migration Success

Although studies on toxicity of Al to fish in seawater are limited, clear negative impacts have been documented in several studies of Al exposure in acidic freshwater on the parr–smolt transformation in Atlantic salmon, and the success of their subsequent downstream migration and osmoregulatory ability in seawater (Saunders et al., 1983; Staurnes et al., 1993, 1996; Poleo and Muniz, 1993; Kroglund and Staurnes, 1999; Magee et al., 2003; Kroglund and Finstad, 2003; Monette et al., 2008, 2010). Crucial to understanding population-level effects, such Al-induced impairments of seawater adaptability and marine survival can ultimately have negative impacts upon the adult return rates to these freshwater rivers (Kroglund and Finstad, 2003; Kroglund et al., 2007). Of particular concern and environmental relevance is the observation that marine osmoregulation can be disrupted even at freshwater Al exposure concentrations that are very low (e.g. 6 μg L^{-1} at pH 5.8) (Kroglund and Finstad, 2003), or short term (e.g. 12 h) (Staurnes et al., 1996), and which may not actually cause significant impairment of the freshwater osmoregulatory ability of these fish (e.g. Monette et al., 2008, 2010).

The hypo-osmoregulatory system of salmon smolts is clearly very sensitive to prior freshwater exposure to Al in acid conditions, which is perhaps not surprising given the major physiological changes occurring during the transition from parr to smolt in such salmonids (Rosseland and Skogheim, 1984; Rosseland et al., 2001; Monette and McCormick, 2008). The most obvious target of this effect of Al is the gill Na$^+$/K$^+$-ATPase activity that is known to increase during smoltification in preparation for the marine migration, and many of the above mentioned studies have shown that this premigratory increase in gill Na$^+$/K$^+$-ATPase activity is either inhibited or even reversed in smolts exposed to acid plus Al (Staurnes et al., 1993, 1995, 1996; Magee et al., 2003; Kroglund et al., 2007; Monette et al., 2008, 2010; Nilsen et al., 2010). However, this is clearly only a partial explanation, as although the Na^+/K^+-ATPase activity is key to powering the branchial ionoregulatory processes in seawater, it is not the only ion transporter of importance for seawater adaptability, and it is also a vital player in gill ion regulation of freshwater fish (Evans et al., 2005; Marshall and Grosell, 2006). It is only recently that cellular and molecular evidence has provided more mechanistic details of this delayed and pervasive effect of freshwater Al on later ability to survive in seawater (Monette et al., 2010; Nilsen et al., 2010).

Salmonids use different isoforms of the α-subunit of Na^+/K^+-ATPase in freshwater and seawater, with $α_{1a}$ predominating in freshwater and $α_{1b}$ in seawater (Richards et al., 2003; Bystriansky et al., 2006; Nilsen et al., 2007; Madsen et al., 2009; McCormick et al., 2009b). Furthermore, this isoform switching is known to begin during the smoltification process in Atlantic salmon, while still in freshwater, in preparation for entering seawater (Nilsen et al., 2007). It appears that this isoform switching is one of the molecular mechanisms of smolting that is impaired by exposure to low levels of Al in acidic freshwater, and instead a total downregulation of all the α-subunits of the Na^+/K^+-ATPase enzyme occurs (Nilsen et al., 2010). Monette et al. (2010) also found a decrease in the mRNA expression of other "seawater" gill transporters caused by Al, such as CFTR-1 (the apical Cl^- channel involved in Cl^- secretion), as well as decreased size and staining intensity of gill mitochondria-rich cells. The precise mechanism of how exposure to Al in freshwater induces such changes is yet to be elucidated, but presumably this relates to accumulation of Al at the gill, which has been repeatedly documented in freshwater-to-seawater transfer studies in Atlantic salmon smolts.

7. NON-ESSENTIALITY OF ALUMINUM

There is little, if any, evidence for the essentiality of Al within biological systems (Wood, 1984, 1985; Williams, 1999; Eichenberger, 1986; Exley, 2003). This probably relates to the low solubility of Al under the anaerobic and circumneutral conditions that were available to primitive biota which would have precluded the evolution of Al-requiring metabolic pathways (Wood, 1984, 1985). While Al has been shown to accumulate over a lifetime within animals (including humans) and plants, there is little evidence of specific biochemical recognition processes for Al, and the accumulation may be more by accident than design (Yokel and Golub, 1997; Jansen et al., 2002; Exley, 2003). However, one recent report has suggested a biological role for nanomolar concentrations of nuclear-associated Al, in biocompacting DNA of eukaryotes ranging from plants to mammals (Lukiw, 2010). This novel idea is sure to receive considerable attention in the near future, not least because of the potential biomedical implications.

8. POTENTIAL FOR BIOCONCENTRATION AND/OR BIOMAGNIFICATION OF ALUMINUM

Adsorption of Al to the gill surface from waterborne exposure to Al is rapid (see Sections 6.2 and 9.1), whereas cellular uptake from the water is slow, but gradual accumulation in internal organs does occur over time (see Section 6.2). Uptake to internal organs (muscle, liver, kidney, and gill) via the diet can occur, but this has only been reported at very high oral doses (10 g Al kg^{-1} dry mass of diet) (Handy, 1993), only a small proportion of this dietary Al was retained, and toxicity was minimal. Lower oral doses via the diet (0.025–2 g kg^{-1}) resulted in no detectable accumulation within the carcass (or vertebrae) of juvenile Atlantic salmon (Poston, 1991). Bioaccumulation and toxicity via the diet are considered very unlikely based on these experimental studies, but also based on the lack of any biomagnification within freshwater invertebrates that are likely to be prey of fish in acidic, Al-rich rivers (Otto and Svensson, 1983; Wren and Stephenson, 1991; Herrmann and Frick, 1995). In fact, Al potentially undergoes the opposite, trophic dilution, up the food chain, as King et al. (1992) observed the rank of Al accumulation to be lowest on fish predators (perch) and highest in the phytoplankton that their zooplankton prey were consuming. In this regard, Al appears to be similar to most other metals (see Wood, Chapter 1, Vol. 31A), with the exceptions of methyl mercury [MeHg(I)] (Kidd and Batchelar, Chapter 5) and various organo-selenium compounds (see Janz, Chapter 7, Vol. 31A).

9. CHARACTERIZATION OF UPTAKE ROUTES

It is important to distinguish between surface adsorption of Al (e.g. onto the gill and skin surfaces, or their mucus layer) and the actual uptake of Al into cells and transport to other internal compartments (blood and other organs). For Al in particular this is a critical issue, because internalization and true cellular uptake are very slow compared to other toxic metals, and yet profound toxicity and even lethality can occur rapidly with only surface adsorption. This distinction is covered below under Section 9.1.

9.1. Gills

Gill (and skin) mucus accumulates Al very rapidly (minutes to hours) (Goossenaerts et al., 1988; Handy and Eddy, 1989; Oughton et al., 1992). Cellular uptake via the gills is slow, but gradual accumulation in internal organs does occur over time (see Section 6.2). The uptake mechanisms are still poorly understood for gill Al. However, it is generally agreed that, at least initially, Al toxicity arises from the actions of bound and precipitated or polymerized Al at the gill surface as described above (although a theoretical mechanism for intracellular toxicity has been suggested) (Exley et al., 1991). From studies using histochemical, X-ray microanalysis, laser microprobe analysis, and [26]Al-isotope techniques, it is apparent that within this initial phase (over the first few minutes to days) almost all the gill Al is found on the branchial surface (Goossenaerts et al., 1988; Oughton et al., 1992), and in particular within mucus-rich areas between gill lamellae (Norrgren et al., 1991). However, following prolonged exposure (from 1 week to 1 year), Al deposits can be found within the gill cells themselves (Karlsson-Norrgren et al., 1986a, b; Youson and Neville, 1987; Evans et al., 1988; Norrgren et al., 1991; Peuranen et al., 2003). The biphasic pattern of total Al concentrations in the gill may therefore represent two separate processes: (1) a rapid accumulation on the branchial surface, and (2) a second, much slower, internalization of Al within the gill cells themselves. The brief "recovery" of total gill Al levels following the initial shock or damage phase may relate to acclimatory changes in gill mucus secretion and gill binding affinity which serve to reduce surface Al and toxicity (see Section 16 for Acclimation). Since internalized Al deposits in the gill are apparently membrane-bound precipitates and hence "non-toxic" (Youson and Neville, 1987), the secondary increase in gill total Al may represent a slower development of cellular detoxification mechanisms that store and isolate the metal before exocytosis (McDonald and Wood, 1993). Thus, the internalization of Al and resultant secondary increase in total gill Al may actually represent a chronic response to additionally protect the gills from Al toxicity. This might also explain why rainbow trout previously acclimated to a low level of Al actually accumulate more total gill Al than control fish when challenged with an acutely lethal [Al] at pH 5.2 (Wilson et al., 1994a). However, Al-acclimated brook trout accumulated less Al under acute high-level exposure (McDonald et al., 1991; Reid et al., 1991), perhaps indicating species differences in this response.

9.2. Other Routes

As previously mentioned, uptake to internal organs via the diet can occur, only at very high oral doses (Poston, 1991; Handy, 1993), but actual uptake of dietary Al into gut tissue has not been documented.

10. CHARACTERIZATION OF INTERNAL HANDLING

10.1. Transport through the Bloodstream

In mammals, approximately 90% of blood Al is associated with plasma transferrin and 7% with citrate (Yokel, 2004; Exley, 2008). However, no such details are yet available on the transport of Al within the blood of fish. Indeed, only a few studies have found increased levels of Al within plasma of Al-exposed fish relative to levels found in control fish (Dietrich and Schlatter, 1989; Allin, 1999).

10.2. Accumulation in Specific Organs

Gill accumulation is fairly well studied, which is mostly rapid, superficial, and associated with mucus, but intracellular deposition occurs more slowly (see Section 9.1). There are a few field and laboratory studies which document internal accumulation in brain, liver, kidney, gonads, heart, white muscle, and scales during chronic exposure (see Section 6.2).

A reduction in whole-body protein translational efficiency (the amount of protein synthesized per unit of ribosomes) was observed following chronic exposure (32 days) to sublethal Al in acidic water, suggesting that Al may have a long-lasting effect on the mechanics of cellular protein synthesis (Wilson et al., 1996). This coincided with the slow accumulation of Al within the internal organs (Lee and Harvey, 1986; Karlsson-Norrgren et al., 1986a; Booth et al., 1988; Witters et al., 1988). Thus, although a cellular mechanism has not been proposed, it is possible that chronic biochemical effects observed within internal tissues are caused directly by elevated tissue Al levels.

10.3. Subcellular Partitioning, Detoxification, and Storage

There have been very few studies on these areas of Al handling in fish, and these few only really focused on the subcellular localization in gills and kidney (Galle et al., 1990; Youson and Neville, 1987; see Section 9.1). There are also no clear data available on the detoxification or storage of Al in fish. Aluminum can be bound to phosphorus in specialized organs of some invertebrates (e.g. digestive gland in pond snails) for immobilization and ultimately excretion (Elangovan et al., 1997; Gensemer and Playle, 1999). However, no comparable mechanism has been identified in fish. Whilst a large proportion of the body Al burden in mammals is known to be associated with skeletal material (Exley, 2008), surprisingly there are few measurements of Al in bone of fish exposed to Al to confirm a similar relationship. Notably, dietary intake of Al does not result in accumulation in bone material of Atlantic salmon (Poston, 1991), even though Al levels in soft tissues such as liver, kidney, gills, and muscle do rise following dietary exposure (Handy, 1996).

11. CHARACTERIZATION OF EXCRETION ROUTES

Virtually nothing is known about excretion routes for Al in fish. Depuration is initially rapid from gills following return to Al-free water after acute waterborne Al exposure. For example, Allin and Wilson (2000) showed that brown trout exposed to a 4 day pulse of Al (35 μg L^{-1} at pH 5.2) rapidly accumulated Al at the gills (to over 60 μg Al g^{-1} (of wet tissue weight), but this had become insignificantly different from the control fish (< 5 μg g^{-1}) within 2 days of terminating the Al exposure. This rapid depuration is probably related to sloughing of largely mucus-bound Al (Playle and Wood, 1991), although rapid turnover of gill tissue and protein during Al exposure could also contribute to this phenomenon (Wilson et al., 1996). However, depuration of Al that has accumulated on the gills of brown trout over a longer period (e.g. 40 days) takes longer to return to levels found in control fish (within 15 days) (S.F. Owen and R.W. Wilson, unpublished data). This would support the idea that Al is internalized into gill cells more slowly than it binds to the gill surface and mucus, but that once Al is internalized, the rate of removal is also slow. Information on the rate of removal of Al from internal organs following return to clean water is lacking, but depuration from internal organs does appear to be slower than for the gills (e.g. > 15 days for head, kidney, and liver), or may not occur at all (e.g. white muscle), or at least there is no detectable change following 25 days in Al-free water (S.F. Owen and R.W. Wilson, unpublished data). Aluminum from a metal-enriched diet can appear in external mucus of rainbow trout, and relative levels of Al in the gills were higher than in the liver or kidney following 42 days of dietary Al intake, suggesting that branchial excretion (possibly via mucus) may be a means of excretion and detoxification from internal stores (Handy, 1996).

12. BEHAVIORAL EFFECTS OF ALUMINUM

12.1. Feeding Behavior, Appetite, and Growth

As mentioned in Section 6.2.3, sublethal Al exposure can cause a large and immediate reduction in appetite, at least during the acute phase of exposure to Al in acid water, in which food intake can be reduced by 50–90% (Lacroix and Townsend, 1987; Wilson et al., 1994a; 1996; Allin and Wilson, 1999, 2000; Brodeur et al., 2001). Poor rates of live food capture have also been documented in response to acid and Al (Farag et al., 1993; Lacroix and Townsend, 1987; Baker et al., 1996). The decline in appetite may be related to hyperglycemia that is induced by hormonal changes associated with a generalized stress response (see Section 6.1.4), with the raised plasma glucose in turn acting as a satiation signal to control feeding (Waiwood et al., 1992; Wilson et al., 1996; Allin and Wilson, 1999), as observed for other toxic metals (e.g. for Cu see Drummond et al., 1973; Lett et al., 1976; for Zn see Farmer et al., 1979). However, Brodeur et al. (2001) observed that Atlantic salmon (*Salmo salar*) showed even higher levels of plasma glucose after the feeding rate had returned to normal, suggesting that this is not the only issue involved in suppression of appetite by Al.

For sublethal Al exposure at pH 5.2, appetite and normal feeding rates can take 10–30 days to return to normal (Wilson et al., 1994a, 1996; Allin and Wilson, 1999, 2000; Brodeur et al., 2001). This response to sudden yet sublethal increases in Al in acidic waters may be important to understand the ecological effects of episodic acid and Al pulses observed in the wild. Such a major reduction in food intake will clearly impact the energy stores available to deal with any stressor. The chronic reductions in growth reported in many studies on fish exposed to sublethal acid plus Al (Sadler and Lynam, 1987, 1988; Reader et al., 1988; Mount et al., 1988a, b; Tam et al., 1998; Ingersoll et al., 1990a, b; Wilson and Wood, 1992; Vuorinen et al., 2003) may therefore be largely if not entirely due to the effect on appetite. These effects on feeding and growth are purely caused by Al as they are consistently absent in fish exposed to acid alone (Wilson and Wood, 1992; Wilson et al., 1994a, 1996). In fact, appetite can be enhanced in response to sublethal acid alone, likely as a response to compensate gill ion losses with enhanced dietary salt intake (e.g. D'Cruz and Wood, 1998).

12.2. Effects on Activity and Swimming Behavior

Exposure to Al in acidic water causes hypoactivity in salmonids (Cleveland et al., 1986; Wilson et al., 1994a; Smith and Haines, 1995; Allin and Wilson, 1999, 2000). This reduction in spontaneous activity includes less time swimming at slow aerobic speeds, fewer anaerobic high speed bursts, and consequently more time spent stationary (Fig. 2.10). The quantification of reduced swimming activity observed by Allin and Wilson (1999, 2000) followed the same trends as those anecdotally documented in previous studies (e.g. Freeman and Everhart, 1971; Schofield and Trojnar, 1980; Ogilvie and Stechey, 1983; Cleveland et al., 1986; Wilson et al., 1994a; Smith

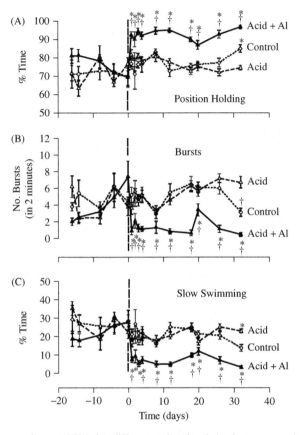

Fig. 2.10. Mean values ± SEM for different swimming behaviors measured in groups of rainbow trout (15 fish per tank and three replicate tanks per treatment) exposed to one of three treatments. Prior to day 0 all fish were maintained in circumneutral, artificial soft water (pH 6.4). Open diamonds joined by a dotted line represent control fish maintained in circumneutral water (pH 6.4) without any added Al. Open triangles joined by a dashed line represent fish exposed to sublethal acid only (pH 5.2) in the absence of Al. Filled triangles joined by a solid line represent fish exposed to acid (pH 5.2) in the presence of 30 μg L^{-1} of Al. Asterisks denote a significant difference from the neutral control group, and daggers denote a significant difference from the acid only group, both at $p < 0.05$. The chronic sublethal exposure to Al caused a sustained reduction in metabolically expensive bursts (sudden movements of at least two body lengths per second, most likely to be anaerobic muscle use) and the time spent cruising at aerobic speeds (slow swimming). From Allin and Wilson (1999).

and Haines, 1995). The high speed burst activity (greater than two body lengths per second) is particularly metabolically costly owing to the oxygen debt incurred from recruitment of the fast anaerobic white muscle (Jayne and Lauder, 1994) compared to aerobic metabolism in red muscle (Duncan and Klekowski, 1975). The reduction in both anaerobic and aerobic swimming activities could be due to direct muscle effects, but the rapidity of the response in some studies (e.g. within hours) (Allin and Wilson, 1999, 2000) suggests that it is directly related to the acute effects of Al on the gill surface and subsequent respiratory impairment by Al (i.e. the capacity for oxygen delivery to the swimming muscle). This has been taken as further evidence of a tradeoff in the energy requirements of damage–repair mechanisms reducing the apparent scope for activity, but in this case revealed in spontaneous activity in social groups of fish rather than forced aerobic swimming performance in a swim tunnel. At the moderate pH used (5.2) in the Al behavior studies of Allin and Wilson, sublethal acid exposure alone had little or no measurable effect on swimming behavior.

It is also known that during sublethal exposure to Al in acid soft water, the reduction in anaerobic burst activity within a social group of brown trout is associated with reduced aggressive attacks initiated by the most dominant fish in each tank (Fig. 2.11). The dominant fish initiating these attacks is thought to cause the majority of the burst activity within the whole group (e.g. causing one or more subordinate fish to escape using anaerobic burst activity). As this is a particularly energetically expensive activity it is not surprising that a respiratory toxicant such as Al which limits aerobic scope (Wilson et al., 1994b) can have a major impact upon the social interactions of a whole group via its effect on the dominant fish (Sloman and Wilson, 2005). Similar responses within social groups have been found during exposure to hypoxia (Kramer, 1987; Sneddon and Yerbury, 2004). The obvious metabolic constraint imposed by hypoxia is presumably acting in the same way as the respiratory toxicity of Al, and both cause reductions in aggressive behavior and spontaneous activity.

12.3. Sensory Effects and Avoidance Behavior (Including Toxic Mixing Zones)

The ability of fish to detect and then avoid high levels of toxicants is obviously a potentially important defense mechanism, provided that refuges can be sought that offer reduced exposure. However, there have been mixed reports about the ability of fish to detect and avoid Al in acidic freshwaters (Gunn and Noakes, 1986; Gunn et al., 1987; Atland and Barlaup, 1996; Exley, 2000). At the same time, toxicants such as Al may influence the sensory ability of fish, for example, to find food or mates and avoid prey (Tierney et al., 2010). Such sensory impacts of Al are not well understood, and only one study has investigated the direct effect of Al on the functions of olfactory sensory neurons (Klaprat et al., 1988). They found that Al at

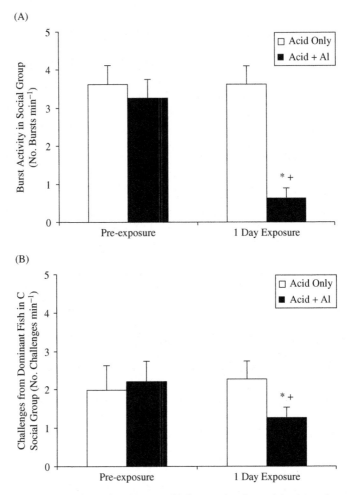

Fig. 2.11. Mean values ± SEM for (A) anaerobic burst swimming activity (more than two body lengths per second) within a social group, and (B) aggressive challenges initiated by the dominant fish in the social group of brown trout both before exposure, and after 1 day exposure to sublethal low pH in the absence of Al (open bars), or in the presence of sublethal Al at 13 μg L^{-1} (solid bars). Asterisks denote a significant difference from the relevant pre-exposure data, and crosses denote a significant difference from the acid-only group at the same time-point, both at $p < 0.05$. Acidic pH alone had no effect, but exposure to Al dramatically reduced both the burst activity within the whole social group and the aggressive challenges initiated by the most dominant fish. From Sloman and Wilson (2005).

quite high levels (20 μM or 540 μg L^{-1}) at pH 4.7 (compared to pH 4.7 alone) substantially impaired the electrical response of the olfactory nerve of rainbow trout to stimulation by L-serine. The repercussions of such olfactory effects on whole organism function are not yet known.

13. MOLECULAR CHARACTERIZATION OF ALUMINUM TRANSPORTERS, STORAGE PROTEINS, AND CHAPERONES

Aluminum has no known biological function in normal health, and no Al-specific transport processes are known. Mechanisms of transport across membranes are poorly understood, and relatively unexplored in fish. However, in mammals there is evidence that approximately 90% of plasma Al is associated with transferrin (the Fe^{3+}-carrying glycoprotein in plasma) and about 7% with citrate (Yokel, 2004; Exley, 2008), and that Al uptake into the brain (via the blood–brain barrier) occurs via transferrin-receptor mediated endocytosis and aluminum citrate uptake by system Xc(–) (an Na^+-independent glutamate transporter) (Nagasawa et al., 2005) as well as by an unidentified organic anion transporter (Yokel, 2006). However, no such detailed studies have been carried out in fish, so the molecular aspects of internal handling of Al remain a mystery in this group of vertebrates.

14. GENOMIC AND PROTEOMIC STUDIES

No large-scale studies have yet been carried out on the general genomic or proteomic effects of Al exposure in fish. However, as mentioned in Section 6.5, several recent studies have shown some effects on mRNA expression of specific genes associated with marine osmoregulatory function in gills of smolts that were exposed to acid and Al in freshwater (Monette et al., 2010; Nilsen et al., 2010). These effects were associated with the accumulation of Al on (or in) gill tissue, but the specificity of these effects on gene expression is not yet known. These responses may simply reflect a generalized stress response within gill tissue suffering from ionregulatory and/or respiratory symptoms associated with gill Al accumulation that were highlighted earlier (Section 6.1.2). It is worth noting that one recent study (Olsvik et al., 2010) examined the liver expression of eight different genes that could be used as markers of apoptosis, oxidative stress, metal stress (metallothionein), and protein degradation, in response to the individual and combined effects of acute (5 h) exposure to gamma irradiation, Al, and Cd in acidified soft water. The combination of Al and Cd (without radiation) did not change the transcription of any of these genes, whereas the gamma radiation alone significantly induced five of these genes. However, when all three toxicants were combined (gamma irradiation, Al, and Cd), none of the eight genes had altered expression, suggesting that Al and Cd worked antagonistically to counteract the molecular responses to gamma-irradiation that on its own resulted in proteins involved in defense mechanisms against cellular free radicals (Olsvik et al., 2010).

15. INTERACTIONS WITH OTHER METALS

Although Al has been identified as the key cause of biotic problems in waters suffering from acidification, it is clearly not the only toxic metal that is solubilized under such conditions, and Al often occurs in concert with other toxic metals where there is potential for it to act additively, less than additively, and synergistically (Gensemer and Playle, 1999). Hickie et al. (1993) exposed rainbow trout alevins and larval fathead minnow to mixtures of toxic metals containing Al, Mn, Fe, Ni, Zn, Cu, and Pb (at a ratio of 75:60:60:12:12:6:6 $\mu g\,L^{-1}$), at a range of pH values. The toxicity of the mixture could largely be explained by the presence of Al, Zn, and Cu alone, and at pH 4.9 the mixture toxicity was accounted for by Al alone. In contrast, Hutchinson and Sprague (1986) found evidence for synergism when examining the toxicity of Al in conjunction with Cu and Zn. Al^{3+} and H^+ may act by a similar mechanism to inhibit branchial ion regulation, and Al^{3+} may even protect against severe low pH (e.g. pH 4) (Neville, 1985; Neville and Campbell, 1988). There is also recent evidence that Al (in combination with Cd) may protect Atlantic salmon against the transcriptional effects of acute (5 h) exposure to gamma-irradiation (4 or 75 mGy) (Salbu et al., 2008; Olsvik et al., 2010) on eight different gene products, although it was not clear whether the primary protective effect was derived from the Cd or Al exposure (see Section 14).

16. KNOWLEDGE GAPS AND FUTURE DIRECTIONS

16.1. Acclimation to Aluminum during Chronic Exposure

The term acclimation is used within the toxicological literature to define an increased resistance to high (usually lethal) levels of a toxicant which develops during prolonged sublethal exposure. Acclimation to Al in acidic soft waters has now been documented in a number of laboratory studies (Orr et al., 1986; McDonald et al., 1991; Wilson et al., 1994a), which may explain the continued presence of fish populations in acidified soft waters containing levels of Al in excess of the thresholds predicted by acute toxicity tests (Wright and Snekvik, 1978; Schofield and Trojnar, 1980; Kelso et al., 1986). The mechanism of acclimation is not precisely understood but is thought to revolve around a damage–repair process which involves biochemical, physiological, and structural changes within the primary target organ, the gills (McDonald et al., 1991; Mueller et al., 1991; McDonald and Wood, 1993; Wilson et al., 1994a, b).

The mechanism of acclimation is intuitively linked to counteracting the mechanism of toxicity and ultimately the recovery of physiological functions. During acclimation the gill binding affinity for Al^{3+} is reduced relative to Ca^{2+} (Reid et al., 1991), which would help to regain control of paracellular ionic permeability and assist the recovery of ion efflux rates. The reduction of active ion uptake by the gills caused by Al likely results from a direct inhibition of branchial ion/acid–base transport proteins such as Na^+/K^+-ATPase and carbonic anhydrase (see Section 6.1.1). However, similar to ionic effluxes, the active uptake of Na^+ also recovers with time (McDonald and Milligan, 1988) and is clearly the result of compensatory mechanisms within the gills. Although increased activity of transport proteins such as Na^+/K^+-ATPase has not been measured during longer term exposures, hyperplasia of gill chloride cells (associated with much of the gills active ion transport capacity) is a prominent feature of chronic sublethal Al exposure (Tietge et al., 1988; Mueller et al., 1991).

Acclimation usually occurs within the first 5–17 days of sublethal exposure (Orr et al., 1986; Wood et al., 1988a, c; McDonald et al., 1991; Wilson and Wood, 1992; Wilson et al., 1994a) and appears to coincide with a pronounced hypertrophy of gill mucous cells. This is considered to be a compensatory rather than pathological response which is directly related to acclimation (McDonald et al., 1991; Mueller et al., 1991; McDonald and Wood, 1993). The purpose of mucocyte hyperplasia is presumably to accelerate the secretion of mucus that preferentially binds and carries away insoluble Al species from the sensitive gill surface (Playle and Wood, 1989, 1990, 1991; McDonald et al., 1991; Mueller et al., 1991; Wilson et al., 1994b). This clearance of precipitated or polymerized Al may reduce the inflammatory response often associated with the respiratory component of Al toxicity (Playle and Wood, 1991; McDonald et al., 1991). However, this is clearly not the sole mechanism of branchial acclimation. Additional responses that may be critical for restoration of normal ionoregulatory function include changes in the affinity of gill surface ligands for Al (Reid et al., 1991). However, more detailed information about the nature of such ligands is currently lacking.

The overall effect of these branchial acclimatory changes is a reduction in the various forms of Al present at the gills. Total gill Al levels generally peak after 2–5 days' exposure to sublethal acid plus Al (McDonald et al., 1991; Wilson and Wood, 1992; Wilson et al., 1994a) and then fall coinciding with the development of acclimation. However, in two separate studies the accumulation of gill Al has been shown to follow a biphasic pattern in which continued exposure (for up to 34 days) leads to gill Al levels undergoing a secondary more gradual increase, which can surpass the initial peak seen prior to acclimation. Reduced toxicity is therefore not simply a function of a reduction in the total level of Al in the gills. What may be of more importance is the distribution of this Al at different sites within and on gills.

16.2. Specificity of Acclimation to Aluminum

The specificity of acclimation to Al (that is, whether it will simultaneously increase a fish's resistance to other metals) is poorly understood. Since the process of acclimation is probably designed to combat the mechanism of toxicity of a metal, it is possible that metals with similar modes of toxic action will elicit similar acclimatory responses and thus may exhibit cross-over resistance (McDonald and Wood, 1993). This is true for Zn and Cd, which both interfere with branchial control of Ca^{2+} balance, and acclimation to Cd has been shown to increase resistance to Zn and vice versa (Kito et al., 1982; Duncan and Klaverkamp, 1983; Thomas et al., 1985; Klaverkamp and Duncan, 1987). Copper has similar effects to Al in freshwater fish in that both can impair respiratory gas exchange during acute exposure (Playle et al., 1989; Wilson and Taylor, 1993) and both interfere with the active influx and passive efflux components of branchial ion transport (Laurén and McDonald, 1985, 1986; Booth et al., 1988; Wood and McDonald, 1987). However, Wilson et al. (1994a) found no evidence for substantial cross-over resistance to Cu in fish acclimated to sublethal acid plus Al over a 34 day period. This implies that acclimation to Al involves changes that are specific to Al rather than the type of physiological toxicity it produces. However, although Cu and Al inhibit Na^+ uptake via a similar if not identical manner (inhibition of transport enzymes) (Wood, 1992), they may exacerbate passive Na^+ effluxes via subtly different mechanisms because Na^+ efflux is independent of the ambient $[Ca^{2+}]$ during Cu exposure (Laurén and McDonald, 1985, 1986), but is Ca^{2+} dependent during Al exposure (Booth et al., 1988; Wood et al., 1988a, c; Playle et al., 1989). This subtle difference between the mechanisms of ionoregulatory toxicities of Al and Cu may relate to their lack of cross-over resistance. No other metals have yet been tested for cross-over resistance with Al. However, one study found that the acute toxicity of beryllium (Be) to perch (*Perca fluviatilis*) and roach (*Rutilus rutilus*) in soft acid water was analogous to that of Al (Jagoe et al., 1993). The similar chemistry, solubility, and aqueous speciation properties of beryllium (Vesely et al., 1989) to Al suggest that it may be a source of similar, as yet relatively unexplored, ecotoxicological problems for fish in soft acid waters. Given these observations, it would be interesting to test the cross-over resistance of Al with Be.

16.3. Impacts of Episodic Exposure and its Moderation by Acclimation

Short-term episodic or pulsed increases in both acidity and Al levels are known to have greater effects on fish than chronic exposure to mild acid and Al levels (see Section 3.2). The physiological impacts of such pulsed exposures in naïve fish (Gunn and Noakes, 1987; Lacroix and Townsend 1987; Cleveland et al., 1991; Reader et al., 1991) are essentially similar to the acute mechanisms of toxicity described in Section 5. However, it has often been demonstrated that chronic exposure to mild acid plus Al conditions in laboratory conditions can induce some degree of acclimation and tolerance to lethal levels of Al (Orr et al., 1986; Wood et al., 1988a, b; Sayer, 1991; Wilson et al., 1994a, b, 1996). There is therefore clear potential for acclimation to low ambient levels in nature to play an important role in the ongoing survival of populations that receive intermittent exposure to potentially lethal levels (Allin and Wilson, 2000), which can be up to an order of magnitude higher than the chronic exposure level (Monteith et al., 1998). Allin and Wilson (2000) addressed this issue and found rainbow trout that were preacclimated to a low level of Al (24 µg L^{-1} at pH 5.2) for 16 days recovered their aerobic and anaerobic swimming behaviors much more quickly when subsequently exposed to a pulse of 36 µg L^{-1} Al (also pH 5.2). In fact, the acclimated fish recovered normal behavior within the time-frame of the pulse itself (4 days). In contrast, Al-naïve fish (previously held in pH 5.2 water only) not only became even more hypoactive at the start of the 4 day Al pulse compared to acclimated fish, but failed to show any recovery of normal behavior even 6 days after the pulse had ceased. There was also significantly more mortality in the Al-naïve group in response to the Al pulse (26% compared to 4% in the preacclimated fish). Intriguingly, the difference between the acclimated and naïve fish cannot be explained by removal of gill Al, as depuration was complete within just 2 days of the pulse ending in the naïve fish, whereas both behavioral and physiological effects (e.g. hematological disturbances) persisted for at least 6 more days without any signs of improvement. This provides further evidence that an acclimation-induced reduction in toxicity is not merely a function of removing Al from the gills (see also Section 16.1). A more detailed understanding is therefore needed of the mechanisms behind acclimation and the survival of populations that receive such episodic surges in Al in the wild.

16.4. Effects of Carbon Dioxide Buildup on Aluminum Toxicity in
       Soft-water Aquaculture Conditions

Hatcheries for rearing Atlantic salmon smolts are often based in soft waters that have low buffer capacity. When water supply is a limiting factor in such aquaculture situations, the accumulation of CO_2 excreted by the fish can have a significant effect on the physiology of fish by itself, but also via its effect on pH. Fivelstad et al. (2003) showed that the pH reduction associated with CO_2 buildup can also influence Al speciation and bioavailability in the incoming water supply, causing fish to accumulate more gill Al and to suffer from the classic symptoms of both ionoregulatory and respiratory toxicity of Al under acidic conditions (see Section 6.1). In these rather specific circumstances, the acidification is caused by the fish's own metabolic CO_2, and the Al is already present (rather than being solubilized from the local geology). The accumulation of CO_2 in such high fish densities in aquaculture is far greater than the CO_2 levels (and related depression in pH) expected to result from anthropogenic CO_2 emissions to the atmosphere in the near future. The degree of freshwater acidification caused by the latter is likely to be too small to cause any significant additional problems of Al toxicity to fish in natural waters.

REFERENCES

Allin, C. J. (1999). *An integrated behavioural and physiological approach to aluminium toxicity in trout.* PhD thesis, University of Exeter.

Allin, C. J., and Wilson, R. W. (1999). Behavioural and metabolic effects of chronic exposure to sublethal aluminum in acidic soft water in juvenile rainbow trout *(Oncorhynchus mykiss).* Can. J. Fish. Aquat. Sci. **56**, 670–678.

Allin, C. J., and Wilson, R. W. (2000). Effects of pre-acclimation to aluminium on the physiology and swimming behaviour of juvenile rainbow trout *(Oncorhynchus mykiss)* during a pulsed exposure. Aquat. Toxicol. **51**, 213–224.

Andersen, D. O. (2006). Labile aluminium chemistry downstream a limestone treated lake and an acid tributary: effects of warm winters and extreme rainstorms. Sci. Total Environ. **366**, 739–748.

Atland, A., and Barlaup, B. T. (1995). Avoidance of toxic mixing zones by Atlantic salmon *(Salmo salar* L.) and brown trout *(Salmo trutta* L.) in the limed river Audna, southern Norway. Environ. Pollut. **90**, 203–208.

Atland, A., and Barlaup, B. T. (1996). Avoidance behaviour of Atlantic salmon *(Salmo salar* L) fry in waters of low pH and elevated aluminum concentration: laboratory experiments. Can. J. Fish. Aquat. Sci. **53**, 1827–1834.

Audet, C., and Wood, C. M. (1993). Branchial morphological and endocrine responses of rainbow trout *(Oncorhynchus mykiss)* during a long term sublethal acid exposure in which acclimation did not occur. Can. J. Fish. Aquat. Sci. **50**, 198–209.

Baker, J., and Schofield, C. (1982). Aluminum toxicity to fish in acidic waters. *Water Air Soil Pollut.* **18**, 289.

Baker, J. P., VanSickle, J., Gagen, G. J., DeWalle, D. R., Sharpe, W. E., Carline, R. F., Baldigo, B. P., Murdoch, P. S., Bath, D. W., Kretser, W. A., Simonin, H. A., and Wigington, P. J. (1996). Episodic acidification of small streams in the North-Eastern United States: effects on fish populations. Ecol. Applic. **6**, 422–437.

Battram, J. C. (1988). The effects of aluminium and low pH on chloride fluxes in the brown trout, *Salmo trutta* L. J. Fish Biol. **32**, 937–947.

Beamish, R. J., Lockhart, W. L., Van Loon, J. C., and Harvey, H. H. (1975). Long-term acidification of a lake and resulting effects on fishes. *Ambio* **4**, 98–102.

Birchall, J. D., Exley, C., Chappell, J. S., and Phillips, M. J. (1989). Acute toxicity of aluminium to fish eliminated in silicon-rich acid waters. *Nature* **338**, 146.

Bjerknes, V., Fyllingen, I., Holtet, L., Teien, H. C., Rosseland, B. O., and Kroglund, F. (2003). Aluminium in acidic river water causes mortality of farmed Atlantic Salmon (*Salmo salar* L.) in Norwegian fjords. *Mar. Chem.* **83**, 169–174.

Bjørnstad, H. E., Oughton, D. H., and Salbu, B. (1992). Determination of aluminium-26 using a low level liquid scintillation spectrometer. *Analyst* **117**, 435–438.

Booth, C. E., McDonald, D. G., Simons, B. P., and Wood, C. M. (1988). Effects of aluminum and low pH on net ion fluxes and ion balance in the brook trout (*Salvelinus fontinalis*). *Can. J. Fish. Aquat. Sci.* **45**, 1563–1574.

Brett, J. R. (1958). Implications and assessments of environmental stress. In: *The Investigation of Fish-Power Problems* (P. A. Larkin, ed.), pp. 69–93. University of British Columbia, Vancouver, BC.

Brett, J. R. (1964). The respiratory metabolism and swimming performance of young sockeye salmon. *J. Fish. Res. Bd Can.* **21**, 1183–1226.

Brodeur, J. C., Okland, F., Finstad, B., Dixon, D. G., and McKinley, R. S. (2001). Effects of subchronic exposure to aluminium in acidic water on bioenergetics of Atlantic salmon (*Salmo salar*). *Ecotoxicol. Environ. Saf.* **49**, 226–234.

Brown, J. A., and Whitehead, C. (1995). Catecholamine release and interrenal response of brown trout, *Salmo trutta*, exposed to aluminium in acidic water. *J. Fish Biol.* **46**, 524–535.

Brown, M. T., Lippiatt, S. M., and Bruland, K. W. (2010). Dissolved Al, particulate Al, and silicic acid in northern Gulf of Alaska coastal waters – glacial–riverine inputs and extreme reactivity. *Mar. Chem.* **122**, 160–175.

Butler, P. J., Day, N., and Namba, K. (1992). Interaction of seasonal temperature and low pH on resting oxygen uptake and swimming performance of adult brown trout, *Salmo trutta*. *J. Exp. Biol.* **165**, 195–212.

Bystriansky, J. S., Richards, J. G., Schulte, P. M., and Ballantyne, J. S. (2006). Reciprocal expression of gill Na^+/K^+-ATPase alpha-subunit isoforms alpha 1a and alpha 1b during seawater acclimation of three salmonid fishes that vary in their salinity tolerance. *J. Exp. Biol.* **209**, 1848–1858.

Cleveland, L., Little, E. E., Hamilton, S. J., Buckler, D. R., and Hunn, J. B. (1986). Interactive toxicity of aluminum and acidity to early life stages of brook trout. *Trans. Am. Fish. Soc.* **115**, 610–620.

Cleveland, L., Little, E. E., Ingersoll, C. G., Wiedmeyer, R. H., and Hunn, J. B. (1991). Sensitivity of brook trout to low pH, low calcium and elevated aluminum concentrations during laboratory pulse exposures. *Aquat. Toxicol.* **19**, 303–318.

Cooke, G. D., Welch, E. B., Peterson, S. A. and Nichols, S. A. (eds) (2005). *Restoration and Management of Lakes and Reservoirs.* Taylor and Francis, Boca Raton, FL.

Cronan, C. S., and Schofield, C. L. (1979). Aluminium leaching response to acid precipitation: effects on high-elevation watersheds in the northeast. *Science* **204**, 304–306.

Cuthbert, A. W., and Maetz, J. (1972). The effects of calcium and magnesium on sodium fluxes through the gills of *Carassius auratus*, L. *J. Physiol.* **221**, 633–643.

D'Cruz, L. M., and Wood, C. M. (1998). The influence of dietary salt and energy on the response to low pH in juvenile rainbow trout. *Physiol. Zool.* **71**, 642–657.

Dietrich, D., and Schlatter, Ch. (1989). Aluminium toxicity to rainbow trout at low pH. *Aquat. Toxicol.* **15**, 197–212.

Dietrich, D., Schlatter, C., Blau, N., and Fischer, M. (1989). Aluminium and acid rain: mitigating effects of NaCl on aluminium toxicity to brown trout (*Salmo trutta* Fario) in acid water. *Toxicol. Environ. Chem.* **19**, 17–23.

Drablos, D., and Tollan, A. (eds) (1980). Proceedings of an International Conference, *Sandfjord, Norway, March 11–14*. SNSF Project, Oslo.

Driscoll, C. T. (1984). A procedure for the fractionation of aqueous aluminum in dilute acidic waters. *Int. J. Environ. Anal. Chem.* **16**, 267–283.

Driscoll, C. T., and Newton, R. M. (1985). Chemical characteristics of Adirondack lakes. *Environ. Sci. Technol.* **19**, 1018–1024.

Driscoll, C. T., and Postek, K. M. (1996). The chemistry of aluminum in surface waters. In: *The Environmental Chemistry of Aluminum* (G. Sposito, ed.), 2nd ed., pp. 363–418. Lewis Publishers, New York.

Driscoll, C. T., and Schecher, W. D. (1988). Aluminum in the environment. In: Metal Ions in Biological Systems, *Vol. 24*. Aluminum and its Role in Biology (H. H. Sigel and A. Sigel, eds), pp. 59–122. Marcel Dekker, New York.

Driscoll, C. T., Lawrence, G. B., Bulger, A. J., Butler, T. J., Cronan, C. S., Eagar, C., Lambert, K. F., Likens, G. E., Stoddard, J. L., and Weathers, K. C. (2001). Acidic deposition in the northeastern United States: sources and inputs, ecosystem effects, and management strategies. *BioScience* **51**, 180–198.

Drummond, R. A., Spoor, W. A., and Olson, G. F. (1973). Some short-term indicators of sublethal effects of copper on brook trout (*Salvelinus fontinalis*). *J. Fish. Res. Bd Can.* **30**, 698–701.

Duncan, A. and Klekowski, R. Z. (1975). Parameters of an energy budget. In: *Methods for Ecological Energetics*. IBP Handbook No. 24. Oxford: A. Brown and Sons.

Duncan, D. A., and Klaverkamp, J. F. (1983). Tolerance and resistance to cadmium in white suckers (*Catostomus commersoni*) previously exposed to cadmium, mercury, zinc or selenium. *Can. J. Fish. Aquat. Sci.* **40**, 128–138.

Dussault, E. B., Playle, R. C., Dixon, D. G., and McKinley, R. S. (2001). Effects of sublethal, acidic aluminum exposure on blood ions and metabolites, cardiac output, heart rate, and stroke volume of rainbow trout. Oncorhynchus mykiss. *Fish Physiol. Biochem.* **25**, 347–357.

Dussault, E. B., Playle, R. C., Dixon, D. G., and McKinley, R. S. (2004). Effects of chronic aluminum exposure on swimming and cardiac performance in rainbow trout. Oncorhynchus mykiss. *Fish Physiol. Biochem.* **30**, 137–148.

Eichenberger, E. (1986). The interrelation between essentiality and toxicity of metals in the aquatic ecosystem. In: Metal Ions in Biological Systems, *Vol. 20*. Concepts on Metal Ion Toxicity (H. Sigel, ed.), pp. 67–100. Marcel Dekker, New York.

Elangovan, R., White, K. N., and McCrohan, C. R. (1997). Bioaccumulation of aluminium in the freshwater snail *Lymnaea stagnalis* at neutral pH. *Environ. Pollut.* **96**, 29–33.

Evans, R. E., Brown, S. B., and Hara, T. J. (1988). The effects of aluminum and acid on the gill morphology in rainbow trout. Salmo gairdneri. *Environ. Biol. Fish.* **22**, 299–311.

Evans, D. H., Piermarini, P. M., and Choe, K. P. (2005). The multifunctional fish gill: dominant site of gas exchange, osmoregulation, acid–base regulation, and excretion of nitrogenous waste. *Physiol. Rev.* **85**, 97–177.

Exley, C. (1996). Aluminium in the brain and heart of the rainbow trout. *J. Fish Biol.* **48**, 706–713.

Exley, C. (2000). Avoidance of aluminum by rainbow trout. *Environ. Toxicol. Chem.* **19**, 933–939.

Exley, C. (2003). A biogeochemical cycle for aluminium. *J. Inorg. Biochem.* **97**, 1–7.

Exley, C. (2008). Aluminium and medicine. In: *Molecular and Supramolecular Bioinorganic Chemistry* (A. L. R. Merce, ed.), pp. 1–24. Nova Science Publishers, New York.

Exley, C., Chappell, J. S., and Birchall, J. D. (1991). A mechanism for acute aluminium toxicity in fish. *J. Theoret. Biol.* **151**, 417.

Exley, C., Wicks, A. J., Hubert, R. B., and Birchall, D. J. (1996). Kinetic constraints in acute aluminium toxicity in the rainbow trout (*Oncorhynchus mykiss*). *J. Theoret. Biol.* **179**, 25–31.

Farag, A. M., Woodward, D. F., Little, E. E, Steadman, B., and Vertucci, F. A. (1993). The effects of low pH and elevated aluminium on Yellowstone cutthroat trout (*Oncorhynchus clarki bouieri*). *Environ. Toxicol. Chem.* **12**, 719–731.

Farmer, G. J., Ashfield, D., and Samant, H. S. (1979). Effects of zinc on juvenile Atlantic salmon *Salmo salar*: acute toxicity, food intake, growth and bioaccumulation. *Environ. Pollut.* **19**, 109–117.

Findlay, D. L., and Kasian, S. E. M. (1986). Phytoplankton community responses to acidification of Lake 223, Experimental Lakes Area, northwestern Ontario. *Water Air Soil Pollut.* **30**, 719–726.

Fivelstad, S., Waagbø, R., Zeitz, S. F., Hosfeld, A. C. D., Olsen, A. B., and Stefansson, S. (2003). A major water quality problem in smolt farms: combined effects of carbon dioxide, reduced pH and aluminium on Atlantic salmon (*Salmo salar* L.) smolts: physiology and growth. *Aquaculture* **215**, 339–357.

Frausto da Silva, J. J. R., and Williams, R. J. P. (1996). *The Biological Chemistry of the Elements* (rev. edn.). Oxford University Press, Oxford.

Freda, J., Sanchez, D. A., and Bergman, H. L. (1991). Shortening of branchial tight junctions in acid exposed rainbow trout (*Oncorhynchus mykiss*). *Can. J. Fish. Aquat. Sci.* **48**, 2028–2033.

Freeman, R. A., and Everhart, W. H. (1971). Toxicity of aluminum hydroxide complexes in neutral and basic media to rainbow trout. *Trans. Am. Fish. Soc.* **100**, 644–658.

Galle, C., Chassardbouchaud, C., Massabuau, J. C., Escaig, F., Boumati, P., Bourges, M., and Pepin, D. (1990). Subcellular localization of acid-rain transported aluminum in gills and kidneys of trout from streams of the Vosges – preliminary results. *Comptes Rend. Acad. Sci. III Life Sci.* **311**, 301–307.

Gensemer, R. W., and Playle, R. C. (1999). The bioavailability and toxicity of aluminum in aquatic environments. *Crit. Rev. Environ. Sci. Technol.* **29**, 315–450.

Goossenaerts, C., Van Grieken, R., Jacob, W., Witters, H., and Vanderborght, O. (1988). A microanalytical study of the gills of aluminium exposed rainbow trout (*Salmo gairdneri*). *Int. J. Environ. Anal. Chem.* **34**, 227–237.

Goss, G. G., and Wood, C. M. (1988). The effects of acid and acid/aluminum exposure on circulating plasma cortisol levels and other blood parameters in the rainbow trout, *Salmo gairdneri*. *J. Fish Biol.* **32**, 63–76.

Gostomski, F. (1990). The toxicity of aluminum to aquatic species in the US. *Environ. Geochem. Health* **12**, 51–54.

Graham, M. S., and Wood, C. M. (1981). Toxicity of environmental acid to the rainbow trout: interactions of water hardness, acid type, and exercise. *Can. J. Zool.* **59**, 1518–1526.

Gundersen, D. T., Bustaman, S., Seim, W. K., and Curtis, L. R. (1994). pH, hardness, and humic acid influence aluminum toxicity to rainbow trout (*Oncorhynchus mykiss*) in weakly alkaline waters. *Can. J. Fish. Aquat. Sci.* **51**, 1345–1355.

Gunn, J. M., and Noakes, D. L. G. (1986). Avoidance of low pH and elevated Al concentrations by brook charr (*Salvelinus fontinalis*) alevins in laboratory tests. *Water Air Soil Pollut.* **30**, 497–503.

Gunn, J. M., and Noakes, D. L. G. (1987). Latent effects of pulse exposure to aluminum and low pH on size, ionic composition, and feeding efficiency of lake trout (*Salvelinus namaycush*) alevins. *Can. J. Fish. Aquat. Sci.* **44**, 1418–1424.

Gunn, J. M., Noakes, D. L. G., and Westlake, G. F. (1987). Behavioral responses of lake char (*Salvelinus namaycush*) embryos to simulated acidic runoff conditions. *Can. J. Zool.* **65**, 2786–2792.

Haines, T. A. (1981). Acidic precipitation and its consequences for aquatic ecosystems: a review. *Trans. Am. Fish. Soc.* **110**, 669–707.

Hamilton, S. J., and Haines, T. A. (1995). Influence of fluoride on aluminum toxicity to Atlantic salmon (*Salmo salar*). *Can. J. Fish. Aquat. Sci.* **52**, 2432–2444.

Handy, R. D. (1993). The accumulation of dietary aluminium by rainbow trout, *Oncorhynchus mykiss*, at high exposure concentrations. *J. Fish Biol.* **42**, 603–606.

Handy, R. D. (1994). Intermittent exposure to aquatic pollutants: assessment, toxicity and sublethal responses in fish and invertebrates. *Comp. Biochem. Physiol. C* **107**, 171–184.

Handy, R. D. (1996). Dietary exposure to toxic metals in fish. In: *Toxicology of Aquatic Pollution.* Society for Experimental Biology Seminar Series, Vol. 57, pp. 29–60. London: Society for Experimental Biology.

Handy, R. D., and Eddy, F. B. (1989). Surface absorption of aluminium by gill tissue and body mucus of rainbow trout, *Salmo gairdneri*, at the onset of episodic exposure. *J. Fish Biol.* **34**, 865–874.

Hargis, J. R. (1976). Ventilation and metabolic rate of young rainbow trout (*Salmo gairdneri*) exposed to sublethal environmental pH. *J. Exp. Zool.* **196**, 39–44.

Havas, M. (1986a). Aluminum chemistry of inland waters. In: *Aluminum in the Canadian Environment* (M. Havas and J. F. Jaworski, eds), pp. 51–77. National Research Council of Canada, Ottawa. Publication No. 24759.

Havas, M. (1986b). Effects of aluminum on aquatic biota. In: *Aluminum in the Canadian Environment* (M. Havas and J. F. Jaworski, eds), pp. 79–127. National Research Council of Canada, Ottawa. Publication No. 24759.

Havas, M., and Jaworski, J. F. (1986). *Aluminum in the Canadian Environment.* National Research Council of Canada, Ottawa, Publication No. 24759

Heath, R. H., Kahl, J. S., and Norton, S. A. (1992). Episodic stream acidification caused by atmospheric deposition of sea salts at Acadia National Park, Maine, United States. *Water Resour. Res.* **28**, 1081–1088.

Heming, T. A., and Blumhagen, K. A. (1988). Plasma acid–base and electrolyte states of rainbow trout exposed to alum (aluminum sulphate) in acidic and alkaline environments. *Aquat. Toxicol.* **12**, 125–139.

Herrmann, J., and Frick, K. (1995). Do stream invertebrates accumulate aluminium at low pH conditions? *Water Air Soil Pollut.* **85**, 407–412.

Hickie, B. E., Hutchinson, N. J., Dixon, D. G., and Hodson, P. V. (1993). Toxicity of trace metal mixtures to alevin rainbow trout (*Oncorhynchus mykiss*) and larval fathead minnow (*Pimephales promelas*) in soft, acidic water. *Can. J. Fish. Aquat. Sci.* **50**, 1348–1355.

Howells, G. D., and Dalziel, T. R. K. (1990). *Restoring Acid Waters: Loch Fleet 1984–1990.* Elsevier Applied Science, London.

Howells, G. D., Brown, D. J. A., and Sadler, K. (1983). Effects of acidity, calcium, and aluminum on fish survival and productivity – a review. *J. Sci. Food Agric.* **34**, 559–570.

Howells, G. D., Dalziel, T. R. K., Reader, J. P., and Solbe, J. F. (1990). EIFAC water quality criteria for European freshwater fish: report on aluminium. *Chem. Ecol.* **4**, 117–173.

Hutchinson, G. E. (1957). A Treatise on Limnology, *Vol. 1, Part 2. Chemistry of Lakes.* John Wiley & Sons, New York.

Hutchinson, N. J., and Sprague, J. B. (1986). Toxicity of trace metal mixtures to American flagfish (*Jordanella floridae*) in soft, acidic water and implications for cultural acidification. *Can. J. Fish. Aquat. Sci.* **43**, 647–655.

Hydes, D. J., and Liss, P. S. (1977). Behavior of dissolved aluminum in estuarine and coastal waters. *Estuar. Coast. Mar. Sci.* **5**, 755–769.

Ingersoll, C. G., Mount, D. R., Gulley, D. D., Lapoint, T. W., and Bergman, H. L. (1990a). Effects of pH, aluminum, and calcium on survival and growth of eggs and fry of brook trout (*Salvelinus fontinalis*). *Can. J. Fish. Aquat. Sci.* **47**, 1580–1592.

Ingersoll, C. G., Gulley, D. D., Mount, D. R., Mueller, M. E., Fernandez, J. D., Hockett, J. R., and Bergman, H. L. (1990b). Aluminum and acid toxicity to two strains of brook trout (*Salvelinus fontinalis*). *Can. J. Fish. Aquat. Sci.* **47**, 1641–1648.

Jagoe, C. H., and Haines, T. A. (1997). Changes in gill morphology of Atlantic salmon (*Salmo salar*) smolts due to addition of acid and aluminum to stream water. *Environ. Pollut.* **97**, 137–146.

Jagoe, C. H., Matey, V. E., Haines, T. A., and Komov, V. T. (1993). Effect of beryllium on fish in acid water is analogous to aluminium toxicity. *Aquat. Toxicol.* **24**, 241–256.

Jansen, S., Watanabe, T., and Smets, E. (2002). Aluminium accumulation in leaves of 127 species in Melastomataceae, with comments on the Order Myrtales. *Ann. Bot.* **90**, 53.

Jansson, M., Persson, G., and Broberg, O. (1986). Phosphorus in acidified lakes: the example of Lake Gårdsjön, Sweden. *Hydrobiologia* **139**, 81–96.

Jayne, B. C., and Lauder, G. V. (1994). How swimming fish use slow and fast muscle fibres: implications for models of vertebrate muscle recruitment. *J. Comp. Physiol. A* **175**, 123–131.

Kahl, J. S., Norton, S. A., MacRae, R. K., Haines, T. A., and Davis, R. B. (1989). The influence of organic acidity on the chemistry of Maine surface waters. *Water Air Soil Pollut.* **46**, 221–234.

Kahl, J. S., Norton, S. A., Haines, T. A., Rochette, E. A., Heath, R. H., and Nodvin, S. C. (1992). Mechanisms of episodic acidification in low-order streams in Maine, USA. *Environ. Pollut.* **78**, 37–44.

Karlsson-Norrgren, L., Dickson, W., Ljungberg, O., and Runn, P. (1986a). Acid water and aluminum exposure: gill lesions and aluminum accumulation in farmed brown trout *Salmo trutta* L. *J. Fish Dis.* **9**, 1–8.

Karlsson-Norrgren, L., Bjorklund, I., Ljungberg, O., and Runn, P. (1986b). Acid water and aluminum exposure: experimentally induced gill lesions in brown trout, *Salmo trutta* L. *J. Fish Dis.* **9**, 11–25.

Kelso, J. R. M., Minns, C. K., Gray, J. E. and Jones, M. L. (1986). Acidification of surface waters in eastern Canada and its relationship to aquatic biota. *Canadian Special Publication of Fisheries and Aquatic Sciences* 87. Ottawa, Canada.

King, S. O., Mach, C. E., and Brezonik, P. L. (1992). Changes in trace metal concentrations in lake water and biota during experimental acidification of Little Rock Lake, Wisconsin, USA. *Environ. Pollut.* **78**, 9–18.

Kito, H., Kazawa, T., Ose, Y., Sato, T., and Ishikawa, T. (1982). Protection by metallothionein against cadmium toxicity. *Comp. Biochem. Physiol.* **73**, 135–139.

Klaprat, D. A., Brown, S. B., and Hara, T. J. (1988). The effect of low pH and aluminum on the olfactory organ of rainbow trout. Salmo gairdneri. Environ. Biol. Fishes **22**, 69–77.

Klaverkamp, J. F., and Duncan, D. A. (1987). Acclimation to cadmium toxicity by white suckers: cadmium binding capacity and metal distribution in gill and liver cytosol. *Environ. Toxicol. Chem.* **6**, 275–289.

Kramer, D. L. (1987). Dissolved oxygen and fish behaviour. *Environ. Biol. Fish.* **18**, 81–92.

Kroglund, F., and Finstad, B. (2003). Low concentrations of inorganic monomeric aluminum impair physiological status and marine survival of Atlantic salmon. *Aquaculture* **222**, 119–133.

Kroglund, F., and Staurnes, M. (1999). Water quality requirements of smolting Atlantic salmon (*Salmo salar*) in limed acid rivers. *Can. J. Fish. Aquat. Sci.* **56**, 2078–2086.

Kroglund, F., Finstad, B., Stefansson, S. O., Nilsen, T. O., Kristensen, T., Rosseland, B. O., Teien, H. C., and Salbu, B. (2007). Exposure to moderate acid water and aluminum reduces Atlantic salmon post-smolt survival. *Aquaculture* **273**, 360–373.

Kroglund, F., Rosseland, B. O., Teien, H. C., Salbu, B., Kristensen, T., and Finstad, B. (2008). Water quality limits for Atlantic salmon (*Salmo salar* L.) exposed to short term reductions in pH and increased aluminum simulating episodes. *Hydrol. Earth Syst. Sci.* **12**, 491–507.

Lacroix, G. L., and Townsend, D. R. (1987). Responses of juvenile Atlantic salmon (*Salmo salar*) to episodic increases in acidity of Nova Scotia rivers. *Can. J. Fish. Aquat. Sci.* **44**, 1475–1484.

Lacroix, G. L., Peterson, R. H., Belfry, C. S., and Martinrobichaud, D. J. (1993). Aluminum dynamics on gills of Atlantic salmon fry in the presence of citrate and effects on integrity of gill structures. *Aquat. Toxicol.* **27**, 373–401.

Laitinen, M., and Valtonen, T. (1995). Cardiovascular, ventilatory and haematological responses of brown trout (*Salmo trutta* L.), to the combined effects of acidity and aluminium in humic water at winter temperatures. *Aquat. Toxicol.* **31**, 99–112.

Laudon, H., Poleo, A. B. S., Vollestad, L. A., and Bishop, K. (2005). Survival of brown trout during spring flood in DOC-rich streams in northern Sweden: the effect of present acid deposition and modelled pre-industrial water quality. *Environ. Pollut.* **135**, 121–130.

Laurén, D. J., and McDonald, D. G. (1985). Effects of copper on branchial ionoregulation in the rainbow trout, *Salmo gairdneri* Richardson. Modulation by water hardness and pH. *J. Comp. Physiol. B* **155**, 635–644.

Laurén, D. J., and McDonald, D. G. (1986). Influence of water hardness, pH, and alkalinity on the mechanisms of copper toxicity in juvenile rainbow trout. Salmo gairdneri. *Can. J. Fish. Aquat. Sci.* **43**, 1488–1496.

LaZerte, B. D., van Loon, G., and Anderson, B. (1997). Aluminum in water. In: *Research Issues in Aluminum Toxicity* (R. A. Yokel and M. S. Golub, eds), pp. 17–45. Taylor and Francis, Washington, DC.

Lee, C., and Harvey, H. H. (1986). Localization of aluminum in tissues of fish. *Water Air Soil Pollut.* **30**, 649–655.

Lett, P. F., Farmer, G. J., and Beamish, F. W. H. (1976). Effect of copper on some aspects of the bioenergetics of rainbow trout (*Salmo gairdneri*). *J. Fish. Res. Bd Can.* **33**, 1335–1342.

Lewis, T. E. (1989). *Environmental Chemistry and Toxicology of Aluminum.* Lewis Publishers, Chelsea, MI.

Lin, H., and Randall, D. J. (1990). The effect of varying water pH on the acidification of expired water in rainbow trout. *J. Exp. Biol.* **149**, 149–160.

Linthurst, R. A., Landers, D. H., Eilers, J. M., Brakke, D. F., Overton, W. S., Meier, E. P., and Crowe, R. E. (1986). Characteristics of lakes in the eastern United States, *Vol. 1.* Population Descriptions and Physico-chemical Relationships. *EPA/600/4-86/007a.* US Environmental Protection Agency, Washington, DC.

Lukiw, W. J. (2010). Evidence supporting a biological role for aluminum in chromatin compaction and epigenetics. *J. Inorg. Biochem.* **104**, 1010–1012.

Lydersen, E. (1990). The solubility and hydrolysis of aqueous aluminium hydroxides in dilute fresh waters at different temperatures. *Nord. Hydrol.* **21**, 195–204.

Lydersen, E., and Löfgren, S. (2002). Potential effects of metals in reacidified limed water bodies in Norway and Sweden. *Environ. Monit. Assess.* **73**, 155–178.

Madsen, S. S., Kiilerich, P., and Tipsmark, C. K. (2009). Multiplicity of expression of Na$^+$, K$^+$-ATPase α-subunit isoforms in the gill of Atlantic salmon (*Salmo salar*): cellular localisation and absolute quantification in response to salinity change. *J. Exp. Biol.* **212**, 78–88.

Magee, J. A., Haines, T. A., Kocik, J. F., Beland, K. F., and McCormick, S. D. (2001). Effects of acidity and aluminum on the physiology and migratory behavior of Atlantic salmon smolts in Maine, USA. *Water Air Soil Pollut.* **130**, 881–886.

Magee, J. A., Obedzinski, M., McCormick, S. D., and Kocik, J. F. (2003). Effects of episodic acidification on Atlantic salmon (*Salmo salar*) smolts. *Can. J. Fish. Aquat. Sci.* **60**, 214–221.

Marshall, W. S., and Grosell, M. (2006). Ion transport, osmoregulation and acid–base balance. In: *Physiology of Fishes* (D. H. Evans and J. B. Claiborne, eds). 3rd edn. CRC Press, Boca Raton, FL.

Mazeaud, M. M., and Mazeaud, F. (1981). Adrenergic responses to stress in fish. In: *Stress and Fish* (A. D. Pickering, ed.), pp. 49–75. Academic Press, London.

McCartney, A. G., Harriman, R., Watt, A. W., Moore, D. W., Taylor, E. M., Collen, P., and Keay, E. J. (2003). Long-term trends in pH, aluminium and dissolved organic carbon in Scottish fresh waters; implications for brown trout (*Salmo trutta*) survival. *Sci. Total Environ.* **310**, 133–141.

McCormick, S. D., Keyes, A., Nislow, K. H., and Monette, M. Y. (2009a). Impacts of episodic acidification on in-stream survival and physiological impairment of Atlantic salmon (*Salmo salar*) smolts. *Can. J. Fish. Aquat. Sci.* **66**, 394–403.

McCormick, S. D., Regish, A. M., and Christensen, A. K. (2009b). Distinct freshwater and seawater isoforms of Na$^+$/K$^+$-ATPase in gill chloride cells of Atlantic salmon. *J. Exp. Biol.* **212**, 3994–4001.

McCormick, S. D., Lerner, D. T., Regish, A. M., O'Dean, M. F., and Monette, M. Y. (2011). Thresholds for short-term acid and aluminum impacts on Atlantic salmon smolts. *Aquaculture* (in press).

McDonald, D. G., and Milligan, C. L. (1988). Sodium transport in the brook trout, *Salvelinus fontinalis* – effects of prolonged low pH exposure in the presence and absence of aluminum. *Can. J. Fish. Aquat. Sci.* **45**, 1606–1613.

McDonald, D. G., and Rogano, M. S. (1986). Branchial mechanisms of ion and acid–base regulation in the freshwater rainbow trout. Salmo gairdneri. Can. J. Zool. **66**, 2699–2708.

McDonald, D. G., and Wood, C. M. (1993). Branchial mechanisms of acclimation to metals in freshwater fish. In: *Fish Ecophysiology* (C. Rankin and F. B. Jensen, eds), pp. 295–319. Chapman and Hall, London.

McDonald, D. G., Wood, C. M., Rhem, R. G., Mueller, M. E., Mount, D. R., and Bergman, H. L. (1991). Nature and time course of acclimation to aluminum in juvenile brook trout (*Salvelinus fontinalis*). 1. Physiology. *Can. J. Fish. Aquat. Sci.* **48**, 2006–2015.

Milligan, C. L., and Wood, C. M. (1982). Disturbances in haematology, fluid volume distribution and circulatory function associated with low environmental pH in the rainbow trout, *Salmo gairdneri*. *J. Exp. Biol.* **99**, 397–415.

Monette, M. Y., and McCormick, S. D. (2008). Impacts of short-term acid and aluminum exposure on Atlantic salmon (*Salmo salar*) physiology: a direct comparison of parr and smolts. *Aquat. Toxicol.* **86**, 216–226.

Monette, M. Y., Bjornsson, B. T., and McCormick, S. D. (2008). Effects of short-term acid and aluminum exposure on the parr–smolt transformation in Atlantic salmon (*Salmo salar*): disruption of seawater tolerance and endocrine status. *Gen. Comp. Endocrinol.* **158**, 122–130.

Monette, M. Y., Yada, T., Matey, V., and McCormick, S. D. (2010). Physiological, molecular, and cellular mechanisms of impaired seawater tolerance following exposure of Atlantic salmon, *Salmo salar*, smolts to acid and aluminum. *Aquat. Toxicol.* **99**, 17–32.

Monteith, D. T., Evans, C., Beaumont, W. R. C., Delinikajtis, C.1998. *The United Kingdom Acid Waters Monitoring Network Data Report for 1997–1998 (Year 10)*. ENISIS Ltd, Report to the Department of the Environment Transport and the Regions (Contract EPG 1:3:92).

Mount, D. R., Ingersoll, C. G., Gulley, D. D., Fernandez, J. D., LaPoint, T. W., and Bergman, H. L. (1988a). Effect of long-term exposure to acid, aluminum, and low calcium on adult brook trout (*Salvelinus fontinalis*). 1. Survival, growth, fecundity, and progeny survival. *Can. J. Fish. Aquat. Sci.* **45**, 1623–1632.

Mount, R., Hockett, J. R., and Gern, W. A. (1988b). Effect of long-term exposure to acid, aluminum, and low calcium on adult brook trout (*Salvelinus fontinalis*). 2. Vitellogenesis and osmoregulation. *Can. J. Fish. Aquat. Sci.* **45**, 1633–1642.

Mueller, M. E., Sanchez, D. A., Bergman, H. L., McDonald, D. G., Rhem, R. G., and Wood, C. M. (1991). Nature and time course of acclimation to aluminum in juvenile brook trout (*Salvelinus fontinalis*). 2. Histology. *Can. J. Fish. Aquat. Sci.* **48**, 2016–2027.

Muniz, I. P., and Lievestad, H. (1980). Acidification – effects on freshwater fish. In: Ecological Impact of Acid Precipitation: Proceedings of an International Conference, *Sandfjord, Norway, March 11–14, 1980* (D. Drablos and A. Tollan, eds), pp. 84–92. SNSF Project, Oslo.

Nagasawa, K., Ito, S., Kakuda, T., Nagai, K., Tamai, I., Tsuji, A., and Fujimoto, S. (2005). Transport mechanism for aluminum citrate at the blood–brain barrier: kinetic evidence implies involvement of system Xc- in immortalized rat brain endothelial cells. *Toxicol. Lett.* **155**, 289–296.

National Academy of Science (2004). *Atlantic Salmon in Maine*. National Academy Press, Washington, DC, Report of the National Research Council of the National Academies.

Neville, C. M. (1985). Physiological response of juvenile rainbow trout, *Salmo gairdneri*, to acid and aluminum – prediction of field responses from laboratory data. *Can. J. Fish. Aquat. Sci.* **42**, 2004–2019.

Neville, C., and Campbell, P. (1988). Possible mechanisms of aluminum toxicity in a dilute, acidic environment to fingerlings and older life stages of salmonids. *Water Air Soil Pollut.* **42**, 311–327.

Nilsen, T. O., Ebbesson, L. O. E., Madsen, S. S., McCormick, S. D., Andersson, E., Björnsson, B. Th., Prunet, P., and Stefansson, S. O. (2007). Differential expression of gill Na^+,K^+-ATPase alpha and beta-subunits, $Na^+,K^+,2Cl^-$ cotransporter and CFTR anion channel in juvenile anadromous and landlocked Atlantic salmon *Salmo salar*. *J. Exp. Biol.* **210**, 2885–2896.

Nilsen, T. O., Ebbesson, L. O. E., Kverneland, O. G., Kroglund, F., Finstad, B., and Stefansson, S. O. (2010).). Effects of acidic water and aluminum exposure on gill Na^+,K^+-ATPase alpha-subunit isoforms, enzyme activity, physiology and return rates in Atlantic salmon (*Salmo salar* L.). *Aquat. Toxicol.* **97**, 250–259.

Nordstrom, D. K., and May, H. M. (1996). Aqueous equilibrium data for mononuclear aluminium species. In: *The Environmental Chemistry of Aluminum* (G. Sposito, ed.), 2nd edn, pp. 39–80. Lewis Publishers, New York.

Norrgren, L., Wicklund Glynn, A., and Malmborg, O. (1991). Accumulation and effects of aluminium in the minnow (*Phoxinus phoxinus* L.) at different pH levels. *J. Fish Biol.* **39**, 833–847.

Ogilvie, D. M., and Stechey, D. M. (1983). Effects of aluminium on respiratory responses and spontaneous activity of rainbow trout. Salmo gairdneri. *Environ. Toxicol. Chem.* **2**, 43–48.

Olsvik, P. A., Heier, L. S., Rosseland, B. O., Teien, H. C., and Salbu, B. (2010). Effects of combined gamma-irradiation and metal (Al plus Cd) exposures in Atlantic salmon (*Salmo salar* L.). *J. Environ. Radioact.* **101**, 230–236.

Orr, P. L., Bradley, R. W., Sprague, J. B., and Hutchinson, N. J. (1986). Acclimation-induced change in toxicity of aluminum to rainbow trout (*Salmo gairdneri*). *Can. J. Fish. Aquat. Sci.* **43**, 243–246.

Otto, C., and Svensson, B. S. (1983). Properties of acid brown water streams in south Sweden. *Arch. Hydrobiol.* **99**, 15–36.

Oughton, D. H., Salbu, B., and Bjornstad, E. (1992). Use of an aluminium-26 tracer to study the deposition of aluminium species on fish gills following mixing of limed and acidic waters. *Analyst* **117**, 619–621.

Paquin, P. R., Gorsuch, J. W., Apte, S., Batley, G. E., Bowles, K. C., Campbell, P. G. C., Delos, C. G., Di Toro, D. M., Dwyer, R. L., Galvez, F., Gensemer, R. W., Goss, G. G., Hogstrand, C., Janssen, C. R., McGeer, J. C., Naddy, R. B., Playle, R. C., Santore, R. C., Schneider, U., Stubblefield, W. A., Wood, C. M., and Wu, K. B. (2002). The biotic ligand model: a historical overview. *Comp. Biochem. Physiol. C* **133**, 3–35.

Persson, G., and Broberg, O. (1985). Nutrient concentrations in the acidified Lake Gårdsjøn: The role of transport and retention of phosphorus, nitrogen and DOC in watershed and lake. *Ecol. Bull. (Stockholm)* **37**, 158–175.

Peuranen, S., Keinänen, M., Tigerstedt, C., and Vuorinen, P. J. (2003). Effects of temperature on the recovery of juvenile grayling (*Thymallus thymallus*) from exposure to Al + Fe. *Aquat. Toxicol.* **65**, 73–84.

Pickering, A. D., and Pottinger, T. G. (1989). Stress responses and disease resistance in salmonid fish: effects of chronic elevation of plasma cortisol. *Fish Physiol. Biochem.* **7**, 253–258.

Playle, R. C. (2004). Using multiple metal–gill binding models and the toxic unit concept to help reconcile multiple-metal toxicity results. *Aquat. Toxicol.* **67**, 359–370.

Playle, R. C. (1987a). Chemical effects of spring and summer alum additions to a small, northwestern Ontario lake. *Water Air Soil Pollut.* **34**, 207–225.

Playle, R. C. (1987b). Methods and feasibility of using aluminum-26 as a biological tracer in low pH waters. *Can. J. Fish. Aquat. Sci.* **44**(Suppl. 1), 260–263.

Playle, R. C., and Wood, C. M. (1989). Water pH and aluminum chemistry in the gill micro-environment of rainbow trout during acid and aluminum exposure. *J. Comp. Physiol. B* **159**, 539–550.

Playle, R. C., and Wood, C. M. (1990). Is precipitation of aluminum fast enough to explain aluminum deposition on fish gills? *Can. J. Fish. Aquat. Sci.* **47**, 1558–1561.

Playle, R. C., and Wood, C. M. (1991). Mechanisms of aluminium extraction and accumulation at the gills of rainbow trout, *Oncorhynchus mykiss* (Walbaum), in acidic soft water. *J. Fish Biol.* **38**, 791–805.

Playle, R. C., Goss, G. G., and Wood, C. M. (1989). Physiological disturbances in rainbow trout (*Salmo gairdneri*) during acid and aluminum exposures in soft water of two calcium concentrations. *Can. J. Zool.* **67**, 314–324.

Playle, R. C., Gensemer, R. W., and Dixon, D. G. (1992). Copper accumulation on gills of fathead minnows: influence of water hardness, complexation and pH of the gill micro-environment. *Environ. Toxicol. Chem.* **11**, 381–391.

Poleo, A. B. S. (1995). Aluminium polymerization – a mechanism of acute toxicity of aqueous aluminium to fish. *Aquat. Toxicol.* **31**, 347–356.

Poléo, A., and Muniz, I. (1993). The effect of aluminium in soft water at low pH and different temperatures on mortality, ventilation frequency and water balance in smoltifying Atlantic salmon, *Salmo salar*. *Environ. Biol. Fish.* **36**, 193–203.

Poléo, A. B. S., and Hytterød, S. (2003). The effect of aluminium in Atlantic salmon (*Salmo salar*) with special emphasis on alkaline water. *J. Inorg. Biochem.* **97**, 89–96.

Poleo, A. B. S., Lydersen, E., and Muniz, I. P. (1991). The influence of temperature on aqueous aluminum chemistry and survival of Atlantic salmon (*Salmo salar* L.) fingerlings. *Aquat. Toxicol.* **21**, 267–278.

Poleo, A. B. S., Lydersen, E., Rosseland, B. O., Kroglund, E., Salbu, B., Vogt, R. D., and Kvellestad, A. (1994). Increased mortality of fish due to changing aluminium chemistry of mixing zones between limed streams and acidic tributaries. *Water Air Soil Pollut.* **75**, 335–351.

Poston, H. A. (1991). Effects of dietary aluminum on growth and composition of young Atlantic salmon. *Prog. Fish Cultur.* **53**, 7–10.

Pottinger, T. G. (1999). The impact of stress on animal reproductive activities. In: *Stress Physiology in Animals* (P. H. M. Balm, ed.), pp. 130–169. CRC Press, Sheffield.

Potts, W. T. W., and Fleming, W. R. (1971). The effects of environmental calcium and ovine prolactin on sodium balance in *Fundulus kansae*. *J. Exp. Biol.* **55**, 63–76.

Randall, D., and Brauner, C. (1991). Effects of environmental factors on exercise in fish. *J. Exp. Biol.* **160**, 113–126.

Rask, M., and Tuunaine, N. P. (1990). Acid-induced changes in fish populations of small Finnish lakes. In: *Acidification in Finland* (P. Kauppi, P. Anttila and K. Kenttamies, eds), pp. 911–927. Springer, Berlin.

Reader, J. P., Dalziel, T. R. K., and Morris, R. (1988). Growth, mineral uptake and skeletal calcium deposition in brown trout, *Salmo trutta* L., yolk-sac fry exposed to aluminium and manganese in soft acid water. *J. Fish Biol.* **32**, 607–624.

Reader, J. P., Dalziel, T. R. K., Morris, R., Sayer, M. D. J., and Dempsey, C. H. (1991). Episodic exposure to acid and aluminium in soft water: survival and recovery of brown trout, *Salmo trutta* L. *J. Fish Biol.* **39**, 181–196.

Reid, S. D., Rhem, R. G., and McDonald, D. G. (1991). Acclimation to sublethal aluminum: modifications of metal–gill surface interactions of juvenile rainbow trout (*Oncorhynchus mykiss*). *Can. J. Fish. Aquat. Sci.* **48**, 1995–2004.

Richards, J. G., Semple, J. W., Bystriansky, J. S., and Schulte, P. M. (2003). Na^+/K^+-ATPase – isoform switching in gills of rainbow trout (*Oncorhynchus mykiss*) during salinity transfer. *J. Exp. Biol.* **206**, 4475–4486.

Rosseland, B. O., and Skogheim, O. K. (1984). A comparative study on salmonid fish species in acid aluminum-rich water II. Physiological stress and mortality of one- and two-year-old fish. *Rep. Inst. Freshwater Res. Drottningholm* **61**, 186–194.

Rosseland, B. O., Eldhuset, T. D., and Staurnes, M. (1990). Environmental effects of aluminium. *Environ. Geochem. Health* **12**, 17–27.

Rosseland, B. O., Blakar, I. A., Bulger, A., Kroglund, F., Kvellstad, A., Lydersen, E., Oughton, D. H., Salbu, B., Staurnes, M., and Vogt, R. (1992). The mixing zone between limed and acidic river waters – complex aluminum chemistry and extreme toxicity for salmonids. *Environ. Pollut.* **78**, 3–8.

Rosseland, B. O., Salbu, B., Kroglund, F., Hansen, T., Teien, H. C. and Havardstun, J. (1998). *Changes in Metal Speciation in the Interface Between Freshwater and Seawater (Estuaries), and the Effects on Atlantic Salmon and Marine Organisms.* Final Report to the Norwegian Research Council, Contract No. 108102/122 [in Norwegian].

Rosseland, B. O., Kroglund, F., Staurnes, M., Hindar, K., and Kvellestad, A. (2001). Tolerance to acid water among strains and life stages of Atlantic salmon (*Salmo salar* L.). *Water Air Soil Pollut.* **130**, 899–904.

Sadler, K., and Lynam, S. (1987). Some effects on the growth of brown trout from exposure to aluminium at different pH levels. *J. Fish Biol.* **31**, 209–219.

Sadler, K., and Lynam, S. (1988). The influence of calcium on aluminium-induced changes in the growth rate and mortality of brown trout, *Salmo trutta* L. *J. Fish Biol.* **33**, 171–179.

Salbu, B., Denbeigh, J., Smith, R. W., Heier, L. S., Teien, H. C., Rosseland, B. O., Oughton, D., Seymour, C. B., and Mothersill, C. (2008). Environmentally relevant mixed exposures to radiation and heavy metals induce measurable stress responses in Atlantic salmon. *Environ. Sci. Technol.* **42**, 3441–3446.

Saunders, R. L., Henderson, E. B., Harmon, P. R., Johnston, C. E., and Eales, J. G. (1983). Effects of low environmental pH on smolting of Atlantic salmon (*Salmo salar*). *Can. J. Fish. Aquat. Sci.* **40**, 1203–1211.

Sayer, M. D. J. (1991). Survival and subsequent development of brown trout *Salmo trutta* L., subjected to episodic exposures to acid, aluminium and copper in soft water during embryonic and larval stages. *J. Fish Biol.* **38**, 969–972.

Sayer, M. D. J., Reader, J. P., and Morris, R. (1991). Effects of six trace metals on calcium fluxes in brown trout (*Salmo trutta* L.) in soft water. *J. Comp. Physiol. B* **161**, 537–542.

Sayer, M. D. J., Reader, J. P., and Dalziel, T. R. K. (1993). Freshwater acidification – effects on the early life stages of fish. *Rev. Fish Biol. Fish.* **3**, 95–132.

Scheuhammer, A. M. (1991). Acidification-related changes in the biogeochemistry and ecotoxicology of mercury, cadmium, lead and aluminum – overview. *Environ. Pollut.* **71**, 87–90.

Schindler, D. W. (1980). Experimental acidification of a whole lake: a test of the oligotrophication hypothesis. In: Ecological Impacts of Acid Precipitation, *Proceedings of an International Conference, Sandefjord, Norway, March 11–14, 1980* (D. Drablos and A. Tollan, eds), pp. 370–374. SNSF Project, Oslo.

Schindler, DW. (1988). Effects of acid rain on freshwater ecosystems. *Science* **239**, 149–157.

Schindler, D. W., Mills, K. H., Malley, D. F., Findlay, D. L., Shearer, J. A., Davies, I. J., Turner, M. A., Linsley, G. A., and Cruikshank, D. R. (1985). Long-term ecosystem stress: the effects of years of experimental acidification on a small lake. *Science* **228**, 1395–1401.

Schindler, D. W., Frost, T. M., Mills, K. H., Chang, P. S. S., Davies, I. J., Findlay, L., Malley, D. F., Shearer, J. A., Turner, M. A., Garrison, P. J., Watras, C. J., Webster, K., Gunn, J. M., Brezonik, P. L., and Swenson, W. A. (1991). Comparisons between experimentally and atmospherically acidified lakes during stress and recovery. *Proc. R. Soc. Edinb. B* **97**, 193–226.

Schofield, C. L., and Trojnar, J. R. (1980). Aluminum toxicity to brook trout (*Salvelinus fontinalis*) in acidified waters. In: *Polluted Rain* (T. Y. Toribara, M. W. Miller and P. E. Morrow, eds), pp. 341–363. Plenum Press, New York.

Shearer, J. A., Fee, E. J., DeBruyn, E. R., and DeClercq, D. R. (1987). Phytoplankton primary production and light attenuation responses to the experimental acidification of a small Canadian Shield lake. *Can. J. Fish. Aquat. Sci.* **44**, 83–90.

Sigel, H., and Sigel, A. (1988). Metal Ions in Biological Systems, *Vol. 24. Aluminum and its Role in Biology*. Marcel Dekker, New York.

Sloman, K., and Wilson, R. W. (2005). Anthropogenic effects on behavioural physiology in fish. In: Fish Physiology, *Vol. 24*. Behaviour – Interactions with Fish Physiology (K. Sloman, R. W. Wilson and S. Balshine, eds), pp. 413–468. Academic Press, New York.

Smith, T. R., and Haines, T. A. (1995). Mortality, growth, swimming activity and gill morphology of brook trout (*Salvelinus fontinalis*) and Atlantic salmon (*Salmo salar*) exposed to low pH with and without aluminium. *Environ. Pollut.* **90**, 33–40.

Sneddon, L. U., and Yerbury, J. (2004). Differences in response to hypoxia in the three-spined stickleback from lotic and lentic localities: dominance and an anaerobic metabolite. *J. Fish Biol.* **64**, 799–804.

Sparling, D. W., and Lowe, T. P. (1996). Environmental hazards of aluminum to plants, invertebrates, fish, and wildlife. *Rev. Environ. Contam. Toxicol.* **145**, 1–127.

Sposito, G. (1989). *The Environmental Chemistry of Aluminum*. CRC Press, Boca Raton, FL.

Sposito, G. (1996). *The Environmental Chemistry of Aluminum* (2nd edn.). Lewis Publishers, Boca Raton, FL.

Staurnes, M., Sigholt, T., and Reite, O. B. (1984). Reduced carbonic anhydrase and Na-K-ATPase activity in gills of salmonids exposed to aluminium-containing acid water. *Experientia* **40**, 226–227.

Staurnes, M., Blix, P., and Reite, O. B. (1993). Effects of acid water and aluminum on parr–smolt transformation and seawater tolerance in Atlantic salmon. Salmo salar. Can. J. Fish. Aquat. Sci. **50**, 1816–1827.

Staurnes, M., Kroglund, F., and Rosseland, B. O. (1995). Water quality requirement of Atlantic salmon (*Salmo salar*) in water undergoing acidification or liming in Norway. *Water Air Soil Pollut.* **85**, 347. 352

Staurnes, M., Hansen, L. P., Fugelli, K., and Haraldstad, Ø. (1996).). Short-term exposure to acid water impairs osmoregulation, seawater tolerance, and subsequent marine survival of smolts of Atlantic salmon (*Salmo salar* L.). Can. J. Fish. Aquat. Sci. **53**, 1695–1704.

Tam, W. H., Fryer, J. N., Ali, I., Dallaire, M. R., and Valentine, B. (1988). Growth inhibition, gluconeogenesis, and morphometric studies of the pituitary and interrenal cells of acid-stressed brook trout (*Salvelinus fontinalis*). Can. J. Fish. Aquat. Sci. **45**, 1197–1211.

Teien, H. C., Salbu, B., Kroglund, F., and Rosseland, B. O. (2004). Transformation of positively charged aluminium-species in unstable mixing zones following liming. *Sci. Total Environ.* **330**, 217–232.

Teien, H. C., Kroglund, F., Atland, A., Rosseland, B. O., and Salbu, B. (2006a). Sodium silicate as alternative to liming-reduced aluminium toxicity for Atlantic salmon (*Salmo salar* L.) in unstable mixing zones. *Sci. Total Environ.* **358**, 151–163.

Teien, H. C., Standring, W. J. F., and Salbu, B. (2006b). Mobilization of river transported colloidal aluminium upon mixing with seawater and subsequent deposition in fish gills. *Sci. Total Environ.* **364**, 149–164.

Teien, H. C., Kroglund, F., Salbu, B., and Rosseland, B. O. (2006c). Gill reactivity of aluminium-species following liming. *Sci. Total Environ.* **358**, 206–220.

Thomas, D. G., Brown, M. W., Shurben, D., del, G., Solbe, J. F., Cryer, A., and Kay, J. (1985). A comparison of the sequestration of cadmium and zinc in the tissue of rainbow trout (*Salmo gairdneri*) following exposure to the metals singly or in combination. *Comp. Biochem. Physiol. C* **82**, 55–62.

Tierney, K. B., Baldwin, D. H., Hara, T. J., Rosse, P. S., Scholz, N. L., and Kennedy, C. J. (2010). Olfactory toxicity in fishes. *Aquat. Toxicol.* **96**, 2–26.

Tietge, J. E., Johnson, R. D., and Bergman, H. L. (1988). Morphometric changes in gill secondary lamellae of brook trout (*Salvelinus fontinalis*) after long-term exposure to acid and aluminum. Can. J. Fish. Aquat. Sci. **45**, 1643–1648.

Tria, J., Butler, E. C. V., Haddad, P. R., and Bowie, A. R. (2007). Determination of aluminium in natural water samples. *Anal. Chim. Acta* **588**, 153–165.

Ultsch, G. R., and Gros, G. (1979). Mucus as a diffusion barrier to oxygen: possible role in O_2 uptake at low pH in carp (*Cyprinus carpio*) gills. *Comp. Biochem. Physiol. A* **62**, 685–689.

USEPA (1998). *Ambient Water Quality Criteria for Aluminum.* US Environmental Protection Agency, Washington, DC, EPA 440/5-86-008.

USEPA (2003). *Response of Surface Water Chemistry to the Clean Air Act Amendments of 1990.* US Environmental Protection Agency, Washington, DC, EPA/620/R-03/001.

USEPA (2009). *National Recommended Water Quality Criteria.* http://www.epa.gov/ost/criteria/wqctable/

Verbost, P. M., Berntssen, M. H. G., Kroglund, F., Lydersen, E., Witters, H. E., Rosseland, B. O., Salbu, B., and Bonga, S. E. W. (1995). The toxic mixing zone of neutral and acidic river water: Acute aluminium toxicity in brown trout (*Salmo trutta* L.). *Water Air Soil Pollut.* **85**, 341–346.

Vesely, J., Benes, P., and Sevcik, K. (1989). Occurrence and speciation of beryllium in acidified freshwaters. *Water Res.* **23**, 711–717.

Vuorinen, P. J., and Vuorinen, M. (1991). Effects of long-term prespawning acid/aluminum exposure on whitefish (*Coregonus wartmanni*) reproduction and blood and plasma parameters. *Finn. Fish. Res.* **12**, 125–133.

Vuorinen, P. J., Vuorinen, M., and Peuranen, S. (1990). Long-term exposure of adult whitefish (*Coregonus wartmanni*) to low pH/aluminum: effects on reproduction, growth, blood composition and gills. In: *Acidification in Finland* (P. Kauppi, P. Anttila and K. Kentt amies, eds), pp. 941–961. Springer, Berlin.

Vuorinen, P. J., Vuorinen, M., Peuranen, S., Rask, M., Lappalainen, A., and Raitaniemi, J. (1992). Reproductive status, blood chemistry, gill histology and growth of perch (*Perca fluviatilis*) in three acidic lakes. *Environ. Pollut.* **78**, 19–27.

Vuorinen, P. J., Keinanen, M., Peuranen, S., and Tigerstedt, C. (2003). Reproduction, blood and plasma parameters and gill histology of vendace (*Coregonus albula* L.) in long-term exposure to acidity and aluminum. *Ecotoxicol. Environ. Saf.* **54**, 255–276.

Waiwood, K. G., and Beamish, F. W. H. (1978). Effects of copper, pH and hardness on the critical swimming performance of rainbow trout (*Salmo gairdneri* Richardson). *Water Res.* **12**, 611–619.

Waiwood, B. A., Haya, K., and Van Eeckhaute, L. (1992). Energy metabolism of hatchery-reared juvenile salmon (*Salmo salar*) exposed to low pH. *Comp. Biochem. Physiol. C* **101**, 49–56.

Walker, R. L., Wood, C. M., and Bergman, H. L. (1988). Effects of low pH and aluminum on ventilation in the brook trout (*Salvelinus fontinalis*). *Can. J. Fish. Aquat. Sci.* **45**, 1614–1622.

Walker, R. L., Wood, C. M., and Bergman, H. L. (1991). Effects of longterm pre-exposure to sublethal concentrations of acid and aluminum on the ventilatory response to aluminum challenge in brook trout (*Salvelinus fontinalis*). *Can. J. Fish. Aquat. Sci.* **48**, 1989–1995.

Walton, R. C., McCrohan, C. R., Livens, F. R., and White, K. N. (2009). Tissue accumulation of aluminium is not a predictor of toxicity in the freshwater snail. *Lymnaea stagnalis. Environ. Pollut.* **157**, 2142–2146.

Walton, R. C., White, K. N., Livens, F., and McCrohan, C. R. (2010). The suitability of gallium as a substitute for aluminum in tracing experiments. *Biometals* **23**, 221–230.

Waring, C. P., and Brown, J. A. (1995). Ionoregulatory and respiratory responses of brown trout, *Salmo trutta*, exposed to lethal and sublethal aluminium in acidic soft waters. *Fish Physiol. Biochem.* **14**, 81–91.

Waring, C. P., and Brown, J. A. (1997). Plasma and tissue thyroxine and triiodothyronine contents in sublethally stressed, aluminum-exposed brown trout (*Salmo trutta*). *Gen. Comp. Endocrinol.* **106**, 120–126.

Waring, C. P., Brown, J. A., Collins, J. E., and Prunet, P. (1996). Plasma prolactin, cortisol, and thyroid responses of the brown trout (*Salmo trutta*) exposed to lethal and sublethal aluminium in acidic soft waters. Gen. Comp. Endocrinol. **102**, 377–385.

Wauer, G., and Teien, H. C. (2010). Risk of acute toxicity for fish during aluminium application to hardwater lakes. *Sci. Total Environ.* **408**, 4020–4025.

White, K. N., Ejim, A. I., Walton, R. C., Brown, A. P., Jugdaohsingh, R., Powell, J. J., and McCrohan, C. R. (2008). Avoidance of aluminum toxicity in freshwater snails involves intracellular silicon–aluminum biointeraction. *Environ. Sci. Technol.* **42**, 2189–2194.

Whitehead, C., and Brown, J. A. (1989). Endocrine responses of brown trout, *Salmo trutta* L., to acid, aluminium and lime dosing in a Welsh hill stream. *J. Fish Biol.* **35**, 59–71.

Wiklander, L. (1975). The role of neutral salts in the ion exchange between acid precipitation and soil. *Geoderma* **14**, 93–105.

Wilkinson, K. J., and Campbell, P. G. C. (1993). Aluminum bioconcentration at the gill surface of juvenile Atlantic salmon in acidic media. *Environ. Toxicol. Chem.* **12**, 2083–2095.

Wilkinson, K. J., Campbell, P. G. C., and Couture, P. (1990). Effect of fluoride complexation on aluminum toxicity towards juvenile Atlantic salmon (*Salmo salar*). *Can. J. Fish. Aquat. Sci.* **47**, 1446–1452.

Willén, E. (1991). Planktonic diatoms – an ecological review. *Algol. Stud.* **62**, 69–106.

Williams, R. J. P. (1999). What is wrong with aluminium? The J. D. Birchall memorial lecture. *J. Inorg. Biochem.* **76**, 81–88.

Wilson, R. W., and Taylor, E. W. (1993). The physiological responses of freshwater rainbow trout, *Oncorhynchus mykiss*, during acutely lethal copper exposure. *J. Comp. Physiol. B* **163**, 38–47.

Wilson, R. W., and Wood, C. M. (1992). Swimming performance, whole body ions, and gill Al accumulation during acclimation to sublethal aluminum in juvenile rainbow trout (*Oncorhynchus mykiss*). *Fish Physiol. Biochem.* **10**, 149–159.

Wilson, R. W., Bergman, H. L., and Wood, C. M. (1994a). Metabolic costs and physiological consequences of acclimation to aluminum in juvenile rainbow trout (*Oncorhynchus mykiss*). 1: Acclimation specificity, resting physiology, feeding, and growth. *Can. J. Fish. Aquat. Sci.* **51**, 527–535.

Wilson, R. W., Bergman, H. L., and Wood, C. M. (1994b). Metabolic costs and physiological consequences of acclimation to aluminum in juvenile rainbow trout (*Oncorhynchus mykiss*). 2: Gill morphology, swimming performance, and aerobic scope. *Can. J. Fish. Aquat. Sci.* **51**, 536–544.

Wilson, R. W., Wood, C. M., and Houlihan, D. F. (1996). Growth and protein turnover during acclimation to acid/aluminum in juvenile rainbow trout (*Oncorhynchus mykiss*). *Can. J. Fish. Aquat. Sci.* **53**, 802–811.

Winter, A. R., Nichols, J. W., and Playle, R. C. (2005). Influence of acidic to basic water pH and natural organic matter on aluminum accumulation by gills of rainbow trout (*Oncorhynchus mykiss*). *Can. J. Fish. Aquat. Sci.* **62**, 2303–2311.

Witters, H. E. (1986). Acute acid exposure of rainbow trout, *Salmo gairdneri* Richardson: effects of aluminium and calcium on ion balance and haematology. *Aquat. Toxicol.* **8**, 197.

Witters, H. E., Vangenechten, J., Van Puymbroeck, S., and Vanderborght, O. (1987). Ionoregulatory and haematological responses of rainbow trout, *Salmo gairdneri* to chronic acid and aluminium stress. *Ann. R. Soc. Zool. Belg.* **117**, 411–421.

Witters, H. E., Vangenechten, J. H. D., Van Puymbroeck, S., and Vanderborght, O. L. J. (1988). Internal or external toxicity of aluminium in fish exposed to acid water. *Communities Report Europe* 965–970.

Witters, H. E., Puymbroek, S., Sande, I., and Vanderborght, O. L. J. (1990). Haematological disturbances and osmotic shifts in rainbow trout, *Oncorhynchus mykiss* (Walbaum) under acid and aluminium exposure. *J. Comp. Physiol. B* **160**, 563–571.

Witters, H. E., Vanpuymbroeck, S., and Vanderborght, O. L. J. (1991). Adrenergic response to physiological disturbances in rainbow trout, *Oncorhynchus mykiss*, exposed to aluminum at acid pH. *Can. J. Fish. Aquat. Sci.* **48**, 414–420.

Witters, H. E., Puymbroeck, S., and Vanderborght, O. L. J. (1992). Branchial and renal ion fluxes and transepithelial electrical potential differences in rainbow trout, *Oncorhynchus mykiss*: effects of aluminium at low pH. *Environ. Biol. Fish.* **34**, 197–206.

Witters, H. E., VanPuymbroeck, S., Stouthart, A., and Bonga, S. E. W. (1996). Physicochemical changes of aluminium in mixing zones: mortality and physiological disturbances in brown trout (*Salmo trutta* L.). *Environ. Toxicol. Chem.* **15**, 986–996.

Wood, J. M. (1984). Microbial strategies in resistance to metal ion toxicity. In: Metal Ions in Biological Systems, *Vol. 18.* Circulation of Metals in the Environment (H Sigel, ed.), pp. 333–351. Marcel Dekker, New York.

Wood, J. M. (1985). Effects of acidification on the mobility of metals and metalloids: an overview. *Environ. Health Perspect.* **63**, 115–119.

Wood, C. M. (1992). Flux measurements as indices of H^+ and metal effects on freshwater fish. *Aquat. Toxicol.* **22**, 239–264.

Wood, C. M., and McDonald, D. G. (1987). The physiology of acid/aluminium stress in trout. *Ann. R. Soc. Zool. Belg.* **117**, 399–410.

Wood, C. M., McDonald, D. G., Booth, C. E., Simons, B. P., Ingersoll, C. G., and Bergman, H. L. (1988a). Physiological evidence of acclimation to acid/aluminum stress in adult brook trout (*Salvelinus fontinalis*). 1. Blood composition and net sodium fluxes. *Can. J. Fish. Aquat. Sci.* **45**, 1587–1596.

Wood, C. M., Simons, B. P., Mount, D. R., and Bergman, H. L. (1988b). Physiological evidence of acclimation to acid/aluminum stress in adult brook trout (*Salvelinus fontinalis*). 2. Blood parameters by chronic cannulation. *Can. J. Fish. Aquat. Sci.* **45**, 1597–1605.

Wood, C. M., Playle, R. C., Simons, B. P., Goss, G. G., and McDonald, D. G. (1988c). Blood gases, acid–base status, ions, and hematology in adult brook trout (*Salvelinus fontinalis*). under acid/aluminum exposure. *Can. J. Fish. Aquat. Sci.* **45**, 1575–1586.

Wren, C. D., and Stephenson, G. L. (1991). The effect of acidification on the accumulation and toxicity of metals to freshwater invertebrates. *Environ. Pollut.* **71**, 205–241.

Wright, R. F. (1983). Acidification of freshwaters in Europe. *Water Qual. Bull.* **8**, 137–142.

Wright, R. F., and Snekvik, E. (1978). Chemistry and fish populations in 700 lakes in southernmost Norway. *Verh. Int. Verein. Limnol.* **20**, 765–775.

Ye, X., and Randall, D. J. (1991). The effect of water pH on swimming performance in rainbow trout (*Salmo gairdneri*, Richardson). *Fish Physiol. Biochem.* **9**, 15–21.

Yokel, R. A. (2004). Aluminum. In: *Elements and their Compounds in the Environment* (E. Merian, M. Anke, M. Inhat and M. Stoeppler, eds), 2nd edn, pp. 635–658. Wiley-VCH, Weinheim.

Yokel, R. A. (2006). Blood–brain barrier flux of aluminum, manganese, iron and other metals suspected to contribute to metal-induced neurodegeneration. *J. Alzheimers Dis.* **10**, 223–253.

Yokel, R. A., and Golub, M. S. (eds) (1997). *Research Issues in Aluminum Toxicity.* Taylor & Francis, Washington, DC.

Youson, J., and Neville, C. (1987). Deposition of aluminum in the gill epithelium of rainbow trout (*Salmo gairdneri* Richardson) subjected to sublethal concentrations of the metal. *Can. J. Zool.* **65**, 647–656.

3

CADMIUM

JAMES C. MCGEER
SOM NIYOGI
D. SCOTT SMITH

Homeostasis and Toxicology of Non-Essential Metals: Volume 31B
FISH PHYSIOLOGY
DOI: 10.1016/S1546-5098(11)31025-4

Cadmium (Cd) is a naturally occurring ubiquitous element of environmental concern. Concentrations of Cd in the geosphere are generally low except for enrichment in association with Zn, Pb, and Cu sulfidic ore deposits, as well as some phosphate rock formations. Uses of Cd include batteries, pigments, stabilizers, coatings, and some alloys. The toxicity of Cd to aquatic species depends on speciation, with the free ion, Cd^{2+} concentration being proportional to bioavailability. Toxicity is reduced via complexation of Cd^{2+} by inorganic and organic anions and through competitive interactions between Ca^{2+} and Cd^{2+} for uptake sites. Cd^{2+} acute toxicity involves disruption of ion homeostasis, particularly Ca, but also Na and Mg. Uptake at the fish gill occurs apically via a lanthanum-sensitive voltage-independent epithelial Ca^{2+} channel located in the mitochondria-rich chloride cells and basolaterally via Ca-ATPase and an Na/Ca exchanger. Chronic exposure also involves ionoregulatory disturbance and disruptions to growth, reproduction, the immune system, endocrine, development, and behavior; gill, liver, and kidney histopathologies also develop. Cadmium accumulates primarily in the kidney and liver. Accumulation has been linked to oxidative damage, which may be a mechanism linking the many different chronic effects. Uptake across gastrointestinal and olfactory surfaces is not well understood but evidence exists for gut Cd uptake via L-type voltage-gated Ca^{2+} channels, a high-affinity Ca^{2+}-ATPase and a divalent metal transporter-1. Acclimation to chronic Cd exposure involves a variety of defense and detoxification processes, such as antioxidants (catalase), metallothionein, glutathione, and heat shock proteins, making the development of bioaccumulation-based impact prediction models for Cd a challenge.

1. INTRODUCTION

Cadmium is a naturally occurring ubiquitous element, but it is also rare and is not found in a pure state in nature (Johnson, 1997). Concern is associated with exposure to Cd; it is considered a potential human carcinogen (group 2B) by the US Environmental Protection Agency (EPA) and a human carcinogen (group 1) by the International Agency for Research on Cancer of the World Health Organization (WHO). While emissions of Cd have generally been declining (ATSDR, 2008), the recent study of Pan et al. (2010), which linked environmental contamination across Europe and exposure via the diet with incidences of prostate and breast cancer, illustrates the ongoing concern for the potential impacts of Cd on human health. In terms of environmental exposure and the potential for ecological effects, water quality guidelines and/or criteria for Cd involve low Cd concentrations relative to other contaminants, indicating the low threshold for effects. Cadmium also readily accumulates in biota.

Cadmium emissions and use are subject to regulation and control in most jurisdictions. In Britain, for example, it has been included as a red-list pollutant to highlight that it is of special concern. Under the European Union (EU) Water Framework Directive, in Canada, in Australia, and in many other countries Cd has been identified as a priority hazardous substance. International agreements that cover Cd include the Helsinki Convention on the Baltic Sea, the Convention for the Protection of the Marine Environment of the North East Atlantic, the Convention on Long-Range Transboundary Air Pollution, and the Basel Convention on the Transboundary Movements of Hazardous Wastes and their Disposal. As such, this element is widely recognized as requiring special concern in terms of the potential for both human health and environmental impacts. This chapter reviews the impact of Cd on aquatic biota with a focus on fish, its presence in the environment, the factors that influence its potential for uptake, the short- and long-term physiological effects, and its internal biodynamics.

2. CHEMICAL SPECIATION IN FRESHWATER AND SEAWATER

Cadmium occurs at low concentrations in aquatic systems. In freshwater total dissolved Cd is usually less than $0.5\ \mu g\ L^{-1}$ and it is even lower in seawater $(0.02\ \mu g\ L^{-1})$ (Pan et al., 2010). In a survey of European streams soluble Cd ranged from $< 0.002\ \mu g\ L^{-1}$ for the most pristine sites to $1.25\ \mu g\ L^{-1}$ for more contaminated sites (Pan et al., 2010). For assessing environmental impacts though, just total metal concentration is not sufficient. Metal speciation information must be taken into account. Speciation corresponds to the distribution of an element among its possible physical and chemical forms. Metal toxicity is dependent on speciation. For a metal to exhibit a toxic response it should first bind to a cellular surface. The availability for a metal to bind to cellular surfaces is dependent on other competing reactions in solution. In a system at equilibrium, the free metal ion activity reflects the chemical reactivity of the metal; it is this reactivity that determines the extent of the metal's reactions with surface cellular sites, and hence its bioavailability. Thus, the amount of free metal should be proportional to toxic response. This concept gave rise to the free ion activity model (Pagenkopf, 1983,) which was eventually generalized into the biotic ligand model (BLM) concept (Di Toro et al., 2001). In the BLM framework the metal interacts (binds) with a biotic ligand and toxicity is proportional to the biotic ligand–metal accumulation (see Section 5).

Accurate toxicity predictions using the BLM approach depend on accurate prediction of solution speciation of Cd. Ligands in solution can decrease Cd toxicity by outcompeting the biotic ligand for Cd and thus decreasing its bioavailability.

2.1. Inorganic Ligands for Cadmium

Cadmium can be complexed in solution by various inorganic anions including chloride, bicarbonate, hydroxide, sulfide, and sulfate. The compilation of equilibrium binding constants by the US National Institute of Science and Technology is an excellent resource for determining the inorganic speciation of Cd (Martell and Smith, 2004). Table 3.1 summarizes equilibrium

Table 3.1
Inorganic complexation constants for cadmium

Ligand	Species	Log K [a]	pK_a [b]	IS [c]
HSO_4^- (bisulfate)	$CdHSO_4^+$	1.08	1.54[d]	0.1
H_2CO_3 (carbonic acid)	$CdCO_3$	3.5	6.3, 10.3	0.1
	$Cd(CO_3)_2^{2-}$	6.3		0.1
	$CdHCO_3^+$	0.9		0.1
	$CdCO_3$ (s, otavite)	-11.24		0.1
Cl^- (chloride)	$CdCl^+$	1.98		0
	$CdCl_2$	2.60		0
	$CdCl_3^-$	1.96		2.0
H_2S (hydrogen sulfide)	$CdHS^+$	7.6	7.02	1.0
	$Cd(HS)_2$	14.6		1.0
	$Cd(HS)_3^-$	16.5		1.0
	$Cd(HS)_4^{2-}$	18.9		1.0
	$CdS(s)$	-7.0		0
OH^- (hydroxide)	$CdOH^+$	3.9	13.997[e]	0
	$CdOH_2$	7.7		0
	$CdOH_3^-$	10.3		0
	$CdOH_4^{2-}$	12		3
	Cd_2OH^{3+}	4.6		0
	$Cd_4OH_4^{4+}$	23.2		0
	$CdOH_2$ (s, beta)	-14.35		0

Values taken from Martell and Smith (2004).
[a] Log K values are presented as the overall formation constants corresponding to the reaction $Cd + nL = CdL_n$. For solid species the log K values correspond to solubility products (reaction $CdL(s) = Cd + L$). Values are for 25°C.
[b] Zero ionic strength values. For multiple-step deprotonation, successive values are listed (i.e. pK_{a1}, pK_{a2}).
[c] IS corresponds to ionic strength in molar units.
[d] 0.1 M ionic strength.
[e] pKw for reaction $H_2O \rightarrow H^+ + OH^-$.

binding constants for Cd with common inorganic ligands in aquatic systems
without correction for ionic strength, for illustration purposes only. Most
geochemical software will include these reactions in their databases, along with
corrections for differences in ionic strength.

Some simple observations can be made from Table 3.1. Considering a
general reaction for Cd where $Cd + L = CdL$, where L is ligand, the
corresponding equilibrium constant and rearrangement can be written:

$$K = \frac{[CdL]}{[Cd][L]} \tag{1}$$

and after rearrangement

$$K[L] = \frac{[CdL]}{[Cd]} \tag{2}$$

Thus, when $[L] = 1/K$, the bound and free forms of Cd are equal.
Arguments of this type are valid when ligand concentrations are much
greater than metal concentrations, so that metal complexed ligand can be
ignored in the mass balance expression. For chloride complexation this
means that at chloride concentrations of $10^{-1.98}$ M approximately half the
Cd will be bound to chloride. This corresponds to approximately 0.01 M
chloride concentration. Thus, in most freshwater chloride complexation can
be ignored, but in seawater chloride complexation is significant. In fact, in
seawater, $CdCl_2$ also forms (using similar arguments as above, $CdCl_2$ will
equal $CdCl^+$ when chloride equals 0.25 M).

With respect to hydroxide formation, cadmium hydroxide will equal free
cadmium at $10^{-3.9}$ concentration of hydroxide, which corresponds to a pH of
10.1. Thus, cadmium hydroxide species only need to be considered at very high
pH values and for normal water conditions their concentration will be
negligible. Polynuclear species (multiple metal atoms) need only be considered
at very high metal concentrations. Similarly, cadmium bisulfate species will
only exist at very high sulfate concentrations.

For reduced systems cadmium bisulfide must be considered. With a pK_a
value around 7, bisulfide is a significant form of sulfide at circumneutral pH
values (at low pH bisulfide will protonate and because H_2S is a volatile
species, sulfur will degas from solution). Sulfide concentrations of $10^{-7.6}$ M
result in significant Cd complexation. For oxic systems this complexation
will not be significant, but for reduced systems, such as sediment, this
complexation is important in the total speciation of Cd. In fact, the insoluble
cadmium sulfide species [CdS(s)] forms in reducing sediment environments.
Sulfide in sediment can be estimated using the acid volatile sulfide (AVS)
technique and measurements of AVS in sediment correlate with Cd toxicity
(Di Toro et al., 1992). Greater AVS results in reduced Cd toxicity.

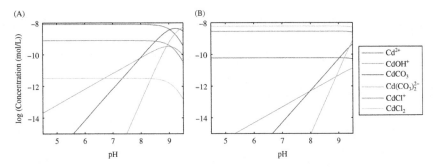

Fig. 3.1. Chemical speciation distribution diagrams for 1 μg L^{-1} Cd in (A) freshwater (1 mM chloride) and (B) saltwater (0.5 M chloride). Speciation was calculated using equilibrium constants from Table 3.1 and equilibrium with atmospheric carbon dioxide. **SEE COLOR PLATE SECTION.**

These observations for speciation, calculated at total Cd of 1 μg L^{-1} (8.90 nmol L^{-1}), atmospheric pressure of carbon dioxide (10$^{-3.5}$ atm) and chloride at 1 mM (i.e. freshwater) and 0.5 M (i.e. seawater), are summarized in Fig. 3.1A and B, respectively. It can be seen that in oxic systems at most pH values, Cd will occur as free cadmium cation (Cd^{2+}). Only at high salinity does chloro complexation become significant. Conversely, for reducing conditions Cd^{2+} will be very low as sulfide complexation is significant in controlling Cd speciation.

2.2. Organic Complexation of Cadmium

Organic ligands existing in natural waters can potentially complex Cd. They include simple organic acids and macromolecular polyelectrolytes, such as humic and fulvic acids, with carboxylic and phenolic binding sites that can complex Cd. In addition to oxygen functionalities, these molecules include a lower concentration of nitrogen and sulfur binding sites (Smith et al., 2002). The Pearson hard/soft classification identified Cd as an intermediate metal (Pearson, 1973). Thus, Cd can bind to both hard (oxygen-containing) and soft (reduced sulfur-containing) binding sites. In addition, Cd can bind to intermediate ligands such as nitrogen-binding sites. Copper is an example of another intermediate metal, while Al and Ag are examples of hard and soft metals, respectively.

Cadmium speciation in the presence of organic ligands can be estimated using the Windermere humic aqueous model (WHAM) (Lofts and Tipping, 2000) or non-ideal competitive adsorption (NICA)–Donnan approaches (Milne et al., 2003). These models have similar abilities to determine Cd speciation in the presence of organic matter but, as with all models, the quality of the input data determines the quality of the output results (Ge et al., 2005). Recent results (Unsworth et al., 2006) have shown that there may be very large discrepancies between the predictions of these two models and measured values, particularly at the low trace metal concentrations encountered in the environment.

The importance of reduced sulfur (i.e. thiol groups) in binding Cd has recently been demonstrated (Karlsson et al., 2007). Binding constants with natural reduced sulfur were found to be very strong (log K values 11.2–11.6) and compared well with stability constants for well-defined thiols. Reduced sulfur complexation can potentially result even in oxic systems; it has been shown that reduced sulfur can be stable in oxic environments at concentrations of tens to hundreds of nanometers (Kramer et al., 2007). In comparison, oxygen-containing functional groups with these same organic materials have log K values in the order of 3.2 and represent weaker binding sites which will become active at higher Cd loadings.

2.3. Particulate Complexation of Cadmium

Minerals and bacterial surfaces can represent significant potential binding sites for Cd (Song et al., 2009). Oxide minerals have terminal oxygen sites on their surface, which can interact with solution and become positively charged (protonated) or negatively charged (deprotonated). The resulting surface charge can attract counterions via electrostatic interactions, as well as forming covalently bonded inner-sphere complexes. Cadmium binding to aluminum and iron oxides can be predicted using the coupled WHAM/surface chemistry assemblage model for particles (SCAMP) routine of Lofts and Tipping (1998, 2000). Hydrous ferric oxide (HFO) is a significant natural mineral that binds metal cations (Dzombak and Morel, 1990). HFO tends to be at least partially negatively charged at its surface at circumneutral pH values. This negative charge can attract positive cations, such as Cd, and decrease their bioavailability in solution (Martinez et al., 2004). Similarly, carboxylic and other functional groups on outer membranes of bacteria can act as complexation sites for Cd (Borrok et al., 2004).

3. SOURCES (NATURAL AND ANTHROPOGENIC) OF CADMIUM AND ECONOMIC IMPORTANCE

The content of Cd in the geosphere is generally low, although relatively high concentrations can occur, for example in association with marine black shales (Thornton, 1995). Naturally occurring Cd enrichment also occurs as greenockite (mineral crystalline form of CdS) in connection with Zn, Pb, and Cu sulfidic ore deposits, and it can also be associated with the apatite in some phosphate rock formations (Mortvedt and Osborn, 1982; WHO, 1992; ATSDR, 2008). While the mean concentrations of Cd in the Earth's crust is 0.2 mg Cd kg^{-1} (see Lalor, 2008), concentrations associated with phosphate rock can be as high as 500 mg Cd kg^{-1} (Alloway, 1995; Pan et al., 2010). Jamaica has widespread and naturally enriched topsoil, with concentrations over 400 mg kg^{-1} that transfer into agricultural crops but without an apparent impact on the local human health (Lalor, 2008). This enrichment has been linked to uptake from marine organisms into birds followed by deposition as guano excretion associated with the extensive rookeries that existed during the late Miocene era (Garrett et al., 2008). Another example of seabirds as a biovector for Cd transferred to land after being bioconcentrated in fish is Devon Island in the Canadian Arctic (Michelutti et al., 2009).

Sources to the environment include the weathering of rock (particularly phosphate rock), volcanic activity, windblown dust, and aerosols from sea spray; as well as anthropogenic sources related to the mining and smelting of Zn, Pb, and Cu ores, use of phosphate fertilizers, burning of fossil fuel, peat, and wood, and the manufacture of cement (Thornton, 1995; International Cadmium Association, 2000; ATSDR, 2008). These latter releases arise from the presence of Cd as an impurity. A source and pathway model developed by Pan et al. (2010) for natural and anthropogenic Cd loadings to environmental compartments is provided in Fig. 3.2.

While mining and smelting activities provide for significant point source loadings to the environment (e.g. Moore and Luoma, 1990; Pan et al., 2010), agricultural activities involving fertilizer use and the application of sewage sludge to fields have resulted in elevated Cd to soils. Estimates given in Panagapko (2007) indicate that approximately one-third of the unintentional release of Cd arises from the production and use of phosphate fertilizers, while Pan et al. (2010) put this value at 56% (Fig. 3.2). Cadmium impurities in fertilizers vary considerably by source (e.g. 10–200 mg Cd kg^{-1} according to the International Cadmium Association, 2000) and once applied it can be taken up into food products or leached into surface water or groundwater (ATSDR, 2008).

Cadmium also has intentional uses in consumer products and industrial processes. Cadmium production is associated with mining of other base metals (e.g. Zn, Pb, Cu; there are no Cd mines) and worldwide production in 2007 is given as 19,900 tonnes (Panagapko, 2007). The manufacture of Cd-containing products and the improper use and disposal of these products at the end of their life cycle provide additional sources of Cd to the environment. Recycling programs, particularly for batteries, have been

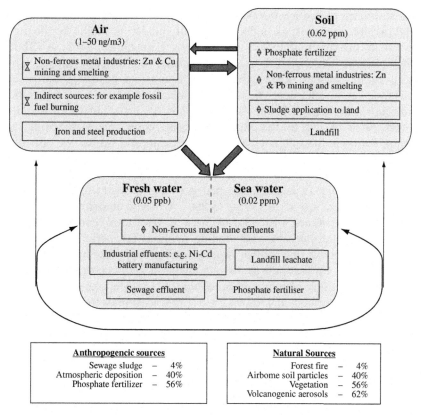

Fig. 3.2. Source pathway model for Cd in the environment, including anthropogenic and natural sources as well as typical concentrations in uncontaminated environmental compartments (air, soil, freshwater, and seawater). Note that natural sources to soil and water due to weathering are not shown (only atmospheric deposition) and within each environmental compartment the notable anthropogenic sources are given with significant sources labeled with symbols. Adapted from Fig. 1 of Pan et al. (2010).

successful in capturing Cd-containing materials and thus lowering the potential for release to the environment.

Uses and applications of Cd have varied considerably over time and currently include NiCd batteries, pigments, stabilizers, coatings, and as a minor constituent in some alloys (International Cadmium Association, 2000). Cadmium is also used in some nanomaterials where the unique optical properties can be exploited (e.g. CdTe or CdS). Battery production accounts for 83% of Cd use (Panagapko, 2007). Use in yellow and red pigments (e.g. in paints, plastics, glass, and ceramics) and as a stabilizer to reduce the degradation due to ultraviolet radiation in polyvinylchloride plastics is declining, but remains a major application for Cd (approximately 15%) (WHO, 1992; Panagapko, 2007; ATSDR, 2008). Minor uses of Cd arise in plating, non-ferrous alloys, and photovoltaic devices, where its inclusion imparts desirable corrosion resistance, or mechanical or electrochemical properties (International Cadmium Association, 2000).

Emissions of Cd to the atmosphere can contribute significantly to contamination in the soil and water compartments and Cd has the potential for long-range transport. Therefore, emission sources (smelters, fossil fuel, and waste incineration) can be local or distant. Canada's National Pollutant Release Inventory (www.ec.gc.ca/npri) provides industry-reported atmospheric releases of Cd at 28,179 kg in 2007. Of these emissions, 95% arose from industrial sources, with 93% of these (88% of total) coming from the non-ferrous metal smelting and refining facilities in operation across the country. It is notable that 70% of emissions to air from this industrial sector (and 62% of all Cd loadings to air in Canada in 2007) came from a single site, the Hudson Bay Mining and Smelting Co. Ltd Metallurgical Complex in Flin Flon, Manitoba, at 17,500 kg Cd. Emissions to air from non-ferrous metal manufacturing (including mines and smelters) reported to the Australian National Pollutant Inventory (www.npi.gov.au) for the year 2008/09 are given at 10,000 kg. Similarly to the Canadian situation, a single site in Australia (Xstrata Plc Mount Isa Mines and Smelter), accounts for 85% of the Cd releases from the sector. These data demonstrate that in spite of the improved technologies and the overall successes in reducing Cd emissions (International Cadmium Association, 2000), significant point sources exist.

Once within the aquatic environment, Cd in solution will react with other constituents, including particulate matter, iron oxides, or clay minerals, and these interactions result in removal from solution into sediments (Thornton, 1995; Lawrence et al., 1996; Skeaff et al., 2002). For example, a multi-year whole-lake dosing experiment demonstrated that less than 1% of the Cd input to the lake remained in the water column, with 93% ending up in the sediment (6% transported downstream) (Lawrence et al., 1996). Over extended periods of continuous Cd input into the aquatic environment, considerable concentrations can build up in the sediment and this reservoir of Cd may enter the food chain via benthic-dwelling organisms, with potential transfer to fish. In cases where very high levels of Cd are found in the sediment, it may be released back into the water column when aquatic loadings are reduced (Stephenson et al., 1996; Bhavsar et al., 2008).

4. A SURVEY OF ACUTE AND CHRONIC AMBIENT WATER QUALITY CRITERIA

Many countries recognize the potential impact of anthropogenic releases of Cd on human and environmental health and consequently have programs for controlling emissions and monitoring for contamination. These measures are varied and include control of uses and applications of Cd-containing products; criteria or guidelines for Cd content in effluents, fertilizers, and sewage sludge; standards for food; and standards or guidelines for air, soil, sediment, and/or water (UNEP, 2008). Concentration-based values for ambient water quality exist in some jurisdictions and because conceptual approaches vary, so do the values (Tables 3.2 and 3.3). For example, some account for the ameliorating effect that Ca and Mg provide on Cd toxicity through the use of toxicity versus hardness relationships (see examples in Table 3.2). The use of hardness (measured as $CaCO_3$ in mg L^{-1}) in equations to adjust criteria or guideline values also recognizes that other water quality characteristics, often correlated with hardness values, can affect the toxicity of metals (USEPA, 2001). Marine values (see example in Table 3.3) are not adjusted for different environmental conditions such as salinity.

The US EPA ambient water quality criteria (AWQC) provide both acute [criteria maximum concentration (CMC)] and chronic [criteria continuous concentration (CCC)] values for both freshwaters and marine waters. Concentration values are given as dissolved metal, defined as passing through a 0.45 μm filter, to account for the binding of Cd to particulate matter and removal of metal from solution. This differs from other countries, such as Canada, where total metal concentrations are the basis for water quality values. The AWQC derivation is based on 95% level of protection for the species for which there are toxicity data. When sufficient data are not available for deriving chronic toxicity criteria, then acute-to-chronic ratio values are applied to the acute data to provide chronic criteria values (USEPA, 2001). Both acute and chronic criteria in freshwater are provided as formulae that account for the effect that hardness has on toxicity (Eqs. 3 and 4). Hardness-adjusted criteria values are calculated as:

$$CMC \text{ (dissolved)} = \exp\{1.0166 \, [\ln \text{ (hardness)}] - 3.924\} \tag{3}$$

and

$$CCC \text{ (dissolved)} = \exp\{0.7409 \, [\ln \text{ (hardness)}] - 4.719\} \tag{4}$$

Values are given with a permissible frequency of exceedance set as once in every 3 years.

Table 3.2
Examples of some of the available national water quality guidelines, criteria, or trigger values
for cadmium concentrations in surface freshwaters ($\mu g \, L^{-1}$)

Country	$\mu g \, L^{-1}$	Details
Argentina[a]	0.2	Aquatic life protection
Mauritius[a]	0.7	Guideline for surface waters
Poland[a]	0.5–5	Highest admissible concentration (varies with classification of water)
Ecuador[a]	1	Water quality criteria, aquatic life protection
Uzbekistan[a]	5	Environmental standards for reservoir water
Moldova[a]	5	Environmental quality standards water of fish facilities
Russia[a]	5	Environmental quality standards water for fishing
Slovakia[a]	5	Qualitative goals for surface water
Iran[a]	10	Water quality standard, natural water
Switzerland[a]	50	Water quality requirement: rivers, streams, and other running water
Australia[b]	0.2	Trigger value for 0–59 hardness
	0.54	Trigger value for 60–119 hardness
	0.84	Trigger value for 120–179 hardness
	1.14	Trigger value for 180–240 hardness
Canada[c]	0.01	Guideline value for hardness of 30
	0.02	Guideline value for hardness of 60
	0.04	Guideline value for hardness of 120
	0.06	Guideline value for hardness of 180
China[a]	1	Category I, high-value waters (headwaters and protected areas)
	5	Category II–IV waters
	10	Category V, agricultural and ordinary sight waters
South Africa[d]	1.8	Guideline acute value for <60 hardness
	2.8	Guideline acute value for 60–119 hardness
	5.1	Guideline acute value for 120–180 hardness
	6.2	Guideline acute value for >180 hardness
	0.15	Guideline chronic value for <60 hardness
	0.19	Guideline chronic value for 60–119 hardness
	0.29	Guideline chronic value for 120–180 hardness
	0.34	Guideline chronic value for >180 hardness
USA[e]	0.6	Criteria maximum concentration (acute) for hardness of 30
	1.3	Criteria maximum concentration (acute) for hardness of 60
	2.6	Criteria maximum concentration (acute) for hardness of 120
	3.9	Criteria maximum concentration (acute) for hardness of 180
	0.11	Criteria continuous concentration (chronic) for hardness of 30
	0.18	Criteria continuous concentration (chronic) for hardness of 60
	0.31	Criteria continuous concentration (chronic) for hardness of 120
	0.42	Criteria continuous concentration (chronic) for hardness of 180

Unless indicated, values are maximum criteria (or standards) for chronic long-term effects.
Some jurisdictions have values that vary with water chemistry and information is included for a
range of hardness (given in mg L^{-1} $CaCO_3$).
[a]UNEP (2008)
[b]ANZECC (2000)
[c]CCME (2007)
[d]CSIR (1996)
[e]USEPA (2001).

Table 3.3

Examples of some of the available national water quality guidelines, criteria, or trigger values for cadmium concentrations in marine waters ($\mu g\ L^{-1}$)

Country	Value	Details
Canada[a]	0.12	Guideline value for aquatic life protection
Australia[b]	0.7	Trigger value (99% protection level)
China[c]	1	Category I, fishery waters and protected area
	5	Category II
	10	Category III/IV, industrial sites, harbors, and development zones
South Africa[d]	4	Target value for protection
Argentina[c]	5	Aquatic life protection
Ecuador[c]	5	Water quality criteria, aquatic life protection
USA[e]	8.8	Criteria continuous concentration (chronic)
Mauritius[c]	20	Coastal water quality guideline
USA[e]	40	Criteria maximum concentration (acute)

Unless indicated, values are maximum criteria (or standards) for chronic long-term effects.
[a]CCME (2007)
[b]ANZECC (2000)
[c]UNEP (2008)
[d]CSIR (1995)
[e]USEPA (2001).

Water quality assessments for Cd in Australia and New Zealand differ from those in the USA and other countries as there is a more explicit application of a risk-based approach. Concentrations are given as trigger values (both in fresh and marine waters) and vary according to four levels of protection (99%, 95%, 90%, or 80% of species). In freshwaters the trigger values are 0.06, 0.2, 0.4, and 0.8 $\mu g\ L^{-1}$, respectively, and in marine waters they are 0.7, 5.5, 14, and 36 $\mu g\ L^{-1}$ for the four levels of biota protection (ANZECC, 2000). Guidance is provided on appropriate circumstances and conditions under which different levels would be used and in this way the approach is similar to that of China (Tables 3.2 and 3.3). Trigger values provide concentrations that should have no significant adverse effects on the aquatic ecosystem and the discussion on how these should be applied also states that elevated concentrations do not mean that effects are probable, simply that a more detailed assessment is required. Detailed assessments include accounting for site-specific (water chemistry) factors that influence bioavailability, and a hardness adjustment equation (Eq. 5) is provided for this purpose:

$$\text{HMTV} = \text{TV} \left(H/30 \right)^{0.89} \tag{5}$$

where HMTV is a hardness-modified trigger value, TV is the trigger value, and H is the hardness in mg L^{-1} $CaCO_3$ (ANZECC, 2000).

The Canadian approach to protection of aquatic life for Cd exposure provides a general value of 0.017 µg L^{-1} for soft waters and also a hardness-based adjustment equation (Eq. 6) (CCME, 2007):

$$\text{Cadmium guideline} = 10^{\{0.86[\log(\text{hardness})]-3.2\}} \tag{6}$$

This value and the hardness equation were published as interim guidelines in 1996 (CCME, 1999). They were based on the study of Porter et al. (1995), which included applying a safety factor of 10 to the lowest of the lowest observed effect concentrations from the chronic studies. Canadian guideline values are designed to protect all forms of aquatic life and all aspects of the aquatic life cycles, including the most sensitive life stage of the most sensitive species over the long term (also known as 100% protection, 100% of the time). The values are exceptionally low compared to those of other countries (Table 3.2). The highly conservative nature of the approach produces values that are unworkable and border on irrelevance as ambient concentrations in uncontaminated environments can be above guidelines. The guidance published with the water quality guidelines does cover the fact that site conditions can influence the response of local biota and that the guidelines do not represent enforceable blanket values for national environmental quality. The Canadian Council of Ministers of the Environment has recently developed a new protocol for deriving water quality guidelines (CCME, 2007). This approach, which has not been applied to Cd yet, develops both acute and chronic values in a tiered approach that facilitates the incorporation of toxicity modifying factors and the development of site-specific adjustment of guideline values with computational tools such as the BLM.

5. MECHANISMS OF TOXICITY

5.1. Acute Toxicity

The toxicity of Cd to aquatic species is generally dependent on concentrations of its bioavailable forms (species), as defined by the total dissolved concentration in combination with the underlying water chemistry. In the context of geochemical speciation (see Section 2), it is particularly the concentration of free Cd^{2+} ions that is generally associated with toxicity (Campbell, 1995; Di Toro et al., 2001). Therefore, as with many other transition metals, complexation reactions that reduce the concentration of Cd^{2+} tend to reduce uptake and decrease toxicity. For example, natural organic matter [NOM; measured as dissolved organic carbon (DOC)] complexation reduced gill binding and acute toxicity in rainbow trout (Niyogi et al., 2008) and fathead minnows (Playle et al., 1993a). Increased concentrations of essential elements such as Ca^{2+} and Mg^{2+} also decrease the accumulation and toxic impact of Cd^{2+} (e.g. Pratap et al., 1989; Tan and Wang, 2009). This effect, which occurs owing to competition for uptake sites, provides insights into the mechanism of acute toxicity of Cd and also explains the basis of the hardness adjustment equations used in water quality criteria and guidelines (see Section 4). In addition to variability induced by water chemistry there are considerable differences in sensitivity among biota. Acute LC50 values for freshwater fish range from 0.5 to 73,500 $\mu g\,L^{-1}$ (USEPA, 2001) and therefore it is not possible to make definitive statements in relation to exposure concentrations that induce acute toxicity.

At extremely high Cd concentrations (0.05–0.1 M, i.e. 5600–11,200 mg L^{-1}), precipitation onto the gill and suffocation occurs (Carpenter, 1927), while high exposure concentrations will result in acute gill epithelium hyperplasia and necrosis (Bilinski and Jonas, 1973). However, these are unlikely to reflect realistic exposure–effect relationships. At more environmentally realistic exposures (e.g. 10 nM, or about 1 $\mu g\,L^{-1}$; see Section 2) the primary acute effect of Cd^{2+} is disruption of ion homeostasis, particularly Ca regulation. Disruption of Ca balance has been demonstrated in numerous fish studies (e.g. Roch and Maly, 1979; Giles, 1984; Reid and McDonald, 1988; Chang et al., 1997; McGeer et al., 2000a; Matsuo et al., 2005) and has been directly associated as the primary cause of acute toxicity (Wood, 2001; Niyogi and Wood, 2004). Calcium homeostasis in freshwater fish depends on gill uptake processes involving basolateral Ca-ATPase and an Na/Ca exchanger and the apical epithelial Ca channel (ECaC) (Shahsavarani et al., 2006), as detailed in Section 8.

The inhibitory effect of Cd^{2+} on branchial Ca transport mechanisms is best characterized for Ca ATPase. Using basolateral membrane vesicles from trout gill tissue, Verbost et al. (1988) showed that there was a direct interaction of Cd^{2+} with the Ca^{2+} transporting ATPase. The work showed that Cd^{2+} had a high affinity for Ca^{2+} binding sites on the ATPase and that there was a direct competition between Ca^{2+} and Cd^{2+} (Verbost et al., 1988), presumably related to the similarities in charge and size of these ions. Recently, Galvez et al. (2006) localized disruption of Ca^{2+} uptake by Cd^{2+} to high-affinity sites in the peanut lectin agglutinin-positive mitochondria-rich (MR) cells in the gills, which leads to Cd accumulation in tissues as well as depletion of plasma Ca concentrations (Roch and Maly, 1979; McGeer et al., 2000a) and cardiac collapse if plasma hypocalcemia ensues.

Cadmium exposure has also been associated with disruption of other physiologically important ions, notably Na^+, and this disruption may also contribute to acute toxicity. Waterborne Cd exposure of rainbow trout at $3 \, \mu g \, L^{-1}$ resulted in significant reductions in whole-body Na (and Ca but not Cl) over the first 4 days of exposure (Hollis et al., 1999; McGeer et al., 2000a). Similarly, a $10 \, \mu g \, L^{-1}$ Cd exposure of tilapia (*Oreochromis mossambicus*) resulted in reductions in plasma Na and Ca (Fu et al., 1990). This loss of Na is likely related to inhibition of uptake, as branchial Na^+/K^+-ATPase activity can be inhibited by Cd exposure (Atli and Canli, 2007). Cadmium-induced acute effects on Na balance do not occur as consistently as those on Ca (e.g. Pelgrom et al., 1994; Hollis et al., 2000a, b). Short-term (3 h) unidirectional Na influx in rainbow trout was not reduced by $5 \, \mu g \, Cd \, L^{-1}$ in soft water (Birceanu et al., 2008).

5.2. Prediction Modeling

The understanding of the mechanisms of acute toxicity and the influence of water chemistry on bioavailability of Cd^{2+} in freshwater has been incorporated into toxicity prediction models known as biotic ligand models (BLMs). The BLM approach is based on the gill surface interaction model for trace metals proposed by Pagenkopf (1983) and provides water chemistry specific estimates of toxicity (Di Toro et al., 2001). The basis for the modeling approach exploits the fact that Cd^{2+} is taken up via ECaC to disrupt a Ca^{2+} binding site on Ca-ATPase and that threshold accumulations can be associated with acute lethality (reviewed by Niyogi and Wood, 2004). An equilibrium modeling approach is applied, using WHAM (Tipping, 1994) to establish the relative concentrations of free and complexed forms of Cd in solution, and then the uptake of Cd^{2+} to the respiratory surface (originally based on the fish gill) (Playle et al., 1993a) is derived. Predicted lethality (LC50) is associated with a threshold accumulation (LA50) (Di Toro et al., 2001).

The estimates of Cd accumulation into the respiratory (biotic ligand) tissue are defined via equilibrium constants that describe the characteristics of both Cd^{2+} binding and the competition for uptake sites by Ca^{2+} and Mg^{2+}. The hardness cations are well established as moderators of Cd toxicity (Calamari et al., 1980; Spry and Wiener, 1991; Davies et al., 1993). In general, the protective effect of Ca^{2+} is stronger than that of Mg^{2+} (Davies et al., 1993; Markich and Jeffree, 1994). The potential for competitive effects from H^+ was suggested by Playle et al. (1993a) based on reduced Cd accumulation at low pH. Toxicity tests in soft water with juvenile steelhead trout are consistent with that hypothesis because as pH increased so did toxicity (40-fold increase between pH 4.7 and 5.7)

(Cusimano et al., 1986). However, Niyogi et al. (2008) more recently showed no variation of short-term gill Cd accumulation in rainbow trout in exposures ranging in pH from 4.8 to 9.5, with no competitive effect of Mg^{2+}, but toxicity mitigation via complexation of Cd^{2+} by DOC. The protection afforded by DOC tends to be less than that observed for other metals such as Cu because Cu complexation to DOC is over 10-fold stronger than that of Cd, as illustrated by conditional equilibrium constants of 9.1 (log $K_{Cu\text{-}DOC}$) and 7.4 (log $K_{Cd\text{-}DOC}$), respectively (Playle et al., 1993b).

The first BLM for Cd was published in 2008 (Niyogi et al., 2008), and was developed based on the experimental results of the effects of various water chemistry variables on short-term (3 h) Cd–gill binding and acute Cd toxicity (96 h LC50) in rainbow trout. The model was successful in accounting for the effects of water chemistry on acute Cd toxicity in trout except for high alkalinity and pH. Another acute Cd BLM for freshwater fish, the HydroQual model, is currently available (version 2.2.3, available at http://www.hydroqual.com/wr_blm.htmlU), which has been developed using the gill-binding characteristics (Log K values) originally cited in Santore et al. (2002). This model provides additional protective effect from Na^+, Mg^{2+}, or H^+, unlike the model developed by Niyogi et al. (2008) which only provides protective effect from Ca^{2+} and DOC. In addition, the Hydroqual model predicts acute toxicity of Cd in two different fish species (rainbow trout and fathead minnow) using a common set of log K values, and species differences in Cd sensitivity is provided through variation in LA50 values.

5.3. Chronic Toxicity

Chronic toxicity values for fish in freshwater range from approximately 0.5 to 160 μg Cd L^{-1} (CCME, 1999; USEPA, 2001) and much less is known on the mechanisms of chronic impact of Cd compared to acute toxicity. Thus, comprehensive prediction models are limited and ecological risk assessments and regulatory approaches tend to be conservative and precautionary as a result. During chronic exposure, Cd distributes via the circulatory system to all tissues (see Section 9) and so understanding the toxicokinetics of chronic accumulation and the links to physiological consequences at an organism or population level are essential to linking exposure with environmental risk (McCarty and Mackay, 1993). Numerous studies have assessed physiological impacts and impaired performance resulting from chronic Cd exposure (see reviews by Sorensen, 1991; USEPA, 2001). A wide variety of processes can be impacted by Cd, for example embryonic development, ionoregulation, energy metabolism, immunity, reproduction, and stress response elements. These impacts can be linked to direct interactions by Cd, for example on Ca-ATPase, or indirectly, for example by the generation of reactive oxygen species (ROS), and are discussed in the following sections.

5.3.1. IONOREGULATORY EFFECTS

Chronic sublethal Cd exposure is unequivocally linked to ionoregulatory disturbance. Larsson et al. (1981) measured plasma Ca declines and Mg increases in flounder (*Platichthys flesus* L.) exposed to concentrations ranging from 5 to 500 µg Cd L^{-1} in brackish water. Previously, these authors had observed similar trends in full-strength seawater (Larsson et al., 1976). In trout in freshwater, chronic Cd exposure to either 3.2 or 6.4 µg L^{-1} over 178 days resulted in reduced plasma Ca, Na, Cl, and K, with an increase in Mg, and these effects were transient and less (not significant) at the lower concentration (Giles, 1984). McGeer et al. (2000a) showed similar results; a transient decrease in whole-body Ca and Na that subsequently returned to control levels during 100 days of exposure to 3 µg Cd L^{-1}. Atlantic salmon (*Salmo salar*) alevins exposed over 3 months showed reduced Ca content at waterborne concentrations of 0.87 µg L^{-1} and greater, while 8 µg Cd L^{-1} was required to cause a reduction in K content (Rombough and Garside, 1984). Tilapia also showed these effects, although higher concentrations were usually required. An exposure concentration of 10 µg L^{-1} over 35 days in low hardness water produced a transitory decline of plasma Ca with increased Mg that had returned to control levels within 2 weeks or less (Pratap et al., 1989). In that study, increasing the water hardness alleviated ionoregulatory effects and dietary Cd exposure produced a similar hypocalcemia (Pratap et al., 1989). Different studies but with the same species (tilapia) at similar exposure concentrations have shown the consistency of these responses (Fu et al., 1989, 1990; Kalay, 2006). Carp also show ionoregulatory disruption with reduced plasma Ca, Na, and Cl (Reynders et al., 2006a). In that study the ionoregulatory effects were associated with growth impacts and the presence of alanine transaminase in the plasma, indicating that increases in hepatocyte membrane permeability had occurred. However, unrealistically high levels of Cd exposure (480 µg L^{-1}) were required to produce this effect and an exposure at 105 µg L^{-1} had no significant effect (Reynders et al., 2006a).

The transient nature of ionoregulatory disturbances is considered part of the classic damage–repair–acclimation scenario described by McDonald and Wood (1993). Acute branchial disruption of Ca balance and to a lesser degree Na, K, and Mg homeostasis occur within the first week (approximately) of exposure and then internal ion levels return to control levels (e.g. Pratap et al., 1989; McGeer et al., 2000a; Kalay, 2006). The repair and recovery process has two aspects: the production and mobilization of metal binding moieties, such as metallothionein (MT) or glutathione (GSH) that sequester and detoxify metal (Mason and Jenkins, 1995; Chowdhury et al., 2005); and the repair and compensation processes that correct the physiological disruption caused by metal accumulation. Metallothionein reduces the impact of Cd, both by direct binding and by countering the ROS that are induced by exposure (Roesijadi, 1992, 1994; Kimura and Itoh, 2008; Wang and Rainbow, 2010) and therefore it has a key role in the physiological transition from the damage phase to repair and acclimation. The re-establishment of ionic homeostasis has been linked to the release of cortisol, elevated prolactin levels, and an increase in the density of chloride cells in the gill (Fu et al., 1989, 1990). Other physiological changes associated with this second phase response include reduced apical metal uptake into the gill and increased clearance from blood to tissues (Hollis et al., 1999; Chowdhury et al., 2003). Overall, the repair and acclimation induced by Cd bioaccumulation result in either a new homeostatic setpoint (e.g. metal uptake and elimination processes) or return to pre-existing setpoints (ion concentrations) as well as increased tolerance to acute challenges.

5.3.2. OXIDATIVE DAMAGE

Accumulation of Cd is known to result in the production of ROS and this is presumed to be one of the key mechanisms of toxic action. ROS include hydrogen peroxide (H_2O_2), hydroxyl radical ($\bullet OH$), singlet oxygen, hydroperoxyl radical ($HO_2\bullet$), superoxide anion radical ($O_2\bullet^-$) and other forms of oxygen-derived species (Stohs and Bagchi, 1995; Livingstone, 2001). Under normal uncontaminated conditions the production of ROS is held in check by antioxidant processes that include specific enzymes such as catalase, glutathione reductase, and superoxide dismutase as well as reducing agents such as GSH, ascorbate, β-carotene and α-tocopherol (Di Giulio et al., 1989; Valavanidis et al., 2006). Exposure to contaminants can disrupt this balance, leading to elevated levels of ROS. Cadmium is known to inhibit the multienzyme mitochondrial electron transport chain (ETC), resulting in excess ROS production and reduced ATP production (Livingstone, 2001). Cadmium appears to bind to a site on complex III of the ETC, impairing normal processes and resulting in an accumulation of semiubiquinones, which in turn leads to the formation of ROS (Wang et al., 2004).

Although the half-life of ROS tends to be extremely short, they are also highly reactive and have the potential for damaging biological molecules. Interactions with lipids result in peroxidation reactions, which disrupt function (for example, degrading lipid bilayer integrity) and create toxic reaction products such as aldehydes and lipid epoxides (Livingstone, 2001;

Valavanidis et al., 2006). Oxidative damage to proteins occurs via carbonylation reactions (Valavanidis et al., 2006). Studies with fish illustrated that oxidative damage occurs and that antioxidant responses can serve as indicators of exposure (reviewed by Martinez-Alvarez et al., 2005). Thomas and Wofford (1993) documented lipid peroxidation in Atlantic croaker (*Micropogonias undulates*) exposed to 1 mg Cd L^{-1} in seawater for a month. Firat et al. (2009) exposed tilapia to 0.1 and 1 mg Cd L^{-1} for up to a month and concluded that antioxidant responses, specifically GSH and glucose-6-phosphate dehydrogenase, contributed to resisting oxidative stress. Testing with Japanese flounder (*Paralichthys olivaceus*) exposed to Cd over 88 days of early life development from embryo to juvenile found that responses varied with stage of development (Cao et al., 2010). In that study, juveniles showed enhanced superoxide dismutase (SOD) activity at exposures of 12 µg L^{-1}, reduced glutathione *S*-transferase activity at 6 µg L^{-1}, and elevated lipid peroxidation as Cd exposure increased, indicating that protective responses were insufficient (Cao et al., 2010). Padmini and Rani (2009) demonstrated that biomarkers of oxidative stress and damage in hepatocytes (elevated lipid peroxidation and protein carbonylation along with lower levels of reduced GSH, SOD, catalase, and glutathione peroxidase) could be used as biomarkers of physiological impairment linked to Cd accumulation in grey mullet (*Mugil cephalus*) from a polluted estuary. Cadmium-induced oxidative stress has also be shown to vary depending on social position, with antioxidant defenses being significantly lower in subordinate tilapia compared to dominant (Almeida et al., 2009). Although ROS and antioxidant responses have not been linked to the damage–repair–acclimation scenario described by McDonald and Wood (1993), the repair–acclimation aspect is consistent with the ROS-induced enhancement of antioxidant defense. It seems relevant to study the time-course of ROS damage and antioxidant responses in relation to sensitive endpoints such as ionoregulation.

5.3.3. GROWTH AND SURVIVAL

Cadmium exposure can also result in whole-organism impacts such as reduced survival and growth, although there is considerable variability across studies. In some studies, growth was an insensitive endpoint. For example, Pickering and Gast (1972) found that survival was a more sensitive endpoint than growth for Cd effects on fathead minnows. Studies with rainbow trout by Hollis et al. (1999, 2000b) and also McGeer et al. (2000a) showed that exposure to 3 µg L^{-1} resulted in mortalities (<10%) but no effect on growth over 30, 30 and 100 days respectively. However, some studies have demonstrated that survival and growth can be sensitive measures of exposure. Cadmium exposure to bull trout (*Salvelinus confluentus*) at $0.8\,\mu g\,Cd\,L^{-1}$ resulted in a 37% mortality within the first 4 days and then over the next 51 days growth was inhibited at that concentration (Hansen et al., 2002). Benaduce et al. (2008), working with silver catfish (*Rhamdia quelen*), noted that concentrations of $4.5\,\mu g\,L^{-1}$ and higher had significant impacts on hatching survival and subsequently the growth of larvae. These impacts were not evident when the alkalinity was increased from 63 to 92 mg $CaCO_3\,L^{-1}$, but spinal column deformities and a transient deformation of barbells during development were observed (Benaduce et al., 2008).

5.3.4. REPRODUCTION

Reproduction appears to be a sensitive endpoint in fish, although there are few comprehensive studies. Benoit et al. (1976) conducted a multi-generational study of Cd effects in brook trout (*Salvelinus fontinalis*) over 3 years at exposure concentrations ranging from 0.06 to $6.4\,\mu g\,Cd\,L^{-1}$. In that study, growth inhibition and mortality in male spawners were noted at $3.4\,\mu g\,Cd\,L^{-1}$ but not at $1.7\,\mu g\,Cd\,L^{-1}$. Brown et al. (1994) exposed rainbow trout to concentrations up to $5.5\,\mu g\,Cd\,L^{-1}$ without effects on growth or survival over 65 weeks, but reproductive development was delayed or failed at an even lower exposure ($1.8\,\mu g\,Cd\,L^{-1}$). Lizardo-Daudt and Kennedy (2008) exposed rainbow trout eggs to Cd and followed responses from shortly after fertilization through to juvenile stages. Hatching was disrupted, being premature at concentrations of 0.05 and $0.25\,\mu g\,Cd\,L^{-1}$ and delayed at $2.5\,\mu g\,Cd\,L^{-1}$. Growth of juveniles was slightly reduced at the highest concentration of $2.5\,\mu g\,Cd\,L^{-1}$ but otherwise unaffected at lower exposures (Lizardo-Daudt and Kennedy, 2008). Juveniles exposed to all concentrations greater than $0.05\,\mu g\,Cd\,L^{-1}$ had elevated levels of either estradiol (female trout) or 11-ketotestosterone (males) (Lizardo-Daudt and Kennedy, 2008). Exposure of adult fathead minnows to $50\,\mu g\,Cd\,L^{-1}$ for 21 days followed by breeding and subsequent characterization of development resulted in no effect on fecundity, fertilization rates, or hatching rate (Sellin et al., 2007). Although these are only a few examples, it appears that early life stages seem to be more sensitive to Cd exposure than juveniles.

5.3.5. EFFECTS ON THE IMMUNE SYSTEM

The disruption caused by bioaccumulated Cd extends to the immune system, developmental abnormalities, and histopathologies. It is unclear whether these represent direct mechanisms of toxicity or secondary (or tertiary) effects resulting from, for example, oxidative damage or ionoregulation disruption. The review of Bols et al. (2001) discusses the effects of contaminants (including Cd) on fish immunity and their value as important measures of health. Iger et al. (1994) characterized responses of mucous cells in carp skin to Cd, showing altered characteristics and a reduction in overall numbers (22 and 560 µg Cd L^{-1}). Chronic exposure to sublethal levels of waterborne Cd was associated with leukopenia in goldfish over a 3 week exposure to Cd at levels ranging from 90 to 445 µg L^{-1} (Murada and Houston, 1988). While such effects are associated with relatively high concentration, there are examples of effects at low exposures. A 2 µg Cd L^{-1} chronic exposure to rainbow trout produced no effects on survival or growth compared to controls, but phagocytic activity and superoxide radical formation were impacted (Zelikoff et al., 1995).

5.3.6. ENDOCRINE DISRUPTION

Cadmium has also been implicated as an endocrine-disrupting substance. The Weybridge meeting in 1996 defined an endocrine disrupter as "an exogenous substance that causes adverse health effects in an intact organism, or its progeny, secondary to changes in endocrine function". In studies that characterize the interaction of Cd on the endocrine system it is not always clear whether the endocrine effects are simply measures of physiological effect or whether they are linked directly to subsequent adverse health effects (see review of Brown et al., 2004). For example, rainbow trout exposed to sublethal concentrations of 25 µg Cd L^{-1} for a month experienced reduced growth and this was associated with endocrine effects, namely inhibited plasma cortisol and thyroxine (T_4) levels (Ricard et al., 1998). This study is sometimes interpreted as demonstrating that Cd is an endocrine-disrupting substance. However, mechanistic links to subsequent adverse health effects were inferred and at low exposure concentrations (10 µg Cd L^{-1}) cortisol levels were elevated rather than inhibited (Ricard et al., 1998). *In vitro* studies (e.g. Brodeur et al., 1998; Lizardo-Daudt et al., 2007) show that interrenal tissue of Cd-exposed trout have inhibited responses to cortisol, and these types of studies are useful contributions that help to develop a mechanistic understanding of Cd and cortisol interactions. Additional discussion of Cd effects on cortisol are provided near the end of Section 11.

Cadmium exposure has been shown to inhibit vitellogenin mRNA levels in rainbow trout hepatocytes (Vetillard and Bailache, 2005) and cultured cells (Isidori et al., 2010) *in vitro*. *In vivo*, Cd has been shown to inhibit induction of vitellogenin (Olsson et al., 1995) and delay oogenesis in brown trout (Brown et al., 1994). The mechanistic link demonstrating the binding of Cd to estrogenic receptors was provided by Nesatyy et al. (2006) and therefore Cd clearly meets the criteria of being an endocrine-disrupting substance. Research delineating Cd effects on steroid receptors may help us to understand some of the mechanisms of Cd toxicity.

5.3.7. HISTOPATHOLOGY

Chronic sublethal exposure to Cd clearly degrades tissue ultrastructure in gills, liver, and kidney. In white seabass (*Lates calcarifer*) exposed to sublethal Cd (800 μg Cd L^{-1}) concentrations for 3 months, gill tissue showed fusion of secondary lamellae as well as epithelial hypertrophy and hyperplasia (Thophon et al., 2003). Accompanying these gill changes were dramatic changes in hepatic histopathologies and renal tubular dilation and necrosis. Similar results were observed in freshwater with silver barb (*Puntius gonionotus*) exposed to 5% of the experimentally derived maximum acceptable toxicant concentration (60 μg Cd L^{-1}), with gill hypertrophy and hyperplasia, hepatic vacuolization, and glomerular as well as tubular damage in the kidney (Wangsongsak et al., 2007). Although these examples involve unrealistically high concentrations they do illustrate the sublethal effects of Cd, as well as the tolerance of some fish species for Cd.

Chronic exposures at more environmentally realistic Cd concentrations demonstrate a wide variety of macrostructural and microstructural impacts. Minnows (*Phoxinus phoxinus* L.) exposed to concentrations of 7.5 μg Cd L^{-1} and higher over 2.5 months of exposure in brackish water (6–7‰) experienced vertebral injuries that impaired swimming (Bengtsson et al., 1975). This was also observed in sublethal exposures with carp in freshwater where vertebral deformities were observed at the lowest exposure concentration (10 μg Cd L^{-1}) after 47 days and were attributed to effects on Ca and P balance (Muramoto, 1981). Larval silver catfish also showed significant skeletal deformities at 4.5 μg Cd L^{-1} (Benaduce, 2008). The effects of Cd exposure (22 μg L^{-1}) over 3 weeks on the ultrastructure of carp skin included a small increase in the number of chloride cells with no changes in the number of mucous or club cells but increases in the number of mucosomes and phagosomes per cell (Iger et al., 1994). Additional changes indicating sublethal disruption included increased pavement cell necrosis, macrophage infiltration of the epidermis, and increased secretion of collagen in the dermis (Iger et al., 1994). Zebrafish exposed to 3 μg Cd L^{-1} showed a thinning of gill epithelia layers and a reduced diffusion distance, but at 10 μg Cd L^{-1} the non-tissue space was enlarged and filled with fluid, thus increasing the lamellar diffusion distance (Karlsson-Norrgren et al., 1985). The potential for impacts under these conditions, such as impaired gas exchange capacity, were exacerbated as exposure continued as significant damage developed, including curling of the secondary lamellae and localized capillary lesions. It would be expected that such damage would impair gas exchange capacity and thus influence aerobic swimming performance, which depends on gill-based oxygen exchange processes, but this was not observed in rainbow trout chronically exposed to sublethal Cd (3 µg Cd L^{-1}) (Hollis et al., 1999; McGeer et al., 2000a).

5.3.8. BEHAVIORAL EFFECTS

Studies have also documented the chronic sublethal effect of Cd on behavior. A detailed discussion of the physiological basis for behavioral effects is provided in Section 11.

6. ESSENTIALITY OF CADMIUM

Cadmium is regarded as a non-essential element and lacks the essential nutrient properties of other transition metals, such as Cu, Zn, Co, Mn, and Mb. An exception is the marine diatom *Thalassiosira weissflogii*, where Cd (and also Co) can substitute for Zn during Zn deficiency and restore normal growth (Price and Morel, 1990). The response is linked to a Cd-based carbonic anhydrase (Lane and Morel, 2001; Lane et al., 2005; Xu et al., 2008) that allows a hydrated Cd ion at the catalytically active metal site. In addition to metalloenyzmes such as carbonic anhydrase, interactions between Cd and other proteins have been documented. For example, Cd interacts with Cu chaperones, can also bind to structural elements such as zinc-finger proteins, and readily binds to metallothioneins (Waldron et al., 2009; Maret, 2010). While the biochemistry of binding to metallochaperones, and to structural and other elements within cells, is not well understood, interactions with Cd do not in any way reflect essentiality. In fact, Cd binding generally results in a loss of function. Cadmium–protein interactions usually occur as a result of the binding affinity similarities among Cd and nutritionally required elements, particularly Zn, Cu, and Ca. Cysteine thiol groups are one of the prominent examples of a functional group for which Cd and essential elements such as Zn and Cu share a high affinity.

7. POTENTIAL FOR BIOCONCENTRATION AND BIOMAGNIFICATION OF CADMIUM

Cadmium readily bioaccumulates and bioconcentrates in aquatic organisms. Tissue Cd concentrations build up at the site of exposure, gills in a waterborne exposure or gastrointestinal tract in a dietborne exposure, and are transferred via the circulation to other tissues. Cadmium accumulates in nearly all tissues and organs, with liver, kidney, and gill (or gut) reaching relatively high levels and muscle tissue being generally much lower (e.g. Karaytug et al., 2007). Hepatic and renal bioaccumulation tends to be relatively slow compared to other nutritionally required metals (McGeer et al., 2000b). Metal accumulation, including discussions of Cd in vertebrates and invertebrates, has been reviewed by Frasier (1979), Sorensen (1991), Rainbow (2002), Vijver et al. (2004), Luoma and Rainbow (2005), and Adams et al. (2011). The conclusions of these reviews are that metal bioaccumulation and internal dynamics vary considerably among species; it is difficult to predict accumulation beyond a particular organism in a particular environment, or group of organisms in similar environments; and accumulated Cd fractionates into metabolically reactive and detoxified pools. Detoxification responses are dynamic and linked to the damage–repair–acclimation process. Therefore, the potential for effects is often not related to total Cd burden because a considerable proportion of the bioaccumulated load will be in a detoxified form. Chronic sublethal waterborne Cd exposure in rainbow trout results in a reduction in branchial Cd uptake over time and this may play a role in reducing the detoxification capacity required as the exposure continues (Hollis et al., 1999).

Cadmium bioaccumulation in and of itself does not mean that deleterious effects will occur. Deriving tissue residue approaches for assessing the potential for impacts is attractive because it provides a way of accounting for different exposure conditions, routes of uptake, and other external factors that influence bioavailability and uptake. Three approaches to assessing tissue residues in relation to effects are discussed here. The relatively simple measure of bioaccumulation potential, the bioconcentration factor (BCF: ratio of accumulated tissue concentration to exposure concentration) is currently considered as a tool for hazard assessment, although its use has been discouraged for metals (Fairbrother et al., 2007). BCFs cannot be applied for assessing Cd (McGeer et al., 2003). To be a representative property of a contaminant, the BCF should be constant across different exposure conditions and species and this is not the case for Cd (Fig. 3.3) or indeed most other metals (Wood, Chapter 1, Vol. 31A). In fact, BCFs decrease as exposure concentrations increases (Fig. 3.30) (McGeer et al., 2003) because Cd uptake follows saturation kinetics.

The critical body residue (CRB) approach has been proposed as a tool for assessing the potential for Cd impacts. A case study applied to Cd is presented in Adams et al. (2011). In brief, the authors found sufficient data on whole-body and tissue burdens in studies where endpoints were measured, and also that it was possible to construct a species sensitivity distribution. But because different taxa have different capacities for detoxification and storage and because tissue accumulation was positively related to exposure time, it was not possible to provide a toxicological threshold or to derive a CRB (Adams et al., 2011). The third approach,

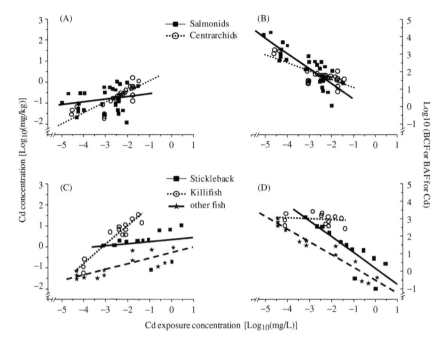

Fig. 3.3. (A, C) Effect of chronic Cd exposure on Cd content in fish and (B, D) associated bioconcentration factor and bioaccumulation factor (BCF/BAF) values for Cd. Adapted from Fig. 3 of McGeer et al. (2003). Data are on a log–log basis and the best fit line from the linear regression analysis is shown for each species group, generally showing that as Cd exposure concentration increases, bioconcentration factors decrease.

which was recently proposed in Adams et al. (2011), is a modified critical residue approach where toxicological endpoints are derived from the sensitive organisms and tissue residue thresholds are derived from resistant organisms. In this approach cause and effect are related, but in different organisms. Effects of Cd accumulation in sensitive organisms can provide toxicologically relevant endpoints but these organisms might not reliably and consistently accumulate Cd. Reliable bioaccumulators living in the same environment can be used to establish exposure–accumulation relationships but they are likely to be resistant to Cd and show few effects unless concentrations are elevated. This model, which is conceptual and awaits validation, accounts for the variability in metal bioavailability across exposure routes and the fact that sensitive organisms account for only a very narrow part of the dose–response curve (Luoma et al., 2010).

As reviewed by Suedel et al. (1994), little evidence exists to suggest that Cd biomagnifies in aquatic systems. Here, a distinction is made between Cd enrichment from one trophic level to the next (trophic transfer, e.g. from prey to predator) and Cd enrichment involving at least three linked trophic levels (biomagnification) within a food web (Leblanc, 1995; Fairbrother et al., 2007). Biomagnification is typically assessed by comparing the ratio of tissue concentrations between successive trophic levels. Ratios consistently greater than one indicate that concentrations are increasing through trophic levels. Calculations are often based on whole-body accumulation, although tissue-specific calculations are also done. Using whole-organism or tissue concentrations does not account for the detoxification and sequestration of Cd that removes the potential for direct toxicity (Wang, 2002; McGeer et al., 2003). Cadmium stored in a detoxified form in a prey organism may be available, however, to predators, although granules generally are not (Wang, 2002; also see Sections 8.2 on uptake and 9.3 on subcellular partitioning).

Ferard et al. (1983) examined the transfer of Cd in an experimental food chain consisting of algae (*Chlorella vulgaris*), zooplankton (*Daphnia magna*), and fish (*Leucaspias delineatus*), and illustrated that Cd concentrations decreased with increasing trophic level. A comprehensive study of Cd accumulation across a gradient of 20 lakes (pristine to contaminated) showed that while trophic transfer enrichment occurred from plankton to macrozooplankton, there was a biodilution between the latter and fish (Chen et al., 2000). Mathews and Fisher (2008b) demonstrated a consistent lack of Cd enrichment moving from phytoplankton to zooplankton to killifish (*Fundulus heteroclitus*) to striped bass (*Morone saxatilis*) by calculating trophic transfer factors as the product of the ingestion rate and the assimilation efficiency of consumed Cd divided by the associated efflux rate constant (Wang, 2002) and then also confirmed this finding in a different four-level food chain (Mathews and Fisher, 2008a). In a chronic exposure, Ng and Wood (2008) noted lower tissues concentrations of Cd in rainbow trout compared to the prey organism (*Lumbriculus variegatus*) contaminated with Cd via waterborne exposure. Trophic transfers were less than 1.0, but trout were still impacted, with dietary toxicity reducing growth rates (Ng and Wood, 2008); these results suggest that trophic transfer potential is not necessarily a useful indicator of effects. These studies illustrate the complex nature of bioaccumulation as influenced by ingestion, assimilation, and species differences in physiological processes such as detoxification, storage, and elimination.

A few examples of Cd biomagnification can be found in peer-reviewed literature (see reviews by Suedel et al., 1994; Wang, 2002). Croteau et al. (2005) provide an interesting example of Cd transfer dynamics within food webs of San Francisco Bay using N and C stable isotopes to establish the trophic position of various biota. They noted a general trend for decreasing concentrations of bioaccumulated Cd from algae to invertebrates (whole body) to fish (liver). In spite of significant variability in tissue concentrations among species, it was possible to describe biomagnification in discrete epiphyte-based food webs under two conditions: either macrophyte-dwelling invertebrates with epiphytic algae as the bottom tier of the food web, or when goby are the first link (Croteau et al., 2005). The authors concluded that Cd biomagnification was clearly demonstrated but that this characterization depended on being able to describe food web dynamics, suggesting that because other studies have not done this, Cd enrichment up the food chain may be more common than thought (Croteau et al., 2005). Another unique example of trophic transfer was observed on Devon Island, where accumulation of Cd within the Arctic marine food chain is transferred to freshwater systems by the upper trophic level organisms (seabirds). Elimination of ingested Cd in and around freshwater ponds by northern fulmars (*Fulmarus glacialis*) has led to enrichment of water and sediments in these oligotrophic systems (Michelutti et al., 2009). This biovector mechanism also serves to transfer nutrients (Keatley et al., 2008) and climate-induced increases in bird populations appear to be altering the equilibrium, resulting in enrichment of nutrients and contaminants (Michelutti et al., 2009). The ecological impacts associated with these two magnification scenarios were not characterized but implied.

8. CHARACTERIZATION OF UPTAKE ROUTES

8.1. Gills

The uptake of Cd in fish occurs primarily via the gills and intestine. The branchial uptake of Cd in fish has been well characterized by several recent studies. Cadmium is known to act like a calcium analogue (see Section 5), and the antagonistic interaction between the uptake of waterborne Ca^{2+} and Cd^{2+} is well documented (Playle et al., 1993a; Wicklund-Glynn et al., 1994; Hollis et al., 2000a; Niyogi and Wood, 2004; Niyogi et al., 2008). The uptake of waterborne Cd^{2+} in fish is also inhibited by Zn^{2+} (Wicklund Glynn, 2001), which shares a common branchial uptake pathway with Ca^{2+} (Hogstrand et al., 1994; Hogstrand, Chapter 3, Vol. 31A). The apical uptake of waterborne Cd^{2+} is believed to occur via a lanthanum-sensitive voltage-independent epithelial Ca^{2+} channel (ECaC) located in the MR chloride cells of the gill epithelium (Verbost et al., 1987, 1989; Wicklund-Glynn et al., 1994; Galvez et al., 2006) (Fig. 3.4A). In rainbow trout, the MR cells of primary interest in this regard are PNA^+ MR cells, which are also thought to be the sites of Ca^{2+} and Cl^- uptake, whereas PNA^- cells are believed to

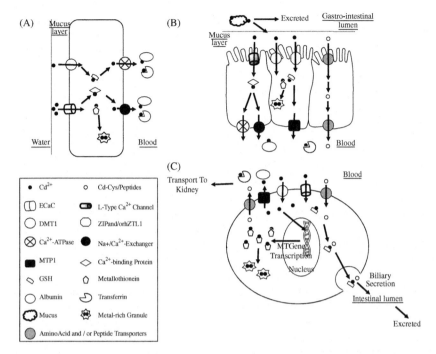

Fig. 3.4. Diagrammatic representation of the mechanisms involved in Cd uptake and handling in three major target organs in fish: (A) gill (mitochondria-rich chloride cell), (B) gastrointestinal tract, and (C) liver (hepatocytes). Note that some of the mechanisms presented in the diagram are hypothesized on the basis of information reported in mammalian systems (see the text for details).

be the sites of Na^+ uptake (PNA is peanut lectin agglutinin, referring to whether or not the cells bind to this diagnostic tool) (Galvez et al., 2006). Recent evidence suggests that apical Cd^{2+} uptake in fish gill may also occur via the Fe^{2+}-transporter divalent metal transporter-1 (DMT-1; also known as Nramp2, DCT1 and Slc11a2) (Bury and Grosell, 2003; Cooper et al., 2007) (Fig. 3.4A). DMT-1 is a proton-coupled apical membrane transporter (H^+/Fe^{2+} symporter), and to date two different isoforms of DMT-1 (Nramp-β and -γ) have been identified in teleosts (Cooper et al., 2007; Kwong et al., 2010). Although Fe^{2+} transport by gill Nramp-β and -γ transcripts expressed in *Xenopus* oocytes is inhibited by Cd, only Nramp-β has been found to mediate apical Cd^{2+} transport (Cooper et al., 2007). The basolateral extrusion of Cd^{2+} in the gill chloride cells is believed to occur via the high-affinity Ca^{2+}-ATPase as well as the Na^+/Ca^{2+} exchanger (Verbost et al., 1989; Flik, 1990) (Fig. 3.4A).

The acute toxicity of waterborne Cd exposure in freshwater fish is generally associated with the impairment of active Ca^{2+} transport by the competitive blockade of apical Ca^{2+} channels and/or selective inhibition of basolateral Ca^{2+}-ATPase transporters by Cd^{2+} in gill chloride cells (Verbost et al., 1987, 1989). The binding of Cd^{2+} to the Ca^{2+} transporters is believed to be the key process in the transfer of Cd into the body across the gill epithelium. At both apical and basolateral membranes, the Ca^{2+} and Cd^{2+} interaction for common binding sites seems to occur, and recent kinetic analyses in rainbow trout indicate at least in the short term, the interactions are by direct competition (Niyogi and Wood, 2004; Niyogi et al., 2008). The characterization of the short-term Cd binding profile of fish gill revealed two different types of Cd^{2+} binding sites: (1) the high-affinity, low-capacity sites that are saturable at a relatively low waterborne Cd concentration range ($\leq 20 \ \mu g \ L^{-1}$), and probably represent the ECaCs and/or Ca^{2+}-ATPases that play a critical role in maintaining Ca^{2+} homeostasis in fish; and (2) the low-affinity, high-capacity sites that become more evident at a waterborne Cd concentration of approximately $\geq 20 \ \mu g \ L^{-1}$, and are less influenced by waterborne calcium and thus thought to be not directly related to branchial Ca^{2+} uptake (Hollis et al., 1997; Birceanu et al., 2008; Niyogi et al., 2008).

The affinity constants (log K) for Ca^{2+} versus Cd^{2+} competition and binding site density (β_{max}) in freshwater fish gill were first determined experimentally by Playle et al. (1993a, b), and subsequently by several other researchers (Hollis et al., 1996, 1997, 1999; Niyogi et al., 2004, 2008). In general, Cd^{2+} binds to fish gills with an affinity more than three orders of magnitude greater than Ca^{2+} (log K_{Cd2+} 7.5–8.6 vs log K_{Ca2+} 3.7–5). In addition to Ca^{2+}, H^+ is believed to compete for the gill–Cd^{2+} binding sites, but with approximately 100-fold lower affinity than Cd^{2+} (log K_{H+} 6.7) (Playle et al., 1993a, b). However, neither Mg^{2+} nor Na^+ appears to have any effect on Cd^{2+} binding to the fish gill (Niyogi et al., 2008). The binding of waterborne Cd^{2+} to DOC, the other important modifier of acute Cd toxicity besides calcium, seems to be about 10-fold weaker relative to that of the gill Cd^{2+} binding (Playle et al., 1993a, b; Niyogi et al., 2008).

The gill–Cd^{2+} binding characteristics (log K and β_{max}) in freshwater fish are not fixed but dynamic, and are influenced by both water (e.g. long-term exposure to variable waterborne calcium and chronic exposure to waterborne Cd) as well as diet quality (e.g. dietary Ca^{2+} and Cd^{2+} concentrations; see Section 8.3 for details). Acclimation to soft water (low Ca^{2+}) leads to a significant increase in β_{max}, with a moderate or no change in log K (Hollis et al., 2000a, b; Niyogi et al., 2004). This increase in β_{max} of gill–Cd^{2+} binding probably results from an upregulation of branchial Ca^{2+} uptake in order to compensate the greater loss of Ca^{2+} from the body in ion-poor soft water. In contrast, acclimation to chronic waterborne Cd exposure results in either a decreased log K with an increased β_{max} or alteration of the gill–Cd^{2+} binding profile from saturable to almost perfect linear (in soft water) (Hollis et al., 1999, 2000a; Szebedinszky et al., 2001; Niyogi et al., 2004). The decrease in log K of gill–Cd^{2+} binding occurs probably to reduce Cd uptake, whereas the increase in β_{max} or linear binding pattern likely reflects an increased Cd pool size in the gill due to increased detoxification capacity.

8.2. Gut

In contrast to the gills, the gastrointestinal transport mechanisms of Cd^{2+} are not very well characterized, although recent investigations provide some important insights. The gastrointestinal uptake of Cd^{2+} occurs in all sections of the intestine (anterior, mid, and posterior) as well as in the stomach. The *in vitro* study of Ojo and Wood (2007) suggested the anterior intestine was the predominant area of luminal Cd^{2+} uptake and found that the total uptake via the stomach is almost equal to that of the mid and posterior intestine. However, very recently Klinck and Wood (2011) reported that the relative order of Cd uptake (high to low) across the gastrointestinal tissues was posterior, anterior, stomach, and mid-intestine. Differences between these two studies are attributed to the fact that the latter (Klinck and Wood, 2011) ensured that gut sacs were filled to a consistent pressure which was representative of *in vivo* tensions. This is an important methodological consideration as stretch-sensitive mechano-gated channels may contribute to Cd uptake and gut tension may influence mucus production and thereby Cd^{2+} availability. Cd^{2+} uptake seems to occur via a calcium-sensitive mechanism in the stomach that has different pharmacological and kinetic properties from that in the gills (Fig. 3.4B). There is evidence for the presence of L-type voltage-gated Ca^{2+} channels in fish enterocytes that are very different from voltage-insensitive Ca^{2+} channels in the gills (Larsson et al., 1998), and Klinck and Wood (2011) concluded that at least part of the Cd uptake in the stomach was via L-type Ca^{2+} channels. Similar Ca^{2+} channels exist in the anterior intestine that may also mediate Cd^{2+} uptake, but in the mid and posterior sections Cd^{2+} transport is not directly calcium sensitive (Wood et al., 2006; Ojo and Wood, 2008; Klinck and Wood, 2011).

In mammals, apical Cd^{2+} transport in the intestine is believed to occur by DMT1 (Elisma and Jumarie, 2001), and by the copper transporter, CTR1 (Lee et al., 2002). The expression of both transporters has been found in the teleost intestine (Mackenzie et al., 2004; Cooper et al., 2006; Kwong et al., 2010; Nadella et al., 2011), and there is circumstantial physiological evidence for the functioning of both (Burke and Handy, 2005; Nadella et al., 2006, 2007; Kwong and Niyogi, 2009; Kwong et al., 2010). Very recently, it has been demonstrated that Cd inhibits intestinal iron (Fe^{2+}) uptake and vice versa through competitive interaction (Kwong and Niyogi, 2009; Kwong et al., 2010), and treatment with an iron-deficient diet increases intestinal DMT1 expression and uptake of Cd^{2+} in fish (Cooper et al., 2006). These findings strongly indicate that Cd^{2+} uptake in the piscine intestine occurs via DMT1, at least in part (Fig. 3.4B). The expression of DMT1 has also been recorded in the fish stomach (Kwong et al., 2010), and treatment with elevated dietary Fe has been found to reduce Cd accumulation in the stomach in addition to the intestine (Kwong et al., 2011), providing indirect evidence of DMT1-driven gastric Cd^{2+} uptake as well.

It has been suggested that luminal Cd^{2+} uptake in mammals can occur via the zinc transporters, the ZIP family of transporters (Girijashanker et al., 2008) and the hZTL1 group (Zalups and Ahmad, 2003). The homologues of ZIP family of transporters have been found to express in the teleost gut (Qiu et al., 2005), and it is likely that teleosts possess hZTL1 homologues as well. Circumstantial evidence of a potential shared Cd^{2+} and Zn^{2+} uptake mechanism in the fish intestine exists (Shears and Fletcher, 1983; Klinck and Wood, 2011; Kwong and Niyogi, unpublished data). Moreover, owing to the great abundance of amino acid and small peptide transporters in the intestinal epithelium, it is also possible that these transporters may participate in the luminal uptake of Cd as conjugates of cysteine or cysteine-containing oligopeptides (Zalups and Ahmad, 2003) (Fig. 3.4B). The basolateral mechanism of gastrointestinal Cd^{2+} transport is largely unknown, though there is evidence that Cd may inhibit a basolateral high-affinity Ca^{2+} ATPase in fish enterocytes (Schoenmakers et al., 1992), similar to its action in gill chloride cells discussed earlier. Schoenmakers et al. (1992) postulated that Cd^{2+} export from the piscine enterocytes could also occur via an Na^+/Ca^{2+} exchange mechanism. In mammalian enterocytes, a membrane transporter known as metal transport protein-1 (MTP1) has been proposed to be involved in basolateral extrusion of Cd^{2+} (Zalups and Ahmad, 2003) (Fig. 3.4B); however, further research is required to characterize this function in fish. Overall, it appears that there are multiple routes for Cd uptake along the gastrointestinal tract of fish.

8.3. Interactions between Uptake via Gill and Gut

Although the Cd^{2+}-binding characteristics (log K and β_{max}) of the fish gut have not been derived yet, recent evidence suggests that the affinity of Cd^{2+} binding in the fish gut is much lower than that in the gills (Klinck et al., 2007). The pre-exposure to dietary Cd modulates Cd^{2+} uptake character-istics via both intestine and gill. The gastrointestinal uptake of Cd^{2+} in fish is a rapid process, and acclimation to elevated dietary Cd via chronic exposure appears to increase the rate of Cd^{2+} uptake in the intestine (Chowdhury et al., 2004). Similarly, acclimation to chronic dietary Cd exposure leads to altered gill–Cd^{2+} binding characteristics (decreased log K and increased β_{max}), which ultimately results in decreased new Cd accumulation in target organs (e.g. gill, liver, and kidney) of fish (Szebedinszky et al., 2001). Calcium is approximately 100-fold greater in fish diets than in most natural freshwaters [typical concentration of calcium: <5 mM (water) vs 500 mM (diet)] because of the high dietary component of bone, scale, carapace, and shell in typical food. Therefore, an interesting recent finding is that waterborne Cd^{2+} uptake across the gills is also sensitive to the calcium level in the diet. Very modest increases (1.5–3-fold) in dietary calcium concentration significantly reduce the affinity (log K) of gill–Cd^{2+} binding, short-term branchial Cd^{2+} uptake, and long-term internal Cd accumulation in various target organs (Zohouri et al., 2001; Baldisserotto et al., 2004a, b, 2005; Franklin et al., 2005; Wood et al., 2006). Taken together, these studies have shown that the protective effect is independent of the anion (Cl^- or CO_3^-) that associates with Ca^{2+} in dietary supplementation, and appears to occur because the fish downregulate branchial active Ca^{2+} uptake pathway, and the reduction in the expression of ECaCs in MR cells probably plays an important role in this regard (Galvez et al., 2007). Thus, by taking more calcium from the diet and less from the water, the fish gains protection against the unintentional uptake of Cd from water and consequent disruption of internal calcium balance. The chronic exposure to elevated dietary calcium also reduces dietary Cd accumulation in the gut (both stomach and intestine), gills, and whole body (Franklin et al., 2005; Klinck et al., 2009; Ng et al., 2009), although the mechanisms of interaction between dietary Ca^{2+} and Cd^{2+} remains unclear at present.

8.4. Other Routes

The additional uptake routes of Cd^{2+} during waterborne exposure in fish are the skin and olfactory epithelium. The uptake of Cd^{2+} through skin is relatively small in magnitude in comparison to that through the gills or intestine, and appears to occur through mechanisms different than that in the gills (Wicklund-Glynn, 2001). Cd^{2+} readily crosses the olfactory epithelium, likely via Ca^{2+} uptake pathways, and accumulates in the olfactory bulb after anterograde axonal transport along the olfactory nerve (Tjälve et al., 1986; Gottofrey and Tjälve, 1991; Tjälve and Henriksson, 1999; Scott et al., 2003; Matz, 2008). This axonal transport is facilitated by metallothionein complexation (Tallkvist et al., 2002). However, Cd does not accumulate in other regions of the brain and does not enter central nervous tissue from the circulation, indicating that it cannot cross the blood–brain barrier or synapses in the olfactory bulb (Evans and Hastings, 1992; Szebedinsky et al., 2001).

9. CHARACTERIZATION OF INTERNAL HANDLING

9.1. Internal Transport

After internalization within the epithelial cells, Cd is sequestered by a number of intracellular low molecular weight thiols (e.g. MTs, GSH), cysteine (Cys) (Olsson and Hogstrand, 1987; Wicklund-Glynn, 1996), and Ca^{2+} binding proteins (e.g. calbindin) (Zalups and Ahmad, 2003) (Fig. 3.4A, B and C). Some Cd is eventually transported across the basolateral membrane into the bloodstream, either in the form of Cd^{2+} or in conjugated forms (e.g. MT–Cd, GSH–Cd–GSH, Cys–Cd–Cys) (Zalups and Ahmad, 2003). In the bloodstream Cd is transported primarily by plasma proteins, albumin, and transferrin, as well as by MTs, albeit in small proportion because of their low concentration in plasma relative to other tissues. However, there are probably differences among fish species in Cd transport in the plasma, since albumin is present in the plasma of salmonids but little or no plasma albumin was found in carp or in elasmobranchs (De Smet et al., 2001; Metcalf and Gemmell, 2005). In addition, Cys, GSH, and several proteins such as ferritin, histidine-rich glycoprotein and γ-globulins are reported to have the ability to serve as carriers of Cd in the plasma in vertebrates including fish (Guthans and Morgan, 1982; Scott and Bradwell, 1983; De Smet et al., 2001). Very recently, Kwong et al. (2011) reported that chronic dietary Cd exposure increases hepatic transferrin expression and plasma transferrin level in rainbow trout, indicating the importance of this Fe-carrier protein in Cd handling during chronic exposure. The plasma clearance of Cd occurs almost twice as rapidly as that of other metals such as Zn, and acclimation to chronic waterborne Cd, but not to dietary Cd, stimulates its rate of plasma clearance (Chowdhury et al., 2003, 2004). The pre-exposure of fish to Cd, via both water and diet, increases Cd burden in red blood cells (RBCs) and thereby influences Cd kinetics in plasma (Chowdhury et al., 2003, 2004), although vertebrate RBCs are not known to be involved in metal transport. This probably results from the induced MT synthesis in the RBCs following Cd exposure. Although there is currently no evidence of Cd-induced MT synthesis in fish RBCs, the increased Cd accumulation in mammalian RBCs in response to Cd exposure has been attributed to the distribution of Cd to RBCs following MT induction in immature RBCs (erythroblasts) (Tanaka et al., 1985).

9.2. Tissue Accumulation Pattern

Cadmium is rapidly absorbed from the plasma by various internal organs in fish, but the highest percent absorption occurs in the liver, followed by the kidney. Chowdhury et al. (2003, 2004) have shown that the concentration of Cd rises more rapidly in the liver than in the kidneys following exposure, although the kidney is the primary organ of long-term Cd accumulation and storage in vertebrates including fish. This implies that some of the Cd accumulated in liver is transported later to the kidney for storage. Thomann et al. (1997) developed a pharmacokinetic model using exposure to Cd to rainbow trout via water and/or diet for 350 days, which indicated that while the Cd concentration in the whole body reaches a steady state in approximately 50 days, the kidney never reaches a steady-state condition. Thomann et al. (1997) also demonstrated that the Cd concentration in the kidney continues to increase even when fish are no longer exposed to Cd (depuration phase), indicating the mobilization of Cd from other target tissues (e.g. liver) to kidney for long-term storage. The absorption of Cd from plasma into various internal organs likely occurs via any of the same transporters/mechanisms described previously for branchial and gastrointestinal uptake of Cd (Fig. 3.4C). In freshwater fish, Cd accumulates maximally in the kidney, gills, liver, and gut, to a lesser extent in the blood, but not significantly in the brain or muscle, although the pattern of accumulation differs depending on the exposure route. The increase in Cd accumulation during waterborne exposure occurs generally in the order kidney > gills > liver > intestine, whereas in dietary exposure the tissue Cd accumulation follows the order intestine > kidney > liver > gill (Szebedinszky et al., 2001; Chowdhury et al., 2005; Kwong et al., 2011). In addition, significant Cd accumulation occurs in the olfactory system of fish exposed to waterborne Cd, but not dietary Cd (Scott et al., 2003). The gills accumulate considerable amounts of Cd during dietary Cd exposure, despite a lack of direct exposure (Szebedinszky et al., 2001; Chowdhury et al., 2004, 2005). Therefore, the ratio of gill to gut accumulation, rather than gill burden alone, appears to be a better indicator of the exposure route. Furthermore, fish acclimated to dietary Cd accumulate a much greater overall body burden of Cd than fish acclimated to waterborne Cd (Szebedinszky et al., 2001).

9.3. Subcellular Partitioning

Subcellular Cd partitioning is a potential indicator of Cd toxicity in aquatic organisms (Wang and Rainbow, 2006). The subcellular distribution of accumulated Cd in fish has been investigated in several recent studies, both in the laboratory and in the field (Giguère et al., 2006; Kraemer et al., 2006; Ng and Wood, 2008; Campbell and Hare, 2009; Ng et al., 2009; Kamunde, 2009; Goto and Wallace, 2010). In general, the partitioning of Cd was analyzed in five separate pools: cellular debris (membrane), metallothionein-like (heat-stable) proteins (MTLPs), heat-sensitive proteins (e.g. enzymes), metal-rich granules (MRGs), and organelle fractions (nucleus, mitochondria, microsomes). Different combinations of the subcellular fractions have been proposed to represent a metabolically active and metal-sensitive fraction (MSF: organelles and heat-sensitive proteins), and a metabolically detoxified metal fraction (MDF: MTLP and MRG). Contrary to the spillover hypothesis, which postulates that low cellular loads of Cd would be entirely sequestered in MDF, these studies reported that Cd partitions into all cellular compartments irrespective of the exposure concentration and time, and tissue burden of Cd. This indicates that there is no threshold below which Cd binding to sensitive cellular compartments does not occur. However, the pattern of subcellular partitioning of Cd in fish is organ specific, as well as being dynamic to the Cd exposure. For example, a shift in Cd distribution from the MSF to the MDF was observed in the intestine, stomach, and RBCs (Ng and Wood, 2008; Ng et al., 2009) in Cd-exposed fish, suggesting an efficient detoxification in these tissues. In contrast, Cd binding to MSF was found to be higher in the liver of fish living in metal-contaminated lakes relative to fish from reference lakes, suggesting potential liver dysfunction in metal-contaminated feral fish populations (Kraemer et al., 2006). Ng et al. (2009) reported that calcium supplementation of a Cd-rich diet increased Cd accumulation in the MDF of RBCs and decreased that in the MSF of stomach, although total tissue Cd remained unaffected. The mechanism(s) by which dietary calcium alters subcellular distribution dynamics of Cd remains to be elucidated.

9.4. Cellular Detoxification

The intracellular detoxification of Cd is primarily mediated by GSH and MTs. GSH, which represents the major non-protein thiols of cells, is able to modify Cd toxicity by altering the rates of metal uptake and elimination (Kang and Enger, 1987; Ochi et al., 1988), and by chelation of metal ions as soon as they enter the cell (Freedman et al., 1989). GSH also functions as an antioxidant by scavenging intracellular ROS. Chronic exposure to Cd has been reported to elevate GSH levels in liver and RBCs, indicating an enhanced detoxification process in fish (Tort et al., 1996; Lange et al., 2002; Firat and Kargin, 2010). MTs provide protection against metal toxicity by lowering the cellular level of free metal ions through sequestration as well as by scavenging intracellular ROS (Roesijadi, 1996). Hollis et al. (2001) and Chowdhury et al. (2005) demonstrated that chronic exposure to Cd via both water and diet produced a clear induction of MT in all tissues where significant Cd accumulation occurred, but it was insufficient to bind all accumulated Cd on a molar binding site basis. This suggests Cd binding to other cellular proteins in addition to the available pool of MT. Potentially,

Cd may compete with the essential metals for binding sites on non-MT proteins and in this scenario may induce cellular damage. Cadmium exposure can enhance oxidative stress in vertebrates including fish by the alteration of GSH turnover and depletion of the MT pool, as well as by the impairment of mitochondrial oxidative phosphorylation (Tort et al., 1996; Stohs et al., 2000; Belyaeva et al., 2001). Organisms, however, possess physiological mechanisms to defend against toxicity and stress, including the expression of heat shock proteins (HSPs), which are representative stress-defense proteins. Recent studies suggest that the induction of HSPs (e.g. HSP70, HSP90) expression plays an important role in the physiological changes related to metabolism and cell protection that occur in Cd-exposed aquatic animals including fish (Matz and Krone, 2007; Choi et al., 2008; Kwong et al., 2011). Another important detoxification process of Cd involves sequestration of MT-bound Cd in lysosomes and/or in MRG (carbonate and phosphate) for long-term storage (Klaassen et al., 1999) (see Fig. 3.4A–C). This process is regarded as the primary Cd detoxification process in some aquatic invertebrates (Wang and Rainbow, 2006). Very recently, it has been documented that some fish inhabiting metal-polluted sites may store large amounts of metals including Cd in MRG, and the increasing Cd body burden in fish may correlate with the accumulation of Cd in this cellular compartment (Goto and Wallace, 2010). These findings indicate that Cd sequestration by MRG may have critical implications for Cd tolerance in fish living in Cd-contaminated habitats. The general responses for acclimation result in an increased tolerance of fish to metals through a damage–repair hypothesis (McDonald and Wood, 1993). For example, there is a disruption of ion regulation immediately after exposure to Cd, followed by a restoration during prolonged exposure (McGeer et al., 2000a). The internal mobilization of metal-binding proteins and thiols (e.g. MT, GSH), increased sequestering by MRG, and induction of HSPs are important factors in the restoration phase through detoxification and storage of Cd in tissues (also see Section 5).

10. CHARACTERIZATION OF EXCRETION ROUTES

Cadmium is known to have a long biological half-life in vertebrates, which reflects the fact that they do not have any efficient pathways for Cd excretion. In fish, the gastrointestinal tract serves as an important barrier for Cd absorption during dietary exposure and a large proportion of ingested Cd is excreted from the body via feces by mucosal sloughing. Chowdhury et al. (2004) demonstrated that only a small fraction (2–6%) of an infused Cd dose into the stomach was internalized across the gut wall, and almost 50% of the infused dose was lost from the body within 24 h. Cadmium is also excreted by fish in a small proportion via bile, urine and gills. The urinary clearance rate of plasma Cd is less than 2% of the overall plasma clearance rate of Cd in fish, indicating a large proportion of plasma Cd is retained in the body tissues, including the kidney (Chowdhury et al., 2004). Acclimation to dietary Cd exposure dramatically increases the urinary excretion rate of Cd (Chowdhury and Wood, 2007). Similarly, acclimation to chronic waterborne and dietary exposure markedly increases hepatobiliary excretion of Cd (Chowdhury et al., 2003, 2004). However, Cd excretion via both the kidney and the bile is very low in relation to Cd uptake and accumulation on a mass-balance basis. A small proportion of Cd is also excreted by the gills, probably through mucosal sloughing (Handy, 1996).

11. BEHAVIORAL EFFECTS OF CADMIUM

While information on the behavioral effects of Cd in fish is limited, a few studies have documented the effects of Cd exposure on a variety of ecologically important behaviors in fish. McNicol and Scherer (1991) measured responses of lake whitefish (*Coregonus clupeaformis*) to Cd in preference–avoidance tests and showed an ability to detect and respond to concentrations as low as 0.2 μg L^{-1}. They also demonstrated a dichotomous response pattern where some fish were repelled and others attracted to Cd, and postulated that contact with Cd might disorient fish (McNicol and Scherer, 1991). Eissa et al. (2010) showed that sublethal exposure to waterborne Cd produced a depression in activity in carp (*Cyprinus carpio*) and the chameleon cichlid (*Australoheros facetum*), but no hypoactivity response was observed in banded astyanax (*Astyanax fasciatus*) exposed under the same conditions. Ellgaard et al. (1978) similarly linked Cd exposure with hypoactivity in bluegill (*Lepomis macrochirus*) and attributed such behavior to the onset of lethality.

Behavioral studies also include the effect of Cd exposure on predator–prey relationships and feeding. For example, exposure of fathead minnows (*Pimephales promelas*) and largemouth bass (*Micropterus salmoides*) to 25 μg L^{-1} Cd in very hard water (349 mg $CaCO_3$ L^{-1} and pH 7.9) for 21 days in a simulated natural environment that included artificial vegetation for cover showed that prey fish (fathead minnows) were more vulnerable to predation as a result of disrupted schooling behavior (Sullivan et al., 1978). Similarly, Scherer et al. (1997) showed that chronic exposure of lake trout (*Salvelinus namaycush*, as predators) and rainbow trout (the prey) altered predator–prey relationships such that prey were more vulnerable and predators less successful. A temporary reduction in food consumption in the first week of exposure of rainbow trout to sublethal Cd (3 μg L^{-1}) was noted by McGeer et al. (2000a). Felten et al. (2008) also showed that feeding, ventilation, and locomotor activity were all reduced by Cd exposure. The latter was the most sensitive and was therefore proposed as an endpoint for ecological assessment.

The olfactory system in fish is an extremely susceptible target for Cd since it is in direct contact with water and known to be a site of significant Cd accumulation (Scott et al., 2003). In fish, olfaction is involved in communication among conspecific and interspecific individuals (Ide et al., 2003; Volpato et al., 2006). Baker and Montgomery (2001) reported that fish were unable to sense adult migratory pheromones following exposure to extremely low levels of waterborne Cd (≤ 1 μg L^{-1}) for 48 h. One of the well-studied aspects of chemical communication in fish is the use of chemical cues to avoid predators. The fish olfactory system is very efficient in detecting alarm substances, a chemical cue released from the epithelial cells of the prey, which elicits typical predator avoidance behaviors of conspecifics, including area avoidance, freezing, shoaling, and shelter use (Lürling and Scheffer, 2007). Exposure to low, environmentally realistic levels of waterborne Cd for 4–7 days impaired the normal antipredator behavior that fish exhibit in response to the alarm substance (Scott et al., 2003; Honda et al., 2008). In contrast, dietary Cd exposure caused no such behavioral impairment (Scott et al., 2003), suggesting a direct effect of waterborne Cd on olfactory function. Similarly, Kusch et al. (2008) demonstrated that exposure to chronic waterborne Cd (2–20 μg L^{-1}) during embryonic and larval development resulted in the long-term impairment of antipredatory behavior in zebrafish (*Danio rerio*), even after the Cd exposure was removed for 2 weeks. Furthermore, Blechinger et al. (2007) reported that zebrafish larvae exposed to sublethal waterborne Cd for 3 h developed deficits in olfactory-dependent predatory avoidance behaviors 4–6 weeks after return to clean water, suggesting permanent damage to the olfactory system.

Waterborne Cd exposure is also known to affect the dominance behaviors in salmonid fish. Dominance hierarchies form between a pair and among groups of salmonid fish living in the confinement or in the natural environment, owing to competition over limited resources such as food or mates. Pairs of juvenile rainbow trout (*Oncorhynchus mykiss*) exposed to a low waterborne Cd concentration (15% of the 96 h LC50) for 24 h displayed significantly lower numbers of aggressive attacks during agnostic encounters, and the Cd-exposed fish exhibited a significantly reduced ability to compete with non-exposed fish (Sloman et al., 2003a).

Furthermore, when groups of trout were exposed to Cd during hierarchy formation, hierarchies developed more quickly than among non-exposed trout (Sloman et al., 2003b).

The impairment of antipredatory and dominance behaviors in fish exposed to waterborne Cd may be attributed to the impairment of endocrine and/or olfactory function in fish. The glucocorticoid stress hormone, cortisol, plays a critical role in modulating complex fish behaviors (Sloman, 2007). Cadmium is known to inhibit adrenocorticotropic hormone (ACTH)-stimulated cortisol secretion from the interrenal cells in fish (Lacroix and Hontela, 2004), and Scott et al. (2003) found that Cd exposure results in the decrease of the characteristic elevation of plasma cortisol observed in rainbow trout exposed to alarm cues. The interrenal tissue of subordinate trout was found to have a reduced sensitivity to ACTH stimulation (Sloman et al., 2002), and thus subordinate fish may have an impaired ability to respond when challenged additionally by Cd exposure.

Recently, Blechinger et al. (2007) and Matz and Krone (2007), using transgenic zebrafish larvae carrying an integrated HSP70-enhanced green fluorescent protein (eGFP) reporter gene, demonstrated that brief exposure (3 h) to a high but sublethal waterborne Cd concentration (0.6–14 mg L^{-1}) activated this gene in the olfactory epithelium. They also observed that the upregulation of HSP70/eGFP was associated with cell death and structural alterations in the olfactory epithelium. Elevated Ca in the water was found to reduce HSP responses and ameliorate the detrimental effects of Cd to the olfactory epithelium, and olfactory sensory function in fish (Matz, 2008). These findings indicate that Cd gains entry to the olfactory epithelium via Ca uptake mechanisms, wherein it causes damage to the olfactory system by affecting neurogenesis, which ultimately leads to sensory impairment.

12. MOLECULAR CHARACTERIZATION OF CADMIUM TRANSPORTERS AND STORAGE PROTEINS

12.1. Epithelial Membrane Transporters

Information on the molecular characteristics of iron (DMT1) and zinc (ZIP, hZtL1) transport proteins, which are believed to be involved in epithelial Cd transport in fish, can be found in the chapters on iron (Bury et al., Chapter 4, Vol. 31A) and zinc (Hogstrand, Chapter 3, Vol. 31A). The apical voltage-independent Ca^{2+} channel (ECaC), involved in branchial Cd uptake, has recently been cloned and characterized in teleosts (Qiu and Hogstrand, 2004; Pan et al., 2005; Shahsavarani et al.,

2006). Mammalian ECaC is a member of the transient receptor potential (TRP) gene family, of which two subfamilies have been identified, TRPV5 and TRPV6. Teleost ECaC is similar to the mammalian TRPV5 and TRPV6. The complete coding region of the ECaC gene of rainbow trout is composed of 2184 nucleotides corresponding to a protein of 727 amino acids with a predicted molecular mass of 82.3 kDa. It has a large domain representative of ankyrin sites as well as an ion transport domain and pore-forming region (Shahsavarani et al., 2006). ECaC expression has been found in both MR and pavement cells of rainbow trout (Shahsavarani et al., 2006); however, its expression in the gastrointestinal tract has yet to be examined.

12.2. Metallothioneins

Metallothioneins (MTs) are the primary Cd binding and storage proteins in vertebrates including fish. Genetic determinants of MTs have been isolated from a number of teleost fish belonging to a wide array of taxa, and many fish species have been shown to possess at least two MT isoforms (MT-A and MT-B) in their genomes (Scudiero et al., 2005). MTs are low molecular weight proteins (60–63 amino acid residues, $\sim$6–7 kDa) with a high cysteine content (16–20 Cys residues) and no aromatic residues, and have a characteristic arrangement of cysteinyl residues in Cys–X–Cys or Cys–X–X–Cys motifs (Capasso et al., 2003). At the genomic level, both isoforms of teleost MTs display a common tripartite structure of the mammalian MTs as well as AT richness in their introns (Chen et al., 2004; Cho et al., 2008), and share a high degree of homology at both the nucleotide and amino acid levels (particularly in the location of Cys residues). This suggests their duplication and divergence from a recent ancestral origin (Knapen et al., 2005). Induction of MTs in fish is mediated by multiple copies of metal-responsive elements (MREs) present in the 5′-regulatory regions of the MT genes (Kling and Olsson, 1995). MREs are activated by a zinc-sensitive, metal-responsive transcription factor (MTF-1) (Palmiter, 1994), but induction of MT gene transcription is known to occur in fish for a range of other metals including Cd (Cheung et al., 2004). Both MT-A and MT-B are known to express in major target organs (e.g. intestine, liver, gill, kidney, and brain), but higher tissue expression of MT-A relative to MT-B has been recorded in fish (Bourdineaud et al., 2006; Kwong et al., 2011). It has been found that these two MT isoforms respond differently in different tissues of fish during chronic exposure to Cd, as MT-A is more sensitive than MT-B in the intestine and liver, whereas the latter is more sensitive in kidney (Kwong et al., 2011).

13. GENOMIC AND PROTEOMIC STUDIES

The application of genomic and proteomic methodologies in under-standing the effects of Cd exposure in fish is in its infancy, although the handful of studies conducted so far have revealed important novel findings. In these studies, the functional genomic and proteomic technologies have been used both to understand the molecular mechanisms responding to Cd uptake and toxicity, and to identify novel genes/proteins to serve as biomarkers of Cd exposure and toxicity. Koskinen et al. (2004) used a cDNA microarray comprising 1273 genes to conduct a transcriptome analysis of rainbow trout fry following short-term exposure (24–96 h) to sublethal concentrations (0.05–0.5 mg L^{-1}) of waterborne Cd. They reported increased expression of genes involved in mitochondrial activities, metabolism of metal ions, and protein biosynthesis, whereas genes related to immune function and stress response were downregulated. Using a custom 500-clone cDNA microarray, Sheader et al. (2006) identified changes in several genes related to oxidative stress in European flounder (*Platichthys flesus*) exposed to Cd through a single intraperitoneal injection. Williams et al. (2006) used a more robust 13,270-clone microarray to study hepatic gene expression in the same species, also exposed to Cd through a single intraperitoneal injection but over a period of 1–16 days. They reported upregulation of genes related to oxidative stress, protein synthesis, transport, and degradation pathways, while genes involved in apoptosis, cell cycle, intracellular transport, immune function, and vitellogenin synthesis were also affected.

A more comprehensive study on the pattern of gene expression in the liver of Cd-exposed carp (*Cyprinus carpio*) was carried out by Reynders et al. (2006b). Various doses of waterborne and dietary Cd were used in acute and subchronic exposure experiments. Expression analysis of approximately 650 liver genes showed that the molecular response to Cd exposure is highly dynamic with time, and a differential response in gene expression was observed at low exposure concentrations (9 µg L^{-1} through water and 9.5 µg g^{-1} dry weight through food) relative to high concentrations (480 µg L^{-1} through water and 144 µg g^{-1} dry weight through food). At low levels of exposure, genes related to energy and lipid metabolism were affected, whereas a more general stress response was observed at high exposure levels. More recently, using a cDNA microarray comprising 582 genes, Walker et al. (2008) investigated the expression of genes that are sensitive to Cd (34 µg L^{-1}), Cu (19 µg L^{-1}), and Ag (11 µg L^{-1}) exposure in a reconstructed trout gill epithelium. They reported that several genes related to GSH cycling, oxidative stress, and intracellular trafficking respond to any of the three metals examined. They also identified genes involved in detoxification that are sensitive only to Cd and Ag [e.g. MT-A, MT-B, voltage-dependent anion channel-3 (VDAC-3)], as well as genes that respond only to Cd and Cu (e.g. apical zinc transporter, ZIP1). Pierron et al. (2009) used quantitative polymerase chain reaction analysis to evaluate the transcriptional responses of several genes related to metal detoxification (MT, HSP70), growth (insulin-like growth factor), aerobic energy metabolism [cytochrome c oxidase (CCO-1)], and antioxidant response (SOD-1) in wild yellow perch (*Perca flavescens*) from clean and metal-contaminated lakes of Rouyn-Noranda and Sudbury regions (Canada). Hepatic Cd accumulation was found to be linked with depressed hepatic expression of CCO-1 and SOD-1, indicating Cd-induced impairment of mitochondrial metabolism in these fish. The changes in the expression of these genes were more pronounced in fish from Rouyn-Noranda lakes relative to Sudbury lakes, reflecting higher levels of Cd contamination in the Rouyn-Noranda lakes. Altogether, these studies highlight the usefulness of genomic tools in unraveling the molecular events and responses to Cd exposure. A number of candidate genes have been identified for further analysis as potential novel biomarkers of Cd exposure and toxicity in fish.

To date, only a couple of studies have reported proteomic changes in Cd-exposed fish. Ling et al. (2009) used a two-dimensional polyacrylamide gel electrophoresis technique to examine the response of the gill proteome in a marine fish (*Paralichthys olivaceus*) after acute waterborne Cd exposure (10 mg L^{-1}) for 24 h. They recorded significant changes in the expression of several proteins, including HSP70 and calcium-binding protein, which were upregulated. In a similar study, Zhu et al. (2006) examined the response of the brain proteome in the same species following 24 h exposure to waterborne Cd (10 mg L^{-1}). Significant changes in the expression of several protein types were observed that included structural proteins, metal-binding proteins, and metabolic enzymes. They recorded a linear decrease in transferrin expression in the brain with increasing Cd concentrations in seawater, suggesting its potential as a biomarker of Cd toxicity in fish. The pathways of Cd metabolism in different fish organs are complex, and many of the Cd-responsive proteins detected in these two studies are yet to be identified. Thus, further research is required to elucidate the specific metabolic pathways for these proteins, and to provide a biomarker profile of Cd exposure and toxicity.

14. INTERACTIONS WITH OTHER METALS

The mechanisms whereby Cd interacts with the transport of essential elements such as Ca, Fe, and Zn into target epithelial cells, and their physiological implications in fish, have already been discussed in the preceding sections (Sections 5, 8, 9, and 10). In addition, a few studies have examined the interactive effects of Cd and Pb, and Cd and Cu in fish. Birceanu et al. (2008) reported that Cd and Pb bind to the gill in a less than additive manner, with Cd inhibiting Pb–gill binding. This likely occurs because Cd outcompetes Pb for gill binding sites, which are likely the apical Ca^{2+} channels and/or high-affinity Ca^{2+}-ATPases (Rogers and Wood, 2004; Rogers et al., 2005). However, Cd and Pb in combination exacerbate disruption of gill-mediated Ca^{2+} and Na^+ uptake in a more than additive manner. Thus, fish exposed to such metal mixtures are likely to be similarly more susceptible to toxicity in a greater than additive manner. Pelgrom et al. (1994) reported that fish exposed to a waterborne Cd and Cu mixture accumulated a lower whole-body Cd burden compared to that in fish exposed to a comparable concentration of waterborne Cd only. However, Pelgrom et al. (1994) found no consistent effect of Cd and Cu coexposure on whole-body Cu burden in fish. These findings indicate that complex mechanisms are involved in Cd and Cu interaction, and Cd and Cu accumulation during combined exposure cannot be predicted by simple addition of the effects of single metal exposures.

McGeer et al. (2007) observed that fish acclimated to chronic sublethal waterborne Cu exposure become more tolerant to acute waterborne Cd exposure. Furthermore, they reported that the short-term gill Cd-binding properties change significantly in Cu-acclimated fish relative to those in Cd-acclimated fish. Both the capacity to accumulate new Cd in the gill and gill–Cd affinity increase following acclimation to Cu, whereas only the former increases in Cd-acclimated fish. These results demonstrate the phenomenon of cross-acclimation to Cd in fish, and also illustrate that knowledge of previous exposure conditions is essential, not only for the metal of concern, but for other metals as well.

15. KNOWLEDGE GAPS AND FUTURE DIRECTIONS

In spite of the wealth of data on the impacts of Cd on aquatic organisms there is much that is not understood. In part, this is because of the broad spectrum of effects that Cd has and its capacity to influence many if not most physiological systems. An integrated understanding of the common mechanisms and interactions by which these different effects are mediated is required. Oxidative damage and induction of defense mechanisms appear to be useful areas of research to pursue in this respect. Chronic toxicity follows a damage–response–acclimation scenario and some physiological responses such as ionoregulatory disruption appear to follow this model, but it is not clear whether ROS-induced effects and responses do. Other models of understanding impacts such as the spillover theory do not appear to be supported.

Development of a common and comprehensive model for Cd impacts and physiological response to Cd exposure would be a valuable addition to the field. Characterization of the molecular mechanisms underlying physiological effects of both dietary and waterborne Cd that could be integrated into this model would contribute to making the understanding comprehensive. The application of genomic and proteomic methodologies in understanding the effects of Cd exposure in fish will undoubtedly be a key aspect of this understanding. Because Cd contamination occurs mostly in combination with other contaminants, understanding mixture effects is one of the future areas of relevant research. The work of Playle (2004) provided a framework for interactions at gill uptake sites among metals. However, the recent study by Birceanu et al. (2008) demonstrates that there are many complexities that await our understanding.

ACKNOWLEDGMENTS

The authors wish to acknowledge the very capable assistance of K. Wood, J. Cunningham, and E.-J. Costa of Wilfrid Laurier University, and R.W. Kwong of the University of Saskatchewan in the preparation of the manuscript. Funding to support the preparation of this chapter was provided by the NSERC Discovery and CRD Programs. This work is dedicated to the memory of Dr. Richard C. Playle: significant aspects of the foundational understanding of the biogeochemistry of metals toxicology, including Cd, were developed through his work at Wilfrid Laurier University. His example continues to provide inspiration there and beyond.

REFERENCES

Adams, W. J., Blust, R., Borgmann, U., Brix, K. V., DeForest, D. K., Green, A. S., Meyer, J., McGeer, J. C., Paquin, P., Rainbow, P., and Wood, C. (2011). Utility of tissue residues for predicting effects of metals on aquatic organisms. *Integr. Environ. Assess. Manag.* **7**, 75–98.

Alloway, B. J. (1995). Cadmium. In: *Heavy Metals in Soils* (B. J. Alloway, ed.), pp. 122–151. Chapman and Hall, London.

Almeida, J. A., Barreto, R. E., Novelli, E. L. B., Castro, F. J., and Moron, S. E. (2009). Oxidative stress biomarkers and aggressive behavior in fish exposed to aquatic cadmium contamination. *Neotrop. Ichthyol.* **7**, 103–108.

ANZECC (2000). *Australian and New Zealand Guidelines for Fresh and Marine Water Quality.* Australian and New Zealand Environment and Conservation Council. http://www.mincos.gov. au/publications/australian_and_new_zealand_guidelines_for_fresh_and_marine_water_quality

Atli, G., and Canli, M. (2007). Enzymatic responses to metal exposures in a freshwater fish *Oreochromis niloticus. Comp. Biochem. Physiol.* C **145**, 282–287.

ATSDR (2008). *Toxicological Profile for Cadmium. Draft for Public Comment.* Agency for Toxic Substances and Disease Registry. http://www.atsdr.cdc.gov/toxprofiles/tp5.html

Baker, C. F., and Montgomery, J. C. (2001). Sensory deficits induced by cadmium in banded kokopu, *Galaxias fasciatus*, juveniles. *Environ. Biol. Fish.* **62**, 455–464.

Baldisserotto, B., Kamunde, C., Matsuo, A., and Wood, C. M. (2004a). A protective effect of dietary calcium against acute waterborne cadmium uptake in rainbow trout. *Aquat. Toxicol.* **67**, 57–73.

Baldisserotto, B., Kamunde, C., Matsuo, A., and Wood, C. M. (2004b). Acute waterborne cadmium uptake in rainbow trout is reduced by dietary calcium carbonate. *Comp. Biochem. Physiol. C* **137**, 363–372.

Baldisserotto, B., Chowdhury, M. J., and Wood, C. M. (2005). Effects of dietary calcium and cadmium on cadmium accumulation, calcium and cadmium uptake from the water, and their interactions in juvenile rainbow trout. *Aquat. Toxicol.* **72**, 99–117.

Belyaeva, E. A., Glazunov, V. V., Nikitina, E. R., and Korotkov, S. M. (2001). Bivalent metal ions modulate Cd^{2+} effects on isolated rat liver mitochondria. *J. Bioenerg. Biomembr.* **33**, 303–318.

Benaduce, A. P. S., Kochhann, D., Flores, E. M. M., Dressler, V. L., and Baldisserotto, B. (2008). Toxicity of cadmium for silver catfish *Rhamdia quelen* (Heptapteridae) embryos and larvae at different alkalinities. *Arch. Environ. Contam. Toxicol.* **54**, 274–282.

Bengtsson, B. E., Carlin, C. H., Larsson, A., and Svanberg, O. (1975). Vertebral damage in minnows, *Phoxinus phoxinus* L., exposed to cadmium. *Ambio* **4**, 166–168.

Benoit, D. A., Leonard, E. N., Christensen, G. M., and Fiandt, J. T. (1976). Toxic effects of cadmium on three generations of brook trout (*Salvelinus fontinalis*). *Trans. Am. Fish. Soc.* **105**, 550–560.

Bhavsar, S. P., Gandhi, N., Diamond, M. L., Lock, A. S., Spiers, G., and De la Torre, M. C. A. (2008). Effects of estimates from different geochemical models on metal fate predicted by coupled speciation-fate models. *Environ. Toxicol. Chem.* **27**, 1020–1030.

Bilinski, E., and Jonas, R. E. E. (1973). Effects of cadmium and copper on the oxidation of lactate by rainbow trout (*Salmo gairdneri*) gills. *J. Fish. Res. Bd Can.* **30**, 1553–1558.

Birceanu, O., Chowdhury, M. J., Gillis, P. L., McGeer, J. C., Wood, C. M., and Wilkie, M. P. (2008). Modes of metal toxicity and impaired branchial ionoregulation in rainbow trout exposed to mixtures of Pb and Cd in soft water. *Aquat. Toxicol.* **89**, 222–231.

Blechinger, S. R., Kusch, R. C., Haugo, K., Matz, C., Chivers, D. P., and Krone, P. H. (2007). Brief embryonic cadmium exposure induces a stress response and cell death in the developing olfactory system followed by long-term olfactory deficits in juvenile zebrafish. *Toxicol. Applied Pharmacol.* **224**, 72–80.

Bols, N. C., Brubacher, J. L., Ganassin, R. C., and Lee, L. E. J. (2001). Ecotoxicology and innate immunity in fish. *Dev. Comp. Immunol.* **25**, 853–873.

Borrok, D., Fein, J. B., and Kulpa, C. F. (2004). Proton and Cd adsorption onto natural bacterial consortia: testing universal adsorption behavior. *Geochim. Cosmochim. Acta* **68**, 3231–3238.

Bourdineaud, J. P., Baudrimont, M., Gonzalez, P., and Moreau, J. L. (2006). Challenging the model for induction of metallothionein gene expression. *Biochimie* **88**, 1787–1792.

Brodeur, J. C., Daniel, C., Ricard, A. C., and Hontela, A. (1998). *In vitro* response to ACTH of the interrenal tissue of rainbow trout (*Oncorhynchus mykiss*) exposed to cadmium. *Aquat. Toxicol.* **42**, 103–113.

Brown, S. B., Adams, B. A., Cyr, D. G., and Eales, J. (2004). Contaminant effects on the teleost fish thyroid. *Environ. Toxicol. Chem.* **23**, 1680–1701.

Brown, V., Shurben, D., Miller, W., and Crane, M. (1994). Cadmium toxicity to rainbow trout *Oncorhynchus mykiss* Walbaum and brown trout *Salmo trutta* L. over extended exposure periods. *Ecotoxicol. Environ. Saf.* **29**, 38–46.

Burke, J., and Handy, R. D. (2005). Sodium-sensitive and -insensitive copper accumulation by isolated intestinal cells of rainbow trout *Oncorhynchus mykiss. J. Exp. Biol.* **208**, 391–407.

Bury, N. R., and Grosell, M. (2003). Waterborne iron acquisition by a freshwater teleost fish, zebrafish *Danio rerio. J. Exp. Biol.* **206**, 3529–3535.

Calamari, D., Marchetti, R., and Vailati, G. (1980). Influence of water hardness on cadmium toxicity to *Salmo gairdneri. Rich. Water Res.* **14**, 1421–1426.

Campbell, P. G. C. (1995). Interactions between trace metals and organisms: critique of the free-ion activity model. In: *Metal Speciation and Bioavailability in Aquatic Systems* (A. Tessier and D. Turner, eds), pp. 45–102. John Wiley and Sons, Chichester.

Campbell, P. G. C., and Hare, L. (2009). Metal detoxification in freshwater animals. Roles of metallothioneins. *Metal Ions Life Sci.* **5**, 239–277.

Cao, L., Huang, W., Liu, J., Yin, X., and Dou, S. (2010). Accumulation and oxidative stress biomarkers in Japanese flounder larvae and juveniles under chronic cadmium exposure. *Comp. Biochem. Physiol. C* **151**, 386–392.

Capasso, C., Carginale, V., Crescenzi, O., Di Maro, D., Parisi, E., Spadaccini, R., and Temussi, P. A. (2003). Solution structure of MT_nc, a novel metallothionein from the Antarctic fish *Notothenia coriiceps. Structure* **11**, 435–443.

Carpenter, K. E. (1927). The lethal action of soluble metallic salts on fishes. *Br. J. Exp. Biol.* **4**, 378–390.

CCME (1999). Canadian Water Quality Guidelines for the Protection of Aquatic Life: Appendix XXI. Canadian Water Quality Guidelines: Updates (May 1996) Cadmium. In: *Canadian Environmental Quality Guidelines 1999*, pp. 1397–1411. Winnipeg: Canadian Council of Ministers of the Environment.

CCME (2007). *Canadian Environmental Quality Guidelines: Canadian Water Quality Guidelines for the Protection of Aquatic Life.* Winnipeg: Canadian Council of Ministers of the Environment. http://ceqg-rcqe.ccme.ca/

Chang, M. H., Lin, H. C., and Hwang, P. P. (1997). Effects of cadmium on the kinetics of calcium uptake in developing tilapia larvae. Oreochromis mossambicus. *Fish Physiol. Biochem.* **16**, 459–470.

Chen, C. Y., Stemberger, R. S., Klaue, B., Blum, J. D., Pickhardt, P. C., and Folt, C. L. (2000). Accumulation of heavy metals in food web components across a gradient of lakes. *Limnol. Oceanogr.* **45**, 1525–1536.

Chen, W. Y., John, J. A. C., Lin, C. H., Lin, H. F., Wu, S. C., and Chang, C. Y. (2004). Expression of metallothionein gene during embryonic and early larval development in zebrafish. *Aquat. Toxicol.* **69**, 215–227.

Cheung, A. P. L., Lam, T. H. J., and Chan, K. M. (2004). Regulation of metallothionein gene expression by heavy metal ions. *Mar. Environ. Res.* **58**, 389–394.

Cho, Y. S., Lee, S. Y., Kim, K. Y., Bang, I. C., Kim, D. S., and Nam, Y. K. (2008). Gene structure and expression of metallothionein during metal exposures in *Hemibarbus mylodon. Ecotoxicol. Environ. Saf.* **71**, 125–137.

Choi, Y. K., Jo, P. G., and Choi, C. Y. (2008). Cadmium affects the expression of heat shock protein 90 and metallothionein mRNA in the Pacific oyster. Crassostrea gigas. *Comp. Biochem. Physiol. C* **147**, 286–292.

Chowdhury, M. J., and Wood, C. M. (2007). Renal function in the freshwater rainbow trout after dietary cadmium acclimation and waterborne cadmium challenge. *Comp. Biochem. Physiol. C* **145**, 321–332.

Chowdhury, M. J., Grosell, M., McDonald, D. G., and Wood, C. M. (2003). Plasma clearance of cadmium and zinc in non-acclimated and metal-acclimated trout. *Aquat. Toxicol.* **64**, 259–275.

Chowdhury, M. J., McDonald, D. G., and Wood, C. M. (2004). Gastrointestinal uptake and fate of cadmium in rainbow trout acclimated to sublethal dietary cadmium. *Aquat. Toxicol.* **69**, 149–163.

Chowdhury, M. J., Baldisserotto, B., and Wood, C. M. (2005). Tissue-specific cadmium and metallothionein levels in rainbow trout chronically acclimated to waterborne or dietary cadmium. *Arch. Environ. Contam. Toxicol.* **48**, 381–390.

Cooper, C. A., Handy, R. D., and Bury, N. R. (2006). The effects of dietary iron concentration on gastrointestinal and branchial assimilation of both iron and cadmium in zebrafish (*Danio rerio*). *Aquat. Toxicol.* **79**, 167–175.

Cooper, C. A., Shayeghi, M., Techau, M. E., Capdevila, D. M., MacKenzie, S., Durrant, C., and Bury, N. R. (2007). Analysis of the rainbow trout solute carrier 11 family reveals iron import $\geq$ pH 7. 4 and a functional isoform lacking transmembrane domains 11 and 12. *FEBS Lett.* **581**, 2599–2604.

Croteau, M.-N., Luoma, S. N., and Stewart, A. R. (2005). Trophic transfer of metals along freshwater food webs: evidence of cadmium biomagnifications in nature. *Limnol. Oceanogr.* **50**, 1511–1519.

CSIR (1995). South African Water Quality Guidelines for Coastal Marine Waters, *Vol. 1.* Natural Environment. Department of Water Affairs and Forestry, Council for Scientific and Industrial Research, Pretoria.

CSIR (1996). South African Water Quality Guidelines, *Vol. 7.* Aquatic Ecosystems. Department of Water Affairs and Forestry, Council for Scientific and Industrial Research, Pretoria.

Cusimano, R. F., Brakke, D. F., and Chapman, G. A. (1986). Effects of pH on the toxicities of cadmium, copper, and zinc to steelhead trout (*Salmo gairdneri*). *Can. J. Fish. Aquat. Sci.* **42**, 1497–1503.

Davies, P. H., Gorman, W. C., Carlson, C. A., and Brinkman, S. F. (1993). Effect of hardness on bioavailability and toxicology of cadmium to rainbow trout. *Chem. Speciat. Bioavail.* **5**, 67–77.

De Smet, H., Blust, R., and Moens, L. (2001). Cadmium-binding to transferrin in the plasma of the common carp *Cyprinus carpio*. *Comp. Biochem. Physiol. C* **128**, 45–53.

Di Giulio, R. T., Washburn, P. C., Wenning, R. J., Winston, G. W., and Jewell, C. S. (1989). Biochemical responses in aquatic animals: a review of determinants of oxidative stress. *Environ. Toxicol. Chem.* **8**, 1103–1123.

Di Toro, D. M., Mahony, J. D., Hansen, D. J., Scott, K. J., Carlson, A. R., and Ankley, G. T. (1992). Acid volatile sulfide predicts the acute toxicity of cadmium and nickel in sediments. Environ. Sci. Technol. **26**, 96–101.

Di Toro, D. M., Allen, H., Bergman, H., Meyer, J., Paquin, P., and Santore, R. (2001). A biotic ligand model of the acute toxicity of metals: I. Technical basis. *Environ. Toxicol. Chem.* **20**, 2383–2396.

Dzombak, D. A., and Morel, F. M. M. (1990). *Surface Complexation Modeling: Hydrous Ferric Oxide.* Wiley-Interscience, New York.

Eissa, B. L., Ossana, N. A., Ferrari, L., and Salibian, A. (2010). Quantitative behavioral parameters as toxicity biomarkers: fish responses to waterborne cadmium. *Arch. Environ. Contam. Toxicol.* **58**, 1032–1039.

Elisma, F., and Jumarie, C. (2001). Evidence for cadmium uptake through Nramp2: metal speciation studies with Caco-2 cells. *Biochem. Biophys. Res. Commun.* **285**, 662–668.

Ellgaard, E. G., Tusa, J. E., and Malizia, A. A., Jr. (1978). Locomotor activity of the bluegill *Lepomis macrochirus*: hyperactivity induced by sublethal concentrations of cadmium, chromium and zinc. *J. Fish Biol.* **12**, 19–23.

Evans, J., and Hastings, L. (1992). Accumulation of Cd(II) in the CNS depending on the route of administration: intraperitoneal, intratracheal, or intranasal. *Fundam. Appl. Toxicol.* **19**, 275–278.

Fairbrother, A., Wenstel, R., Sappington, K., and Wood, W. (2007). Framework for metals risk assessment. *Ecotoxicol. Environ. Saf.* **68**, 145–227.

Felten, V., Charmantier, G., Mons, R., Geffard, A., Rousselle, P., Coquery, M., Garric, J., and Geffard, O. (2008). Physiological and behavioural responses of *Gammarus pulex* (Crustacea: Amphipoda) exposed to cadmium. *Aquat. Toxicol.* **86**, 413–425.

Ferard, J. F., Jouany, J. M., Truhaut, R., and Vasseur, P. (1983). Accumulation of cadmium in a freshwater food chain experimental model. *Ecotoxicol. Environ. Saf.* **7**, 43–52.

Firat, Ö., and Kargin, F. (2010). Individual and combined effects of heavy metals on serum biochemistry of Nile tilapia *Oreochromis niloticus*. *Arch. Environ. Contam. Toxicol.* **58**, 151–157.

Firat, Ö., Cogun, H. Y., Aslanyavrusu, S., and Kargin, F. (2009). Antioxidant responses and metal accumulation in tissues of Nile tilapia *Oreochromis niloticus* under Zn, Cd and Zn plus Cd exposures. *J. Appl. Toxicol.* **29**, 295–301.

Flik, G. (1990). Hypocalcin physiology. Prog. *Clin. Biol. Res.* **342**, 578–585.

Franklin, N. M., Glover, C. N., Nicol, J. A., and Wood, C. M. (2005). Calcium/cadmium interactions at uptake surfaces in rainbow trout: waterborne versus dietary routes of exposure. *Environ. Toxicol. Chem.* **24**, 2954–2964.

Frazier, J. M. (1979). Bioaccumulation of cadmium in marine organisms. *Environ. Health Perspect.* **28**, 75–79.

Freedman, J. H., Ciriolo, M. R., and Peisach, J. (1989). The role of glutathione in copper metabolism and toxicity. *J. Biol. Chem.* **264**, 5598–5605.

Fu, H., Lock, R. A. C., and Wendelaar Bonga, S. E. (1989). Effect of cadmium on prolactin cell activity and plasma electrolytes in the freshwater teleost *Oreochromis mossambicus*. *Aquat. Toxicol.* **14**, 295–306.

Fu, H., Steinebach, O. M., van den Hamer, C. J. A., Balm, P. H. M., and Lock, R. A. C. (1990). Involvement of cortisol and metallothionein-like proteins in the physiological responses of tilapia (*Oreochromis mossambicus*) to sublethal cadmium stress. *Aquat. Toxicol.* **16**, 257–269.

Galvez, F., Wong, D., and Wood, C. M. (2006). Cadmium and calcium uptake in isolated mitochondria-rich cell populations from the gills of the freshwater rainbow trout. *Am. J. Physiol. Regul. Integr. Comp. Physiol.* **291**, 170–176.

Galvez, F., Franklin, N. M., Tuttle, R. B., and Wood, C. M. (2007). Interactions of waterborne and dietary cadmium on the expression of calcium transporters in the gills of rainbow trout: influence of dietary calcium supplementation. *Aquat. Toxicol.* **84**, 208–214.

Garrett, R. G., Porter, A. R. D., Hunt, P. A., and Lalor, G. C. (2008). The presence of anomalous trace element levels in present day Jamaican soils and the geochemistry of Late-Miocene or Pliocene phosphorites. *Appl. Geochem.* **23**, 822–834.

Ge, Y., MacDonald, D., Sauvé, S., and Hendershot, W. (2005). Modeling of Cd and Pb speciation in soil solutions by WinHumicV and NICA-Donnan model. *Environ. Model. Softw.* **20**, 353–359.

Giguère, A., Campbell, P. G. C., Hare, L., and Couture, P. (2006). Sub-cellular partitioning of cadmium and zinc in indigenous yellow perch (*Perca flavescens*) sampled along a polymetallic gradient. *Aquat. Toxicol.* **77**, 178–189.

Giles, M. A. (1984). Electrolyte and water balance in plasma and urine of rainbow trout (*Salmo gairdneri*) during chronic exposure to cadmium. *Can. J. Fish. Aquat. Sci.* **41**, 1678–1685.

Girijashanker, K., He, L., Soleimani, M., Reed, J. M., Li, H., Liu, Z., Wang, B., Dalton, T. P., and Nebert, D. W. (2008). Slc39a14 gene encodes ZIP14, a metal/bicarbonate symporter: similarities to the ZIP8 transporter. *Mol. Pharmacol.* **73**, 1413–1423.

Goto, D., and Wallace, W. G. (2010). Metal intracellular partitioning as a detoxification mechanism for mummichogs (*Fundulus heteroclitus*) living in metal-polluted salt marshes. *Mar. Environ. Res.* **69**, 163–171.

Gottofrey, J., and Tjalve, H. (1991). Axonal transport of cadmium in the olfactory nerve of the pike. *Pharmacol. Toxicol.* **69**, 242–252.

Guthans, S. L., and Morgan, W. T. (1982). The interaction of zinc, nickel and cadmium with serum albumin and histidine-rich glycoprotein assessed by equilibrium dialysis and immunoadsorbent chromatography. *Arch. Biochem. Biophys.* **218**, 320–328.

Handy, R. D. (1996). Dietary exposure to toxic metals in fish. In: *Toxicology of Aquatic Pollution: Physiological, Cellular and Molecular Approaches* (E. W. Taylor, ed.), pp. 29–60. Cambridge University Press, Cambridge.

Hansen, J. A., Welsh, P. G., Lipton, J., and Suedkamp, M. J. (2002). The effects of long-term cadmium exposure on the growth and survival of juvenile bull trout (*Salvelinus confluentus*). *Aquat. Toxicol.* **58**, 165–174.

Hogstrand, C., Wilson, R. W., Polgar, D., and Wood, C. M. (1994). Effects of zinc on the kinetics of branchial calcium uptake in freshwater rainbow trout during adaptation to waterborne zinc. *J. Exp. Biol.* **186**, 55–73.

Hollis, L., Burnison, K., and Playle, R. C. (1996). Does the age of metal-dissolved organic carbon complexes influence binding of metals to fish gills? *Aquat. Toxicol.* **35**, 253–264.

Hollis, L., Muench, L., and Playle, R. C. (1997). Influence of dissolved organic matter on copper binding, and calcium on cadmium binding, by gills of rainbow trout. *J. Fish Biol.* **50**, 703–720.

Hollis, L., McGeer, J. C., McDonald, D. G., and Wood, C. M. (1999). Cadmium accumulation, gill Cd binding, acclimation, and physiological effects during long term sublethal Cd exposure in rainbow trout. *Aquat. Toxicol.* **46**, 101–119.

Hollis, L., McGeer, J. C., McDonald, D. G., and Wood, C. M. (2000a). Effects of long term sublethal Cd exposure in rainbow trout during soft water exposure: implications for biotic ligand modelling. *Aquat. Toxicol.* **51**, 93–105.

Hollis, L., McGeer, J. C., McDonald, D. G., and Wood, C. M. (2000b). Protective effects of calcium against chronic waterborne cadmium exposure to juvenile rainbow trout. *Environ. Toxicol. Chem.* **19**, 2725–2734.

Hollis, L., Hogstrand, C., and Wood, C. M. (2001). Tissue specific cadmium accumulation, metallothionein induction, and tissue zinc and copper levels during chronic sublethal cadmium exposure in juvenile rainbow trout. *Arch. Environ. Contam. Toxicol.* **41**, 468–474.

Honda, R. T., Fernandes-de-Castilho, M., and Val, A. L. (2008). Cadmium-induced disruption of environmental exploration and chemical communication in matrinxã. Brycon amazonicus. *Aquat. Toxicol.* **89**, 204–206.

Ide, L. M., Urbinati, E. C., and Hoffmann, A. (2003). The role of olfaction in the behavioural and physiological responses to conspecific skin extract in *Brycon cephalus*. *J. Fish Biol.* **63**, 332–343.

Iger, Y., Lock, R. A. C., ver der Meij, J. C. A., and Wendelaar Bonga, S. E. (1994). Effects of water-borne cadmium on the skin of the common carp (*Cyprinus carpio*). *Arch. Environ. Contam. Toxicol.* **26**, 342–350.

International Cadmium Association (2000). *Cadmium Products, The Issues and Answers.* http://www.cadmium.org/environmental.html

Isidori, M., Cangiano, M., Palermo, F. A., and Parrella, A. (2010). E-screen and vitellogenin assay for the detection of the estrogenic activity of alkylphenols and trace elements. *Comp. Biochem. Physiol. C* **152**, 51–56.

Johnson, B. T. (1997). *Cadmium Contamination of Food.* UCD Extoxnet FAQ Team. http://extoxnet.orst.edu/faqs/foodcon/cadmium.htm

Kalay, M. (2006). The effect of cadmium on the levels of Na^+, K^+, Ca^{++} and Mg^{++} in serum of *Tilapia nilotica* (Linnaeus, 1758). *Ekoloji* **59**, 1–7.

Kamunde, C. (2009). Early subcellular partitioning of cadmium in gill and liver of rainbow trout (*Oncorhynchus mykiss*) following low-to-near-lethal waterborne cadmium exposure. *Aquat. Toxicol.* **91**, 291–301.

Kang, Y. J., and Enger, M. D. (1987). Effect of cellular glutathione depletion on cadmium-induced cytotoxicity in human lung carcinoma cells. *Cell Biol. Toxicol.* **3**, 347–360.

Karaytug, S., Erdem, C., and Cicik, B. (2007). Accumulation of cadmium in the gill, liver, kidney, spleen, muscle and brain tissues of *Cyprinus carpio.* *Ekoloji* **63**, 16–22.

Karlsson, T., Elgh-Dalgren, K., Björn, E., and Skyllberg, U. (2007). Complexation of cadmium to sulfur and oxygen functional groups in an organic soil. *Geochim. Cosmochim. Acta* **71**, 604–614.

Karlsson-Norrgren, L., Runn, P., Haux, C., and Forlin, L. (1985). Cadmium-induced changes in gill morphology of zebrafish, *Brachydanio rerio* (Hamilton–Buchanan), and rainbow trout, *Salmo gairdneri* Richardson. *J. Fish Biol.* **27**, 81–95.

Keatley, B. E., Douglas, M. S. V., Blais, J. M., Mallory, M. L., and Smol, J. P. (2008). Impacts of seabird-derived nutrients on water quality and diatom assemblages from Cape Vera, Devon Island, Canadian high Arctic. *Hydrobiology* **621**, 191–205.

Kimura, T., and Itoh, N. (2008). Function of metallothionein in gene expression and signal transduction: newly found protective role of metallothionein. *J. Health Sci.* **54**, 251–260.

Klaassen, C. D., Liu, J., and Choudhuri, S. (1999). Metallothionein: an intracellular protein to protect against cadmium toxicity. *Annu. Rev. Pharmacol. Toxicol.* **39**, 267–294.

Klinck, J. S., and Wood, C. M. (2011). *In vitro* characterization of cadmium transport along the gastro-intestinal tract of freshwater rainbow trout (*Oncorhynchus mykiss*). *Aquat. Toxicol.* **102**, 58–72.

Klinck, J. S., Green, W. W., Mirza, R. S., Nadella, S. R., Chowdhury, M. J., Wood, C. M., and Pyle, G. G. (2007). Branchial cadmium and copper binding and intestinal cadmium uptake in wild yellow perch (*Perca flavescens*) from clean and metal-contaminated lakes. *Aquat. Toxicol.* **84**, 198–207.

Klinck, J. S., Ng, T. Y. T., and Wood, C. M. (2009). Cadmium accumulation and in vitro analysis of calcium and cadmium transport functions in the gastro-intestinal tract of trout following chronic dietary cadmium and calcium feeding. *Comp. Biochem. Physiol. C* **150**, 349–360.

Kling, P., and Olsson, P. E. (1995). Regulation of the rainbow trout metallothionein-A gene. *Mar. Environ. Res.* **39**, 117–120.

Knapen, D., Redeker, E. S., Inácio, I., De Coen, W., Verheyen, E., and Blust, R. (2005). New metallothionein mRNAs in *Gobio gobio* reveal at least three gene duplication events in cyprinid metallothionein evolution. *Comp. Biochem. Physiol. C* **140**, 347–355.

Koskinen, H., Pehkonen, P., Vehniäinen, E., Krasnov, A., Rexroad, C., Afanasyev, S., Mölsa, H., and Oikari, A. (2004). Response of rainbow trout transcriptome to model chemical contaminants. Biochem. *Biophys Res. Commun.* **320**, 745–753.

Kraemer, L. D., Campbell, P. G. C., and Hare, L. (2006). Seasonal variations in hepatic Cd and Cu concentrations and in the sub-cellular distribution of these metals in juvenile yellow perch (*Perca flavescens*). *Environ. Pollut.* **142**, 313–325.

Kramer, J. R., Bell, R. A., and Smith, D. S. (2007). Determination of sulfide ligands and associations with natural organic matter. *Appl. Geochem.* **22**, 1606–1611.

Kusch, R. C., Krone, P. H., and Chivers, D. P. (2008). Chronic exposure to low concentrations of waterborne cadmium during embryonic and larval development results in the long-term hindrance of antipredator behavior in zebrafish. *Environ. Toxicol. Chem.* **27**, 705–710.

Kwong, R. W. M., and Niyogi, S. (2009). The interactions of iron with other divalent metals in the intestinal tract of a freshwater teleost, rainbow trout (*Oncorhynchus mykiss*). *Comp. Biochem. Physiol. C* **150**, 442–449.

Kwong, R. W. M., Andrés, J. A., and Niyogi, S. (2010). Molecular evidence and physiological characterization of iron absorption in isolated enterocytes of rainbow trout (*Oncorhynchus mykiss*): implications for dietary cadmium and lead absorption. *Aquat. Toxicol.* **99**, 343–350.

Kwong, R. W. M., Andrés, J. A., and Niyogi, S. (2011). Effects of dietary cadmium exposure on tissue-specific cadmium accumulation, iron status and expression of iron-handling and stress-inducible genes in rainbow trout: influence of elevated dietary iron. *Aquat. Toxicol.* **102**, 1–9.

Lacroix, A., and Hontela, A. (2004). A comparative assessment of the adrenotoxic effects of cadmium in two teleost species, rainbow trout, *Oncorhynchus mykiss*, and yellow perch, *Perca flavescens. Aquat. Toxicol.* **67**, 13–21.

Lalor, G. C. (2008). Review of cadmium transfers from soil to humans and its health effects in the Jamaican environment. *Sci. Total Environ.* **400**, 162–172.

Lane, T. W., and Morel, F. M. M. (2001). A biological function for cadmium in marine diatoms. *Proc. Natl. Acad. Sci. U.S.A.* **97**, 4627–4631.

Lane, T. W., Saito, M. A., George, G. N., Pickering, I. J., Prince, R. C., and Morel, F. M. M. (2005). A cadmium enzyme from a marine diatom. *Nature* **345**, 42.

Lange, A., Ausseil, O., and Segner, H. (2002). Alterations of tissue glutathione levels and metallothionein mRNA in rainbow trout during single and combined exposure to cadmium and zinc. *Comp. Biochem. Physiol. C* **131**, 231–243.

Larsson, A., Bengtsson, B.-E., and Svanberg, O. (1976). Some haematological and biochemical effects of cadmium on fish. In: *Effects of Pollutants on Aquatic Organisms* (A. M. P. Lockwood, ed.), pp. 35–45. Cambridge University Press, New York.

Larsson, A., Bengtsson, B.-E., and Haux, C. (1981). Disturbed ion balance in flounder, *Platichthys flesus* L. exposed to sublethal level of cadmium. Aquat. Toxicol. **1**, 19–35.

Larsson, D., Lundgren, T., and Sundell, K. (1998). Ca^{2+} uptake through voltage-gated L-type Ca^{2+} channels by polarized enterocytes from Atlantic cod *Gadus morhua. J. Membr. Biol.* **164**, 229–237.

Lawrence, S. G., Holoka, M. H., Hunt, R. V., and Hesslein, R. H. (1996). Multi-year experimental additions of cadmium to a lake epilimnion and resulting water column cadmium concentrations. *Can. J. Fish. Aquat. Sci.* **53**, 1876–1887.

Leblanc, G. A. (1995). Trophic-level differences in the bioconcentration of chemicals: implications for assessing environmental biomagnifications. *Environ. Sci. Technol.* **29**, 154–160.

Lee, J., Peña, M. M. O., Nose, Y., and Thiele, D. J. (2002). Biochemical characterization of the human copper transporter Ctr1. *J. Biol. Chem.* **277**, 4380–4387.

Ling, X. P., Zhu, J. Y., Huang, L., and Huang, H. Q. (2009). Proteomic changes in response to acute cadmium toxicity in gill tissue of *Paralichthys olivaceus. Environ. Toxicol. Pharmacol.* **27**, 212–218.

Livingstone, D. R. (2001). Contaminant-stimulated reactive oxygen species production and oxidative damage in aquatic organisms. *Mar. Pollut. Bull.* **42**, 656–666.

Lizardo-Daudt, H. M., and Kennedy, C. (2008). Effects of cadmium chloride on the development of rainbow trout *Oncorhynchus mykiss* early life stages. *J. Fish Biol.* **73**, 702–718.

Lizardo-Daudt, H. M., Bains, O. S., Singh, C. R., and Kennedy, C. (2007). Biosynthetic capacity of rainbow trout (*Oncorhynchus mykiss*) interrenal tissue after cadmium exposure. *Arch. Environ. Contam. Toxicol.* **52**, 90–96.

Lofts, S., and Tipping, E. (1998). An assemblage model for cation binding by natural particulate matter. *Geochim. Cosmochim. Acta* **62**, 2609–2625.

Lofts, S., and Tipping, E. (2000). Solid–solution metal partitioning in the Humber rivers: application of WHAM and SCAMP. *Sci. Total Environ.* **251**, 381–399.

Luoma, S. N., and Rainbow, P. S. (2005). Why is metal bioaccumulation so variable? Biodynamics as a unifying concept. *Environ. Sci. Technol.* **39**, 1921–1931.

Luoma, S. N., Cain, D. J., and Rainbow, P. S. (2010). Calibrating biomonitors to ecological disturbance: a new technique for explaining metal effects in natural waters. *Integr. Environ. Assess. Manag.* **6**, 199–209.

Lürling, M., and Scheffer, M. (2007). Info-disruption: pollution and the transfer of chemical information between organisms. *Trends Ecol. Evol.* **22**, 374–379.

Mackenzie, N. C., Brito, M., Reyes, A. E., and Allende, M. L. (2004). Cloning, expression pattern and essentiality of the high-affinity copper transporter 1 (ctr1) gene in zebrafish. *Gene* **328**, 113–120.

Maret, W. (2010). Metalloproteomics, metalloproteomes, and the annotation of metalloproteins. *Metallomics* **2**, 117–125.

Markich, S. J., and Jeffree, R. A. (1994). Absorption of divalent trace metals as analogues of calcium by Australian freshwater bivalves: an explanation of how water hardness reduces metal toxicity. *Aquat. Toxicol.* **29**, 257–290.

Martell, A. E. and Smith, R. M. (2004). *NIST standard reference database 46 version 8.0.* Database software developed by R. J. Motekaitis, Gaithersburg, MD.

Martinez, R. E., Ferris, F. G., and Pedersen, K. (2004). Cadmium complexation by bacteriogenic iron oxides from a subterranean environment. *J. Colloid Interface Sci.* **275**, 82–89.

Martinez-Alvarez, R. M., Morales, A. E., and Sanz, A. (2005). Antioxidant defenses in fish: biotic and abiotic factors. *Rev. Fish Biol. Fish.* **15**, 75–88.

Mason, A. Z., and Jenkins, K. D. (1995). Metal detoxification in aquatic organisms. In: *Metal Speciation and Bioavailability in Aquatic Systems* (A. Tessier and D. Turner, eds), pp. 479–608. John Wiley and Sons, Chichester.

Mathews, T., and Fisher, N. S. (2008a). Trophic transfer of seven trace metals in a four-step marine food chain. *Mar. Ecol. Prog. Ser.* **367**, 23–33.

Mathews, T., and Fisher, N. S. (2008b). Evaluating the trophic transfer of cadmium, polonium, and methylmercury in an estuarine food chain. *Environ. Toxicol. Chem.* **27**, 1093–1101.

Matsuo, A. Y., Wood, C. M., and Val, A. L. (2005). Effects of copper and cadmium on ion transport and gill metal binding in the Amazonian teleost tambaqui (*Colossoma macropomum*) in extremely soft water. *Aquat. Toxicol.* **74**, 351–364.

Matz, C. J. (2008). *The Effects of Cadmium on the Olfactory System of Larval Zebrafish.* PhD Thesis, University of Saskatchewan.

Matz, C. J., and Krone, P. H. (2007). Cell death, stress-responsive transgene activation, and deficits in the olfactory system of larval zebrafish following cadmium exposure. *Environ. Sci. Technol.* **41**, 5143–5148.

McCarty, L. S., and Mackay, D. (1993). Enhancing ecotoxicological modeling and assessment. Body residues and modes of toxic action. *Environ. Sci. Technol.* **27**, 1718–1728.

McDonald, D. G., and Wood, C. M. (1993). Branchial mechanisms of acclimation to metals in freshwater fish. In: *Fish Ecophysiology* (J. C. Rankin and F. B. Jensen, eds), pp. 297–321, London: Chapman.

McGeer, J. C., Szebedinszky, C., McDonald, D. G., and Wood, C. M. (2000a). Effects of chronic sublethal exposure to waterborne Cu, Cd or Zn in rainbow trout. 1: Iono-regulatory disturbance and metabolic costs. *Aquat. Toxicol.* **50**, 231–243.

McGeer, J. C., Szebedinszky, C., McDonald, D. G., and Wood, C. M. (2000b). Effects of chronic sublethal exposure to waterborne Cu, Cd or Zn in rainbow trout. 2: Tissue specific metal accumulation. *Aquat. Toxicol.* **50**, 245–256.

McGeer, J. C., Brix, K. V., Skeaff, J. M., DeForest, D. K., Brigham, S. I., Adams, W. J., and Green, A. (2003). Inverse relationship between bioconcentration factor and exposure concentration for metals: implications for hazard assessment of metals in the aquatic environment. *Environ. Toxicol. Chem.* **22**, 1017–1037.

McGeer, J. C., Nadella, S., Alsop, D. H., Hollis, L., Taylor, L. N., McDonald, D. G., and Wood, C. M. (2007). Influence of acclimation and cross-acclimation of metals on acute Cd toxicity and Cd uptake and distribution in rainbow trout (*Oncorhynchus mykiss*). *Aquat. Toxicol.* **84**, 190–197.

McNicol, R. E., and Scherer, E. (1991). Behavioural responses of lake whitefish (*Coregonus clupeaformis*) to cadmium during preference–avoidance testing. *Environ. Toxicol. Chem.* **10**, 225–234.

Metcalf, V. J., and Gemmell, N. J. (2005). Fatty acid transport in cartilaginous fish: absence of albumin and possible utilization of lipoproteins. *Fish Physiol. Biochem.* **31**, 55–64.

Michelutti, N., Keatley, B. E., Brimble, S., Blais, J. M., Liu, H., Douglas, M. S. V., Mallory, M. L., Macdonald, R. W., and Smol, J. P. (2009). Seabird-driven shifts in Arctic pond ecosystems. *Proc. R. Soc. B* **276**, 591–596.

Milne, C. J., Kinniburgh, D. G., van Riemsdijk, W. H., and Tipping, E. (2003). Generic NICA–Donnan model parameters for metal–ion binding by humic substances. *Environ. Sci. Technol.* **37**, 958–971.

Moore, J. N., and Luoma, S. N. (1990). Hazardous wastes from large-scale metal extraction. A case study. *Environ. Sci. Technol.* **24**, 1278–1285.

Mortvedt, J. J., and Osborn, G. (1982). Studies on the chemical form of cadmium contaminants in phosphate fertilizers. *Soil Sci.* **134**, 185–192.

Murada, A., and Houston, A. H. (1988). Leucocytes and leucopoietic capacity in goldfish, *Carassius auratus*, exposed to sublethal levels of cadmium. *Aquat. Toxicol.* **13**, 141–154.

Muramoto, S. (1981). Vertebral column damage and decrease of calcium concentrations in fish exposed experimentally to cadmium. *Environ. Pollut. Ser. A* **24**, 125–133.

Nadella, S. R., Grosell, M., and Wood, C. M. (2006). Physical characterization of high-affinity gastrointestinal Cu transport in vitro in freshwater rainbow trout *Oncorhynchus mykiss*. *J. Comp. Physiol. B* **176**, 793–806.

Nadella, S. R., Grosell, M., and Wood, C. M. (2007). Mechanisms of dietary Cu uptake in freshwater rainbow trout: evidence for Na-assisted Cu transport and a specific metal carrier in the intestine. *J. Comp. Physiol. B* **177**, 433–446.

Nadella, S. R., Hung, C. C., and Wood, C. M. (2011). Mechanistic characterization of gastric copper transport in rainbow trout. *J. Comp. Physiol. B* **181**, 27–41.

Nesatyy, V. J., Ammann, A. A., Rutishauser, B. V., and Suter, M. J. F. (2006). Effect of cadmium on the interaction of 17 beta-estradiol with the rainbow trout estrogen receptor. *Environ. Sci. Technol.* **40**, 1358–1363.

Ng, T. Y. T., and Wood, C. M. (2008). Trophic transfer and dietary toxicity of Cd from the oligochaete to the rainbow trout. *Aquat. Toxicol.* **87**, 47–59.

Ng, T. Y. T., Klinck, J. S., and Wood, C. M. (2009). Does dietary Ca protect against toxicity of a low dietborne Cd exposure to the rainbow trout? *Aquat. Toxicol.* **91**, 75–86.

Niyogi, S., and Wood, C. M. (2004). Kinetic analyses of waterborne Ca and Cd transport and their interactions in the gills of rainbow trout (*Oncorhynchus mykiss*) and yellow perch (*Perca flavescens*), two species differing greatly in acute waterborne Cd sensitivity. *J. Comp. Physiol.* B **174**, 243–253.

Niyogi, S., Couture, P., Pyle, G., McDonald, D. G., and Wood, C. M. (2004). Acute cadmium biotic ligand model characteristics of laboratory-reared and wild yellow perch (*Perca flavescens*) relative to rainbow trout (*Oncorhynchus mykiss*). *Can. J. Fish. Aquat. Sci.* **61**, 942–953.

Niyogi, S., Kent, R., and Wood, C. M. (2008). Effects of water chemistry variables on gill binding and acute toxicity of cadmium in rainbow trout (*Oncorhynchus mykiss*): a biotic ligand model (BLM) approach. *Comp. Biochem. Physiol.* C **148**, 305–314.

Ochi, T., Otsuka, F., Takahashi, K., and Ohsawa, M. (1988). Glutathione and metallothioneins as cellular defense against cadmium toxicity in cultured Chinese hamster cells. *Chem. Biol. Interact.* **65**, 1–14.

Ojo, A. A., and Wood, C. M. (2007). In vitro analysis of the bioavailability of six metals via the gastro-intestinal tract of the rainbow trout (*Oncorhynchus mykiss*). *Aquat. Toxicol.* **83**, 10–23.

Ojo, A. A., and Wood, C. M. (2008). In vitro characterization of cadmium and zinc uptake via the gastro-intestinal tract of the rainbow trout (*Oncorhynchus mykiss*): interactive effects and the influence of calcium. *Aquat. Toxicol.* **89**, 55–64.

Olsson, P. E., and Hogstrand, C. (1987). Subcellular distribution and binding of cadmium to metallothionein in tissues of rainbow trout after exposure to [109]Cd in water. *Environ. Toxicol. Chem.* **6**, 867–874.

Olsson, P. E., Kling, P., Petterson, C., and Silversand, C. (1995). Interaction of cadmium and oestradiol-17 beta on metallothionein and vitellogenin synthesis in rainbow trout (*Oncorhynchus mykiss*). *Biochem J* **307**, 197–203.

Padmini, E., and Rani, M. U. (2009). Evaluation of oxidative stress biomarkers in hepatocytes of grey mullet inhabiting natural and polluted estuaries. *Sci. Total Environ.* **407**, 4533–4541.

Pagenkopf, G. K. (1983). Gill surface interaction model for trace-metal toxicity to fishes: role of complexation, pH, and water chemistry. *Environ. Sci. Technol.* **17**, 342–346.

Palmiter, R. D. (1994). Regulation of metallothionein genes by heavy metals appears to be mediated by a zinc-sensitive inhibitor that interacts with a constitutively active transcription factor, MTF-1. *Proc. Natl. Acad. Sci. U.S.A* **91**, 1219–1223.

Pan, J., Plant, J. A., Voulvoulis, N., Oates, C. J., and Ihlenfeld, C. (2010). Cadmium levels in Europe: implications for human health. *Environ. Geochem. Health* **32**, 1–12.

Pan, T. C., Liao, B. K., Huang, C. J., Lin, L. Y., and Hwang, P. P. (2005). Epithelial Ca^{2+} channel expression and Ca^{2+} uptake in developing zebrafish. *Am. J. Physiol. Regul. Integr. Comp. Physiol.* **289**, 1202–1211.

Panagapko, D. (2007). *Mineral and Metal Commodity Reviews: Cadmium.* Natural Resources Canada. http://www.nrcan.gc.ca/smm-mms/busi-indu/cmy-amc/content/2007/15.pdf

Pearson, R. G. (ed.) (1973). *Hard and Soft Acids and Bases.* Dowden Hutchinson and Ross, Stroudsburg, PA.

Pelgrom, S. M. G. J., Lamers, L. P. M., Garritsen, J. A. M., Pels, B. M., Lock, R. A. C., Balm, P. H. M., and Wendelaar Bonga, S. E. (1994). Interactions between copper and cadmium during single and combined exposure in juvenile tilapia *Oreochromis mossambicus*: influence of feeding condition on whole body metal accumulation and the effect of the metals on tissue water and ion content. *Aquat. Toxicol.* **30**, 117–135.

Pickering, Q. H., and Gast, M. H. (1972). Acute and chronic toxicity of cadmium to the fathead minnow (*Pimephales promelas*). *J. Fish. Res. Bd Can.* **29**, 1099–1106.

Pierron, F., Bourret, V., St-Cyr, J., Campbell, P. G. C., Bernatchez, L., and Couture, P. (2009). Transciptional responses to environmental metal exposure in wild yellow perch (*Perca flavescens*) collected in lakes with differing environmental metal concentrations (Cd, Cu, Ni). *Ecotoxicology* **18**, 620–631.

Playle, R. C. (2004). Using multiple metal–gill binding models and the toxic unit concept to help reconcile multiple-metal toxicity results. *Aquat. Toxicol.* **67**, 359–370.

Playle, R. C., Dixon, D. G., and Burnison, K. (1993a). Copper and cadmium binding to fish gills: modification by dissolved organic carbon and synthetic ligands. *Can. J. Fish. Aquat. Sci.* **50**, 2667–2677.

Playle, R. C., Dixon, D. G., and Burnison, K. (1993b). Copper and cadmium binding to fish gills: estimates of metal–gill stability constants and modelling of metal accumulation. *Can. J. Fish. Aquat. Sci.* **50**, 2678–2687.

Porter, E. L., Kent, R. A., Andersen, D. E., Keenleyside, K. A., Milne, D., Cureton, P., Smith, S. L., Drouillard, K. G., and MacDonald, D. D. (1995). Development of proposed Canadian Environmental Quality Guidelines for cadmium. *J. Geochem. Explor.* **52**, 205–219.

Pratap, H. B., Fu, H., Lock, R. A. C., and Wendelaar Bonga, S. E. (1989). Effect of waterborne and dietary cadmium in relation to water calcium levels. *Arch. Environ. Contam. Toxicol.* **18**, 568–575.

Price, N. M., and Morel, F. M. M. (1990). Cadmium and cobalt substitution for zinc in a marine diatom. *Nature* **344**, 658–660.

Qiu, A., and Hogstrand, C. (2004). Functional characterisation and genomic analysis of an epithelial calcium channel (ECaC) from pufferfish, *Fugu rubripes*. *Gene* **342**, 113–123.

Qiu, A., Shayeghi, M., and Hogstrand, C. (2005). Molecular cloning and functional characterization of a high-affinity zinc importer (DrZIP1) from zebrafish (*Danio rerio*). *Biochem. J* **388**, 745–754.

Rainbow, P. S. (2002). Trace metal concentrations in aquatic invertebrates: why and so what? *Environ. Pollut.* **120**, 497–507.

Reid, S. D., and McDonald, D. G. (1988). Effects of cadmium, copper, and low pH on ion fluxes in the rainbow trout. Salmo gairdneri. *Can. J. Fish. Aquat. Sci.* **45**, 244–253.

Reynders, H., Van Campenhout, K., Bervoets, L., De Coen, W. M., and Blust, R. (2006a). Dynamics of cadmium accumulation and effects in common carp (*Cyprinus carpio*) during simultaneous exposure to water and food (*Tubifex tubifex*). *Environ. Toxicol. Chem.* **25**, 1558–1567.

Reynders, H., van der Ven, K., Moens, L. N., van Remortel, P., De Coen, W. M., and Blust, R. (2006b). Patterns of gene expression in carp liver after exposure to a mixture of waterborne and dietary cadmium using a custom-made microarray. *Aquat. Toxicol.* **80**, 180–193.

Ricard, A. C., Daniel, C., Anderson., P., and Hontela, A. (1998). Effects of subchronic exposure to cadmium chloride on endocrine and metabolic functions in rainbow trout *Oncorhynchus mykiss*. *Arch. Environ. Contam. Toxicol.* **34**, 377–381.

Roch, M., and Maly, E. J. (1979). Relationship of cadmium-induced hypocalcemia with mortality in rainbow trout (*Salmo gairdneri*) and the influence of temperature on toxicity. *J. Fish. Res. Bd Can.* **36**, 1297–1303.

Roesijadi, G. (1992). Metallothioneins in metal regulation and toxicity in aquatic animals. *Aquat. Toxicol.* **22**, 81–114.

Roesijadi, G. (1994). Metallothionein induction as a measure of response to metal exposure in aquatic animals. *Environ. Health Perspect.* **102**, 91–95.

Roesijadi, G. (1996). Metallothionein and its role in toxic metal regulation. *Comp. Biochem. Physiol. C* **113**, 117–123.

Rogers, J. T., and Wood, C. M. (2004). Characterization of branchial lead–calcium interaction in the freshwater rainbow trout *Oncorhynchus mykiss. J. Exp. Biol.* **207**, 813–825.

Rogers, J. T., Patel, M., Gilmour, K. M., and Wood, C. M. (2005). Mechanisms behind Pb-induced disruption of Na^+ and Cl^- balance in rainbow trout (*Oncorhynchus mykiss*). *Am. J. Physiol. Regul. Integr. Comp. Physiol.* **289**, 463–472.

Rombough, P. J., and Garside, E. T. (1984). Disturbed ion balance in alevins of Atlantic salmon *Salmo salar* chronically exposed to sublethal concentrations of cadmium. *Can. J. Zool.* **62**, 1443–1450.

Santore, R. C., Mathew, R., Paquin, P. R., and Di Toro, D. (2002). Application of the biotic ligand model to predicting zinc toxicity to rainbow trout, fathead minnow, and *Daphnia magna. Comp. Biochem. Physiol.* **133**, 271–285.

Scherer, E., McNicol, R. E., and Evans, R. E. (1997). Impairment of lake trout foraging by chronic exposure to cadmium: a black-box experiment. *Aquat. Toxicol.* **37**, 1–7.

Schoenmakers, T. J. M., Klaren, P. H. M., Flik, G., Lock, R. A. C., Pang, P. K. T., and Wendelaar Bonga, S. E. (1992). Actions of cadmium on basolateral plasma membrane proteins involved in calcium uptake by fish intestine. *J. Membr. Biol.* **127**, 161–172.

Scott, B. J., and Bradwell, A. R. (1983). Identification of the serum binding proteins for iron, zinc, cadmium, nickel, and calcium. *Clin. Chem.* **29**, 629–633.

Scott, G. R., Sloman, K. A., Rouleau, C., and Wood, C. M. (2003). Cadmium disrupts behavioural and physiological responses to alarm substance in juvenile rainbow trout (*Oncorhynchus mykiss*). *J. Exp. Biol.* **206**, 1779–1790.

Scudiero, R., Temussi, P. A., and Parisi, E. (2005). Fish and mammalian metallothioneins: a comparative study. *Gene* **345**, 21–26.

Sellin, M. K., Eidem, T. M., and Kolok, A. S. (2007). Cadmium exposure in fathead minnows: are there sex-specific differences in mortality, reproductive success, and Cd accumulation? *Arch. Environ. Contam. Toxicol.* **52**, 535–540.

Shahsavarani, A., McNeill, B., Galvez, F., Wood, C. M., Goss, G. G., Hwang, P. P., and Perry, S. F. (2006). Characterization of a branchial epithelial calcium channel (ECaC) in freshwater rainbow trout (*Oncorhynchus mykiss*). *J. Exp. Biol.* **209**, 1928–1943.

Sheader, D. L., Williams, T. D., Lyons, B. P., and Chipman, J. K. (2006). Oxidative stress response of European flounder (*Platichthys flesus*) to cadmium determined by a custom cDNA microarray. *Mar. Environ. Res.* **62**, 33–44.

Shears, M. A., and Fletcher, G. L. (1983). Regulation of Zn^{2+} uptake from the gastrointestinal tract of a marine teleost, the winter flounder (*Pseudopleuronectes americanus*). *Can. J. Fish. Aquat. Sci.* **40**, 197–205.

Skeaff, J. M., Dubreuil, A. A., and Brigham, S. I. (2002). The concept of persistence as applied to metals for aquatic hazard identification. *Environ. Toxicol. Chem.* **21**, 2581–2590.

Sloman, K. A. (2007). Effects of trace metals on salmonid fish: the role of social hierarchies. *Appl. Anim. Behav. Sci.* **104**, 326–345.

Sloman, K. A., Montpetit, C. J., and Gilmour, K. M. (2002). Modulation of catecholamine release and cortisol secretion by social interactions in the rainbow trout. Oncorhynchus mykiss. *Gen. Comp. Endocrinol.* **127**, 136–146.

Sloman, K. A., Baker, D. W., Ho, C. G., McDonald, D. G., and Wood, C. M. (2003a). The effects of trace metal exposure on agonistic encounters in juvenile rainbow trout. Oncorhynchus mykiss. *Aquat. Toxicol.* **63**, 187–196.

Sloman, K. A., Scott, G. R., Diao, Z., Rouleau, C., Wood, C. M., and McDonald, D. G. (2003b). Cadmium affects the social behaviour of rainbow trout. Oncorhynchus mykiss. *Aquat. Toxicol.* **65**, 171–185.

Smith, D. S., Bell, R. A., and Kramer, J. R. (2002). Metal speciation in natural waters with emphasis on reduced sulfur groups as strong metal binding sites. *Comp. Biochem. Physiol. C* **133**, 65–74.

Song, Y., Swedlund, P. J., Singal, N., and Swift, S. (2009). Cadmium (II) speciation in complex aquatic systems: a study with ferrihydrite, bacteria and an organic ligand. *Environ. Sci. Technol.* **43**, 7430–7436.

Sorensen, E. M. B. (1991). Cadmium. In: *Metal Poisoning in Fish*, pp. 175–234. Boca Raton, FL: CRC Press.

Spry, D. J., and Wiener, J. G. (1991). Metal bioavailability and toxicity to fish in low-alkalinity lakes: a critical review. *Environ. Pollut.* **71**, 243–304.

Stephenson, M., Bendell Young, L., Bird, G. A., Brunskill, G. J., Curtis, P. J., Fairchild, W. L., Holoka, M. H., Hunt, R. V., Lawrence, S. G., Motycka, M. F., Schwartz, W. J., Turner, M. A., and Wilkinson, P. (1996). Sedimentation of experimentally added cadmium and Cd-109 in Lake 382, Experimental Lakes Area, Canada. *Can. J. Fish. Aquat. Sci.* **53**, 1888–1902.

Stohs, S. J., and Bagchi, D. (1995). Oxidative mechanisms in the toxicity of metal ions. *Free Rad. Biol. Med.* **18**, 321–336.

Stohs, S. J., Bagchi, D., Hassoun, E., and Bagchi, M. (2000). Oxidative mechanisms in the toxicity of chromium and cadmium ions. *J. Environ. Pathol. Toxicol. Oncol.* **19**, 201–213.

Suedel, B. C., Boraczek, J. A., Peddicord, R. K., Clifford, P. A., and Dillon, T. M. (1994). Trophic transfer and biomagnification potential of contaminants in aquatic ecosystems. *Rev. Environ. Contam. Toxicol.* **136**, 22–89.

Sullivan, J. F., Atchison, G. J., Kolar, D. J., and McIntosh, A. W. (1978). Changes in the predator–prey behavior of fathead minnows (*Pimephales promelas*) and largemouth bass (*Micropterus salmoides*) caused by cadmium. *J. Fish. Res. Bd Can.* **35**, 446–451.

Szebedinszky, C. S., McGeer, J. C., McDonald, D. G., and Wood, C. M. (2001). Effects of chronic Cd exposure via the diet or water on internal organ-specific distribution and subsequent gill Cd uptake kinetics in juvenile rainbow trout (*Oncorhynchus mykiss*). *Environ. Toxicol. Chem.* **20**, 597–607.

Tallkvist, J., Persson, E., Henriksson, J., and Tjälve, H. (2002). Cadmium–metallothionein interactions in the olfactory pathways of rats and pikes. *Toxicol. Sci.* **67**, 108–113.

Tan, Q.-G., and Wang, W.-X. (2009). The influence of ambient and body calcium on cadmium and zinc accumulation in *Daphnia magna. Environ. Toxicol. Chem.* **27**, 1605–1613.

Tanaka, K., Min, K.-S., Onosaka, S., Fukuhara, C., and Ueda, M. (1985). The origin of metallothionein in red blood cells. *Toxicol. Appl. Pharmacol.* **78**, 63–68.

Thomann, R. V., Shkreli, F., and Harrison, S. (1997). A pharmacokinetic model of cadmium in rainbow trout. *Environ. Toxicol. Chem.* **16**, 2268–2274.

Thomas, P., and Wofford, H. W. (1993). Effects of cadmium and Aroclor 1254 on lipid peroxidation, glutathione peroxidase activity, and selected antioxidants in Atlantic croaker tissues. *Aquat. Toxicol.* **27**, 159–177.

Thophon, S., Kruatrachue, M., Upatham, E. S., Pokethitiyook, P., Sapaphong, S., and Jaritkhuan, S. (2003). Histopathological alterations of white seabass, *Lates calcarifer*, in acute and subchronic cadmium exposure. *Environ. Pollut.* **121**, 307–320.

Thornton, I. (1995). *Metals in the Global Environment: Facts and Misconceptions*. International Council on Metals and the Environment (ICME), Ottawa, http://www.icmm.com/library

Tipping, E. (1994). WHAM – a chemical-equilibrium model and computer code for waters, sediments, and soils incorporating a discrete site electrostatic model of ion-binding by humic acid. *Comput. Geosci.* **20**, 973–1023.

Tjalve, H., and Henriksson, J. (1999). Uptake of metals in the brain via olfactory pathways. *Neurotoxicology* **20**, 181–196.

Tjalve, H., Gottofrey, J., and Bjorklund, I. (1986). Tissue disposition of ^{109}Cd^{2+} in the brown trout (*Salmo trutta*) studied by autoradiography and impulse counting. *Toxicol. Environ. Chem.* **12**, 31–45.

Tort, L., Kargacin, B., Torres, P., Giralt, M., and Hidalgo, J. (1996). The effect of cadmium exposure and stress on plasma cortisol, metallothionein levels and oxidative status in rainbow trout (*Oncorhynchus mykiss*) liver. *Comp. Biochem. Physiol. C* **114**, 29–34.

UNEP (2008). Draft Final Review of Scientific Information on Cadmium. Appendix: Overview of existing and future national actions, including legislation, relevant to cadmium. *November 2008 Version.* United Nations Environment Programme, Geneva.

Unsworth, E. R., Warnken, K. W., Zhang, H., Davison, W., Black, F., Buffle, J., Cao, H., Cleven, R., Galceran, J., Gunkel, P., Kalis, E., Kistler, D., van Leeuwen, H. P., Martin, M., Noël, S., Nur, Y., Odzak, N., Puy, J., van Riemsdijk, W., Sigg, L., Temminghoff, E., Tercier-Waeber, M., Toepperwien, S., Town, R. M., Weng, L., and Xue, H. (2006). Model predictions of metal speciation in freshwaters compared to measurements by in situ techniques. *Environ. Sci. Technol.* **40**, 1942–1949.

USEPA (2001). *Update of Ambient Water Quality Criteria for Cadmium.* United States Environmental Protection Agency, Washington, DC, EPA-822-R-01-001

Valavanidis, A., Vlahogianni, T., Dassenakis, M., and Scoullos, M. (2006). Molecular biomarkers of oxidative stress in aquatic organisms in relation to toxic environmental pollutants. *Ecotoxicol. Environ. Saf.* **64**, 178–189.

Verbost, P. M., Flik, G., Lock, R. A., and Wendelaar Bonga, S. E. (1987). Cadmium inhibition of Ca^{2+} uptake in rainbow trout gills. *Am. J. Physiol.* **253**, 216–221.

Verbost, P. M., Flik, G., Lock, R. A. C., and Wendelaar Bonga, S. E. (1988). Cadmium inhibits plasma-membrane calcium-transport. *J. Membr. Biol.* **102**, 97–104.

Verbost, P. M., Van Rooij, J., Flik, G., Lock, R. A. C., and Wendelaar Bonga, S. E. (1989). The movement of cadmium through freshwater trout branchial epithelium and its interference with calcium transport. *J. Exp. Biol.* **145**, 185–197.

Vetillard, A., and Bailache, T. (2005). Cadmium: an endocrine disrupter that affects gene expression in the liver and brain of juvenile rainbow trout. *Biol. Reprod.* **72**, 119–126.

Vijver, M. G., Van Gestel, C. A. M., Lanno, R. P., Van Straalen, N. M., and Peijnenburg, W. J. G. M. (2004). Internal metal sequestration and its ecotoxicological relevance: a review. *Environ. Sci. Technol.* **38**, 4705–4712.

Volpato, G. L., Castro, A. L. S., Gonçalves-de-Freitas, E., Giaquinto, P. C., Fernandes-de-Castilho, M., Pereira-da-Silva, E., and Jordão, L. C. (2006). Comunicação química em peixes. In: *Tópicos Especiais em Biologia Aquática e Aqüicultura* (J. E. P. Cyprino and E. C. Urbinati, eds), pp. 15–52. Sociedade Brasileira de Aqüicultura e Biologia Aquática, Jaboticabal, São Paulo.

Waldron, K. J., Rutherford, J. C., Ford, D., and Robinson, N. J. (2009). Metalloproteins and metal sensing. *Nature* **460**, 823–830.

Walker, P. A., Kille, P., Hurley, A., Bury, N. R., and Hogstrand, C. (2008). An *in vitro* method to assess toxicity of waterborne metals to fish. *Toxicol. Appl. Pharmacol.* **230**, 67–77.

Wang, W.-X. (2002). Interactions of trace metals and different marine food chains. *Mar. Ecol. Prog. Ser.* **243**, 295–309.

Wang, W. X., and Rainbow, P. S. (2006). Subcellular partitioning and the prediction of cadmium toxicity to aquatic organisms. *Environ. Chem.* **3**, 395–399.

Wang, W.-X., and Rainbow, P. S. (2010). Significance of metallothioneins in metal accumulation kinetics in marine animals. *Comp. Biochem. Physiol.* C **152**, 1–8.

Wang, Y., Fang, J., Leonard, S. S., and Rao, K. M. K. (2004). Cadmium inhibits the electron transfer chain and induces reactive oxygen species. *Free Rad. Biol. Med.* **36**, 1434–1443.

Wangsongsak, A., Utarnpongsa, S., Kruatrachue, M., Ponglikitmongkol, M., Pokethitiyook, P., and Sumranwanich, T. (2007). Alterations of organ histopathologies and metallothionein mRNA expression in silver barb, *Puntius gonionotus* during subchronic cadmium exposure. *J. Environ. Sci.* **19**, 1341–1348.

WHO (1992). *Environmental Health Criteria 134 – Cadmium International Programme on Chemical Safety (IPCS) Monograph.* World Health Organization, Geneva.

Wicklund-Glynn, A. W. (1996). The concentration dependency of branchial intracellular cadmium distribution and influx in the zebrafish (*Brachydanio rerio*). *Aquat. Toxicol.* **35**, 47–58.

Wicklund Glynn, A. (2001). The influence of zinc on apical uptake of cadmium in the gills and cadmium influx to the circulatory system in zebrafish (*Danio rerio*). *Comp. Biochem. Physiol.* C **128**, 165–172.

Wicklund Glynn, A., Norrgren, L., and Mussener, A. (1994). Differences in uptake of inorganic mercury and cadmium in the gills of the zebrafish, *Brachydanio rerio*. *Aquat. Toxicol.* **30**, 13–26.

Williams, T. D., Diab, A. M., George, S. G., Godfrey, R. E., Sabine, V., Conesa, A., Minchin, S. D., Watts, P. C., and Chipman, J. K. (2006). Development of the GENIPOL European flounder (*Platichthys flesus*) microarray and determination of temporal transcriptional responses to cadmium at low dose. *Environ. Sci. Technol.* **40**, 6479–6488.

Wood, C. M. (2001). Toxic responses of the gill. In: Target Organ Toxicity in Marine and Freshwater Teleosts, *Vol. 1.* Organs (D Schlenk and W. H Benson, eds), pp. 1–89. Taylor and Francis, New York.

Wood, C. M., Franklin, N. M., and Niyogi, S. (2006). The protective role of dietary calcium against cadmium uptake and toxicity in freshwater fish: an important role for the stomach. *Environ. Chem.* **3**, 389–394.

Xu, Y., Feng, L., Jeffrey, P. D., Shi, Y., and Morel, F. M. M. (2008). Structure and metal exchange in the cadmium carbonic anhydrase of marine diatoms. *Nature* **452**, 56–61.

Zalups, R. K., and Ahmad, S. (2003). Molecular handling of cadmium in transporting epithelia. *Toxicol. Appl. Pharmacol.* **186**, 163–188.

Zelikoff, J. T., Bowser, D., Squibb, K. S., and Frenkel, K. (1995). Immunotoxicity of low level cadmium exposure in fish: an alternative animal model for immunotoxicological studies. *J. Toxicol. Environ. Health* **45**, 235–248.

Zhu, J. Y., Huang, H. Q., Bao, X. D., Lin, Q. M., and Cai, Z. (2006). Acute toxicity profile of cadmium revealed by proteomics in brain tissue of *Paralichthys olivaceus*: potential role of transferrin in cadmium toxicity. *Aquat. Toxicol.* **78**, 127–135.

Zohouri, M. A., Pyle, G. G., and Wood, C. M. (2001). Dietary Ca inhibits waterborne Cd uptake in Cd-exposed rainbow trout. Oncorhynchus mykiss. *Comp. Biochem. Physiol.* C **130**, 347–356.

4

LEAD

EDWARD M. MAGER

Homeostasis and Toxicology of Non-Essential Metals: Volume 31B
FISH PHYSIOLOGY

Among the non-essential metals, few have rivaled the significance of lead (Pb) within human history in terms of both utility and toxicity. Much of this can be attributed to the use of Pb in plumbing applications, lead-based paints, and leaded gasoline. Today, the primary use for Pb is in the production of batteries and most concern for Pb entering aquatic environments is from point-source discharges related to Pb mining and industrial processing. The bioavailability of Pb to aquatic organisms will depend largely on the pH, alkalinity, hardness, and natural organic matter content of the receiving water. In fish, acute Pb toxicity is due to respiratory asphyxiation under extreme concentrations and the disruption of ionoregulatory homeostasis under more environmentally relevant concentrations. Chronic effects are similar to those in humans, involving primarily hematological and neurological dysfunction. Lead is taken up and elicits its effects by substituting for calcium and potentially other essential divalent cations, such as iron and zinc. Accordingly, Pb is predominantly found within the calcified hard tissues of the skeleton and scales, but also concentrates to a large extent within the blood, gill, and kidney. While great strides have been made in characterizing the uptake, accumulation, and toxicity of Pb in fish, there remains much to be learned regarding the specific transport pathways and internal handling of Pb, as well as the mechanisms of Pb excretion.

1. CHEMICAL SPECIATION IN FRESHWATER AND SEAWATER

Lead (Pb) is a class B, post-transition metal (atomic number 82) that exists predominantly in its divalent oxidative state. The most prevalent and economically important mineral of Pb is galena (PbS), with cerussite ($PbCO_3$) and anglesite ($PbSO_4$) among the most substantial of the other Pb deposits. In natural aquatic environments at or above neutral pH, Pb is readily complexed and most inorganic salts of Pb are poorly soluble with the exception of nitrate, chlorate, and chloride salts. In contrast, Pb salts tend to be quite soluble under acidic conditions. Lead can also form stable organic compounds such as tetraethyl Pb, once a common antiknock additive in gasoline. Such influences of water chemistry affect not only the chemical form (speciation) of Pb, but also its toxicity. In general, and for the purposes of this chapter, the most toxic form of a metal is assumed to be the free ionic form (Pb^{2+}), although it remains to be seen whether other species (e.g. hydroxides and/or carbonates) contribute to Pb toxicity in fish.

1.1. Freshwater

Under typical freshwater conditions, pH, alkalinity and the concentration and quality of natural organic matter (NOM) will represent the parameters of greatest importance for Pb speciation. As shown in Table 4.1, changes in pH and alkalinity have a profound effect on the relative contributions of the free Pb^{2+} ion, Pb carbonate ($PbCO_3^2$), and Pb monohydroxide ($PbOH^+$) species. This occurs in large part because, as for other divalent metals, Pb complexes strongly with CO_3^{2-} and OH^- ions. Accordingly, in waters of high pH and alkalinity the Pb carbonato and hydroxo species will dominate, while in waters of low pH and alkalinity a far greater percentage of ionic Pb^{2+} will prevail. The transition between these two states is quite sharp, leading to a dramatic shift in the speciation and solubility of Pb as pH approaches neutrality (i.e. between pH 6.5 and 7.5). While a low pH will correspond with greater free Pb^{2+} concentrations, and therefore greater toxicity of Pb, toxicity may be mitigated to some degree by the competitive interactions between Pb^{2+} and H^+ for binding at the gill. Of course, at some point the acidity itself will contribute to toxicity. Conversely, the toxicity of Pb will tend to be much less at alkaline pH as the free Pb^{2+} ions are removed by complexation with CO_3^{2-} and OH^- ions. Although water hardness (i.e. Ca^{2+} and Mg^{2+}) may contribute a significant effect on Pb solubility in very hard conditions, pH appears to be the dominant factor in determining Pb solubility in most waters typically employed for toxicity testing. A sigmoidal relationship has been recently described characterizing the solubility of Pb as a function of pH (R. Blust,

Table 4.1

Estimated inorganic speciation of lead in conditions representative of typical natural waters at 25°C: two examples are provided for freshwater to illustrate the influences of pH and alkalinity

Major Pb species	FW (pH 6.0) (%)	FW (pH 9.0) (%)	SW (pH 8.2) (%)
Free	86	<1	3
CO_3	7	95	41
Cl	1	<1	47
OH	2	5	9
SO_4	4	<1	1

Data from Turner et al. (1981).
FW: freshwater; SW: seawater.

personal communication): log (Pb solubility) in $mg\,L^{-1} = -0.9705 + (1.316 + 0.9705)/(1 + 10\,((6.962 - pH) \times (-3.926)))$.

Indeed, the solubility of Pb, especially with regard to the kinetics of Pb precipitation, is currently receiving much attention in the field of aquatic toxicology. Recent experiments have shown that several days may be required for dissolved Pb concentrations to reach equilibrium (R. Blust, personal communication). This is an issue of particular concern during acute toxicity tests which are performed over short durations (e.g. 48–96 h) and require high concentrations of Pb. Thus, it is now recommended that total and dissolved Pb concentrations are repeatedly measured to confirm that an apparent equilibrium has been reached before exposures are initiated, and that both total and dissolved Pb concentrations measured during the exposures are reported.

Broadly characterized as a poorly defined complex mixture of particulate, colloidal, and dissolved organic carbon (DOC; e.g. humic substances), NOM putatively affords protection against metal toxicity owing to an abundance of various high-affinity ligands. In both natural waters (Taillefert et al., 2000) and experimental waters using Aldrich humic acid as a DOC surrogate (Mager et al., 2010b), NOM has been estimated to complex the vast majority of Pb at environmentally relevant concentrations. It is important to note, however, that the protective nature of DOC is dependent on its quality, with darker allochthonous sources imparting greater protection than lighter autochthonous DOC, a measure of which can be reasonably estimated by a simple spectrophotometric analysis of aromaticity (Richards et al., 2001).

Speciation effects attributed to changes in one parameter are often complicated by concurrent changes in other parameters. For example, hardness and pH often co-vary with alkalinity (as $CaCO_3$ equivalents), leading to an inherent difficulty in elucidating the relative effects of each on the speciation, and therefore toxicity, of Pb. Hardness can also influence Pb speciation through cation competition for binding both inorganic and organic ligands, although most important for toxicity is likely the competition between Ca^{2+} and Pb^{2+} for binding at the gill (see Section 5.1.1.2). While such issues may pose significant challenges to modeling efforts, the problem can be approached systematically by empirically determining the binding affinities of each of the relevant chemical species for various inorganic, organic, and biotic ligands.

In an elegant study aimed at developing a model to predict the acute toxicity of Pb to rainbow trout (*Oncorhynchus mykiss*), Macdonald et al. (2002) calculated Pb^{2+}–gill binding affinities in low ionic waters supplemented with various complexing ligands or competing cations. Their results revealed a conditional log equilibrium constant (log K) for Pb^{2+}–gill binding of 6.0, indicating a binding strength lower than that calculated for Pb binding to $(CO_3^{2-})_2$ (9.0), CO_3^{2-} (6.4), and organic matter (8.4), but higher than that for Cl_2^{2-} (2.0) and Cl^- (1.4). In addition, binding affinity at the gill was greater for Pb^{2+} (6.0) than for Ca^{2+} (4.0), Mg^{2+} (4.0), H^+ (4.0), and Na^+ (3.5). With the exception of Na^+ and Cl^-, each of these parameters correlated well with acute Pb toxicity. These findings were in good agreement with a later study by Rogers and Wood (2004), which led to a similar value (log $K_i = 6.3$) for Pb^{2+} inhibition of active Ca^{2+} uptake at the rainbow trout gill. Thus, there is clear evidence that Ca^{2+} and Pb^{2+} compete for ligands at the gill and therefore increased Ca^{2+} (hardness), as with alkalinity and DOC, will likely afford protection against acute Pb toxicity. Indeed, results from a number of toxicity studies confirm the protective influence of these parameters during acute Pb exposures to fish (Table 4.2).

Up to this point, this section has examined the influence of pH and alkalinity, DOC, and the competing effects of other cations on the speciation and toxicity of Pb. The freshwater examples in Table 4.1, however, ignore another important parameter for Pb speciation in the way of particle adsorption and deposition. Such particles may be organic (e.g. phytoplankton and biological debris) and inorganic (mineral formations) in nature. The amount of Pb adsorbed onto particles will depend largely on the Pb and particle loads, as well as the pH and hardness, with typically less adsorption occurring as pH decreases and water hardness increases (USEPA, 1979; Gao et al., 2003). While it has been reported that on average 90% of Pb in relatively unpolluted United States East Coast rivers is complexed with particulate matter (Windom et al., 1991), this percentage may be far less (e.g. ~20–30%) in those receiving anthropogenic Pb input (Muller and Sigg, 1990). Once settled, Pb complexation may shift to binding with reactive sulfides as Pb is liberated from the deposited particles in reduced anoxic regions near the sediment–water interface. Lead can also bind to hydrous iron oxide, hydrous manganese oxide, and other solid phases in oxic sediments where acid-volatile sulfides are not present.

1.2. Seawater

While the parameters in freshwater environments can vary significantly, the overall constancy of the ocean allows for a more generalized description with regard to the influence of water chemistry on Pb speciation. For the most part, the parameters influencing speciation in freshwater will also be important for Pb speciation in seawater; nonetheless, there are some specific points worth considering. Perhaps the most notable difference in seawater is that the influence of Cl^- on Pb speciation is far greater owing to the much higher concentration than that found in freshwater (Table 4.1). Secondly, with respect to DOC, Pb may be one of only two metals (the other being Cu)

Table 4.2

Representative summary of acute toxic effect levels (LC50) for waterborne Pb exposures to freshwater fish in test waters of varying composition

Species	Test duration (chronic)	LC50 (LC20) ($\mu g\ L^{-1}$) dissolved Pb	pH	Hardness ($mg\ L^{-1}$)	DOC (or TOC) ($mg\ L^{-1}$)	Reference
Oncorhynchus mykiss	96 h	1170	6.9	32	NR	Davies et al. (1976)
Oncorhynchus mykiss	96 h	1320	8.2	385	NR	Davies et al. (1976)
Oncorhynchus mykiss	96 h	1470	8.8	290	NR	Davies et al. (1976)
Oncorhynchus mykiss	96 h	1000	7.9–8.0	140	3.0	Rogers et al. (2003)
Pimephales promelas	48 h	114–610	8.1	20–30	< 1.0	Diamond et al. (1997)
Pimephales promelas	96 h (30 days)	52 (22)	7.4	19	1.2	Grosell et al. (2006)
Pimephales promelas	96 h (30 days)	174–524 (47–51)	7.2–7.3	47–218	1.2	Grosell et al. (2006)
Pimephales promelas	96 h (30 days)	372–1656 (189–1729)	7.8–7.9	21–25	2.0–10.5	Grosell et al. (2006)
Pimephales promelas	96 h	178–439	7.4–7.6	19–24	0.9–1.1	Mager et al. (2010b)
Pimephales promelas	96 h	744–1719	7.4–7.7	33–309	1.1–1.2	Mager et al. (2010b)
Pimephales promelas	96 h	608–3294	7.5–7.6	20–22	1.4–5.0	Mager et al. (2010b)
Pimephales promelas	96 h	698–816	7.5–8.3	22–23	1.1–1.2	Mager et al. (2010b)
Pimephales promelas	96 h (30 days)	162–790 (41–149)	5.4–8.3	22–30	1.2–1.6	Mager et al. (2010b)
Pimephales promelas	96 h	810	6.3	280–300	NR	Schubauer-Berigan et al. (1993)
Pimephales promelas	96 h	> 5400	7.1	280–300	NR	Schubauer-Berigan et al. (1993)
Pimephales promelas	96 h	> 5400	8.3	280–300	NR	Schubauer-Berigan et al. (1993)
Pimephales promelas	96 h	2100	7.4–7.7	44	2.0 (TOC)	Spehar and Fiandt (1986)

In cases where similar chronic Pb exposures were performed, the test durations and effect values (LC20) are included in parentheses for comparison. LC50 and LC20: concentrations at which an effect of 50% or 20% mortality is elicited, respectively; DOC: dissolved organic carbon; TOC: total organic carbon; NR: not reported.

significantly complexed by humic acid in the marine environment (Turner et al., 1981), despite the fact that, in general, humic acid will have less metal buffering capacity in seawater owing to the higher concentrations of competing cations. It has been estimated that 50–70% of Pb in the open ocean is organically bound, with inorganic compounds accounting for a vast majority of the balance (Reuer and Weiss, 2002). Finally, the process of Pb scavenging and settling in oceans is similar to that in lakes (albeit on different scales), with Pb entering from the surface, decreasing in concentration with depth due to settling of Pb-adhered particles, and ultimately complexing with reduced species near the sediment.

The speciation of Pb within estuaries is more complex owing to conditions of intermediate and changing salinity. Generally speaking, however, it is reasonable to expect that under most estuarine conditions Pb solubility and toxicity will decrease with increasing salinity, a hypothesis supported by a Pb accumulation study with the estuarine fish *Gillichthys mirabilis* (Somero et al., 1977a; see Section 5.2). In addition, estuaries are commonly used as dumping grounds for effluents and sewage treatment plant outflows. Aside from contributing metals and other pollutants to the environment, such effluents may contain high concentrations of nutrients (e.g. phosphates) that will not only influence Pb speciation, but may also contribute to environmental stress for fish by driving conditions to hypoxic or anoxic states.

2. SOURCES (NATURAL AND ANTHROPOGENIC) OF LEAD AND ECONOMIC IMPORTANCE

2.1. Sources of Lead to the Environment

Lead can enter a body of water from a variety of aquatic, atmospheric, and terrestrial routes. Both natural and anthropogenic sources contribute Pb to the environment, but the atmospheric dispersal of Pb, primarily from its use as a fuel additive, has made it a pervasive and persistent pollutant worldwide. In fact, estimates indicate that atmospheric Pb deposition has increased up to 1000-fold since prehistoric times (Renberg et al., 2000). Much of this is attributed to the > 7 million tonnes of Pb burned as a gasoline additive between 1926 and 1985 in the USA alone (Nriagu, 1990). In response to the growing awareness of the hazards of Pb pollution, the USA phased out the use of leaded gasoline, beginning in the 1970s under the Clean Air Act. This effort has been regarded as one of the great successes in environmental legislation, having dramatically improved air quality with respect to Pb (Fig. 4.1). As evident from an analysis of sediment cores

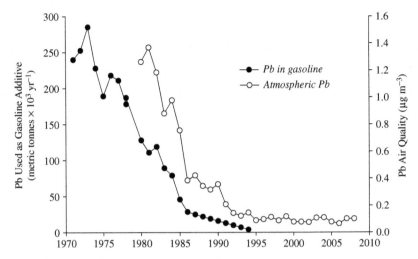

Fig. 4.1. Graph illustrating parallel trends in improved air quality in the USA and the phasing out of leaded gasoline following enactment of the Clean Air Act in 1970. Source: USEPA (1986, 2010).

from various lakes across the USA (Mahler et al., 2006), a corresponding reduction in the atmospheric input of Pb to the aquatic environment has followed. However, the use of leaded gasoline persists in many developing countries today, primarily in Africa and Asia.

Except in areas of high Pb mineralization, natural weathering contributes relatively minor amounts of Pb to aquatic environments on a local scale compared to anthropogenic sources. Total anthropogenic input of Pb to aquatic ecosystems has been estimated at 138,000 t year^{-1} (Nriagu and Pacyna, 1988), although most of this (around 100,000 t) was attributed to atmospheric fallout (Table 4.3). Metal processing and manufacturing, sewage sludge, and domestic wastewater represented the other major anthropogenic discharges (Nriagu and Pacyna, 1988). Considering the input of Pb to the atmosphere, anthropogenic emissions have been estimated at nearly 300,000 t (Nriagu and Pacyna, 1988), far exceeding the 12,000 t estimated for natural sources (Nriagu, 1989). However, owing to the decline in the use of leaded fuels, more recent estimates of mobile sources were about 30% of the 1983 amounts (OECD, 1993). Natural sources of Pb to the atmosphere include volcanoes, wild forest fires, and seasalt spray, whereas anthropogenic input arises largely from mobile sources (e.g. leaded fuel, break and engine wear, battery leaks), combustion of coal, oil, and wood, and various processes in metal production and manufacturing (Nriagu and Pacyna, 1988; Nriagu, 1989).

Table 4.3
Worldwide anthropogenic Pb input to aquatic ecosystems in 1983 (tonnes year^{-1})

Source	Range (median)
Atmospheric fallout	87,000–113,000
Metal manufacturing	2500–22,000
Dumping of sewage sludge	2900–16,000
Domestic wastewater, central	900–7200
Smelting and refining, non-ferrous metals	1000–6000
Domestic wastewater, non-central	600–4800
Chemical manufacturing	400–3000
Smelting and refining, iron and steel	1400–2800
Base metal mining and dressing	250–2500
Steam electric	240–1200
Pulp and paper manufacturing	10–900
Petroleum product manufacturing	0–120
Total anthropogenic input	97,000–180,000 (138,000)

Data from Nriagu and Pacyna (1988).

2.2. Economic Importance

Lead has a number of unique chemical and physical properties that have attracted its use for a variety of applications throughout human history. The desire for silver, along with the discovery of cupellation (a process by which Pb is separated from lead–silver alloys), ushered in the original period of major Pb production some 5000 years ago (Settle and Patterson, 1980). Demand for Pb rose sharply during the Roman Empire and again during the industrial revolution. The latter saw a dramatic rise in Pb production, with worldwide levels increasing from 749,000 t in 1900 to current estimates of nearly 4 million t produced annually (ILZSG, 2010; USGS, 2010a). Aside from its use as a fuel additive, other major applications have gained public notoriety in recent years. Because Pb is quite malleable and highly resistant to water corrosion and discoloration, it has been a natural choice in the manufacture of paints and pipes. Both of these applications have led to inadvertent human consumption. Some have even speculated that the neurotoxic effects of Pb, arising from the use of leaden pipes, food and drink containers, and the use of Pb as a dietary sweetener, may have contributed to the fall of the Roman Empire (Gilfillan, 1965).

Today, the major application for Pb by far is in the production of lead-acid batteries. The United States Geological Survey (USGS) estimated that of the nearly 1.5 million tonnes of Pb consumed in the USA in 2008, almost 90% was used in producing batteries (USGS, 2010b). The remainder was consumed in the production of ammunition (5%) and radiation shielding (3%), followed by various other applications including sheet lead, solder, and fishing weights. Approximately 60% of Pb used for industrial purposes is in the form of Pb alloys such as lead–antimony and lead–calcium, among many others (Prengaman, 2003). The amount of Pb mined worldwide in 2008 was 3.8 million tonnes, with China producing nearly 39%, followed by Australia (17%), the USA (11%), Peru (9%), and Mexico (3%) (USGS, 2010b). China also ranked highest for consumption of Pb (37%), followed by the USA (18%), with Germany and the Republic of Korea accounting for 4% each (USGS, 2010b).

3. ENVIRONMENTAL SITUATIONS OF CONCERN

The anthropogenic input of Pb to the environment can occur at virtually any stage, from its extraction and processing, to product usage and wear, and ultimately as consumer waste. While natural background concentrations of Pb have been estimated in the range of approximately $1–23 \, ng \, L^{-1}$ in surface waters (Settle and Patterson, 1980; Flegal et al., 1989; Windom et al., 1991) and $1.3 \, mg \, kg^{-1}$ in sediment (Flegal et al., 1987), much higher concentrations can be found in areas receiving anthropogenic input (e.g. several hundred $\mu g \, L^{-1}$ dissolved Pb, or $mg \, L^{-1}$ range for total Pb; see Table 4.4). In this section, two of the major sources leading to environmental situations of concern for Pb will be discussed, namely mining and smelting of Pb ores and shooting ranges.

3.1. Lead Mining and Smelting

The historical mining districts of Wales and England provide some of the oldest and most extreme examples of environmental situations of concern. Lead mining in these areas, which pre-dates the Roman occupation, peaked during the mid-nineteenth century until declining rapidly towards the late nineteenth century. Extraction and processing inefficiencies, combined with a lack of environmental policy, resulted in a trail of long-lasting and devastating impacts on the resident biota. In the most heavily polluted river, the Ystwyth (once home to a thriving fish community), nearly a century passed before fish returned to distant regions downstream of the mine (Davies, 1987). In fact, many areas throughout the region remained devoid of vegetation as late as the 1980s (Davies, 1987) and significant leaching of Pb from abandoned mine tailings in the Ystwyth catchment continue to this day, with reports of Pb concentrations as high as $130 \, g \, kg^{-1}$ and $320 \, \mu g \, L^{-1}$ in sediments and surface waters, respectively (Palumbo-Roe et al., 2009).

Within the USA, the abundant Pb deposits known as the Old and New Lead Belts of Missouri are among the most heavily exploited. Operations

Table 4.4

Lead concentrations from stream food webs in mining-disturbed areas of Missouri and the western USA (total and dissolved Pb concentrations in $\mu g\ L^{-1}$; all others are in $\mu g\ g^{-1}$ Pb, dry weight)

	Water (total Pb)	Water (dissolved Pb)	Sediments	Plant biomass/biofilm	Invertebrates	Fish
Animas River, CO, USA (Besser et al., 2001)						
Reference streams	<1.8	<0.2	6	9	0.8–2	~0.1[a]
Mining-disturbed areas	0.9–8.6	<0.1–6.9	69–1820	207–2023	9–228	~0.7–4[a]
Boulder River, MT, USA (Farag et al., 2007)						
Reference streams	0.4 (colloidal)	0.3–0.4	10–26	12	1.0–1.6	0.1–2.1[b]
Mining-disturbed areas	0.1–44 (colloidal)	0.1–2	27–1100	32–1100	3–38	0.4–1.7[b]
Coeur d'Alene River, ID, USA (Farag et al., 1998; Clark, 2002)						
Reference streams	2–20	0.01–2	10–203	18–279	9–12	<0.3–1.6[c]
Mining-disturbed areas	6–2000	2–50	253–9187	169–26,425	34–3900	15–128[c]
New Lead Belt, MO, USA (Besser et al., 2007; Brumbaugh et al., 2007)						
Reference streams	NR	<0.01–1.6	4.3–20	1.7–50	0.1–3.4	~0.2–4[d]
Mining-disturbed areas	NR	0.02–1.7	7–827	10–779	1.5–106	0.4–31[d]

[a]Liver
[b]gill, liver, whole fish
[c]gill, kidney, whole fish
[d]whole fish.
NR: not reported.

at the Old Lead Belt (OLB) ceased in 1972, yet drainage from the mining areas continues to impact area streams (Czarnezki, 1985; Schmitt et al., 1993). While much of the contamination from the OLB has also been attributed to inefficient technologies and lack of environmental regulation, the New Lead Belt to the west (also known as the Viburnum Trend) was mined with modern equipment under regulatory guidance until its depletion in the 1980s. Nevertheless, several reports have documented elevated Pb concentrations in area streams (Table 4.4) as well as Pb effects in fish (Schmitt et al., 1993; Besser et al., 2007; Schmitt et al., 2007a). Today, the expansion of potentially exploitable Pb deposits to the south is being pursued, although the implications for surrounding resources including a National Park have given rise to public concern (Besser et al., 2007; Schmitt et al., 2007a, b). Other examples where Pb mining or smelting operations have affected fish include the Spring and Neosho Rivers in Oklahoma (Schmitt et al., 2005), various watersheds of western Montana (Schmitt et al., 2002; Farag et al., 2007), areas along the Columbia River in Trail, British Columbia (Schmitt et al., 2002), the Animas River in Colorado (Besser et al., 2001) and perhaps the most extreme example, the Coeur d'Alene River Basin in Idaho (Farag et al., 1998; Farag et al., 1999), where total Pb concentrations as high as 2 mg L^{-1} have been reported (Clark, 2002) (Table 4.4).

3.2. Shooting Ranges

Although generally considered a greater threat to waterfowl, Pb entering the environment from recreational activities such as hunting, fishing, and outdoor shooting ranges (i.e. as Pb shot and fishing weights) may impact nearby surface waters and the fish therein. Of these activities, shooting ranges are likely of greatest concern owing to the accumulation of Pb shot in confined areas near the backstops. The amount of Pb deposited at shooting ranges is difficult to determine, but global estimates may reach as high as tens to perhaps hundreds of thousands of tonnes annually (Vantelon et al., 2005). Spent Pb from ranges can enter surface waters and groundwaters either through weathering at the soil surface or by direct transport to nearby streams or lakes by storm water runoff. The degree and speed of Pb seepage into the groundwater will depend largely on the type (e.g. sandy versus clay) and chemistry (e.g. pH and organic matter) of soil through which it passes, as well as the amount of rainfall received (USEPA, 2005).

The potential for Pb contamination to surface waters due to Pb shot was well illustrated by a study of the streams draining a nearby shooting range in Southwest Virginia (Craig et al., 1999). Lead concentrations were as high as 473 μg L^{-1} closest to the backstop compared to 0.5 μg L^{-1} just upstream. However, contamination appeared confined to the near vicinity of the backstop as Pb concentrations quickly fell downstream. Although effects on biota were not examined in this study, Pb accumulation and biochemical effects have been documented in fish collected from a stream draining a nearby shooting range in Norway, with total Pb concentrations ranging from 15–45 μg L^{-1} (Heier et al., 2009).

4. A SURVEY OF ACUTE AND CHRONIC AMBIENT WATER QUALITY CRITERIA IN VARIOUS JURISDICTIONS IN FRESHWATER AND SEAWATER

Table 4.5 provides a survey of ambient water quality criteria (AWQC) values for Pb in the USA, Canada, European Union (EU), Australia/New

Table 4.5
Survey of water quality criteria across various jurisdictions: acute and chronic values are provided where applicable

Jurisdiction	Reference	Acute ($\mu g\ L^{-1}$)	Chronic ($\mu g\ L^{-1}$)	Notes
Australia/	ANZECC		2	Hardness 20 mg L^{-1}
New Zealand	(2000)[a]		6.5	Hardness 50 mg L^{-1}
			37.8	Hardness 200 mg L^{-1}
			4.4	Seawater
Canada	CCME (2008)		1	Hardness 0–60 mg L^{-1}
			2	Hardness 60–120 mg L^{-1}
			4	Hardness 120–180 mg L^{-1}
			7	Hardness >180 mg L^{-1}
European Union	CEC (2006)		7.2	All surface waters
South Africa	CSIR (1996)[b]	4	0.5	Hardness 0–60 mg L^{-1}
		7	1	Hardness 60–120 mg L^{-1}
		13	2	Hardness 120–180 mg L^{-1}
		16	2.4	Hardness >180 mg L^{-1}
	CSIR (1995)		12	Seawater
United States	USEPA (1985)	10.8	0.4	Hardness 20 mg L^{-1}
		30.1	1.2	Hardness 50 mg L^{-1}
		136	5.3	Hardness 200 mg L^{-1}
		210	8.1	Seawater

Several examples are listed for jurisdictions allowing site-specific adjustments to criteria based on water hardness.
[a]The values listed represent a targeted protection of 95% of species; however, protection levels (i.e. trigger values) may be adjusted at the discretion of local authorities.
[b]Adjustments are also made to account for increasing Pb toxicity as dissolved oxygen concentrations decrease. The following correction factors are applied for the percentage of oxygen saturation: 100% = 1.0; 80% = 0.95; 60% = 0.85; 40% = 0.71.

Zealand, and South Africa, revealing fairly similar levels of protection across jurisdictions. Of those listed, only the USA and South Africa provide separate guidelines for protection against acute and chronic Pb toxicity. In most cases, site-specific adjustments to Pb AWQC may be made based on the hardness (i.e. Ca^{2+} and Mg^{2+}) of the receiving water. South Africa also provides guidance on accounting for differences in local dissolved oxygen concentrations. Although AWQC values are typically established with the goal of protecting a given percentage of resident species (e.g. 100% in Canada and 95% in the USA), in Australia and New Zealand the protection levels (referred to as "trigger values") may be adjusted at the discretion of local authorities.

While the importance of water chemistry parameters other than hardness (e.g. DOC, pH, and alkalinity) has long been recognized by regulatory agencies, it was not until the advent of the biotic ligand model (BLM) that an approach became available to potentially account for all factors that may affect metal toxicity. In essence, the BLM is a chemical equilibrium model that accounts for the mitigating effects of: (1) cations that compete with the metal for binding to a biotic ligand (e.g. gill), and (2) complexation with DOC or various inorganic species that render the metal unavailable for uptake (Paquin et al., 2002). To date, regional risk assessments in the USA and Europe have employed BLMs for metals such as Cu, Ni, and Zn. The parameterization for an acute freshwater BLM for Pb was recently completed (Mager et al. 2010b) and BLMs for other high-priority metals are currently in various stages of development. As the future appears to focus on a BLM-based approach to establishing AWQC for metals, additional efforts will be needed to develop BLMs for chronic Pb exposures as well as for Pb exposures in seawater.

5. MECHANISMS OF TOXICITY

5.1. Freshwater

5.1.1. ACUTE TOXICITY

5.1.1.1. Mucus production. As early as the 1920s, it was observed that fish exposed to lethal concentrations of Pb displayed a thin coat of mucus covering the body, a thicker "pad" of mucus within the operculum, and irregular patterns of ventilation and swimming (or lack thereof) indicative of respiratory distress (Carpenter, 1927). Following transfer of Pb-exposed fish to a weak ammonium sulfide solution, Carpenter (1927) noted a black coloration of the mucus film attributed to precipitated Pb sulfide. It was therefore suggested, and later substantiated by Westfall (1945), that the cause of acute mortality was suffocation due to the inability to clear a Pb-induced coagulation of mucus at the gill. This coagulation effect of Pb appears preventable if sufficient amounts of Ca^{2+} are initially present, although if Ca^{2+} is added subsequent to Pb exposure this effect was not observed (Jones, 1938), suggesting either very slow exchange or irreversible binding of these competing cations within the mucus.

It is important to note, however, that the concentrations of Pb to which fish were exposed in these early experiments were relatively high (e.g. 20 to > 100 mg L^{-1}), likely exceeding the solubility limit, and well above the range of concentrations typically sufficient to cause acute mortality (Table 4.2). A generalized role for mucus production and sloughing has since been proposed as a routine defensive measure against a variety of stress responses including those caused by toxicant exposure, experimental handling, and disease (see Shephard, 1994, for review). While a less severe production of mucus in response to more moderate Pb concentrations may not lead to death, it will increase the diffusive distance for gas exchange at the respiratory surface. As discussed below, this effect will add to the respiratory distress caused by other hematological mechanisms of Pb toxicity as well as other potential morphological alterations to the gill (e.g. swelling, filament clubbing) as commonly observed during metal exposures (Wood, 2001). In addition, Pb has been shown to alter the fluidity of mucus, which may in turn affect the hydrodynamic resistance, as well as ionoregulatory and gas exchange functions at the gill (Varanasi et al., 1975).

5.1.1.2. Ionoregulatory effects. Surprisingly little was known regarding the mechanism of acute toxicity of Pb at environmentally relevant concentrations until the past decade. A series of reports by Rogers and colleagues clearly details the detrimental effects of acute Pb exposure on Ca^{2+}, Na^+, and Cl^- homeostasis in rainbow trout (Rogers et al., 2003, 2005; Rogers and Wood, 2004). Their findings indicate that hypocalcemia is the primary cause of acute Pb toxicity, with effects on Na^+ and Cl^- balance representing secondary contributions (Rogers et al., 2003). Using kinetic and pharmacological approaches, it was determined that hypocalcemia occurs as a result of Ca^{2+} uptake inhibition due initially to competitive interactions with Pb at shared voltage-independent Ca^{2+} channels on the apical surface of the gill epithelium (Rogers and Wood, 2004). Once accumulated inside the cell, a secondary non-competitive inhibition to Ca^{2+} influx arises from a Pb-induced reduction in Ca^{2+}-ATPase activity in the basolateral membrane, thereby preventing transport of Ca^{2+} to the blood (Rogers and Wood, 2004). Although Ca^{2+} efflux at the gill did not contribute significantly to the observed hypocalcemia, subsequent work demonstrated increased urinary Ca^{2+}, Mg^{2+}, and glucose excretion due most likely to Pb inhibition of active tubular reabsorption by the kidney (Patel et al., 2006). Thus, it appears that hypocalcemia is the result of disrupted Ca^{2+} homeostasis due to Pb interactions at both the gill and kidney.

With respect to Na^+ and Cl^- balance, the mechanism of Pb-induced disruption was attributed in part to rapid non-competitive inhibition of branchial carbonic anhydrase (CA) activity and, similar to the Ca^{2+}-ATPase, a more slowly developing inhibition of Na^+/K^+-ATPase activity in the basolateral membrane (Rogers et al., 2005). The observed rapid inhibition of CA by Pb is consistent with the findings of a subsequent study of the inhibitory effects of Pb on two human erythrocyte CA isozymes, CA-I and CA-II, although in the case of CA-II inhibition was non-competitive (Ekinci et al., 2007). In contrast to Ca^{2+}, Na^+ efflux at the gill also contributed to net Na^+ loss during latter portions of the exposures

(24–48 h). It should be noted that the above mechanisms (i.e. hypocalcemia with secondary Na^+ and Cl^- effects) were determined from Pb exposures in hard water, and that recent evidence suggests that reduced Na^+ influx (hyponatremia) may represent the primary toxic mechanism in soft water (Birceanu et al., 2008). Finally, such ionoregulatory effects appear less significant with respect to chronic Pb toxicity. While it has been shown that Ca^{2+}, Na^+, and Cl^- homeostasis were also disrupted during the early stages of chronic Pb exposures to fathead minnows, levels eventually recovered, suggesting a potential acclimation response to sublethal Pb concentrations (Grosell et al., 2006).

5.1.2. Chronic toxicity

5.1.2.1. Growth and development. One of the most striking effects of chronic Pb exposure is the development of lordoscoliosis (spinal curvature in both frontal and sagittal planes), as shown to occur in rainbow trout exposed to 8–2300 µg L^{-1} Pb (Davies et al., 1976; Hodson et al., 1978) and brook trout (*Salvelinus fontinalis*) exposed to 119–474 µg L^{-1} Pb (Holcombe et al., 1976). Lead residues in kidney and liver of brook trout exposed to the lowest concentration (119 µg L^{-1}) were approximately 200 and 60 µg g^{-1}, respectively. Given that Pb burdens had reached an apparent equilibrium within these tissues, it was suggested that they might serve in an environmental monitoring capacity as indicators of potential scoliosis (Holcombe et al., 1976). Spinal curvature is usually preceded by a characteristic black discoloration of the caudal–peduncle region (also called black tail effect, a precursor to scoliosis), with later stages often accompanied by erosion of the caudal fin. In the early stages, the curvature is reversible if the fish is anesthetized, but calcification over time leads to permanence of the curved state (Holcombe et al., 1976). Other effects observed from these studies included hyperactivity, muscular atrophy, loss of equilibrium, hemorrhaging at the base of the caudal fin, and in severe cases, paralysis (Davies et al., 1976; Holcombe et al., 1976). In light of the known neurotoxicity of Pb (see Section 5.1.2.3) it would seem these effects are likely due to damage of motor neurons and the sympathetic nerves controlling caudal pigment cells (Davies et al., 1976). Although the mechanisms have yet to be confirmed, it is clear that the severely impaired movement imparted by these effects ultimately contributes to reduced reproduction by limiting spawning mobility (Holcombe et al., 1976).

It is interesting to note that incidence and time of onset of these effects were closely linked to the stage at which Pb exposure was initiated. In experiments with both brook trout and rainbow trout, the earlier the exposure was initiated (e.g. egg or sac fry), the earlier the onset of effects was observed and the lower the concentration of Pb that was needed to induce effects (Davies et al., 1976; Holcombe et al., 1976; Hodson et al., 1979). Hodson et al. (1982) subsequently showed that growth rate also affected timing of black-tail development, with symptoms occurring earlier in faster growing fish. A greater imbalance between growth rate and neuronal repair in faster growing fish was proposed as a potential mechanism. Consistent with the protective effects of hardness and alkalinity, onset of lordoscoliosis was delayed considerably in hard water (353 mg L^{-1} CaCO$_3$), occurring some 4–5 months later compared to fish reared in soft water (28 mg L^{-1} CaCO$_3$) (Davies et al., 1976).

Reports of growth reductions in fish attributable to Pb exposures have been inconsistent. For example, in studies demonstrating Pb-induced growth reduction in fathead minnows, the effective concentrations were typically above the 96 h LC50 value (Grosell et al., 2006; Mager et al., 2010b). With salmonids, reduced growth has been reported as both a sensitive and an insensitive response to waterborne Pb exposure. Holcombe et al. (1976) found reduced growth in the third generation of Pb exposed brook trout, with the same lowest observed effect concentration (LOEC) for growth reduction as for lordoscoliosis and mortality. From two ~60 day early life stage (ELS) Pb exposures of rainbow trout, growth and mortality LOECs were the same in the first test, but the LOEC for reduced growth was about seven times lower than that for increased mortality in the second (Mebane et al., 2008). In 30 day tests with rainbow trout, reduced growth was observed, but not increased mortality (Burden et al. 1998). In contrast, Sauter et al. (1976) found no reductions in growth after 60 day ELS exposures, nor did Davies et al. (1976) following 19 month exposures of rainbow trout. Although the mechanisms of reduced growth are not entirely clear, one of the primary reasons likely relates to reduced feeding ability (e.g. prey capture) and/or appetite owing to the putative neurological effects described below. In support of this, Pb has been shown to affect feeding ability in fathead minnows (Weber et al., 1991; Mager et al., 2010a) and killifish (*Fundulus heteroclitis*) (Weis and Weis, 1998; Weis et al., 2003). However, other mechanisms, such as increased metabolic costs and/or hormonal dysregulation, may also contribute to reduced growth.

5.1.2.2. Hematological effects. The characteristic hematological effects of Pb are anemia and the basophilic stippling of erythrocytes. The latter arises from ribosomal aggregations and may indicate impaired protein synthesis, most important of which for erythrocytes is the manufacture of hemoglobin (Hb). The anemic response is manifested in the form of microcytic and hypochromic erythrocytes, suggesting a loss of mature erythrocytes and inhibition of Hb synthesis, respectively. The effects are largely attributed to inhibition of the enzyme δ-aminolevulinic acid dehydratase (ALAD, also called porphobilinogen synthase), which catalyzes a key step in the heme synthesis pathway, condensing two δ-aminolevulinic acid (ALA) molecules to form a single porphobilinogen molecule. The mechanism of inhibition appears to stem from a dysfunctional conformation imparted by Pb displacement of Zn from cysteine (Cys) binding sites in the protein (Magyar et al., 2005). Inhibition is highly specific to Pb (Jackim, 1973; Hodson et al., 1977), a feature that, combined with the simplicity of the enzyme assay, has made ALAD a common biomarker of Pb exposure in fish. The inhibition results not only in reduced heme synthesis, but also in accumulation of ALA. Studies have linked ALA accumulation to the production of reactive oxygen species (ROS) via oxidative interactions with oxyHb (Monteiro et al., 1989) and ferritin (Oteiza et al., 1995). Hence, Pb inhibition of ALAD likely contributes to both reduced heme synthesis and ROS-mediated hemolysis. Lead can also induce the formation of free radicals such as hydrogen peroxide and singlet oxygen via lipid peroxidation events (Sanders et al., 2009), providing additional pathways by which Pb may elevate ROS and ultimately contribute to a hemolytic anemia.

Numerous studies using common hematological indices (e.g. hematocrit, Hb concentration, and red blood cell counts) have demonstrated anemia in a variety of fish following waterborne Pb exposure, including rainbow trout (Hodson et al., 1978; Johansson-Sjobeck and Larsson, 1979), goldfish (*Carassius carassius*) (Fantin et al., 1988), rosy barb (*Barbus conchonius*) (Tewari et al., 1987), catfish (*Ameiurus nebolosus*) (Dawson, 1935) tilapia (*Oreochromis hornorum* and *aureus*) (Tabche et al., 1990) (Allen, 1993), tench (*Tinca tinca*) (Shah, 2006), gourami (*Colisa fasciatus*) (Srivastava and Mishra, 1979), and grey mullet (*Mugil auratus*) (Krajnovic-Ozretic and Ozretic, 1980). However, the effects appear to vary with species, and compensatory responses attributed to erythropoiesis and splenic release of erythrocytes have been reported (Hodson et al., 1978; Johansson-Sjobeck and Larsson, 1979). The role of dietary Pb exposure with respect to anemia appears less of a concern, as exposures to concentrations as high as 500 μg L^{-1} Pb elicited no discernible change in Hb concentration collected from the blood of rainbow trout (Alves and Wood, 2006). Given the lack of consistent effects and potential for acclimation, the environmental relevance of Pb-induced anemia in fish is uncertain. Lead has been shown to reduce maximum swim velocity (Weber and Dingel, 1997) and stamina (Adams, 1975) in fish, indicating a potential impairment in oxygen carrying capacity. However, these findings could reflect effects other than, or in addition to, anemia, such as reduced oxygen uptake at the gill due to an increased diffusive distance imparted by mucus secretion or morphological modifications (hypertrophy, hyperplasia). In addition, impaired neurological function could affect swim performance by altering neuronal transmission and muscular coordination.

5.1.2.3. Neurological effects. In mammals, Pb impairs both the central and peripheral nervous systems by a variety of mechanisms involving synaptogenesis, cell differentiation, neurotransmission, and neuronal death. During development, Pb impacts cell differentiation and the timed programming of cell–cell connections, effects that ultimately lead to a dysfunctional alteration in the neuronal circuitry (Silbergeld, 1992). Within the central nervous system, Pb appears to principally target the hippocampus, impairing the synaptic plasticity putatively responsible for memory and learning in the mammalian brain. This is believed to occur by Pb-induced changes in N-methyl-D-aspartate receptor (NMDAR) subunit expression and NMDAR-mediated Ca^{2+} signaling (Toscano and Guilarte, 2005). Indeed, the ability of Pb to affect a variety of neurotransmitter systems is likely related to its ability to mimic Ca^{2+} in ion transport and intracellular signaling pathways or by altering intracellular Ca^{2+} homeostasis (Goyer, 1996). For example, Pb competes with Ca^{2+} at presynaptic voltage-gated Ca^{2+} channels (Bernal et al., 1997) and affects the activities of Ca^{2+}-dependent kinases, such as protein kinase A and C, Ca^{2+}/calmodulin-dependent protein kinase (CAMK), and mitogen-activated protein kinase (MAPK) (Toscano and Guilarte, 2005). In addition, Pb-induced increases in circulating ALA may cause indirect neurotoxic effects by inhibiting the release of γ-aminobutyric acid (GABA) and/or by competing at its receptors (Anderson et al., 1996). The mechanisms of Pb-induced neuronal cell death are not fully understood, but may involve the disruption of intracellular Ca^{2+} homeostasis leading to apoptosis and/or a ROS-mediated cytotoxicity similar to that implicated in hemolysis (Sanders et al., 2009). Peripheral neuropathy may be associated with demyelination due to Pb-induced Schwann cell degeneration and the potential subsequent degeneration of exposed axons (Goyer and Clarkson, 2001).

The mechanisms of Pb neurotoxicity in fish have received far less consideration. However, from the limited number of studies available it would seem that the effects are consistent with those observed in mammals. Lead has been found to affect various neurotransmitter systems in fish, including serotonergic (Spieler et al., 1995; Rademacher et al., 2003; Sloman et al., 2005) dopaminergic (Spieler et al., 1995; Rademacher et al., 2001), and noradrenergic (Spieler et al., 1995) pathways. In addition, Pb increased brain endocannabinoid levels in male fathead minnows, likely accentuating Pb-induced effects on neurotransmitter release (Rademacher et al., 2005). As in mammals, Pb was found to induce neuronal injury in the hippo-campus, as well as the optic tectum, of the ornate wrasse (*Thalassoma pavo*), suggesting that Pb affects memory and visuomotor function in fish (Giusi et al., 2008). Finally, there is recent evidence that Pb activates the MAPKs, extracellular signal-regulated kinases (ERK1/2) and p38^{MAPK}, in the brains of Brazilian catfish (*Rhamdia quelen*) (Leal et al., 2006). These MAPKs are believed instrumental in the signaling pathways required for learning and memory in mammals (Toscano and Guilarte, 2005), and the Pb-induced effects are similar to those described for the catfish.

5.2. Seawater

While the mechanism(s) of Pb toxicity to fish in seawater has received little attention, the effects are likely to be comparable with those of freshwater fish. Considering the Ca^{2+} uptake mechanisms are similar among freshwater and marine teleosts, it seems reasonable to expect that hypocalcemia may contribute to acute Pb toxicity regardless of the environment. However, the high alkalinity and Ca^{2+} content of seawater will limit Pb solubility and bioavailability, thus potentially preventing the ionic Pb^{2+} fraction from reaching levels sufficient to inhibit Ca^{2+} influx. From Pb exposures to the estuarine fish *Gillichthys mirabilis*, Somero et al. (1977a) found that Pb accumulation rates were inversely proportional to salinity, as might be expected owing to increasing cation competition with greater salinity. Still, it remains to be seen to what extent ionoregulatory effects of Pb may play a role in the acute toxicity of marine fish. Nevertheless, once taken up, there is little reason to suspect that the hematological and neurological effects of Pb will differ among freshwater or marine fish.

6. NON-ESSENTIALITY OF LEAD

Lead has no known biological function and there exists no evidence that it is required, or otherwise beneficial, for life. Furthermore, Pb is toxic even at low doses. These features are in direct contrast with essential metals which are required to sustain life and furthermore improve health with increasing levels up to an optimal threshold. As discussed throughout this chapter, perhaps the most convincing evidence to the non-essentiality of Pb is the fact that it is taken up, and imparts its dysfunctional effects, by substituting for essential elements such as Ca^{2+} and Zn.

7. POTENTIAL FOR BIOCONCENTRATION AND BIOMAGNIFICATION OF LEAD

7.1. Bioconcentration

It is well known that fish are able to concentrate Pb from the water within various tissues. However, the degree to which Pb will bioconcentrate depends on a number of factors including developmental stage, water quality, Pb concentration, and variability within and among populations and/or species. Table 4.6 lists bioconcentration factors (BCFs) for different tissues from fathead minnows exposed to either 28 or 105 μg L^{-1} Pb for 300 days. Clearly, the tissues that bioconcentrate Pb the most are the kidney, gill, and intestine, regardless of ambient Pb concentration. These findings are consistent with the developmental Pb accumulation observed within various tissues across two generations of brook trout exposed to 119–474 μg L^{-1} Pb (Holcombe et al., 1976).

It is important to note from Table 4.6 that, despite a 3.75-fold increase in waterborne Pb concentration, tissue and whole-body BCFs not only failed to increase proportionately, but actually decreased in most cases. A study by Grosell et al. (2006) revealed a wide range of BCFs in fathead minnows (spanning three orders of magnitude from 10^2 to 10^5) that were negatively correlated with ambient Pb concentrations (Fig. 4.2). This inverse relationship between Pb BCF and exposure concentrations represents a trend that is shared across taxa, although the greatest rates of change appear to occur within fish (McGeer et al., 2003). Also evident from Table 4.6 is a declining trend in whole-body BCFs as development progresses. This likely reflects a growth dilution effect owing to the fact that most of the accumulated Pb is

Table 4.6
Summary of lead bioconcentration factors (BCFs) in whole-body and individual tissue samples of *Pimephales promelas*

	Waterborne [Pb]		Fold difference
	28 μg L^{-1}	105 μg L^{-1}	3.75
Whole body			
Day 4	1269	1045	0.82
Day 30	1453	1003	0.69
Day 150	522	752	1.4
Day 300	402	375	0.93
Day 300 tissue			
Gill	1509	957	0.63
Brain	156	69	0.44
Testes	32	7.0	0.22
Ovaries	17	5.7	0.33
Intestine	783	505	0.64
Liver	212	74	0.35
Kidney	1949	974	0.50
Carcass	399	383	0.96

Data from Mager et al. (2010a).

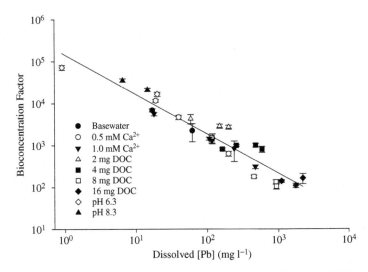

Fig. 4.2. Bioconcentration of Pb by *Pimephales promelas* as a function of dissolved Pb concentration following 30 days of waterborne exposure in freshwater media modified for hardness (as $CaSO_4$), pH, and dissolved organic carbon (DOC). Basewater represents the unmodified tap water supply (deionized water and dechlorinated Virginia Key tap water mixed at a 2:1 ratio). Source: Grosell et al. (2006).

sequestered in bone (see Section 9.3), which contributes less proportional mass as the fish grows (Mager et al., 2008, 2010a).

The effect of growth on Pb accumulation or the converse, effects of Pb tissue burdens on growth, in studies of feral fish is less evident, further illustrating the challenges associated with interpreting BCFs. Hodson et al. (1982) found a negative correlation between whole-body Pb burden and weight in feral rainbow smelt (*Osmerus mordax*) but not feral lake trout (*Salvelinus namaycush*). Furthermore, no relationship between size and whole-body Pb accumulation was observed by Vinikour et al. (1980) in a study of various species of feral fish. However, large individual variation was noted in the latter report and it is likely that the initial period of rapid accumulation in early development was missed. Species-specific differences during development (e.g. habitat selection) could also account for the lack of consistent Pb accumulation trends in feral fish. In light of the above examples, and considering the inverse relationship between Pb BCF and exposure concentrations, the use of a default BCF for regulatory purposes is not recommended (for a critical analysis of BCFs see McGeer et al., 2003). Finally, while much of the focus in the literature has centered on freshwater fish, the same trends and issues for BCFs are likely to apply for marine species, if not to a greater extent given the low solubility of Pb and higher concentrations of competing cations in seawater.

7.2. Biomagnification

It is clear that Pb does not biomagnify along the food web (Settle and Patterson, 1980; Demayo et al., 1982; Farag et al., 1998). As just mentioned, this is likely due to the fact that most accumulated Pb is sequestered in calcified tissues (e.g. bone) and is scarcely found in the large and far more readily assimilated tissue masses of the muscles. Furthermore, at the subcellular level Pb is partitioned to metal rich granules (see Section 9.4), which putatively represent a largely non-trophically available fraction in potential prey (Goto and Wallace, 2009). [Pb]/[Ca^{2+}] ratios have been shown to decrease with increasing trophic levels, indicating a potential biopurification process owing to greater uptake of Ca^{2+} relative to Pb (Patterson, 1980). This was illustrated by more than an order of magnitude decline in [Pb]/[Ca^{2+}] ratios from plants to carnivores in both terrestrial and marine ecosystems (Patterson, 1980). Thus, given the above, it would appear that Pb does not biomagnify owing to the combination of its low trophic bioavailability and sequential biopurification by Ca^{2+}. Although Pb does not biomagnify within the food web, some trophic transfer assuredly takes place for some species (Table 4.4). Indeed, in situations of low waterborne Pb concentrations, as often occurs in the wild, the bulk of Pb accumulation may occur via trophic transfer as opposed to uptake across the gill (Farag et al., 1994).

8. CHARACTERIZATION OF UPTAKE ROUTES

As Pb has no known biological function, its uptake, regardless of the route, must be attributed in large part to the mimicry of other, nutritive, divalent cations. Among these, the transport pathways for Ca^{2+} appear most susceptible, although those for iron (Fe) may also be important, particularly within the gut. In light of the propensity for Pb to substitute for Zn within the enzyme, ALAD (see Section 5.1.2.2), and potentially other Zn binding proteins (Godwin, 2001), it would seem that Zn pathways may represent additional means by which Pb may be taken up. However, the subject has received surprisingly little attention, even within mammalian systems. Therefore, the following sections will focus on the pathways for which there is substantial evidence implicating their involvement in Pb uptake, namely those for Ca^{2+} and Fe.

8.1. Gills

At the gill, uptake of Pb appears to follow the same route(s) as for Ca^{2+}. Transport of Ca^{2+} across the gill is believed to occur via chloride cells (also called mitochondria-rich cells) in a similar fashion in both freshwater and seawater, involving passive apical entry through voltage-insensitive channels and basolateral extrusion via a high-affinity Ca^{2+}-ATPase and/or an Na^+/Ca^{2+} exchanger (see Marshall, 2002, for review). One of the first studies examining the specific influence of Ca^{2+} on Pb uptake in fish by Varanasi and Gmur (1978) provided indirect evidence for shared branchial uptake pathways. Their work revealed reduced waterborne Pb uptake by coho salmon (*Oncorhynchus kisutch*) fed a high-Ca^{2+} diet, potentially reflecting a reduced rate of Ca^{2+} uptake by the gill. In support of this mechanism, decreased branchial uptake rates for Ca^{2+} have been reported for rainbow trout fed a high-Ca^{2+} diet (Baldisserotto et al., 2005) and Rogers and Wood (2004) showed that $CaCl_2$ injections reduced both Ca^{2+} and Pb uptake at the gill. Lead gill accumulation has also been shown to decrease with increasing ambient Ca^{2+} concentrations and, conversely, Pb has been shown to competitively inhibit Ca^{2+} influx (Macdonald et al., 2002; Rogers and Wood, 2004). In addition, Pb uptake was inhibited by known voltage-independent Ca^{2+} channel competitors [lanthanum (La), Cd, and Zn], but not by the voltage-dependent Ca^{2+} channel blockers nifedipine and verapamil (Rogers and Wood, 2004). These findings clearly illustrate a competitive Pb–Ca^{2+} interaction at shared binding sites on the gill and strongly indicate that Pb uptake occurs via a voltage-independent Ca^{2+} channel. However, there is evidence of a second low-affinity/high-capacity population of branchial Pb binding sites demonstrating an apparent non-competitive interaction with Ca^{2+} (Rogers and Wood, 2004; Birceanu et al., 2008), although the Pb concentrations needed for binding to these sites appear to exceed the range of environmental relevance in most cases (e.g. mg Pb L^{-1} range).

While the fate of Pb following entry inside the cell is less clear, some is assuredly transported, potentially via Ca^{2+} binding proteins (CaBPs), to the basolateral membrane for extrusion to the circulation. However, there exists little information as to the exact mechanism(s) by which this extrusion might occur. As previously mentioned, exposure to a high concentration of Pb (~1 mg L^{-1}) resulted in a delayed non-competitive inhibition of the branchial Ca^{2+}-ATPase in rainbow trout, suggesting that sufficient Pb accumulation in the gill must occur for inhibition to take place (Rogers and Wood, 2004). The significance of inhibition at lower concentrations of Pb exposure, however, was not examined. Thus, although the potential role for the Ca^{2+}-ATPase in basolateral Pb extrusion is uncertain given the inhibitory influence of Pb at relatively high concentrations, it does not preclude that some transport occurs via this mechanism, particularly at lower, more environmentally relevant Pb concentrations. Alternatively or additionally, transport through an Na^+/Ca^{2+} exchange mechanism could contribute to basolateral Pb extrusion.

8.2. Gut

In many ways, characterizing toxicant exposure and uptake at the gill is a more amenable proposition than for that at the gut. For example, differences in drinking rates, dietary composition, nutritional status, and varying chemical environments along the gastrointestinal tract pose significant challenges to assessing dietary Pb uptake. Clearly, Pb is taken up by the gut of fish, as demonstrated by a number of studies (Crespo et al., 1986; Mount et al., 1994; Farag et al., 1999; Alves et al., 2006; Alves and Wood, 2006; Ojo and Wood, 2007; Dai et al., 2009); however, the transport mechanisms are virtually unknown and the factors that influence uptake are just beginning to be understood. Nevertheless, some speculations can be made with respect to potential Pb uptake routes using parallels drawn between studies conducted in fish and higher vertebrates. Of particular note, the evidence seems to indicate that Pb may gain entry across the gut by following the uptake pathways for Ca^{2+} and/or Fe.

An early indication that Pb uptake across the gut could follow Fe pathways was provided by a study demonstrating enhanced intestinal Pb absorption by rats deficient in Fe (Six and Goyer, 1972). As in mammals, Fe acquisition by fish is believed to occur predominantly via an Fe^{2+}/H^+ symporter known as divalent metal transporter-1 (DMT1). The name was derived from its ability to transport a range of divalent metals in addition to Fe, such as Cu, Ni, Mn, Co, Cd, and Pb in DMT1-expressing *Xenopus* oocytes (Gunshin et al., 1997). Subsequent confirmation of Pb uptake via DMT1 was demonstrated in transfected yeast and human embryonic kidney fibroblasts (HEK293 cells) expressing rat DMT1 (Bannon et al., 2002). In mammals, uptake of Fe and other metals occurs in the acidic environment of the duodenum, where DMT1 has been shown to be abundantly expressed on the apical surfaces of enterocytes in the luminal villi (Canonne-Hergaux et al., 1999). While it would appear that DMT1 is expressed in the fish intestine (Bury and Grosell, 2003; Sibthorpe et al., 2004), a detailed characterization as to its localization and function in the fish gut remains lacking. However, the evolutionary conservation of DMT1 sequences across vertebrate species including fish (Dorschner and

Phillips, 1999; Saeij et al., 1999) suggests a similar functional role in Fe, and thus potentially Pb, uptake in the fish intestine. Recently, Kwong and Niyogi (2009) showed that Pb significantly inhibited Fe uptake rates across the mucosal epithelium in all sections of rainbow trout intestine *in vitro*, supporting a role for DMT1 in Pb intestinal uptake. However, an excess of luminal Fe was without effect on intestinal Pb accumulation, suggesting a greater affinity for Pb to the Fe uptake mechanism and/or the potential for alternative Pb transport pathways, such as those for Ca^{2+} (Kwong and Niyogi, 2009).

Studies of both freshwater and seawater species indicate that fish take up Ca^{2+} across enterocytes of the anterior regions of the intestine in a manner similar to that by chloride cells at the gill (see Flik and Verbost, 1993, for review). In addition, recent evidence indicates that the stomach is of great importance to dietary Ca^{2+} uptake in rainbow trout (Bucking and Wood, 2007). In support of shared uptake routes by Ca^{2+} and Pb along the gut, Alves and Wood (2006) revealed clear protective effects of dietary Ca^{2+} supplements in juvenile rainbow trout chronically fed diets containing Pb. These findings are consistent with results from experiments with mammals. For example, Six and Goyer (1970) showed that considerably less Pb was accumulated by rats fed Pb mixed with a normal Ca^{2+} diet compared to Pb with a diet low in Ca^{2+}. Although such evidence is circumstantial, it strongly indicates that Pb uptake along the gastrointestinal tract via the Ca^{2+} route(s) is at least partially involved.

Using an *in vitro* gut sac technique, Ojo and Wood (2007) characterized the uptake of Pb (as well as Cu, Zn, Ni, and Cd) from luminal saline in the stomach, anterior, mid, and posterior intestine of rainbow trout. Their findings revealed the highest rates for Pb transport, mucosal binding, and epithelial accumulation were exhibited by the mid intestine, followed by the posterior intestine, with typically much lower rates in the anterior intestine and stomach. This pattern would seem somewhat difficult to reconcile with what might be expected if Pb were to follow the Fe and/or Ca^{2+} uptake routes in the stomach and anterior intestine. The more acidic environments of these segments favor the bioavailability of metals, particularly in the case of Fe, which utilizes a H^+ symport mechanism, although it should be noted that all segments of the intestine demonstrated a similar capacity for transporting Fe (Kwong and Niyogi, 2009). However, Pb speciation under *in vivo* conditions will vary markedly as it moves down the gut owing to the presence of chyme and the progressive change from acidic to more alkaline environments. Indeed, Pb may behave like Ca^{2+} in chyme, presenting greater levels of exposure within the stomach and anterior intestine (Bucking and Wood, 2007), a situation that contrasts with the gut sac studies of

Ojo and Wood in which the same luminal Pb concentrations were used in all sections of the gastrointestinal tract. Furthermore, in the intestines of marine teleosts, which secrete bicarbonate for osmoregulatory purposes and potentially for postprandial acid–base regulation (Wilson et al., 2002; Taylor and Grosell, 2009), bioavailability may be limited by precipitation of lead carbonates. Thus, differences in Pb speciation and concentration (due to absorption and/or extrusion) under *in vivo* conditions compared to the standardized conditions of the *in vitro* studies may help to explain the apparent discrepancies.

8.3. Other Routes

8.3.1. Skin

The cutaneous uptake of Pb has received little attention as efforts have focused predominately on uptake by the gill and intestine. This is perhaps not surprising given that the major ionoregulatory responses to the external medium are regulated by the gill and the assumption that the skin represents a protective barrier that is generally impervious to charged, non-lipophilic xenobiotics such as metals. Nevertheless, Pb may gain entry across the skin in a manner similar to that described for branchial uptake; that is, via Ca^{2+} pathways across chloride cells, often found in greatest density in the skin covering regions adjacent to the gill such as the jaw and operculum. Indeed, uptake of Ca^{2+} has been correlated with the number of chloride cells in the skin (Marshall et al., 1992; McCormick et al., 1992), although it should be noted that La was ineffective at inhibiting Ca^{2+} influx across *in vitro* skin preparations (Marshall et al., 1992). This is in contrast to the effect of La at the gill, suggesting potential alternative uptake mechanisms at the different locations. In any event, uptake of Pb via the skin would likely be of greatest significance during the early life stages when chloride cells are most abundant and widespread, as has been demonstrated for various teleost larvae, presumably as a means of meeting the osmoregulatory needs of the fish until the gill, kidney, and intestine develop (Kaneko et al., 2002). In adult fish, the role of the skin in Ca^{2+} transport is considered to be of minor importance in most cases (Flik and Verbost, 1993). Notable exceptions include gobies and blennies, which rely to a greater extent on cutaneous chloride cells for ion transport (Marshall and Grosell, 2006). In addition, Perry and Wood (1985) showed that the skin and gill contributed equally to Ca^{2+} uptake by rainbow trout (~244 g) under control conditions. Hence, if Pb is indeed being taken up by chloride cells in the integument, the degree would likely be strongly influenced by ambient Ca^{2+} levels.

9. CHARACTERIZATION OF INTERNAL HANDLING

9.1. Biotransformation

Perhaps the most notable biotransformation enzyme for inorganic Pb is glutathione *S*-transferase (GST), which catalyzes a thioester-type conjugation between the tripeptide, glutathione (GSH), and electrophilic xenobiotics or metabolites. Indeed, studies have reported increased GST mRNA expression (Daggett et al., 1998), protein expression (Di Ilio et al., 1989), and enzyme activity (Daggett et al., 1998) following Pb exposure in rats. It has been suggested that GST may facilitate conjugation of GSH and Pb as a means of limiting its reactivity and thereby reducing toxicity (Daggett et al., 1998). As Pb exposure can lead to the production of ROS (see Section 5.1.2.2), GST induction may also serve to detoxify these harmful species. In support of this hypothesis, a toxicogenomic analysis of Pb-exposed fathead minnows revealed increased GST and G6PD mRNA expression, potentially indicating a coordinated defense response against ROS-mediated hemolytic anemia (Mager et al., 2008; see Section 13).

The biotransformation of alkyllead compounds such as tetraethyllead or tetramethyllead takes place via an oxidative dealkylation reaction catalyzed by cytochrome p450 (ATSDR, 2007). The process, which requires NADPH and oxygen, occurs in virtually all tissues, but primarily in liver microsomes where p450 enzyme levels are greatest (Parkinson, 2001). While the putative aim of biotransformation is to decrease xenobiotic toxicity by facilitating elimination, in some cases, as with alkyllead compounds, the process forms products that are as toxic as, or even more toxic, than the original compound. The known metabolites arising from the dealkylation of tetraethyllead in mammals are triethyllead and inorganic Pb, both of which exert potent neurotoxicity, but it is the lipophilic nature of the former that gives rise to the persistent toxic effects of tetraethyllead poisoning (Bolanowska and Wisniewska-Knypl, 1971).

9.2. Transport through the Bloodstream

Of the Pb entering the circulation, the vast majority (~99%) becomes rapidly bound to erythrocytes (Manton and Cook, 1984). It has been shown by *in vitro* analysis that most of the Pb in erythrocytes is cytoplasmic and not membrane bound (Simons, 1993). The predominant mode of Pb entry into erythrocytes appears to occur via an anion exchanger known as solute carrier 4A1 (SLC4A1) or band 3 (Bannon et al., 2000). While it has been demonstrated that erythrocytes are capable of extruding Pb via a Ca^{2+}-ATPase (Simons, 1988), cytoplasmic levels remain high due to binding of various cellular constituents. From human studies, it was determined that the primary contributor is ALAD, accounting for up to 81% of protein-bound Pb within erythrocytes (Bergdahl et al., 1998), apparently by occupying the thiol-rich Zn-binding sites of the protein (Magyar et al., 2005). Most of the remaining fraction was attributed to a 45 kDa protein, potentially pyrimidine-5-nucleotidase, another known target of Pb inhibition (Paglia et al., 1977), and approximately 1% bound to an undefined protein of < 10 kDa (Bergdahl et al., 1998). There was no evidence that Pb was bound to Hb.

The minor fraction (~1%) of Pb not taken up by erythrocytes remains in the plasma, where it is estimated that approximately 40–75% is protein bound, mainly to albumin, although immunoglobulins may also contribute, with most of the rest bound to small molecular weight sulfhydryl complexes (ATSDR, 2007). In humans, the binding capacity of erythrocytes appears to reach saturation at a blood Pb concentration of about 400–500 $\mu g\ L^{-1}$, after which serum levels rapidly increase (Manton and Cook, 1984). In contrast, it was recently reported that Pb in rainbow trout plasma remains constant up to a concentration of about 1000 $\mu g\ L^{-1}$ Pb, indicating that the capacity of erythrocytes to bind Pb may be higher in some fish than in humans (Alves et al., 2006).

9.3. Accumulation

Aside from blood, the predominant tissues to which Pb is distributed in fish are bone, gill, kidney, spleen, and intestine. As discussed in Section 7, the highest Pb concentrations are often found in the gill and kidney, but as these tissues are of relatively little mass, their overall contribution to whole-body Pb accumulation may be rather small. Indeed, consistent with the notion of Pb as a Ca^{2+} analogue, the vast majority of Pb is sequestered in bone. The accumulation of Pb within scales may occur from adsorption/absorption from direct exposure to waterborne Pb (Coello and Khan, 1996) or from the incorporation of internal sources (Sauer and Watabe, 1989; Farrell et al., 2000). Accumulation of Pb in the spleen is also quite substantial and is likely related to its role in erythrocyte storage, production, and removal. Conversely, Pb is typically found at relatively low levels in the liver and muscle.

In general, Pb accumulation appears to occur rapidly within target tissues, followed by either an apparent stabilization or a mild continued increase over time. A representative analysis of the internal distribution of Pb over time was recently provided from studies of Pb-exposed fathead minnows. From an initial 30 day study by Grosell et al. (2006), results demonstrated a rapid Pb accumulation by the gill and intestine with a more gradual increase in Pb accumulation by the carcass (Fig. 4.3A). Using a stepwise digestion procedure, it was determined that the vast majority of Pb within the carcass was associated with bone (Grosell et al., 2006). On a percentage basis, the bone and intestine accumulated nearly 40% each at day 4, but the trends deviated in opposite directions thereafter, with bone representing nearly 80% and intestine less than 10% by day 30 (Fig. 4.3B).

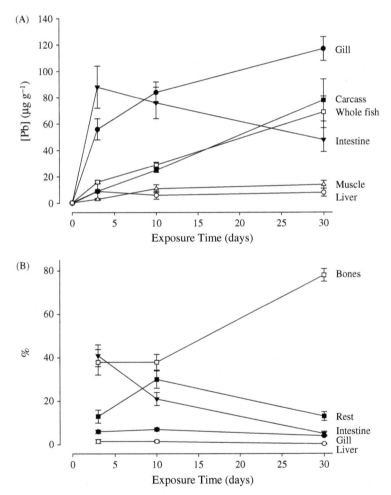

Fig. 4.3. Time-course of (A) Pb accumulation (B) and internal distribution in *Pimephales promelas* during 30 days of waterborne exposure to 26 μg Pb L^{-1}. Significant Pb accumulation was observed in all examined tissues during the 30 days of exposure. Source: Grosell et al. (2006).

The gill maintained a fairly consistent percentage of approximately 2.5–5% throughout the exposure. A subsequent experiment by Mager et al. (2010a) showed that these trends continued to at least 300 days of Pb exposure, at which time the percentage of Pb in bone had reached nearly 95% of total whole-body Pb. Although the kidney was not analyzed separately in the 30 day fathead minnow study, Pb kidney levels from the 300 day study revealed an accumulation similar to that of the gill in relative and absolute amounts. In all cases, the liver was found to accumulate a relatively small amount of Pb, expressed either on a per mass basis or as the percentage of whole-body Pb.

While some variation is to be expected, for the most part these findings are consistent with temporal Pb tissue distributions found in other teleosts, including brook trout (Holcombe et al., 1976), rainbow trout (Farag et al., 1994), coho salmon (Reichert et al., 1979), sunfish (Merlini and Pozzi, 1977), and longjaw mudskipper (*Gillichthys mirabilis*) (Somero et al., 1977a), as well as an elasmobranch, the spotted dogfish (*Scyliorhinus canicula*) (De Boeck et al., 2010). In the case of the dogfish, relatively high concentrations were also found in the skin and rectal gland (De Boeck et al., 2010). Given that the rectal gland regulates Na^+ and Cl^+ balance, the mechanism of Pb accumulation within this tissue may involve Na^+ mimicry, and/or or non-specific binding by Pb, along a Na^+ transport pathway (e.g. Na^+/K^+-ATPase).

Studies of dietary Pb accumulation in rainbow trout (Farag et al., 1994; Alves et al., 2006) and coho salmon (Varanasi and Gmur, 1978) indicate that, while in general the target tissues are consistent between waterborne and dietary exposures, Pb appears to accumulate to a greater extent in the gastrointestinal tract during dietary exposures, with the gill and kidney assuming less of a Pb burden.

On a final note, it is perhaps surprising that Pb appears to accumulate very little within nervous tissue (Holcombe et al., 1976; Hodson et al., 1978; Sloman et al., 2005; Mager et al., 2010a) given its putative neurological effects in mammals and fish. This evidence may indicate a low threshold of toxicity by the nervous system and/or reflect secondary effects by mediators induced during Pb exposure (e.g. ALA).

9.4. Subcellular Partitioning

Goto and Wallace (2009) studied the subcellular partitioning of Pb, Cd, Cu, and Zn in killifish collected from various metal-polluted salt marshes in New York, USA. Subcellular fractions from whole-body homogenates were divided into organelles, metal-rich granules (MRGs), heat-denatured proteins, heat-stable proteins, and cellular debris (CD). Although there were substantial differences in partitioning among the populations, Pb was generally found mostly within the MRG (~40–60%) and CD (~30–40%) fractions from the polluted sites. Localization of Pb to more specific subcellular regions and organelles can be gleaned from studies in mammalian cell types. It is clear that much of the Pb entering target cells becomes sequestered within cytoplasmic and nuclear inclusion bodies (discussed in the following section). In fact, approximately 50% of Pb within the kidney of Pb-exposed rats has been attributed to these formations (Goyer et al., 1970), a value similar to the corresponding MRG fraction of the killifish study above. In addition, Pb appears to target specific organelles including the mitochondria and endoplasmic reticulum. Reichert et al. (1979) found that between 25 and 28% of Pb in the cells of the kidney and liver of Pb-exposed coho salmon was deposited in the mitochondrial fraction. Mitochondria are quite sensitive to Pb, which can elicit a number of effects leading to inhibited respiration (Scott et al., 1971; Goyer, 1983). In addition, Pb has been implicated in the impairment of proper protein synthesis and transport in the endoplasmic reticulum of astrocytes (Qian and Tiffany-Castiglioni, 2003). Lead is also likely found within lysosomes due to scavenging of Pb-bound components.

9.5. Detoxification and Storage Mechanisms

Once taken up by an organism, the amount of free Pb that exists is presumably very small as most of it is bound by proteins and other small sulfhydryl containing compounds such as GSH (see Section 9.1). Toxicity will relate to the proportion of Pb bound to detoxifying versus non-detoxifying cellular constituents. A classical characteristic feature of Pb exposure in mammals is the presence of insoluble Pb-protein aggregates known as inclusion bodies, found most commonly in cells of the kidney and nervous system (Goyer, 1983), but also reported within other known targets for Pb such as bone (Hsu et al., 1973). Inclusion bodies are found in the cytoplasm as well as the nucleus, where they persist likely without threat of proteolytic attack (Nolan and Shaikh, 1992). Cytoplasmic inclusion bodies, in contrast, have a short half-life, presumably due to increased lysosomal degradation (Spit et al., 1981) and transport to the nucleus by way of invagination of nuclear membranes (Goyer, 1983). Aside from largely consisting of a protein, or proteins, rich in acidic amino acids, little is known regarding their composition. Evidence has indicated that metallothionein (MT), α-synuclein (Zuo et al., 2009), and a cleavage product of α_2-microglobulin (Fowler, 1998) are critical to inclusion body formation, although their exact roles remain to be elucidated. It should be noted, however, that Pb is a poor inducer of MT compared to other metals such as Cd and Zn (Waalkes and Klaassen, 1985). Thus, in the case of Pb the role of MT may be limited to inclusion body formation as opposed to a more general means of sequestration as described for other metals. Nevertheless, it appears that the main function of inclusion bodies is to sequester intracellular Pb to reduce exposure to critical cell constituents including enzymes and organelles such as the mitochondria and endoplasmic reticulum as mentioned previously.

It is often assumed that bone tissue represents a large reservoir for sequestering Pb in an inert capacity, thereby serving as a mode of detoxification. While this may be true to some extent, there are some points worth considering. In mammals, skeletal Pb may be mobilized in response to Ca^{2+} status, leading to potentially adverse effects in other tissues (Pounds et al., 1991). A release of accumulated Pb from bone and/or scales may similarly occur in fish. Indeed, it has been shown that Ca^{2+} can be mobilized from the scales of river running Atlantic salmon during sexual maturation, presumably to provide Ca^{2+} to the growing gonads (Persson et al., 1998), and from the scales of rainbow trout during periods of starvation (Persson et al., 1997). In addition, the notion that Pb may exist in an inert form while residing within bone may not be entirely accurate. Several lines of evidence suggest that Pb affects various aspects of bone physiology, including bone formation, resorption, and fracture healing, and may cause necrosis (Pounds et al., 1991; Carmouche et al., 2005). The significance of such effects in fish, however, remains largely unknown.

9.6. Homeostatic Controls

As Pb is a non-essential metal, a discussion of its homeostatic controls is limited primarily to its depuration and induced acclimation responses. Several studies have examined the tissue-specific depuration of Pb in fish (Holcombe et al., 1976; Somero et al., 1977a; Varanasi and Gmur, 1978). In rainbow trout exposed to 119 µg L^{-1} for 105 weeks, a 12 week depuration period led to reductions in tissue Pb burdens of gill, kidney, and liver by 70%, 74%, and 78%, respectively (Holcombe et al., 1976). Reduced Pb burdens for these tissues were likewise reported for coho salmon (Varanasi and Gmur, 1978), and also for gill and liver in longjaw mudskipper (Somero et al., 1977a), during periods of depuration following chronic Pb exposures. However, magnitudes differed somewhat, likely because of variations in exposure and depuration conditions and/or species-specific differences. In both latter species, Pb continued to accumulate during the depuration period in the skeleton, and to a lesser extent, the skin. Furthermore, Pb was also found to accumulate in the spleen during depuration, likely reflecting the role of erythrocytes in the transfer and removal of Pb from other tissues (Somero et al., 1997a). The amount of Pb accumulation within skeleton and skin during depuration appeared to decrease in the presence of elevated Ca^{2+}, suggesting a Ca^{2+}-enhanced Pb turnover rate within these tissues (Varanasi and Gmur, 1978).

Considering that some of the major effects of Pb toxicity are attributed to ionoregulatory and hematological alterations, several lines of evidence seem to indicate that fish are able to acclimate, at least to some extent, to Pb during chronic exposures. For example, fathead minnows have been shown to recover from disturbances to Na^+, K^+, and Ca^{2+} homeostasis during 30 day exposures to waterborne Pb (Grosell et al., 2006). In addition, some reports indicate recovery of hematological parameters during Pb exposures owing to either increased erythropoiesis and/or splenic release of erythrocytes (Hodson et al., 1978; Johansson-Sjobeck and Larsson, 1979). The other major effects of Pb, due to neurological impairment, may be less prone to acclimation. For example, although black tail may be reversed if exposures are stopped in time, once calcification occurs, the effects are permanent. Furthermore, while killifish recovered from various Pb-induced behavioral impairments following cessation of Pb exposure (Weis and Weis, 1998), effects on goldfish cognition were found to persist for several weeks afterwards (Weir and Hine, 1970). Clearly, more research is needed to fully characterize the relationships of the durations and doses of Pb exposure to which fish may be able to acclimate.

10. CHARACTERIZATION OF EXCRETION ROUTES

10.1. Gills

Accumulation of Pb by the gills during dietary exposures suggests that a branchial role for Pb excretion may exist (Alves and Wood, 2006), although it is unclear as to exactly how this might occur. A study by Varanasi and Markey (1978) demonstrated that saltwater-adapted coho salmon injected intravenously with Pb nitrate (400 µg kg^{-1}) showed a small but significant amount of Pb (4.4 µg L^{-1}) in the epidermal mucus secreted during a postinjection time course of 3–384 h. Thus, it is possible that the gills may act in a similar fashion to that of the skin, to excrete Pb within mucus, thereby implicating a potential role for branchial goblet cells in the elimination of Pb. However, the true significance of the gill in the excretion of Pb remains largely unknown.

10.2. Kidney

The renal excretion of Pb by fish appears related to an inhibition of active reabsorption along proximal tubule I, with the notable exception of aglomerular teleosts which lack this part of the nephron. Urinary Pb excretion was recently examined during acute (96 h) Pb exposures to rainbow trout at a concentration of 1.2 mg L^{-1} dissolved Pb (Patel et al., 2006). As expected, Pb accumulated in the kidney over time, peaking at 96 h with a tissue burden approaching 30 $\mu g\,g^{-1}$ Pb wet weight. At 24–96 h of exposure, urinary Pb concentrations ranged from 60 to 95 $\mu g\,L^{-1}$ Pb, which corresponded to an excretion rate of approximately 0.08–0.12 μg Pb $kg^{-1}\,h^{-1}$. As discussed in Section 9.2, Pb will exist primarily in complexed forms (e.g. to albumin or GSH) within plasma as it is delivered to the kidney, with a small percentage in the free ionic form, although the relative proportions entering the nephron will likely depend on glomerular filtration and dose. Tubular reabsorption of these forms could thus potentially occur by endocytosis or following the active transport pathways for Ca^{2+} and Mg^{2+}, respectively. Given that protein reabsorption was unaffected during the rainbow trout exposures, while net reabsorption efficiencies for Ca^{2+}, Mg^{2+}, and Pb were progressively decreased as indicated by clearance ratio analyses, enhanced urinary Pb excretion was attributed to inhibition of active tubular reabsorption. As noted previously, the concurrent decrease in Ca^{2+} reabsorption provides an additional mechanism for the hypocalcemia observed during acute waterborne Pb exposures (see Section 5.1.1.2).

10.3. Liver/Bile

Studies in mammals have distinctly revealed the importance of biliary excretion in the elimination of Pb, although the relative contribution appears to vary among species and with the route of exposure (e.g. inhalation versus ingestion) (Klaassen and Shoeman, 1974; Cikrt et al., 1983; DeMichele, 1984). An examination of biliary Pb excretion in fish has not been reported; however, Pb accumulates within the fish liver, albeit to an apparent lesser degree than the kidney, thus suggesting hepatic transfer to the bile is highly probable. Recent evidence for the biliary excretion of Ca^{2+} in rainbow trout (Bucking and Wood, 2007) provides further indirect support that clearance through the bile likely represents a significant route of Pb elimination in fish. Another important role for bile during Pb exposures will involve the clearance of accumulating bilirubin, a toxic breakdown product of Hb arising from the loss of erythrocytes. Nevertheless, further studies are needed to directly assess the significance of Pb excretion via the bile in fish.

10.4. Gut

While the processes have not been well characterized pertaining to Pb specifically, there exists a variety of potential mechanisms by which a toxicant can be excreted by the gut, including both biliary and non-biliary pathways, such as active secretion, efflux transport and rapid exfoliation of intestinal cells (Kleinow and James, 2001). Given the large binding capacity of epithelial mucus demonstrated *in vitro* (Ojo and Wood, 2007), it would seem that mucosal sloughing represents a first line of defense against Pb absorption by the fish gut. With respect to absorbed Pb, the gut likely functions to eliminate Pb secreted within the bile as well as that potentially secreted across the intestinal epithelium. The results of Bucking and Wood (2007) suggest that, although minor, some Ca^{2+} may be secreted by the mid and posterior intestine of rainbow trout during digestion. Thus, given the propensity of Pb to follow Ca^{2+} transport, Pb may similarly be secreted to a minor extent in these regions. In the end, aside from the elimination of incompletely absorbed dietary Pb, and presumably Pb excreted within the bile, the role of the gut in piscine Pb excretion is poorly defined at this time.

11. BEHAVIORAL EFFECTS OF LEAD

Although the literature is sparse on the behavioral effects of Pb in fish, it appears that Pb can impair both cognitive and sensorimotor function. Clearly, Pb affects motor function, as illustrated by the common observances of erratic swimming, hyperactivity and hyperventilation, trembling and muscle spasms (Davies et al., 1976; Holcombe et al., 1976; Somero et al., 1977b; Giusi et al., 2008). With respect to cognition, a study by Weir and Hine (1970) assessed the effects of Pb, Se, Hg, and As on the ability of goldfish (*Carassius auratas*) to maintain a conditioned response (i.e. escape an area to avoid a mild electric shock). Lead elicited the second most sensitive threshold (behind Hg) at which a significant effect was observed, occurring at $1/1570$ (70 μg L^{-1}) of the 7 day LC50. The average behavioral impairment at this concentration was approximately 10%, whereas at the highest dose tested (10 mg L^{-1}) impairment reached nearly 70%. Lead has also been shown to impair fish feeding ability on live prey, as assessed by feeding duration (Weber et al., 1991; Weis and Weis, 1998; Mager et al., 2010a), number of feeding miscues (Weber et al., 1991; Weis and Weis, 1998), choice of prey size (Nyman, 1981; Weber, 1996), and changes in reaction distance to prey (Nyman, 1981). Weis and Weis (1998) further found that Pb reduced spontaneous activity, swimming performance, and the ability to avoid predation. Finally, Weber (1993) observed Pb-induced alterations to the reproductive behavior of fathead minnows, including a reduction in time spent on nest preparation and maintenance by males, reduced oviposition by females, and increased interspawn periods. While the effects just described may have important ecological implications, more work will be necessary to more clearly define dose responses as well as threshold levels at which Pb impairment to various behaviors occur, particularly in the context of ambient water quality.

12. MOLECULAR CHARACTERIZATION OF LEAD TRANSPORTERS, STORAGE PROTEINS, AND CHAPERONES

As discussed in Section 8, there is strong evidence that Pb transport occurs principally via the Ca^{2+} and Fe pathways. Although the specific Ca^{2+} channels through which Pb is taken up have yet to be identified, recent discovery of an epithelial Ca^{2+} channel (ECaC) in several species of fish (Qiu and Hogstrand, 2004; Pan et al., 2005; Shahsavarani et al., 2006) presents a likely candidate for branchial uptake. ECaC is a voltage-independent channel belonging to the vanilloid subfamily of the transient receptor potential (TRP) superfamily that appears to function by forming a tetrameric pore that is highly specific for Ca^{2+} (Hoenderop et al., 2003). The predicted protein sequences from fish share similarities of around 80–90% and range in size from 709 to 727 amino acids. Although ECaC appears to be expressed to a minor extent in all tissues examined, it is predominantly expressed within gills (Qui and Hogstrand, 2004; Pan et al., 2005; Shahsavarani et al., 2006). By immunohistochemical analysis, ECaC was localized to pavement cells in addition to chloride cells in rainbow trout gills, thus challenging the notion that only chloride cells mediate Ca^{2+} uptake (Shahsavarani et al., 2006).

In contrast to the voltage-independent nature of ECaC, evidence indicates that the apical Ca^{2+} channel of enterocytes is an L-type voltage-dependent channel (Larsson et al., 1998). However, further molecular characterization of this channel is currently unavailable. An additional uptake mechanism for Pb along the gastrointestinal tract, and potentially kidney, presumably occurs by way of the Fe^{2+}/H^+ symporter, DMT1. Also referred to as the natural resistance-associated macrophage protein 2 (NRAMP2) and divalent cation transporter-1 (DCT1), DMT1 belongs to the large family of highly conserved solute carrier 11 (Slc11) transport proteins. The deduced protein sequences of DMT1s cloned from *Takifugu rubripes* (Sibthorpe et al., 2004), carp (Saeij et al., 1999), and rainbow trout (Dorschner and Phillips, 1999) exhibit shared similarities of about 90%, with sizes ranging from 548 to 558 amino acids. Analyses of mRNA tissue distributions reveal that DMT1 is ubiquitously expressed, with highest levels found in the kidney, intestine, and gonads (Gunshin et al., 1997; Sibthorpe et al., 2004).

Little is known regarding the identity of the specific storage proteins and chaperones for Pb. However, as Pb appears to follow uptake via Fe and/or Ca^{2+} apical transport mechanisms, it seems reasonable to expect that transcellular transport and basolateral extrusion would follow similar routes involving mimicry of these cations. Thus, one may speculate that CaBPs such as calmodulin, and basolateral transporters/exchangers such as ferroportin, Ca^{2+}-ATPase, and the Na^+/Ca^{2+} exchanger, represent likely candidates warranting further investigation.

13. GENOMIC STUDIES

A time-course toxicogenomic analysis was recently performed using fathead minnows exposed to approximately 35 μg L^{-1} Pb for 150 days in waters modified to examine the influences of hardness (as $CaSO_4$) and DOC (as humic acid) on Pb accumulation and toxicity (Mager et al., 2008). Initial microarray analyses (GEO accession no. GSE8404) of fish sampled from the basewater controls with and without Pb revealed four genes exhibiting pronounced Pb-induced responses: GST-α, G6PD, β-globin and ferritin heavy chain. These genes were linked in biochemical pathways supporting an ROS-mediated hemolytic anemia within Pb-exposed fish (Fig. 4.4). In particular, the closely matched coinduction of GST and G6PD (Fig. 4.5) suggested recruitment of the pentose phosphate shunt employed by erythrocytes to combat oxidative stress (Lachant et al., 1984). In addition, an upregulation in ferritin transcription may have reflected an increase in free Fe^{2+} concentrations arising from an ALA-mediated displacement of Fe^{2+} from ferritin (Oteiza et al., 1995; Rocha et al., 2003). Finally, a decrease in β-globin mRNA levels was observed, thus further supporting a potential hemolytic loss of erythrocytes.

Quantitative polymerase chain reaction (qPCR) was subsequently used to examine whether DOC and Ca^{2+} afforded protection against Pb-induced transcriptional responses. Except for β-globin, changes in mRNA expression largely paralleled the influence of water chemistry on whole-body Pb accumulation, with reduced responses by GST, G6PD, and ferritin observed in water supplemented with DOC but not Ca^{2+} (Fig. 4.5). Although additional genomic studies are needed to more fully characterize the molecular responses of Pb exposure in fish, it is likely that these genes could form the basis of a

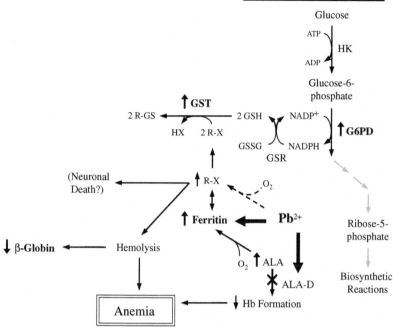

Fig. 4.4. Lead-induced genes from *Pimephales promelas* identified by microarray analysis (bold) are linked in pathways with potential roles in hematological and neurological dysfunction. Lead inhibition of δ-aminolevulinic acid dehydratase (ALAD) results in (1) impaired heme synthesis and (2) accumulation of aminolevulinic acid (ALA), which promotes release of reactive oxygen species (R–X) from ferritin, thereby contributing to oxidative stress and potentially hemolysis and neuronal death. Ferritin and the pentose phosphate shunt enzymes, glutathione S-transferase alpha (GST) and glucose-6-phosphate dehydrogenase (G6PD), increase as compensatory and detoxification responses, respectively. HK: hexokinase; Hb: hemoglobin. Source: Mager et al. (2008).

sensitive Pb-specific profile for future use in assessing Pb exposure and accumulation. Ideally, mRNA expression profiles from early life-stage Pb exposures would be used to predict chronic outcomes of ecological relevance. However, such utility will demand the phenotypic anchoring of transcriptional responses across multiple levels of biological organization.

14. INTERACTIONS WITH OTHER METALS

While most laboratory studies focus on a single metal, exposures in the environment commonly occur in combination with other metals and

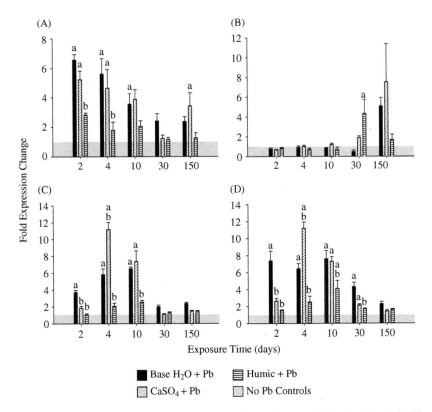

Fig. 4.5. Time-course quantitative polymerase chain reaction analysis of whole-body Pb-induced mRNA expression changes in (A) ferritin, (B) β-globin, (C) G6PD, and (D) GST from fathead minnows (*Pimephales promelas*) exposed to ~35 µg L^{-1} Pb for 150 days in waters modified to examine the influences of hardness (as CaSO$_4$) and dissolved organic carbon (as humic acid). All data were normalized to the housekeeping gene, elongation factor-1α (EF1α), and expressed relative to treatment-matched controls. Statistically significant difference from: [a]corresponding control or [b]basewater+Pb. Source: Mager et al. (2008).

toxicants. A recent study by Komjarova and Blust (2009) examined the interactions between Pb, Cd, Cu, Ni, and Zn uptake by zebrafish (*Danio rerio*) during mixed waterborne exposures. Their findings revealed that in most cases Pb accumulation rates in both the gills and whole body increased with increasing concentrations of the other metals. The influence of Cu was one exception, eliciting a more complex accumulation pattern of an initial increase and then decrease in Pb accumulation rate. A previous study using neon tetras (*Paracheirodon innesi*) demonstrated that Pb and Cu facilitate the uptake of one another (i.e. synergistic effect) during simultaneous exposures (Tao et al., 1999). The interactions between Pb and Cd have also been examined in soft water acclimated rainbow trout (Birceanu et al., 2008). By performing stepwise increases for each metal in the presence of 100 nM (nominal) of the other (measured concentrations of 18 µg L^{-1} Pb or 9 µg L^{-1} Cd), it was found that Pb minimally influenced Cd–gill binding, whereas Cd reduced Pb–gill binding by up to 50% during concurrent Pb exposures up to approximately 73 µg L^{-1}. In addition, Pb–Cd mixtures inhibited Ca^{2+} and Na$^+$ influx in a synergistic manner, a finding consistent with hypocalcemia and hyponatremia as the putative mechanisms for acute Pb toxicity (see Section 5.1.1.2). Finally, with respect to Pb–Zn mixtures, results from several studies suggest that the inhibitory effect of Pb on ALAD activity may be ameliorated by concurrent exposures due to Zn reactivation (Schmitt et al., 1984, 1993, 2005; Rodrigues et al., 1989). Hence, during chronic exposures, the combined effects of these two metals may be less than additive.

15. KNOWLEDGE GAPS AND FUTURE DIRECTIONS

While much has been learned regarding the influence of water chemistry on Pb bioavailability, the uptake and internal distribution of Pb, and the mechanisms of acute and chronic Pb toxicity, there remain many areas warranting further research. Importantly, a complete characterization of the Pb uptake mechanisms, particularly those involved with the transcellular transport and basolateral extrusion, remains to be elucidated for both the gill and gut. Although the target tissues for Pb are well understood, details regarding the internal handling of Pb at the molecular and subcellular levels are sparse. Perhaps least is known about the mechanisms of Pb excretion by fish. With respect to the mechanisms of Pb toxicity, the relevance of anemia and neurological dysfunction in the wild remains uncertain in many respects. For example, the hematological responses to Pb appear variable and it would seem that many fish have the capacity to acclimate and compensate for Pb-induced effects during chronic exposures. Further studies utilizing a combined analysis of hematological effects, swimming respirometry, and gill morphology during acute and chronic Pb exposures should more accurately assess the environmental relevance of anemia to Pb toxicity in fish. Finally, although the genomic responses of Pb exposure are just beginning to be examined in fish, future efforts such as concentration-dependent and/or tissue-specific microarrays, as well as proteomic analyses, should help to elucidate the underlying toxic mechanisms of Pb and provide far more sensitive biomarkers of exposure and effects.

REFERENCES

Adams, E. S. (1975). Effects of lead and hydrocarbons from snowmobile exhaust on brook trout (*Salvelinus fontinalis*). *Trans. Am. Fish. Soc.* **104**, 363–373.

Allen, P. (1993). Effects of acute exposure to cadmium (II) chloride and lead (II) chloride on the haematological profile of *Oreochromis aureus* (Steindachner). *Comp. Biochem. Physiol. C* **105**, 213–217.

Alves, L. C., and Wood, C. M. (2006). The chronic effects of dietary lead in freshwater juvenile rainbow trout (*Oncorhynchus mykiss*) fed elevated calcium diets. *Aquat. Toxicol.* **78**, 217–232.

Alves, L. C., Glover, C. N., and Wood, C. M. (2006). Dietary Pb accumulation in juvenile freshwater rainbow trout (*Oncorhynchus mykiss*). *Arch. Environ. Contam. Toxicol.* **51**, 615–625.

Anderson, A. C., Pueschel, S. M., and Linakis, J. G. (1996). Pathophysiology of lead poisoning. In: *Lead Poisoning in Children* (S.M. Pueschel, J.G. Linakis and A.C. Anderson, eds), pp. 75–96. P.H. Brookes, Baltimore, MD.

ANZECC (2000). *Australian and New Zealand Guidelines for Fresh and Marine Water Quality.* Canberra: Australian and New Zealand Environment and Conservation Council.

ATSDR (2007). *Toxicological Profile for Lead.* Agency for Toxic Substances and Disease Registry. Atlanta, GA: US Department of Health and Human Services.

Baldisserotto, B., Chowdhury, M. J., and Wood, C. M. (2005). Effects of dietary calcium and cadmium on cadmium accumulation, calcium and cadmium uptake from the water, and their interactions in juvenile rainbow trout. *Aquat. Toxicol.* **72**, 99–117.

Bannon, D. I., Olivi, L., and Bressler, J. (2000). The role of anion exchange in the uptake of Pb by human erythrocytes and Madin–Darby canine kidney cells. *Toxicology* **147**, 101–107.

Bannon, D. I., Portnoy, M. E., Olivi, L., Lees, P. S., Culotta, V. C., and Bressler, J. P. (2002). Uptake of lead and iron by divalent metal transporter 1 in yeast and mammalian cells. *Biochem. Biophys. Res. Commun.* **298**, 978–984.

Bergdahl, I. A., Sheveleva, M., Schutz, A., Artamonova, V. G., and Skerfving, S. (1998). Plasma and blood lead in humans: capacity-limited binding to delta-aminolevulinic acid dehydratase and other lead-binding components. *Toxicol. Sci.* **46**, 247–253.

Bernal, J., Lee, J. H., Cribbs, L. L., and Perez-Reyes, E. (1997). Full reversal of Pb^{++} block of L-type Ca^{++} channels requires treatment with heavy metal antidotes. *J. Pharmacol. Exp. Ther.* **282**, 172–180.

Besser, J. M., Brumbaugh, W. G., May, T. W., Church, S. E., and Kimball, B. A. (2001). Bioavailability of metals in stream food webs and hazards to brook trout (*Salvelinus fontinalis*) in the upper Animas River watershed, Colorado. *Arch. Environ. Contam. Toxicol.* **40**, 49–59.

Besser, J. M., Brumbaugh, W. G., May, T. W., and Schmitt, C. J. (2007). Biomonitoring of lead, zinc, and cadmium in streams draining lead-mining and non-mining areas, southeast Missouri, USA. *Environ. Monit. Assess.* **129**, 227–241.

Birceanu, O., Chowdhury, M. J., Gillis, P. L., McGeer, J. C., Wood, C. M., and Wilkie, M. P. (2008). Modes of metal toxicity and impaired branchial ionoregulation in rainbow trout exposed to mixtures of Pb and Cd in soft water. *Aquat. Toxicol.* **89**, 222–231.

Bolanowska, W., and Wisniewska-Knypl, J. M. (1971). Dealkylation of tetraethyllead in the homogenates of rat and rabbit tissues. *Biochem. Pharmacol.* **20**, 2108–2110.

Brumbaugh, W. G., May, T. W., Besser, J. M., Allert, A. L. and Schmitt, C. J. (2007). *Assessment of Elemental Concentrations in Streams of the New Lead Belt in Southeastern Missouri, 2002–05.* USGS Scientific Investigations Report 2007-5057. Reston, VA: US Geological Survey.

Bucking, C., and Wood, C. M. (2007). Gastrointestinal transport of Ca^{2+} and Mg^{2+} during the digestion of a single meal in the freshwater rainbow trout. *J. Comp. Physiol. B* **177**, 349–360.

Burden, V. M., Sandheinrich, M. B., and Caldwell, C. A. (1998). Effects of lead on the growth and delta-aminolevulinic acid dehydratase activity of juvenile rainbow trout. Oncorhynchus mykiss. *Environ. Pollut.* **101**, 285–289.

Bury, N., and Grosell, M. (2003). Iron acquisition by teleost fish. *Comp. Biochem. Physiol. C Toxicol. Pharmacol.* **135**, 97–105.

Canonne-Hergaux, F., Gruenheid, S., Ponka, P., and Gros, P. (1999). Cellular and subcellular localization of the Nramp2 iron transporter in the intestinal brush border and regulation by dietary iron. *Blood* **93**, 4406–4417.

Carmouche, J. J., Puzas, J. E., Zhang, X., Tiyapatanaputi, P., Cory-Slechta, D. A., Gelein, R., Zuscik, M., Rosier, R. N., Boyce, B. F., O'Keefe, R. J., and Scharz, E. M. (2005). Lead exposure inhibits fracture healing and is associated with increased chondrogenesis, delay in cartilage mineralization, and a decrease in osteoprogenitor frequency. *Environ. Health Perspect.* **113**, 749–755.

Carpenter, K. E. (1927). The lethal action of soluble metallic salts on fishes. *J. Exp. Biol.* **4**, 378–390.

CCME (2008). *Canadian Water Quality Guidelines.* Ottawa: Canadian Council of Ministers of the Environment.

CEC (2006). *Proposal for a Directive of the European Parliament and of the Council on Environmental Quality Standards in the Field of Water Policy and Amending Directive 2000/60/EC.* Brussels: Commission of the European Union.

Cikrt, M., Lepsi, P., and Tichy, M. (1983). Biliary excretion of lead in rats drinking lead-containing water. *Toxicol. Lett.* **16**, 139–143.

Clark, G. M. (2002). *Occurrence and Transport of Cadmium, Lead, and Zinc in the Spokane River Basin, Idaho and Washington, Water Years 1999–2001.* Water-Resources Investigation Report 02–4183. Boise, ID: US Geological Survey.

Coello, W. F., and Khan, M. A. (1996). Protection against heavy metal toxicity by mucus and scales in fish. *Arch. Environ. Contam. Toxicol.* **30**, 319–326.

Craig, J. R., Rimstidt, J. D., Bonnaffon, C. A., Collins, T. K., and Scanlon, P. F. (1999). Surface water transport of lead at a shooting range. *Bull. Environ. Contam. Toxicol.* **63**, 312–319.

Crespo, S., Nonnotte, G., Colin, D. A., Leray, C., Nonnotte, L., and Aubree, A. (1986). Morphological and functional alterations induced in trout intestine by dietary cadmium and lead. *J. Fish Biol.* **28**, 69–80.

CSIR (1995). *South African Water Quality Guidelines for Coastal Marine Waters.* Department of Water Affairs and Forestry, Council for Scientific and Industrial Research, Pretoria.

CSIR (1996). *South African Water Quality Guidelines.* Department of Water Affairs and Forestry, Council for Scientific and Industrial Research, Pretoria.

Czarnezki, J. M. (1985). Accumulation of lead in fish from Missouri streams impacted by lead mining. *Bull. Environ. Contam. Toxicol.* **34**, 736–745.

Daggett, D. A., Oberley, T. D., Nelson, S. A., Wright, L. S., Kornguth, S. E., and Siegel, F. L. (1998). Effects of lead on rat kidney and liver: GST expression and oxidative stress. *Toxicology* **128**, 191–206.

Dai, W., Du, H., Fu, L., Jin, C., Xu, Z., and Liu, H. (2009). Effects of dietary Pb on accumulation, histopathology, and digestive enzyme activities in the digestive system of tilapia (*Oreochromis niloticus*). *Biol. Trace Elem. Res.* **127**, 124–131.

Davies, B. E. (1987). Consequences of environmental contamination by lead mining in Wales. *Hydrobiologia* **149**, 213–220.

Davies, P. H., Goettl, J. P., Sinley, J. R., and Smith, N. F. (1976). Acute and chronic toxicity of lead to rainbow trout *Salmo gairdneri*, in hard and soft water. *Water Res.* **10**, 199–206.

Dawson, A. B. (1935). The hemopoietic response in the catfish (*Ameiurus nebulosus*) to chronic lead poisoning. *Biol. Bull.* **68**, 335–346.

De Boeck, G., Eyckmans, M., Lardon, I., Bobbaers, R., Sinha, A. K., and Blust, R. (2010). Metal accumulation and metallothionein induction in the spotted dogfish *Scyliorhinus canicula*. *Comp. Biochem. Physiol. A* **155**, 503–508.

Demayo, A., Taylor, M. C., and Taylor, K. W. (1982). Toxic effects of lead and lead compounds on human health, aquatic life, wildlife plants, and livestock. *Crit. Rev. Environ. Contr.* **1**, 257–305.

DeMichele, S. J. (1984). Nutrition of lead. *Comp. Biochem. Physiol. A* **78**, 401–408.

Diamond, J. M., Koplish, D. E., McMahon, J., and Rost, R. (1997). Evaluation of the water-effect ratio procedure for metals in a riverine system. *Environ. Toxicol. Chem.* **16**, 509–520.

Di Ilio, C., Aceto, A., Columbano, A., Ledda-Columbano, G. M., and Federici, G. (1989). Induction of rat liver glutathione transferase subunit 7 by lead nitrate. *Cancer Lett.* **46**, 167–171.

Dorschner, M. O., and Phillips, R. B. (1999). Comparative analysis of two Nramp loci from rainbow trout. *DNA Cell Biol.* **18**, 573–583.

Ekinci, D., Beydemir, S., and Kufrevloglu, O. I. (2007). *In vitro* inhibitory effects of some heavy metals on human erythrocyte carbonic anhydrases. *J. Enzyme Inhib. Med. Chem.* **22**, 745–750.

Fantin, A. M. B., Trevisan, P., and Pederzoli, A. (1988). Effects of acute experimental pollution by lead on some haematological parameters in *Carassius carassius* (L.) var. *auratus*. *Boll. Zool.* **55**, 251–255.

Farag, A. M., Boese, C. J., Woodward, D. F., and Bergman, H. L. (1994). Physiological changes and tissue metal accumulation in rainbow trout exposed to foodborne and waterborne metals. *Environ. Toxicol. Chem.* **13**, 2021–2029.

Farag, A. M., Woodward, D. F., Goldstein, J. N., Brumbaugh, W., and Meyer, J. S. (1998). Concentrations of metals associated with mining waste in sediments, biofilm, benthic macroinvertebrates, and fish from the Coeur d'Alene River basin, Idaho. *Arch. Environ. Contam. Toxicol.* **34**, 119–127.

Farag, A. M., Woodward, D. F., Brumbaugh, W. G., Goldstein, J. N., McConnell, E., Hogstrand, C., and Barrows, F. T. (1999). Dietary effects of metals-contaminated invertebrates from the Coeur d'Alene River, Idaho, on cutthroat trout. *Trans. Am. Fish. Soc.* **128**, 578–592.

Farag, A. M., Nimick, D. A., Kimball, B. A., Church, S. E., Harper, D. D., and Brumbaugh, W. G. (2007). Concentrations of metals in water, sediment, biofilm, benthic macroinvertebrates, and fish in the Boulder River watershed, Montana, and the role of colloids in metal uptake. *Arch. Environ. Contam. Toxicol.* **52**, 397–409.

Farrell, A. P., Hodaly, A. H., and Wang, S. (2000). Metal analysis of scales taken from arctic grayling. *Arch. Environ. Contam. Toxicol.* **39**, 515–522.

Flegal, A. R., Rosman, K. J. R., and Stephenson, M. D. (1987). Isotope systematics of contaminant leads in Monterey Bay. *Environ. Sci. Technol.* **21**, 1075–1079.

Flegal, A. R., Nriagu, J. O., Niemeyer, S., and Coale, K. H. (1989). Isotopic tracers of lead contamination in the Great Lakes. *Nature* **21**, 1075–1079.

Flik, G., and Verbost, P. M. (1993). Calcium transport in fish gills and intestine. *J. Exp. Biol.* **184**, 17–29.

Fowler, B. A. (1998). Roles of lead-binding proteins in mediating lead bioavailability. *Environ. Health Perspect.* **106**(Suppl. 6), 1585–1587.

Gao, Y., Kan, A., and Tomson, M. B. (2003). Critical evaluation of desorption phenomena of heavy metals from natural sediments. *Environ. Sci. Technol.* **37**, 5566–5573.

Gilfillan, S. C. (1965). Lead poisoning and the fall of Rome. *J. Occup. Med.* **7**, 53–60.

Giusi, G., Alo, R., Crudo, M., Facciolo, R. M., and Canonaco, M. (2008). Specific cerebral heat shock proteins and histamine receptor cross-talking mechanisms promote distinct lead-dependent neurotoxic responses in teleosts. *Toxicol. Appl. Pharmacol.* **227**, 248–256.

Godwin, H. A. (2001). The biological chemistry of lead. *Curr. Opin. Chem. Biol.* **5**, 223–227.

Goto, D., and Wallace, W. G. (2009). Metal intracellular partitioning as a detoxification mechanism for mummichogs (*Fundulus heteroclitus*) living in metal-polluted salt marshes. *Mar. Environ. Res.* **69**, 163–171.

Goyer, R. A. (1983). Intracellular sites of toxic metals. *Neurotoxicology* **4**, 147–156.

Goyer, R. A. (1996). Results of lead research: prenatal exposure and neurological consequences. *Environ. Health Perspect.* **104**, 1050–1054.

Goyer, R. A., and Clarkson, T. W. (2001). Toxic effects of metals. In: *Casarett and Doull's Toxicology: The Basic Science of Poisons* (C.D. Klaassen, ed.), pp. 811–867. McGraw-Hill, New York.

Goyer, R. A., May, P., Cates, M. M., and Krigman, M. R. (1970). Lead and protein content of isolated intranuclear inclusion bodies from kidneys of lead-poisoned rats. *Lab. Invest.* **22**, 245–251.

Grosell, M., Gerdes, R., and Brix, K. V. (2006). Influence of Ca, humic acid and pH on lead accumulation and toxicity in the fathead minnow during prolonged water-borne lead exposure. *Comp. Biochem. Physiol. C* **143**, 473–483.

Gunshin, H., Mackenzie, B., Berger, U. V., Gunshin, Y., Romero, M. F., Boron, W. F., Nussberger, S., Gollan, J. L., and Hediger, M. A. (1997). Cloning and characterization of a mammalian proton-coupled metal-ion transporter. *Nature* **388**, 482–488.

Heier, L. S., Lien, I. B., Stromseng, A. E., Ljones, M., Rosseland, B. O., Tollefsen, K. E., and Salbu, B. (2009). Speciation of lead, copper, zinc and antimony in water draining a shooting range–time dependent metal accumulation and biomarker responses in brown trout (*Salmo trutta* L.). *Sci. Total Environ.* **407**, 4047–4055.

Hodson, P. V., Blunt, B. R., Spry, J., and Austen, K. (1977). Evaluation of erythrocyte delta-aminolevulinic acid dehydratase activity as a short-term indicator in fish of a harmful exposure to lead. *J. Fish. Res. Bd Can.* **34**, 501–508.

Hodson, P. V., Blunt, B. R., and Spry, D. J. (1978). Chronic toxicity of water-borne and dietary lead to rainbow trout (*Salmo gairdneri*) in Lake Ontario water. *Water Res.* **12**, 869–878.

Hodson, P. V., Blunt, B. R., Jensen, D., and Morgan, S. (1979). Effect of fish age on predicted and observed chronic toxicity of lead to rainbow trout in Lake Ontario water. *J. Great Lakes Res.* **5**, 84–89.

Hodson, P. V., Dixon, D. G., Spry, D. J., Whittle, D. M., and Sprague, J. B. (1982). Effect of growth rate and size of fish on rate of intoxication by waterborne lead. *Can. J. Fish. Aquat. Sci.* **39**, 1243–1251.

Hoenderop, J. G., Voets, T., Hoefs, S., Weidema, F., Prenen, J., Nilius, B., and Bindels, R. J. (2003). Homo- and heterotetrameric architecture of the epithelial Ca^{2+} channels TRPV5 and TRPV6. *EMBO J.* **22**, 776–785.

Holcombe, G. W., Benoit, D. A., Leonard, E. N., and McKim, J. M. (1976). Long-term effects of lead exposure on three generations of brook trout (*Salvelinus fontinalis*). *J. Fish. Res. Bd Can.* **33**, 1731–1741.

Hsu, F. S., Krook, L., Shively, J. N., Duncan, J. R., and Pond, W. G. (1973). Lead inclusion bodies in osteoclasts. *Science* **181**, 447–448.

ILZSG (2010). http://www.ilzsg.org/static/statistics.aspx.

Jackim, E. (1973). Influence of lead and other metals on fish delta-aminolevulinic acid dehydrase activity. *J. Fish. Res. Bd Can.* **30**, 560–562.

Johansson-Sjobeck, M., and Larsson, A. (1979). Effects of inorganic lead on delta-aminolevulinic acid dehydratase activity and hematological variables in the rainbow trout, *Salmo gairdnerii*. *Arch. Environ. Contam. Toxicol.* **8**, 419–431.

Jones, J. R. E. (1938). The relative toxicity of salts of lead, zinc and copper to the stickleback (*Gasterosteus aculeatus* L.) and the effect of calcium on the toxicity of lead and zinc salts. *J. Exp. Biol.* **15**, 394–407.

Kaneko, T., Shiraishi, K., Katoh, F., Hasegawa, A., and Hiroi, H. (2002). Chloride cells during early life stages of fish and their functional differentiation. *Fish. Sci.* **68**, 1–9.

Klaassen, C. D., and Shoeman, D. W. (1974). Biliary excretion of lead in rats, rabbits, and dogs. *Toxicol. Appl. Pharmacol.* **29**, 434–446.

Kleinow, K. M., and James, M. O. (2001). Response of the teleost gastrointestinal system to xenobiotics. In: Target Organ Toxicity in Marine and Freshwater Teleosts, *Vol. 1.* Organs (D. Schlenk and W.H. Benson, eds), pp. 269–362. Taylor & Francis, London.

Komjarova, I., and Blust, R. (2009). Multimetal interactions between Cd, Cu, Ni, Pb, and Zn uptake from water in the zebrafish *Danio rerio*. *Environ. Sci. Technol.* **43**, 7225–7229.

Krajnovic-Ozretic, M., and Ozretic, B. (1980). The ALA-D activity test in lead-exposed grey mullet *Mugil auratus*. *Mar. Ecol. Prog. Ser.* **3**, 187–191.

Kwong, R. W., and Niyogi, S. (2009). The interactions of iron with other divalent metals in the intestinal tract of a freshwater teleost, rainbow trout (*Oncorhynchus mykiss*). *Comp. Biochem. Physiol. C* **150**, 442–449.

Lachant, N. A., Tomoda, A., and Tanaka, K. R. (1984). Inhibition of the pentose phosphate shunt by lead: a potential mechanism for hemolysis in lead poisoning. *Blood* **63**, 518–524.

Larsson, D., Lundgren, T., and Sundell, K. (1998). Ca^{2+} uptake through voltage-gated L-type Ca^{2+} channels by polarized enterocytes from Atlantic cod *Gadus morhua*. *J. Membr. Biol.* **164**, 229–237.

Leal, R. B., Ribeiro, S. J., Posser, T., Cordova, F. M., Rigon, A. P., Zaniboni Filho, E., and Bainy, A. C. (2006). Modulation of ERK1/2 and p38(MAPK) by lead in the cerebellum of Brazilian catfish *Rhamdia quelen*. *Aquat. Toxicol.* **77**, 98–104.

Macdonald, A., Silk, L., Schwartz, M., and Playle, R. C. (2002). A lead–gill binding model to predict acute lead toxicity to rainbow trout (*Oncorhynchus mykiss*). *Comp. Biochem. Physiol. C* **133**, 227–242.

Mager, E. M., Wintz, H., Vulpe, C. D., Brix, K. V., and Grosell, M. (2008). Toxicogenomics of water chemistry influence on chronic lead exposure to the fathead minnow (*Pimephales promelas*). *Aquat. Toxicol.* **87**, 200–209.

Mager, E. M., Brix, K. V., and Grosell, M. (2010a). Influence of bicarbonate and humic acid on effects of chronic waterborne lead exposure to the fathead minnow (*Pimephales promelas*). *Aquat. Toxicol.* **96**, 135–144.

Mager, E. M., Esbaugh, A. J., Brix, K. V., Ryan, A. C., and Grosell, M. (2010b). Influences of water chemistry on the acute toxicity of lead to *Pimephales promelas* and *Ceriodaphnia dubia*. *Comp. Biochem. Physiol.* **153**, 82–90.

Magyar, J. S., Weng, T. C., Stern, C. M., Dye, D. F., Rous, B. W., Payne, J. C., Bridgewater, B. M., Mijovilovich, A., Parkin, G., Zaleski, J. M., Penner-Hahn, J. E., and Godwin, H. A. (2005). Reexamination of lead(II) coordination preferences in sulfur-rich sites: implications for a critical mechanism of lead poisoning. *J. Am. Chem. Soc.* **127**, 9495–9505.

Mahler, B. J., Van Metre, P. C., and Callender, E. (2006). Trends in metals in urban and reference lake sediments across the United States, 1970 to 2001. *Environ. Toxicol. Chem.* **25**, 1698–1709.

Manton, W. I., and Cook, J. D. (1984). High accuracy (stable isotope dilution) measurements of lead in serum and cerebrospinal fluid. *Br. J. Ind. Med.* **41**, 313–319.

Marshall, W. S. (2002). Na(+), Cl(-), Ca(2+) and Zn(2+) transport by fish gills: retrospective review and prospective synthesis. *J. Exp. Zool.* **293**, 264–283.

Marshall, W. S., and Grosell, M. (2006). Ion transport, osmoregulation, and acid–base balance. In: *The Physiology of Fishes* (D.H. Evans and J.B. Claiborne, eds), pp. 177–230. CRC Press, Boca Raton, FL.

Marshall, W. S., Bryson, S. E., and Wood, C. M. (1992). Calcium transport by isolated skin of rainbow trout. *J. Exp. Biol.* **166**, 297–316.

McCormick, S. D., Hasegawa, S., and Hirano, T. (1992). Calcium uptake in the skin of a freshwater teleost. *Proc. Natl. Acad. Sci. U.S.A.* **89**, 3635–3638.

McGeer, J. C., Brix, K. V., Skeaff, J. M., DeForest, D. K., Brigham, S. I., Adams, W. J., and Green, A. (2003). Inverse relationship between bioconcentration factor and exposure concentration for metals: implications for hazard assessment of metals in the aquatic environment. *Environ. Toxicol. Chem.* **22**, 1017–1037.

Mebane, C. A., Hennessy, D. P., and Dillon, F. S. (2008). Developing acute-to-chronic toxicity ratios for lead, cadmium, and zinc using rainbow trout, a mayfly, and a midge. *Water Air Soil Pollut.* **188**, 41–66.

Merlini, M., and Pozzi, G. (1977). Lead and freshwater fishes: Part I – Lead accumulation and water pH. *Environ. Pollut.* **12**, 167–172.

Monteiro, H. P., Abdalla, D. S., Augusto, O., and Bechara, E. J. (1989). Free radical generation during delta-aminolevulinic acid autoxidation: induction by hemoglobin and connections with porphyrinpathies. *Arch. Biochem. Biophys.* **271**, 206–216.

Mount, D. R., Barth, A. K., Garrison, T. D., Barten, K. A., and Hockett, J. R. (1994). Dietary and waterborne exposure of rainbow trout (*Oncorhynchus mykiss*) to copper, cadmium, lead and zinc using a live diet. *Environ. Toxicol. Chem.* **13**, 2031–2041.

Muller, B., and Sigg, L. (1990). Interaction of trace metals with natural particle surfaces: comparison between adsorption experiments and field measurements. *Aquat. Sci.* **52**, 79–92.

Nolan, C. V., and Shaikh, Z. A. (1992). Lead nephrotoxicity and associated disorders: biochemical mechanisms. *Toxicology* **73**, 127–146.

Nriagu, J. O. (1989). A global assessment of natural sources of atmospheric trace metals. *Nature* **338**, 47–49.

Nriagu, J. O. (1990). The rise and fall of leaded gasoline. *Sci. Total Environ.* **92**, 13–28.

Nriagu, J. O., and Pacyna, J. M. (1988). Quantitative assessment of worldwide contamination of air, water and soils by trace metals. *Nature* **333**, 134–139.

Nyman, H. G. (1981). Sublethal effects of lead (Pb) on size selective predation by fish: Applications on the ecosystem level. *Verh. Intern. Verein. Limnol.* **21**, 1126–1130.

OECD (1993). *Risk Reduction Monograph No. 1: Lead Background and National Experience with Reducing Risk*. Paris: Organisation for Economic Co-operation and Development.

Ojo, A. A., and Wood, C. M. (2007). *In vitro* analysis of the bioavailability of six metals via the gastro-intestinal tract of the rainbow trout (*Oncorhynchus mykiss*). *Aquat. Toxicol.* **83**, 10–23.

Oteiza, P. I., Kleinman, C. G., Demasi, M., and Bechara, E. J. (1995). 5-Aminolevulinic acid induces iron release from ferritin. *Arch. Biochem. Biophys.* **316**, 607–611.

Paglia, D. E., Valentine, W. N., and Fink, K. (1977). Lead poisoning. Further observations on erythrocyte pyrimidine-nucleotidase deficiency and intracellular accumulation of pyrimidine nucleotides. *J. Clin. Invest.* **60**, 1362–1366.

Palumbo-Roe, B., Klinck, B., Banks, V., and Quigley, S. (2009). Prediction of the long-term performance of abandoned lead zinc mine tailings in a Welsh catchment. *J. Geochem. Explor.* **100**, 169–181.

Pan, T.-C., Liao, B.-K., Huang, C.-J., Lin, L.-Y., and Hwang, P.-P. (2005). Epithelial Ca^{2+} channel expression and Ca^{2+} uptake in developing zebrafish. *Am. J. Physiol. Regul. Integr. Comp. Physiol.* **289**, R1202–R1211.

Paquin, P. R., Gorsuch, J. W., Apte, S. C., Batley, G. E., Bowles, K. C., Campbell, P. G. C., Delos, C. G., Di Toro, D. M., Dwyer, R. L., Galvez, F., Gensemer, R. W., Goss, G. G.,

Hostrand, U., Janssen, C. R., McGeer, J. C., Naddy, R. B., Playle, R. C., Santore, R. C., Schneider, U., Stubblefield, W. A., Wood, C. M., and Wu, K. B. (2002). The biotic ligand model: a historical overview. *Comp. Biochem. Physiol. C* **133**, 3–36.

Parkinson, A. (2001). Biotransformation of xenobiotics. In: *Casarett and Doull's Toxicology: The Basic Science of Poisons* (C.D. Klaassen, ed.), pp. 133–224. McGraw-Hill, New York.

Patel, M., Rogers, J. T., Pane, E. F., and Wood, C. M. (2006). Renal responses to acute lead waterborne exposure in the freshwater rainbow trout (*Oncorhynchus mykiss*). *Aquat. Toxicol.* **80**, 362–371.

Patterson, C. C. (1980). An alternative perspective – lead pollution in the human environment: origin, extent, and significance. In *Lead in the Human Environment*. Washington, DC: National Research Council, Committee on Lead in the Environment.

Perry, S. F., and Wood, C. M. (1985). Kinetics of branchial calcium uptake in the rainbow trout: effects of acclimation to various external calcium levels. *J. Exp. Biol.* **116**, 411–433.

Persson, P., Johannsson, S., Takagi, Y., and Bjornsson, B. (1997). Estradiol–17β and nutritional status affect calcium balance, scale and bone resorption, and bone formation in rainbow trout, *Oncorhynchus mykiss*. *J. Comp. Physiol. B.* **167**, 468–473.

Persson, P., Sundell, K., Bjornsson, B., and Lundqvist, H. (1998). Calcium metabolism and osmoregulation during sexual maturation of river running Atlantic salmon. *J. Fish Biol.* **54**, 669–684.

Pounds, J. G., Long, G. J., and Rosen, J. F. (1991). Cellular and molecular toxicity of lead in bone. *Environ. Health Perspect.* **91**, 17–32.

Prengaman, R. D. (2003). *Lead Alloys*. Wiley-VCH, New York.

Qian, Y., and Tiffany-Castiglioni, E. (2003). Lead-induced endoplasmic reticulum (ER) stress responses in the nervous system. *Neurochem. Res.* **28**, 153–162.

Qiu, A., and Hogstrand, C. (2004). Functional characterisation and genomic analysis of an epithelial calcium channel (ECaC) from pufferfish, *Fugu rubripes*. *Gene* **342**, 113–123.

Rademacher, D. J., Steinpreis, R. E., and Weber, D. N. (2001). Short-term exposure to dietary Pb and/or DMSA affects dopamine and dopamine metabolite levels in the medulla, optic tectum, and cerebellum of rainbow trout (*Oncorhynchus mykiss*). *Pharmacol. Biochem. Behav.* **70**, 199–207.

Rademacher, D. J., Steinpreis, R. E., and Weber, D. N. (2003). Effects of dietary lead and/or dimercaptosuccinic acid exposure on regional serotonin and serotonin metabolite content in rainbow trout (*Oncorhynchus mykiss*). *Neurosci. Lett.* **339**, 156–160.

Rademacher, D. J., Weber, D. N., and Hillard, C. J. (2005). Waterborne lead exposure affects brain endocannabinoid content in male but not female fathead minnows (*Pimephales promelas*). *Neurotoxicology* **26**, 9–15.

Reichert, W. L., Federighi, D. A., and Malins, D. C. (1979). Uptake and metabolism of lead and cadmium in coho salmon (*Oncorhynchus kisutch*). *Comp. Biochem. Physiol. C* **63**, 229–234.

Renberg, I., Brannvall, M.-J., Bindler, R., and Emteryd, O. (2000). Atmospheric lead pollution history during four millennia (2000 BC to 2000 AD) in Sweden. *Ambio* **29**, 150–156.

Reuer, M. K., and Weiss, D. J. (2002). Anthropogenic lead dynamics in the terrestrial and marine environment. *Philos. Trans. A Math. Phys. Eng. Sci.* **360**, 2889–2904.

Richards, J. G., Curtis, H. J., Burnison, B. K., and Playle, R. C. (2001). Effects of natural organic matter source on reducing metal toxicity to rainbow trout (*Oncorhynchus mykiss*) and on metal binding to their gills. *Environ. Toxicol. Chem.* **20**, 1159–1166.

Rocha, M. E., Dutra, F., Bandy, B., Baldini, R. L., Gomes, S. L., Faljoni-Alario, A., Liria, C. W., Miranda, M. T., and Bechara, E. J. (2003). Oxidative damage to ferritin by 5-aminolevulinic acid. *Arch. Biochem. Biophys.* **409**, 349–356.

Rodrigues, A. L., Bellinaso, M. L., and Dick, T. (1989). Effect of some metal ions on blood and liver delta-aminolevulinate dehydratase of *Pimelodus maculatus* (Pisces, Pimelodidae). *Comp. Biochem. Physiol. B* **94**, 65–69.

Rogers, J. T., and Wood, C. M. (2004). Characterization of branchial lead–calcium interaction in the freshwater rainbow trout *Oncorhynchus mykiss*. *J. Exp. Biol.* **207**, 813–825.

Rogers, J. T., Richards, J. G., and Wood, C. M. (2003). Ionoregulatory disruption as the acute toxic mechanism for lead in the rainbow trout (*Oncorhynchus mykiss*). *Aquat. Toxicol.* **64**, 215–234.

Rogers, J. T., Patel, M., Gilmour, K. M., and Wood, C. M. (2005). Mechanisms behind Pb-induced disruption of Na^+ and Cl^- balance in rainbow trout (*Oncorhynchus mykiss*). *Am. J. Physiol. Regul. Integr. Comp. Physiol.* **289**, R463–R472.

Saeij, J. P., Wiegertjes, G. F., and Stet, R. J. (1999). Identification and characterization of a fish natural resistance-associated macrophage protein (NRAMP) cDNA. *Immunogenetics* **50**, 60–66.

Sanders, T., Liu, Y., Buchner, V., and Tchounwou, P. B. (2009). Neurotoxic effects and biomarkers of lead exposure: a review. *Rev. Environ. Health* **24**, 15–45.

Sauer, G. R., and Watabe, N. (1989). Temporal and metal-specific patterns in the accumulation of heavy metals by the scales of *Fundulus heteroclitus*. *Aquat. Toxicol.* **14**, 233–248.

Sauter, S., Buxton, K. S., Macek, K. J. and Petrocelli, S. R. (1976). *Effects of Exposure to Heavy Metals on Selected Freshwater Fish: Toxicity of Copper, Cadmium, Chromium and Lead to Eggs and Fry of Seven Fish Species*. EPA–600/3–76–105. Duluth, MN: USEPA Office of Research and Development.

Schmitt, C. J., Dwyer, F. J., and Finger, S. E. (1984). Bioavailability of Pb and Zn from mine tailings as indicated by erythrocyte delta-aminolevulinic acid dehydratase (ALA-D) activity in suckers (Pisces: Catostomidae). *Can. J. Fish. Aquat. Sci.* **41**, 1030–1040.

Schmitt, C. J., Wildhaber, M. L., Hunn, J. B., Nash, T., Tieger, M. N., and Steadman, B. L. (1993). Biomonitoring of lead-contaminated Missouri streams with an assay for erythrocyte delta-aminolevulinic acid dehydratase activity in fish blood. *Arch. Environ. Contam. Toxicol.* **25**, 464–475.

Schmitt, C. J., Caldwell, C. A., Olsen, B., Serdar, D., and Coffey, M. (2002). Inhibition of erythrocyte delta-aminolevulinic acid dehydratase (ALAD) activity in fish from waters affected by lead smelters. *Environ. Monit. Assess.* **77**, 99–119.

Schmitt, C. J., Whyte, J. J., Brumbaugh, W., and Tillitt, D. E. (2005). Biochemical effects of lead, zinc, and cadmium from mining on fish in the Tri-states District of Northeastern Oklahoma, USA. *Environ. Toxicol. Chem.* **24**, 1483–1495.

Schmitt, C. J., Brumbaugh, W. G., and May, T. W. (2007a). Accumulation of metals in fish from lead–zinc mining areas of southeastern Missouri, USA. *Ecotoxicol. Environ. Saf.* **67**, 14–30.

Schmitt, C. J., Whyte, J. J., Roberts, A. P., Annis, M. L., May, T. W., and Tillitt, D. E. (2007b). Biomarkers of metals exposure in fish from lead–zinc mining areas of southeastern Missouri, USA. *Ecotoxicol. Environ. Saf.* **67**, 31–47.

Schubauer-Berigan, M. K., Dierkes, J. R., Monson, P. D., and Ankley, G. T. (1993). pH-dependent toxicity of Cd, Cu, Ni, Pb and Zn to *Ceriodaphnia dubia*, *Pimephales promelas*, *Hyalella azteca*, and *Lumbriculus variegatus*. *Environ. Toxicol. Chem.* **12**, 1261–1266.

Scott, K. M., Hwang, K. M., Jurkowitz, M., and Brierley, G. P. (1971). Ion transport by heart mitochondria. 23. The effects of lead on mitochondrial reactions. *Arch. Biochem. Biophys.* **147**, 557–567.

Settle, D. M., and Patterson, C. C. (1980). Lead in albacore: guide to lead pollution in Americans. *Science* **207**, 1167–1176.

Shah, S. L. (2006). Hematological parameters in tench *Tinca tinca* after short term exposure to lead. *J. Appl. Toxicol.* **26**, 223–228.

Shahsavarani, A., McNeill, B., Galvez, F., Wood, C. M., Goss, G. G., Hwang, P. P., and Perry, S. F. (2006). Characterization of a branchial epithelial calcium channel (ECaC) in freshwater rainbow trout (*Oncorhynchus mykiss*). *J. Exp. Biol.* **209**, 1928–1943.

Shephard, K. L. (1994). Functions for fish mucus. *Rev. Fish Biol. Fish.* **4**, 401–429.

Sibthorpe, D., Baker, A. M., Gilmartin, B. J., Blackwell, J. M., and White, J. K. (2004). Comparative analysis of two slc11 (Nramp) loci in *Takifugu rubripes*. *DNA Cell Biol.* **23**, 45–58.

Silbergeld, E. K. (1992). Mechanisms of lead neurotoxicity, or looking beyond the lamppost. *FASEB J.* **6**, 3201–3206.

Simons, T. J. (1988). Active transport of lead by the calcium pump in human red cell ghosts. *J. Physiol.* **405**, 105–113.

Simons, T. J. (1993). Lead transport and binding by human erythrocytes *in vitro*. *Pflugers Arch.* **423**, 307–313.

Six, K. M., and Goyer, R. A. (1970). Experimental enhancement of lead toxicity by low dietary calcium. *J. Lab. Clin. Med.* **76**, 933–942.

Six, K. M., and Goyer, R. A. (1972). The influence of iron deficiency on tissue content and toxicity of ingested lead in the rat. *J. Lab. Clin. Med.* **79**, 128–136.

Sloman, K. A., Lepage, O., Rogers, J. T., Wood, C. M., and Winberg, S. (2005). Socially-mediated differences in brain monoamines in rainbow trout: effects of trace metal contaminants. *Aquat. Toxicol.* **71**, 237–247.

Somero, G. N., Chow, T. J., Yancey, P. H., and Snyder, C. B. (1977a). Lead accumulation rates in tissues of the estuarine teleost fish, *Gillichthys mirabilis*: salinity and temperature effects. *Arch. Environ. Contam. Toxicol.* **6**, 337–348.

Somero, G. N., Yancey, P. H., Chow, T. J., and Snyder, C. B. (1977b). Lead effects on tissue and whole organism respiration of the estuarine teleost fish, *Gillichthys mirabilis*. *Arch. Environ. Contam. Toxicol.* **6**, 349–354.

Spehar, R. L., and Fiandt, J. (1986). Acute and chronic effects of water quality criteria-based metal mixtures on three aquatic species. *Environ. Toxicol. Chem.* **5**, 917–931.

Spieler, R. E., Russo, A. C., and Weber, D. N. (1995). Waterborne lead affects circadian variations of brain neurotransmitters in fathead minnows. *Bull. Environ. Contam. Toxicol.* **55**, 412–418.

Spit, B. J., Wibowo, A. A. E., Feron, V. J., and Zielhius, R. L. (1981). Ultrastructural changes in the kidneys of rabbits treated with lead acetate. *Arch. Toxicol.* **49**, 85–91.

Srivastava, A. K., and Mishra, S. (1979). Blood dyscrasia in a teleost, *Colisa fasciatus* after acute exposure to sublethal concentrations of lead. *J. Fish Biol.* **14**, 199–203.

Tabche, L. M., Martinez, C. M., and Sanchez-Hidalgo, E. (1990). Comparative study of toxic lead effect on gill and haemoglobin of tilapia fish. *J. Appl. Toxicol.* **10**, 193–195.

Taillefert, M., Lienemann, C.-P., Gaillard, J.-F., and Perret, D. (2000). Speciation, reactivity, and cycling of Fe and Pb in a meromictic Lake (Paul Lake, MI). *Geochim. Cosmochim. Acta* **64**, 169–183.

Tao, S., Liang, T., Cao, J., Dawson, R. W., and Liu, C. (1999). Synergistic effect of copper and lead uptake by fish. *Ecotoxicol. Environ. Saf.* **44**, 190–195.

Taylor, J. R., and Grosell, M. (2009). The intestinal response to feeding in seawater gulf toadfish, *Opsanus beta*, includes elevated base secretion and increased epithelial oxygen consumption. *J. Exp. Biol.* **212**, 3873–3881.

Tewari, H., Gill, T. S., and Pant, J. (1987). Impact of chronic lead poisoning on the hematological and biochemical profiles of a fish, *Barbus conchonius* (Ham). *Bull. Environ. Contam. Toxicol.* **38**, 748–752.

Toscano, C. D., and Guilarte, T. R. (2005). Lead neurotoxicity: from exposure to molecular effects. *Brain. Res. Brain Res. Rev.* **49**, 529–554.

Turner, D. R., Whitfield, M., and Dickson, A. G. (1981). The equilibrium speciation of dissolved components in freshwater and seawater at 25°C and 1 atm pressure. *Geochim. Cosmochim. Acta* **45**, 855–881.

USEPA (1979). Water-related Environmental Fate of 129 Priority Pollutants, *Vol. I.* Introduction and Technical Background, Metals and Inorganics, Pesticides and PCBs.. Office of Water Planning and Standards, Washington, DC.

USEPA (1985). *Guidelines for Deriving Numerical National Water Quality Criteria for the Protection of Aquatic Organisms and Their Uses.* Office of Water, Washington, DC.

USEPA (1986). *Air Quality Criteria for Lead*, Vol. II of IV. Research Triangle Park, NC: Environmental Criteria and Assessment Office.

USEPA (2005). *Best Management Practices for Lead at Outdoor Shooting Ranges.* Division of Enforcement and Compliance Assistance, New York.

USEPA (2010). http://www.epa.gov/air/airtrends/lead.html.

USGS (2010a). *Lead statistics*, in Kelly, T. D., and Matos, G. R., comps., Historical statistics for mineral and materials commodities in the United States. U.S. Geological Survey Data Series 140. http://pubs.usgs.gov/ds/2005/140/.

USGS (2010b). *2008 Minerals Yearbook: Lead* [advance release]. US Geological Survey, US Department of the Interior. http://minerals.usgs.gov/minerals/pubs/commodity/lead/myb1-2008-lead.pdf

Vantelon, D., Lanzirotti, A., Scheinost, A. C., and Kretzschmar, R. (2005). Spatial distribution and speciation of lead around corroding bullets in a shooting range soil studied by micro-X-ray fluorescence and absorption spectroscopy. *Environ. Sci. Technol.* **39**, 4808–4815.

Varanasi, U., and Gmur, D. J. (1978). Influence of water-borne and dietary calcium on uptake and retention of lead by coho salmon (*Oncorhynchus kisutch*). *Toxicol. Appl. Pharmacol.* **46**, 65–75.

Varanasi, U., and Markey, D. (1978). Uptake and release of lead and cadmium in skin and mucus of coho salmon (*Oncorhynchus kisutch*). *Comp. Biochem. Physiol. C* **60**, 187–191.

Varanasi, U., Robisch, P. A., and Malins, D. C. (1975). Structural alterations in fish epidermal mucus produced by water-borne lead and mercury. *Nature* **258**, 431–432.

Vinikour, W. S., Goldstein, R. M., and Anderson, R. V. (1980). Bioconcentration patterns of zinc, copper, cadmium and lead in selected fish species from the Fox River, Illinois. *Bull. Environ. Contam. Toxicol.* **24**, 727–734.

Waalkes, M. P., and Klaassen, C. D. (1985). Concentration of metallothionein in major organs of rats after administration of various metals. *Fundam. Appl. Toxicol.* **5**, 473–477.

Weber, D. N. (1993). Exposure to sublethal levels of waterborne lead alters reproductive behavior patterns in fathead minnows (*Pimephales promelas*). *Neurotoxicology* **14**, 347–358.

Weber, D. N. (1996). Lead-induced metabolic imablances and feeding alterations in juvenile fathead minnows (*Pimephales promelas*). *Environ. Toxicol. Water Qual.* **11**, 45–51.

Weber, D. N., and Dingel, W. M. (1997). Alterations in neurobehavioral responses in fishes exposed to lead and lead-chelating agents. *Am. Zool.* **37**, 354–362.

Weber, D. N., Russo, A. C., Seale, D. B., and Spieler, R. E. (1991). Waterborne lead affects feeding neurotransmitter levels of juvenile fathead minnows (*Pimephales promelas*). *Aquat. Toxicol.* **21**, 71–80.

Weir, P. A., and Hine, C. H. (1970). Effects of various metals on behavior of conditioned goldfish. *Arch. Environ. Health* **20**, 45–51.

Weis, J. S., and Weis, P. (1998). Effects of exposure to lead on behavior of mummichog (*Fundulus heteroclitus* L.) larvae. *J. Exp. Mar. Biol. Ecol.* **222**, 1–10.

Weis, J. S., Samson, J., Zhou, T., Skurnick, J., and Weis, P. (2003). Evaluating prey capture by larval mummichogs (*Fundulus heteroclitus*) as a potential biomarker for contaminants. *Mar. Environ. Res.* **55**, 27–38.

Westfall, B. A. (1945). Coagulation film anoxia in fishes. *Ecology* **26**, 283–287.

Wilson, R. W., Wilson, J. M., and Grosell, M. (2002). Intestinal bicarbonate secretion by marine teleost fish – why and how? *Biochim. Biophys. Acta Biomembr.* **1566**, 182–193.

Windom, H. L., Byrd, T., Smith, R. G., and Huan, F. (1991). Inadequacy of NAS-QUAN data for assessing metal trends in the nation's rivers. *Environ. Sci. Technol.* **25**, 1137–1142.

Wood, C. M. (2001). Toxic responses of the gill. In: Target Organ Toxicity in Marine and Freshwater Teleosts, *Vol. 1.* Organs (D. Schlenk and W.H. Benson, eds), pp. 1–89. Taylor & Francis, London.

Zuo, P., Qu, W., Cooper, R. N., Goyer, R. A., Diwan, B. A., and Waalkes, M. P. (2009). Potential role of α-synuclein and metallothionein in lead-induced inclusion body formation. *Toxicol. Sci.* **111**, 100–108.

5

MERCURY

KAREN KIDD

KATHARINA BATCHELAR

Homeostasis and Toxicology of Non-Essential Metals: Volume 31B
FISH PHYSIOLOGY

Anthropogenic use of mercury (Hg) has had local and global consequences, with emissions to the atmosphere from fossil fuel combustion or waste incineration and subsequent long-range transport and deposition raising background levels by two to four times in the global environment. There are three main forms of Hg in the aquatic environment, inorganic Hg°, Hg^{2+} [Hg(II)], and organic CH_3Hg^+ [methylmercury or MeHg(I)], with MeHg(I) being of particular concern because of its potent neurotoxicity. Waterborne or dietborne exposures of fish to Hg(II) and MeHg(I) affect their growth, development, and reproduction, even in remote ecosystems. This chapter reviews the current understanding of the sources and cycling of Hg in the aquatic environment, and then focuses on the toxicity and fate of Hg(II) and MeHg(I) in fish.

1. INTRODUCTION

Our understanding of how fish are exposed to mercury (Hg) and its effects on their health is based on studies going back to the 1950s. They were motivated by the increasing recognition that Hg, especially its organic form methylmercury [CH_3Hg^+ or MeHg(I)], has adverse effects on wildlife and human health and that fish consumption is a main route of exposure to this metal. After several mass poisonings occurred in the 1950s through 1970s (e.g. Minamata Bay, Japan), it became critical to understand how Hg was getting into fish consumed by humans and wildlife and to determine whether it was affecting the fish themselves, the two main foci of this chapter.

Mercury's symbol originates from the Latin word that means "liquid silver" or "quick silver", *hydrargyrus*. It is a non-transition metal found in group 12 (IIB) of the periodic table, with unique properties: Hg is liquid under ambient temperatures and its weak Hg–Hg bonds result in low melting (-38.9°C) and boiling (356.58°C) points and a higher volatility (0.0002 Pa) than other metals. It has three main oxidation states (0, +1, +2, and +3) and seven stable isotopes (^{196}Hg, ^{198}Hg, ^{199}Hg, ^{200}Hg, ^{201}Hg, ^{202}Hg, ^{204}Hg) with ^{202}Hg and ^{200}Hg being the two most common in the environment.

About 0.05 ppm of the Earth's crust is made of Hg (Parsons and Percival, 2005) and the element is present in a variety of different forms (Rytuba, 2005). Most Hg-bearing rocks have concentrations below 100 ppb ($\mu g\ kg^{-1}$); the exception is black shale which can contain Hg at concentrations up to 1000 ppb (Rytuba, 2005). It occurs in the Earth's crust as native Hg and as Hg-bearing minerals, including the common form cinnabar (HgS). Mercury deposits have been accessed by humans for centuries because the metal's unique properties make it useful for a wide variety of industrial (e.g. gold mining, chlor-alkali plants), medicinal (e.g. preservative in vaccines), artistic (e.g. pigments), and agricultural (e.g. pesticides) applications.

2. CHEMICAL SPECIATION IN WATER

2.1. Speciation in Freshwater

Mercury is present in freshwaters in three main forms – inorganic [Hg(II)], gaseous elemental (Hg^0) and organic [MeHg(I)] – and the latter two forms, as well as total Hg (THg; sum of all Hg in oxidized sample), are most commonly measured. On their own, these forms of Hg have low water solubility, but are much more soluble when complexed with organic [dissolved organic carbon (DOC)] and inorganic (e.g. sulfate, chloride) ligands. The presence of ligands and the pH and redox of the water strongly influence the speciation of both Hg(II) and MeHg(I). Inorganic Hg binds predominantly to DOC (i.e. with their carboxylic acids and thiols) and Cl^- or OH^- ions in oxic, near neutral waters. Under reducing, anoxic conditions, reduced sulfur species (mainly sulfides) are the dominant inorganic ligands. Both MeHg(I) and THg partition equally between truly dissolved (< 10 kDa), colloidal (0.4 μm–10 kDa) and particulate (> 0.4 μm) phases, and [MeHg(I)] are positively related to aqueous organic carbon, an important ligand for this form (Babiarz et al., 2001).

Mercury concentrations in fresh waters are variable, and accurate measurements require clean techniques to minimize sample contamination (Gill and Bruland, 1990). Unfiltered THg concentrations are typically from 0.3 to 8 ng L^{-1} at pristine sites and generally higher in waters with more DOC; in contrast, THg concentration is much higher (up to 450 μg L^{-1}) at sites where mining or industrial discharges have occurred or at sites with ores high in Hg (Babiarz et al., 2001; Rytuba, 2003; Wiener et al., 2003). MeHg(I) concentrations are usually between 0.1 and 5% of THg concentrations (0.04 and 0.8 ng L^{-1} in oxic, unimpacted waters), but can be 1–2 ng L^{-1} (occasionally 20–30 ng L^{-1}) in waters downstream of mine drainages, industrial discharges, or high-Hg mineral deposits (Babiarz et al., 2001; Rytuba, 2003; Wiener et al., 2003).

2.2. Speciation in Seawater

Cycling of Hg in oceans is complex and not completely understood, but shares many similar processes to those in freshwaters (Mason and Gill, 2005). Mercury is present in both dissolved and particulate-bound forms and includes Hg^0, Hg(II), MeHg(I), and dimethylmercury (Me_2Hg), a form not found in freshwaters. Inorganic (mainly Cl^- in oxic waters, but also SH^-) and organic (mainly thiol-containing) ligands bind Hg(II); MeHg(I) is also found bound to both inorganic and organic ligands. It is generally believed that organic complexes dominate in higher DOC waters, and inorganic complexes with Cl$^-$ (HgCl$_4^{2-}$, HgCl$_3^-$, CH$_3$HgCl) will dominate in lower DOC waters. In addition, there is such strong binding of Hg(II) to Cl$^-$ that concentrations of free Hg(II) are very low ($<2.0 \times 10^{-1}$gL^{-1}). Similarly, the free metal ion concentration for MeHg(I) is less than 1% of the total dissolved fraction (Mason and Gill, 2005).

The main source of Hg to the ocean is from atmospheric deposition of wet and dry forms (Mason and Gill, 2005). THg concentrations are low and range from 40 to 802 pg L^{-1} ($\sim$0.2 to 4.0 pM), with the higher concentrations found near the coast likely due to riverine inputs (Mason and Gill, 2005). Between 5 and 35% of THg is MeHg(I) and measured concentrations of MeHg(I) range from 5 to 144 pg L^{-1} (25 to 670 fM) (Mason and Gill, 2005).

2.3. Methylation of Mercury

Fish are exposed to both Hg(II) and MeHg(I) from water and diet, but food is their main source of MeHg(I) (Hall et al., 1997). Methylation of Hg and demethylation of MeHg(I) are key processes that determine the relative abundance of Hg(II) versus MeHg(I). Aqueous concentrations of MeHg(I) are an important predictor of those in fish tissues because it is this form that is bioconcentrated into lower trophic levels and then biomagnified up through the food web (Wiener et al., 2003). Both biotic and abiotic mechanisms convert Hg(II) to MeHg(I) [and also MeHg(I) to Hg(II)], but methylation by bacteria (predominantly sulfate reducers) is believed to be the dominant process and one that occurs at the oxic–anoxic interface of sediments in both freshwaters and saltwaters. In oceans, demethylation of Me$_2$Hg also contributes to the pool of MeHg(I). Concentrations of MeHg(I) tend to be highest in lakes and streams draining wetlands because they provide conditions favorable for methylation and the DOC important in transporting MeHg(I) to downstream locations. More evidence suggests that the percentage of MeHg(I) in waters is driven not by concentrations of THg, but by conditions favorable for methylation (availability of substrate, temperature, etc.).

3. SOURCES OF MERCURY AND ECONOMIC IMPORTANCE

Mercury was one of the first metals known to humans and has been mined, extracted, and used for millennia. The amalgamation of gold and silver from ores drove its large-scale use from the sixteenth to the twentieth century and has led to a number of highly polluted sites. It was commonly used in chlor-alkali plants to produce chlorine for the production of chlorinated compounds or for the bleaching of pulp, and to generate NaOH for water treatment and numerous industrial applications. Mercury is also an effective pesticide and was used as an active ingredient in chemicals to control weeds, insects, fungi, and bacteria. Finally, scientific instruments (e.g. electrodes, vacuum pumps) as well as household items such as fluorescent bulbs, batteries, computers, and switches contain or contained Hg (Parsons and Percival, 2005).

Mercury in surface waters comes from a combination of anthropogenic and natural sources. Metal mining and smelting, combustion of fossil fuels such as in coal-fired generating stations, production of cement, and garbage incineration have increased Hg loads to the atmosphere (Pacyna et al., 2010), and ultimately to surface waters through local and long-range atmospheric transport and deposition. Direct contamination of waters has also occurred from accidental and intentional release of effluents from chlor-alkali plants and mines. Natural emissions of Hg come from weathering of soils and rocks, volcanic eruptions and geothermal releases, and forest fires, and can enter surface waters through surface runoff or atmospheric transport and deposition (Rytuba, 2003).

Although challenging to determine the relative contributions of natural versus anthropogenic sources, it is well established that human activities have increased environmental concentrations across the globe between two- and four-fold (Lindberg et al., 2007). It is currently believed that anthropogenic and natural (mainly oceans, soils, and terrestrial biogenic sources) emissions contribute equally to the atmospheric pool of Hg (Lindberg et al., 2007). Overall, it is estimated that human activities release over 2300 tonnes each year to the atmosphere, with almost half from coal combustion (Pacyna et al., 2010).

4. ENVIRONMENTAL SITUATIONS OF CONCERN

Surface waters can be both acutely and chronically toxic to fish, although the latter is probably much more common. Though rare, mine drainage or industrial discharges can elevate Hg concentrations in surface waters to levels that are known to be acutely toxic to fish (Rytuba, 2003; Wiener et al., 2003). It is more likely that toxicity in wild fish is related to chronic exposure to MeHg(I) via the diet. This form is effectively biomagnified up aquatic food webs, accumulating to levels in piscivorous fish (Wiener et al., 2003) that have been correlated with adverse effects in wild populations (Sandheinrich and Wiener, 2010).

5. A SURVEY OF ACUTE AND CHRONIC AMBIENT WATER QUALITY CRITERIA FOR FRESHWATER AND SEAWATER

Several countries have water quality criteria for Hg in both freshwaters and saltwaters and these criteria vary by jurisdiction and the form of Hg (Table 5.1). Guidelines are common for Hg(II) or THg and rare for MeHg(I), likely reflecting the difficulties in measuring the latter form in waters. For example, the Canadian Council of Ministers of the Environment (CCME) (2007) guidelines for Hg(II) are 0.026 and 0.016 µg L^{-1} in fresh and marine waters, respectively. Their interim guideline for MeHg(I) (0.004 µg L^{-1}), one of the few available, may not be protective of higher trophic level fish because the CCME (1991) protocol used to derive the guidelines does not address exposure via the diet. No marine water quality guideline currently exists for MeHg(I) because of a lack of data to derive one (CCME, 2007).

To determine no-effect whole-body residues, Beckvar et al. (2005) compiled studies that measured both effects and tissue concentrations. After comparing various methods for deriving a protective concentration (see manuscript for details), they recommended a tissue threshold effect level of 0.2 µg Hg g^{-1} wet weight (ww) for adults and juveniles, and speculated, owing to limited data, that a whole-body residue of 0.02 µg Hg g^{-1} ww would be protective for eggs, larvae, or fry.

6. MECHANISMS OF TOXICITY

Considerable literature exists on the acute and chronic toxicity of Hg(II) (mainly as HgCl$_2$) and MeHg(I) (mainly as MeHgCl) on fish. However, most is based on aqueous exposures. Recent reviews have identified the need to better understand the effects of dietborne Hg, and to improve our understanding of how tissue concentrations are related to toxicity (Beckvar et al., 2005). In addition, it is important to note that fish-eating fish are exposed mainly to MeHg complexed to cysteine (Lemes and Wang, 2009). Laboratory studies using MeHgCl to examine dietborne toxicity may therefore not completely reflect what is occurring in nature.

6.1. Acute Toxicity of Inorganic Mercury

Acute toxicity of aqueous HgCl$_2$ to fish has been studied since the 1950s (e.g. Doudoroff and Katz, 1953). Inorganic Hg is a potent neurotoxicant and causes loss of equilibrium, inactivity, respiratory distress, and ultimately death in fish exposed to high concentrations for 24–96 h (Wobeser, 1975a).

Table 5.1

Mercury ambient water quality guidelines or criteria for the protection of aquatic life in fresh and marine waters

Jurisdiction	Hg form	Acute [Hg] ($\mu g\ L^{-1}$)	Chronic [Hg] ($\mu g\ L^{-1}$)	Target [Hg] ($\mu g\ L^{-1}$)	Notes	Reference
Freshwater						
USA	MeHg(I)	–	–	–	0.3 mg kg^{-1} in organism, based on consumption of 0.0175 kg day^{-1}	USEPA (2009)
Canada	Hg(II)	–	–	0.026		CCME (2007)
Australia/New Zealand	Hg(II)	–	–	0.06	Trigger values for 99% level of protection (% species)	ANZECC and ARMCANZ (2000)
Australia/New Zealand	Hg(II)	–	–	0.6	Trigger values for 95% level of protection (% species)	ANZECC and ARMCANZ (2000)
Australia/New Zealand	Hg(II)	–	–	1.9	Trigger values for 90% level of protection (% species)	ANZECC and ARMCANZ (2000)
Canada	MeHg(I)	–	–	0.004		CCME (2007)
South Africa	THg	1.7	0.08	0.04	Target water quality range	CSIR (1996)
USA	THg	1.4	0.77	–		USEPA (2009)
Marine						
Canada	Hg(II)	–	–	0.016		CCME (2007)
Australia/New Zealand	Hg(II)	–	–	0.1	Trigger values for 99% level of protection (% species)	ANZECC and ARMCANZ (2000)
Australia/New Zealand	Hg(II)	–	–	0.4	Trigger values for 95% level of protection (% species)	ANZECC and ARMCANZ (2000)
Australia/New Zealand	Hg(II)	–	–	0.7	Trigger values for 90% level of protection (% species)	ANZECC and ARMCANZ (2000)
Canada	MeHg(I)	–	–	–		CCME (2007)
South Africa	THg	–	–	0.3	Target value for South African coastal zone: primary consumers	CSIR (1995)
South Africa	THg	–	–	0.3	Target value for South African coastal zone: secondary consumers	CSIR (1995)
USA	THg	1.8	0.94	1		USEPA (2009)

One-day exposure to $100 \, \mu g \, L^{-1}$ decreases oxygen consumption in gill, liver, and muscle of carp by up to 60% (Radhakrishnaiah et al., 1993). Its neurotoxicity is due in part to the suppression of acetylcholinesterase activity; for example, enzyme levels decrease in brain, gill, and liver of fish exposed to $180 \, \mu g \, L^{-1}$ for 2 days (Gill et al., 1990). However, some acute effects, such as lethargy, can be reversed by moving fish to clean water (Wobeser, 1975a).

Elevated acute exposures also affect a number of fish tissues. Mucus production increases on both gills and skin (Wobeser, 1975a), and hemorrhaging, hyperplasia, edema, and necrosis have been found in gill tissues of several species (Jagoe et al., 1996; Ribeiro et al., 2000, 2002; Giari et al., 2008). Aqueous Hg(II) also causes severe necrosis of brain tissues, congestion of liver blood vessels, and damage to kidney tubules and hepatocytes (Wobeser, 1975a; Krishnani et al., 2003).

Waterborne Hg(II) interferes with the function of superficial sensory organs and can mask environmental signals critical for finding prey, avoiding predation, and reproduction (Tierney et al., 2010). Short-term exposures ($<2 \, h$) can decrease the electrical signal transmission from the olfactory epithelium to the brain in rainbow trout (*Oncorhynchus mykiss*) (Hara et al., 1976). Mercury interferes with the chemoreceptors in fish; as examples, it interferes with the binding of L-serine, which has a strong odor, to olfactory receptors (Rehnberg and Schreck, 1986) and of L-alanine to taste receptors (Zelson and Cagan, 1979). In addition, exposures to sublethal levels can lead to complete degeneration of the tastebuds (Pevzner et al., 1986).

Enzymes and substrates involved in both aerobic and anaerobic metabolism are disrupted by Hg(II). Short-term, elevated exposures inhibited acid phosphatase activity in liver, gill, and kidney, and increased its levels in gonads of rosy barb (Gill et al., 1990). Succinate dehydrogenase (SDH) is suppressed in gill, liver, and muscle of common carp (*Cyprinus carpio*) exposed to $100 \, \mu g \, L^{-1}$ for up to 30 days (Radhakrishnaiah et al., 1993). In contrast, lactate dehydrogenase (LDH) levels increased (along with the substrates pyruvate and lactate) up to 100% in all tissues (but especially gill) of exposed fish. These responses indicate that Hg(II) exposure stimulates anaerobic glycolysis and suppresses oxidative metabolic pathways. Cellular oxidative metabolism is impaired because Hg(II) affects the membrane integrity of mitochondria, disrupting ion flow and electron transport chains needed for oxidative phosphorylation (see Zaman and Pardini, 1996, for more detail).

Acute aqueous exposures to Hg(II) can affect the expression of antioxidant enzymes or non-enzymatic scavenger molecules. For example, juvenile matrinxã (*Brycon amazonicus*) exposed for 96 h at $150 \, \mu g \, L^{-1}$ showed significant increases in liver, gill, and muscle superoxide dismutase

(SOD), catalase (CAT), and glutathione S-transferase (GST) activities, in glutathione peroxidase (GSH-Px) in both gill and muscle, and in glutathione reductase (GR) in gill tissues (Monteiro et al., 2010). Production of the scavenger molecules glutathione (GSH) and metallothionein (MT) also increased after Hg exposure in liver, gill, and heart tissues (Monteiro et al., 2010).

Plasma ions are higher in fish exposed to lethal and sublethal concentrations of Hg(II), most likely because it damages kidney tissues and affects this organ's ability to balance ions. In a study by Hilmy et al. (1987), 4–7 day exposures of catfish (*Clarias lazera*) to 100, 220, and 400 μg L^{-1} increased plasma concentrations of Na$^+$, K$^+$, Mg^{2+}, and Ca^{2+}.

Gametes, fertilization, and embryonic development are also affected by short-term exposures to Hg(II). Very short (5 min at 50 μg L^{-1}) exposures of mummichog (*Fundulus heteroclitus*) sperm reduced their motility but did not affect fertilization success for similarly exposed eggs (Khan and Weis, 1987a).

As with other elements, sensitivity to Hg(II) varies across species and is typically higher for smaller than larger bodied fish and for earlier than later life stages. For freshwater species, 96 h LC50 values were highest in catfish (507 μg L^{-1}) and lowest for golden shiner (16.75 μg L^{-1}) at similar temperatures (Table 5.2). Tissue concentrations of > 0.04 μg g^{-1} ww are lethal to eggs and larvae of rainbow trout (Birge et al., 1979). Within one species, e.g. catla (*Catla catla*) or mrigal (*Cirrhinus mrigala*), Hg(II) is more toxic at warmer than colder water temperatures (Table 5.2). Fewer tests have been done on saltwater species, but those tested showed similar sensitivities to Hg(II) (96 h LC50 of 36–200 μg L^{-1}) (Table 5.2), and the LC50 was lower in smaller (85 μg L^{-1}) than in larger (200 μg L^{-1}) sea bass (*Lates calcarifer*) (Krishnani et al., 2003).

6.2. Chronic Toxicity of Inorganic Mercury

6.2.1. REPRODUCTION

Gonad growth is impaired in fish exposed to concentrations of HgCl$_2$ found in highly contaminated waters. For example, 10 μg L^{-1} stopped testes growth in spotted murrel (*Channa punctatus*) (Ram and Sathyanesan, 1983), and mean gonadosomatic index (GSI) was significantly reduced after male air-breathing catfish (*Clarias batrachus*) were exposed for 6 months to 50 μg L^{-1} (Kirubagaran and Joy, 1992).

Within the gonads, development of sperm and eggs is delayed or arrested by Hg(II). At exposures of 10 μg L^{-1} or 50 μg L^{-1}, HgCl$_2$ stopped spermatogenesis in spotted murrel (Ram and Sathyanesan, 1983; Ram and Joy, 1988) and air-breathing catfish (Kirubagaran and Joy, 1992), and Leydig cells were

Table 5.2

Acute lethal concentration 50 (LC50) of aqueous inorganic and organic mercury to freshwater and marine fish species

Common name	Species name	Life stage	Average size (mm)	Test conditions	Hg concentration (µg L^{-1})	Hg form	Test duration (h)	Temperature (°C)	Reference
Freshwater, inorganic Hg									
Banded killifish	*Fundulus diaphanus*	–	≤200	Static	110	HgCl$_2$	96	28	Rehwoldt et al. (1972)
Pumpkinseed	*Lepomis gibbosus*	–	≤200	Static	300	HgCl$_2$	96	28	Rehwoldt et al. (1972)
White perch	*Roccus americanus*	–	≤200	Static	220	HgCl$_2$	96	28	Rehwoldt et al. (1972)
Common white sucker	*Catostomus commersoni*	–	121 ± 10	Flow	687	HgCl$_2$	96	12.1	Duncan and Klaverkamp (1983)
Carp	*Cyprinus carpio*	–	≤200	Static	180	HgCl$_2$	96	28	Rehwoldt et al. (1972)
Striped bass	*Roccus saxatilis*	–	≤200	Static	90	HgCl$_2$	96	28	Rehwoldt et al. (1972)
American eel	*Anguilla rostrata*	–	≤200	Static	140	HgCl$_2$	96	28	Rehwoldt et al. (1972)
Rainbow trout	*Oncorhyncus mykiss*	Fingerling	40–60	Static–renewal	900	HgCl$_2$	24	10	Wobeser (1975a)
Matrinxa	*Brycon amazonicus*	Juvenile	115.5 ± 8.3	Static	150	HgCl$_2$	96	24 ± 2	Monteiro et al. (2010)
Largemouth bass	*Micropterus salmoides*	Embryo–larval	–	Static–renewal	140	HgCl$_2$	192	19–22	Birge et al. (1979)
Largemouth bass	*Micropterus salmoides*	Embryo–larval	–	Flow	5.3	HgCl$_2$	192	17–18	Birge et al. (1979)

Common name	Species	Life stage		Test type		Compound			Reference
Rainbow trout	*Oncorhynchus mykiss*	Embryo–larval	–	Static–renewal	4.7	$HgCl_2$	672	12–14	Birge et al. (1979)
Rainbow trout	*Oncorhynchus mykiss*	Embryo–larval	–	Flow	<0.1	$HgCl_2$	672	13–14	Birge et al. (1979)
Channel catfish	*Ictalurus punctatus*	Embryo–larval	–	Static–renewal	30	$HgCl_2$	240	22–24	Birge et al. (1979)
Channel catfish	*Ictalurus punctatus*	Embryo–larval	–	Flow	0.3	$HgCl_2$	240	28–29	Birge et al. (1979)
Goldfish	*Carassius auratus*	Embryo–larval	–	Static–renewal	121.9	$HgCl_2$	192	19–22	Birge et al. (1979)
Goldfish	*Carassius auratus*	Embryo–larval	–	Flow	0.7	$HgCl_2$	192	24–25	Birge et al. (1979)
Bluegill	*Lepomis macrochirus*	Embryo–larval	–	Static–renewal	88.7	$HgCl_2$	192	19–22	Birge et al. (1979)
Redear sunfish	*Lepomis microphus*	Embryo–larval	–	Static–renewal	137.2	$HgCl_2$	192	19–22	Birge et al. (1979)
Air-breathing catfish	*Clarias batrachus*	Adults	–	Static–renewal	507	$HgCl_2$	96	20 ± 2	Kirubagaran and Joy (1988a)
Fathead minnow	*Pimephales promelas*	Juvenile	3 months old	Flow–through	168	$HgCl_2$	96	23–24	Snarski and Olson (1982)
Fathead minnow	*Pimephales promelas*	Juvenile	3 months old	Flow–through	112	$HgCl_2$	120	23–24	Snarski and Olson (1982)
Fathead minnow	*Pimephales promelas*	Juvenile	3 months old	Flow–through	84	$HgCl_2$	144	23–24	Snarski and Olson (1982)
Fathead minnow	*Pimephales promelas*	Juvenile	3 months old	Flow–through	74	$HgCl_2$	168	23–24	Snarski and Olson (1982)
Catla	*Catla catla*	Fingerling	37	Static	43.92	$HgSO_4$	96	35	Kumar and Gupta (2006)
Catla	*Catla catla*	Fingerling	37	Static	52.75	$HgSO_4$	96	16	Kumar and Gupta (2006)
Rohu	*Labeo rohita*	Fingerling	54	Static	194.6	$HgSO_4$	96	35	Kumar and Gupta (2006)

(Continued)

Table 5.2 (*continued*)

Common name	Species name	Life stage	Average size (mm)	Test conditions	Hg concentration ($\mu g\ L^{-1}$)	Hg form	Test duration (h)	Temperature (°C)	Reference
Rohu	*Labeo rohita*	Fingerling	54	Static	228.2	$HgSO_4$	96	16	Kumar and Gupta (2006)
Mrigal	*Cirrhinus mrigala*	Fingerling	58	Static	268.1	$HgSO_4$	96	35	Kumar and Gupta (2006)
Mrigal	*Cirrhinus mrigala*	Fingerling	58	Static	308.5	$HgSO_4$	96	16	Kumar and Gupta (2006)
Catla	*Catla catla*	Fingerling	37	Static	47.7	$HgSO_4$	96	35	Kumar and Gupta (2006)
Catla	*Catla catla*	Fingerling	37	Static	72.7	$HgSO_4$	96	16	Kumar and Gupta (2006)
Rohu	*Labeo rohita*	Fingerling	54	Static	222.1	$HgSO_4$	96	35	Kumar and Gupta (2006)
Rohu	*Labeo rohita*	Fingerling	54	Static	278.3	$HgSO_4$	96	16	Kumar and Gupta (2006)
Mrigal	*Cirrhinus mrigala*	Fingerling	58	Static	281.4	$HgSO_4$	96	35	Kumar and Gupta (2006)
Mrigal	*Cirrhinus mrigala*	Fingerling	58	Static	312.9	$HgSO_4$	96	16	Kumar and Gupta (2006)
Mosquitofish	*Gambusia affinis*	–	–	Static	52.62	$HgCl_2$	96	21.4 ± 0.59	McCrary and Heagler (1997)
Golden shiner	*Notemigonus crysoleucas*	–	–	Static	16.75	$HgCl_2$	96	21.4 ± 0.59	McCrary and Heagler (1997)
Marine, inorganic Hg									
Tidewater silverside	*Menidia peninsulae*	Larvae, age 26 days	–	Static	71	$HgCl_2$	96	26	Mayer (1987)

Common name	Species	Life stage	Age/size	Test type	LC50	Form	Duration (h)	Temp (°C)	Reference
Spot	*Leiostomus xanthurus*	Adult	–	Static	36	HgCl$_2$	96	26	Mayer (1987)
Seabass	*Lates calcarifer*	Fry	11 ± 3	Static–renewal	85	HgCl$_2$	96	28 ± 2	Krishnani et al. (2003)
Seabass	*Lates calcarifer*	Fry	24 ± 4	Static–renewal	200	HgCl$_2$	96	28 ± 2	Krishnani et al. (2003)
Mummichog	*Fundulus heteroclitus*	Eggs	4–8 cell stage	Static–renewal	67.4	HgCl$_2$	96	–	Sharp and Neff (1982)
Freshwater, organic Hg									
Rainbow trout	*Oncorhynchus mykiss*	Fry	2–7 days after hatching	Static–24 h renewal	24	MeHgCl	96	10	Wobeser (1975a)
Rainbow trout	*Oncorhynchus mykiss*	Fingerling	40–60	Static–24 h renewal	42	MeHgCl	96	10	Wobeser (1975a)
Brook trout	*Salvelinus fontinalis*	Juvenile	–	Flow	84	MeHgCl	96	12	McKim et al. (1976)
Brook trout	*Salvelinus fontinalis*	Yearling	–	Flow	65	MeHgCl	96	12	McKim et al. (1976)
Fathead minnow	*Pimephales promelas*	Embryos	–	Flow	39	MeHgCl	96	24 ± 0.5	Devlin (2006)
Fathead minnow	*Pimephales promelas*	Embryos	–	Flow	42	MeHgCl	72	24 ± 0.5	Devlin (2006)
Fathead minnow	*Pimephales promelas*	Embryos	–	Flow	71	MeHgCl	48	24 ± 0.5	Devlin (2006)
Fathead minnow	*Pimephales promelas*	Embryos	–	Flow	221	MeHgCl	24	24 ± 0.5	Devlin (2006)
Air-breathing catfish	*Clarias batrachus*	Adults	–	Static–renewal	430	MeHgCl	96	20 ± 2	Kirubagaran and Joy (1988a)
Blue gourami	*Trichogaster* sp.	Adults	–	–	89.5	MeHgCl	96	26–28	Roales and Perlmutter (1974)

(Continued)

Table 5.2 (*continued*)

Common name	Species name	Life stage	Average size (mm)	Test conditions	Hg concentration ($\mu g\ L^{-1}$)	Hg form	Test duration (h)	Temperature (°C)	Reference
Blue gourami	*Trichogaster* sp.	Adults	–	–	94.2	MeHgCl	48	26–28	Roales and Perlmutter (1974)
Blue gourami	*Trichogaster* sp.	Adults	–	–	123	MeHgCl	24	26–28	Roales and Perlmutter (1974)
Marine, organic Hg									
Mummichog	*Fundulus heteroclitus*	Eggs	–	–	1700.0	MeHgCl	20 min	–	Khan and Weis (1987c)
Mummichog	*Fundulus heteroclitus*	Eggs	–	–	700.0	MeHgCl	20 min	–	Khan and Weis (1987c)
Mummichog	*Fundulus heteroclitus*	Juvenile	24–45	–	210.0	MeHgCl	96	–	Khan and Weis (1987c)
Mummichog	*Fundulus heteroclitus*	Juvenile	24–45	–	190.0	MeHgCl	96	–	Khan and Weis (1987c)
Mummichog	*Fundulus heteroclitus*	Eggs	4–8 cell	Static–renewal	51.1	MeHgCl	96	–	Sharp and Neff (1982)
Mummichog	*Fundulus heteroclitus*	Embryos	1 day	Static–renewal	72.7	MeHgCl	96	–	Sharp and Neff (1982)

inactive (no nuclei) and atrophied (Ram and Joy, 1988). No vitellogenic oocytes were found in female spotted murrel exposed to Hg(II) over a 6 month period (10 μg L^{-1}), whereas eggs were well developed in control females (Ram and Joy, 1988).

The effects of Hg(II) on gamete development are undoubtedly linked to its broader disruption of the endocrine system and hormone production (Crump and Trudeau, 2009). For example, gonadotrophs in the pituitary of spotted murrel exposed for 6 months to 0.01 mg L^{-1} were smaller in size and did not appear to be active (Ram and Sathyanesan, 1983). Synthesis of sex steroids involves 3β-hydroxy-Δ$_5$-steroid dehydrogenase (3βHSD) and activity of this enzyme in testicular tissues was partially inhibited by Hg (II) (Kirubaragarn and Joy, 1988b). Cholesterol, the key substrate for steroid synthesis, was decreased in testes of air-breathing catfish exposed to Hg(II), which may be owing to decreased mobilization of cholesterol from liver to gonads (Kirubagaran and Joy, 1992).

Spawning is also affected by aqueous exposures to Hg(II). For example, Snarski and Olson (1982) showed that concentrations of at least 1.02 μg L^{-1} stopped spawning of fathead minnow (*Pimephales promelas*), and lower concentrations reduced the numbers of spawning pairs. Egg production by adults exposed to 0.26 or 0.5 μg L^{-1} was 46% and 54% lower, respectively, than controls. In another study, fertilization and hatching success were reduced when either sperm from males or eggs from female steelhead trout (*Oncorhynchus mykiss*) exposed to 1 μg HgCl$_2$ L^{-1} over 4 months were used (Birge et al., 1979).

6.2.2. GROWTH

Chronic Hg exposure can interfere with the growth of fish and lead to emaciation. After 41 weeks, reduced growth was observed in female fathead minnow exposed to ≥0.26 μg L^{-1} and males exposed to 3.69 μg L^{-1}. Second generation offspring in this study had lower growth at concentrations at or above 1.02 μg L^{-1}, and stunting and scoliosis at concentrations >2.01 μg L^{-1} (Snarski and Olson, 1982). However, 4 month exposures of Atlantic salmon parr (*Salmo salar*) to a diet of up to 100 mg kg^{-1} dry weight (dw) did not affect their final weight, condition, or specific growth rates (Berntssen et al., 2003).

Fish exposed to Hg(II) also show changes in the tissue storage of energy, proteins, and vitamins. For example, featherback (*Notopterus notopterus*) exposed for 30 days to 44 or 88 μg L^{-1} showed increased total lipid content in liver (but decreases in ovaries), and decreases in proteins and vitamins A, C, and D in muscle and ovary tissues (Verma and Tonk, 1983). Mercury exposure also decreased tissue glycogen (Verma and Tonk, 1983) and cholesterol, fatty acids, and phospholipids (Kirubagaran and Joy, 1992; but see Berntssen et al., 2004).

6.2.3. OXIDATIVE STRESS

It is well known that Hg(II) generates H_2O_2 and OH radicals, though the mechanisms for this are not well understood. Readers are referred to a review by Zaman and Pardini (1996) on the numerous oxidative stresses caused by Hg(II) [and by MeHg(I)].

Inorganic Hg either stimulates or impairs the production of several antioxidant enzymes that protect against reactive oxygen species (ROS). For example, Atlantic salmon parr fed 10 and 100 mg kg^{-1} Hg dw had elevated SOD in liver and kidney and GSH-Px in liver (100 mg kg^{-1} only). In contrast, GSH-Px was suppressed in kidney at the highest exposure (Berntssen et al., 2003).

Inorganic Hg exposure from either the water or the diet also causes lipid oxidation. Dose-dependent increases in thiobarbituric acid reactive substances (TBARS), a measure of lipid peroxidation, were four times higher in the kidney of Atlantic salmon after dietborne exposures (Berntssen et al., 2003). Lipid peroxidative products have been found in liver and kidney of spotted murrel exposed to 5 µg L^{-1} Hg(II) for 30 days (Rana et al., 1995), and in gills, liver, white muscle, and heart tissues of matrinxã exposed over 4 days to 150 µg L^{-1} (Monteiro et al., 2010).

6.2.4. OSMOREGULATION

Several osmoregulatory processes are affected by Hg(II). Waterborne exposure of fish to Hg(II) causes extensive damage to the gill tissues and affects the numbers and ultrastructure of chloride cells critical to osmoregulation (Giari et al., 2008). In addition, the enzymes needed to regulate cellular ion levels are impaired after Hg(II) exposure. For example, the activities of both Ca^{2+}-ATPase and Na^+/K^+-ATPase were depressed in the brains of fish exposed to 27 µg L^{-1} (0.1 µM) (Verma et al., 1983).

6.2.5. DIGESTION

Effects of Hg exposure on fish growth may be linked to its impairment of digestion and absorption of essential nutrients. Sloughing of the epithelial cells of the intestine and regurgitation of food was seen after 3 week exposures to very high dietborne concentrations (Handy and Penrice, 1993). The apparent digestibility of protein and glycogen decreased in Atlantic salmon exposed to 100 µg kg^{-1} in their feed (Berntssen et al., 2004). Similarly, aqueous $HgCl_2$ reduced the absorption of essential amino acids by intestinal tissues of toadfish (*Opsanus tau*) and mummichog by up to 67% (Farmanfarmaian and Socci, 1984).

6.2.6. Nervous system

As described for its acute toxicity, Hg(II) is a potent neurotoxicant and affects both the central nervous system and peripheral sensory organs of fish via a number of mechanisms. Brain tissues of water-exposed fish show changes in lipid stores as well as increases in lipid peroxidation (up to 30 days at 200 μg L^{-1}) (Bano and Hasan, 1989). In longer dietary studies, brain tissues had a proliferation of astrocytes, likely in response to injury and to maintain normal function (Atlantic salmon fed 100 mg kg^{-1}) (Berntssen et al., 2003). Finally, neurons were smaller, degenerated and inactive in spotted murrel exposed to 10 μg L^{-1} over a 6 month period (Ram and Joy, 1988).

Secretion of neurotransmitters and production of the enzymes that metabolize them are also affected by Hg(II). Six month exposures to up to 0.03 mg L^{-1} Hg(II) decreased serotonin [5-hydroxytryptamine (5-HT)] levels in hypothalamus of Nile tilapia (*Oreochromis mossambicus*) (Tsai et al., 1995). Monoamine oxidase (MAO) was depressed after either aqueous (Ram and Joy, 1988) or dietary (Berntssen et al., 2003) exposures. Similarly, several studies have shown that acetylcholinesterase activity is impaired by Hg(II) (e.g. Suresh et al., 1992).

6.2.7. Histopathology and cellular effects

Several external tissues can be damaged during aqueous or dietborne exposures to Hg(II). Dietary exposures (10 mg kg^{-1}) caused a proliferation of intestinal cells, followed by pathologies of the intestinal wall in Atlantic salmon (Berntssen et al., 2004). Waterborne exposures also cause structural damage to the intestine, including degeneration of the villi, vacuolation, and nuclear apoptosis of the epithelial cells (Bano and Hasan, 1990; Banerjee and Bhattacharya, 1995). Other tissues exposed directly to Hg from water were damaged, including olfactory organs and epidermal, epithelial, and lateral line cells (Gardner, 1975).

Once Hg has been taken up into fish, damage to internal organs is also common at higher exposures. Aqueous exposures for 30 days to 200 μg L^{-1} had numerous effects on kidney tissues including smaller Bowman capsules, necrosis of epithelial cells, and infiltration of tissues by inflammatory cells (Bano and Hasan, 1990). Some increases in melanomacrophages were seen in kidney of fish fed 10 mg kg^{-1} dw for 42 days (Handy and Penrice, 1993). Livers in fish from waterborne exposures showed hypertrophy, degeneration of parenchyma, and necrosis of cells (Bano and Hasan, 1990).

6.2.8. Immune system

Several studies have shown that Hg(II) changes both the external and internal innate immune response in fish. Aqueous exposures increase mucus production by gill tissues but it remains speculative whether this response is protective against infection or whether excessive mucus facilitates microbial entry into the vascular system (Bols et al., 2001). Internally, several humoral factors are affected by this metal. Lysozyme, an enzyme that is important in attacking bacteria, increased in plasma of exposed rainbow trout (Sanchez-Dardon et al., 1999). Cellular responses include reductions in the ability of macrophages produced by the head kidney to phagocytose particles (Voccia et al., 1994). Another immune response affected by Hg(II) is the "respiratory burst" or the production of ROS by phagocytes (Bols et al., 2001).

6.2.9. BIOCHEMISTRY/PHYSIOLOGY

Mercury binds strongly to ligands, especially those with R-SH or R-S-S-R groups, including cysteinyl and histidyl side-chains of proteins, amino, and carboxyl groups on enzymes, phosphates, purines, pteridines, and porphyrins (Vallee and Ulmer, 1972). This affinity means it can bind to molecules and interfere with a large number of biochemical processes such as the function of enzymes, the structure of nucleic acids, and the pathways of oxidative phosphorylation (see review by Vallee and Ulmer, 1972). In addition, it interferes with membrane structure and function owing to its affinity for phosphoryl groups found on cell membranes (Vallee and Ulmer, 1972).

Studies have shown that glucocorticoid steroid hormone synthesis is disrupted by Hg(II), which may be due to interferences with the enzymes used or substrates needed for its production. *In vivo* plasma cortisol levels were 54% of controls in air-breathing catfish exposed to Hg(II) (50 $\mu g\,L^{-1}$) for 180 days (Kirubagaran and Joy, 1991).

6.3. Acute Toxicity of Methylmercury

Mechanisms of acute toxicity of MeHg(I) are similar to those described for Hg(II), with few exceptions. MeHg(I) readily passes the blood–brain barrier and dietary exposures result in higher brain contamination than do similar exposures to Hg(II) (Berntssen et al., 2004). Hence, MeHg(I) is a potent neurotoxicant and exposed fish become inactive and lethargic. There is also damage to gill tissues including swelling, hyperplasia, necrosis, vacuolation, and eventual separation of the epithelial cells from the lamellae and necrosis of pillar and mucus cells (e.g. $> 10\,\mu g\,L^{-1}$, Wobeser, 1975b; up to $1\,\mu g\,L^{-1}$, Liao et al., 2007). Ventilation rates also declined by 40% (Stinson and Mallatt, 1989). However, acute toxicity of MeHg(I) is not caused by excess gill mucus production, as observed in fish exposed to $HgCl_2$ (Wobeser, 1975a). Ribeiro et al. (2002) found an increase in heterochromatin in nuclei after 12 h and widespread necrosis, smaller nuclei, growth of connective tissues, and more phagocytes in the liver 18 days after dosing Arctic charr (*Salvelinus alpinus*) once with dietborne MeHg(I).

Permeability of cell membranes and their ability to regulate ions are affected by MeHg(I). For example, freshwater lamprey (*Petromyzon marinus*) exposed for 2–4 h to 180 µg L^{-1} showed greater efflux of Na$^+$ and Cl$^-$ (by 30 and 22%, respectively) across the gill epithelia, likely due to changes in the permeability of the cell membranes (e.g. via the opening of channels) to ions and other small molecules (Stinson and Mallatt, 1989).

Decreases in growth have been observed in fish exposed to MeHg(I) and may be due to suppression of thyroid hormone production. Kirubagaran and Joy (1994) exposed air-breathing catfish to concentrations of 125 µg L^{-1} and found suppression of both T$_4$ and T$_3$ by up to 60% after 2 weeks; some recovery of T$_3$ and to a lesser extent T$_4$ was seen after fish were moved to clean water for a minimum of 2 weeks. The authors believe the decrease in T$_3$ was related directly to declines in T$_4$ production, and not to the suppression of the deiodinase enzyme, because ratios of T$_4$/T$_3$ were similar across treatments and controls.

As for inorganic Hg, life stage and species are important determinants of MeHg(I) toxicity and larger fish tend to be less sensitive than smaller fish of the same species. For example, MeHg(I) had much higher toxicity (24 µg L^{-1} 96 h LC50) for rainbow trout fry than for fingerlings (42 µg L^{-1} 96 h LC50). This study also found that MeHg(I) was seven times more toxic than HgCl$_2$ to fingerlings (Wobeser, 1975a). Others have also shown higher aqueous toxicities for MeHg(I) compared to HgCl$_2$ (Table 5.2). The 96 h LC50 values for MeHg(I) range from 24 to 221 µg L^{-1} for freshwater species (except the air-breathing catfish at 430 µg L^{-1}) and from 190 to 1700 µg L^{-1} for different life stages of the saltwater species mummichog (Table 5.2).

6.4. Chronic Toxicity of Methylmercury

The literature on chronic toxicity of MeHg(I) in fish has increased substantially over the past decade and it is becoming increasingly evident that effects are occurring at much lower exposures than previously believed. In a review, Sandheinrich and Wiener (2010) show that effects on biochemical processes, tissues, or reproduction are occurring in laboratory-cultured or wild fish with muscle or whole-body concentrations ranging from 0.5 to 1.2 µg g^{-1} or 0.3 to 0.7 µg g^{-1} ww, respectively (see Tables 1 and 2 in their paper). Similarly, tissue concentrations of >0.2 µg g^{-1} ww have adverse effects on reproduction, behavior, and development in both fry and adults (Beckvar et al., 2005). These concentrations (as THg) are common in wild fish (Wiener et al., 2003).

6.4.1. NERVOUS SYSTEM

As in mammals, MeHg(I) is a potent neurotoxicant in fish and effects on this system have been observed at the organismal, tissue, biochemical, and physiological levels. Brain lesions have been observed in Atlantic salmon fed 10 mg kg^{-1} dw MeHg(I) (Berntssen et al., 2003); these fish had vacuolation in brain tissues, with greater effects seen in cerebellum, medulla, and cerebrum of fish, as well as degeneration of neurons. Early life-stage exposure of mummichog to MeHg(I) also resulted in abnormal optical development (Weis and Weis, 1977) and enhanced depolarization of retinal bipolar neurons (Weber et al., 2008). Given that MeHg(I) affects peripheral sensory organs as well as the central nervous system, these fish will likely also have impaired abilities to catch and ingest prey and avoid predation.

6.4.2. REPRODUCTION

There are various mechanisms through which MeHg(I) exposure affects or likely affects reproduction in fish, including behavior and impairment at the biochemical through tissue levels. For a comprehensive review on this subject, readers are referred to Crump and Trudeau (2009).

Several studies have shown that gonad development is impaired in fish exposed to MeHg(I). Male air-breathing catfish exposed for 6 months to aqueous MeHg(I) had GSIs that were 57% of controls (Kirubargaran and Joy, 1992). Similarly, both male walleye (*Stizostedion vitreum*) fed 1.0 mg kg^{-1} ww for 6 months (Friedmann et al., 1996) and female fathead minnow fed 3.93 mg kg^{-1} dw for 8 months (40%) (Drevnick and Sandheinrich, 2003) showed decreases in GSI. In addition, lower egg production and instantaneous rates of reproduction were observed for this species (Hammerschmidt et al., 2002).

Development of spermatocytes and oocytes is delayed in fish exposed to MeHg(I). Methylmercury affects sperm cell development in air-breathing catfish (Kirubagaran and Joy, 1992). Testes of walleye exposed via the diet (0.1 or 1.0 mg kg^{-1} ww) showed altered morphology, including atrophy (Friedmann et al., 1996). Methylmercury causes delays in oocyte development, hypertrophy, vacuolation, necrosis, apoptosis, and atresia in this tissue in some (Kirubagaran and Joy, 1988c; Drevnick et al., 2006b) but not all (Friedmann et al., 1996) studies.

As has been seen for Hg(II), MeHg(I) also affects tissue energy stores in fish. For example, 3–6 month exposures of air-breathing catfish to 40 μg L^{-1} of MeHg(I) decreased levels of all lipids (cholesterol, fatty acids, and phospholipids) in the testes, likely because of an impairment in mobilization of lipids from other tissues into the gonads of these males (Kirubagaran and

Joy, 1992). Testicular lipids are critical for spermatogenesis and reductions in tissue stores likely contribute to the reproductive delays seen in other laboratory experiments.

The delays in gamete development described above are undoubtedly linked to the impacts of MeHg(I) on sex steroid or vitellogenin production. Dietary MeHg(I) exposure decreased 17β-estradiol in female fathead minnow and testosterone in male fathead minnow by up to four-fold when compared to controls (Drevnick and Sandheinrich, 2003; Drevnick et al., 2006b). These changes are likely related to the effects of MeHg(I) on enzymes and substrates needed for steroidogenesis; for example, this form of Hg decreases 3βHSD (Kirubagaran and Joy, 1988b), and changes cholesterol metabolism (e.g. Kirubagaran and Joy, 1992) in testicular tissues. The mRNA of vitellogenin was downregulated in female fish fed MeHg(I) (Klaper et al., 2006), suggesting that production of this protein would also be impaired.

Although only a handful of studies have examined how MeHg(I) affects reproductive behaviors, it is clear that it does delay or stop some that are critical for reproduction. Timing of spawning was delayed in fathead minnow fed 0.88 mg kg^{-1} dw MeHg(I) from the juvenile life stage and this effect has been linked to concurrent delays in gonad maturation (Hammerschmidt et al., 2002). Sandheinrich and Miller (2006) also saw a 10-fold decline in the time that exposed male fathead minnow spent spawning.

Fertilization and hatching success are reduced either for gametes from exposed parents or because eggs or sperm are affected directly by MeHg(I) in the water. Mercury in the diet of females is deposited into eggs (Hammerschmidt and Sandheinrich, 2005), and egg MeHg(I) is higher in larger, older females and in those from more contaminated environments (Latif et al., 2001). Fertilization success is decreased when eggs have MeHg (I) >0.01 μg g^{-1} dw (Matta et al., 2001). This study also found fewer female offspring at the moderate exposures but more females than expected at the highest parental exposures, indicating that MeHg(I) has transgenerational effects. Waterborne MeHg(I) decreases sperm motility, and fertilization and hatching success in a number of species (Khan and Weis, 1987b; Latif et al., 2001), even at low, environmentally relevant concentrations (0.1–7.8 ng L^{-1}) (Latif et al., 2001). Survival of embryos is affected more by MeHg(I) than by Hg(II) and effects are greater when exposure occurs immediately after fertilization rather than several days into embryo development (Sharp and Neff, 1980). In several experiments, decreases in spawning success (numbers of pairs producing a clutch of eggs) were found for fathead minnow fed MeHg(I)-contaminated diets (Hammerschmidt et al., 2002; Drevnick and Sandheinrich, 2003; Sandheinrich and Miller, 2006).

6.4.3. GROWTH

Growth is impaired for some but not all fish exposed to MeHg(I) in the laboratory and the magnitude of these effects is dependent on exposure concentrations and life stage. Rainbow trout fingerlings fed diets contaminated with 16 or 24 mg MeHg(I) kg^{-1} ww (but not at exposures of 4 and 8 mg kg^{-1}) over 105 days had low weight gains (Wobeser, 1975b). Juvenile male walleye fed MeHg(I) (1.0 mg kg^{-1} ww) for 6 months were much shorter than controls or than those with lower exposures (0.1 mg kg^{-1}), and concentrations of Hg in tissues of fish from the highest treatment were inversely related to total length or weight; however, juvenile females fed either the high or low MeHg(I) diets did not show impairments in growth (Friedmann et al., 1996). Atlantic salmon parr fed MeHg(I) (at 5 or 10 mg kg^{-1} dw) over 4 months did not show any decreases in final weight, condition, or specific growth rate (Berntssen et al., 2003). Similarly, walleye larvae exposed to low ng L^{-1} concentrations for 8 days had similar lengths to the controls (Latif et al., 2001).

6.4.4. OXIDATIVE STRESS

As was found for inorganic Hg, MeHg(I) exposure also results in oxidative damage via the production of ROS and an activation, or sometimes suppression, of the redox defense system in fish (see review by Zamman and Pardini, 1996). Lipid peroxidative products (TBARS) increased seven-fold in the brain, but not liver or kidney, of fish fed the highest concentration of MeHg(I) (10 mg kg^{-1} dw) (Berntssen et al., 2003). Within liver cells exposed *in vitro* to MeHg(I), elevated levels of hydrogen peroxide and superoxide anions were measured along with increased lipid peroxidation and protein carboxylation [by 85% and 24% in 540 µg L^{-1} (2.5 µM), respectively] (Neto et al., 2008). Again, in the study by Berntssen et al. (2003), Atlantic salmon parr fed 10 mg kg^{-1} (dw), but not 5 mg kg^{-1}, had increased SOD and GSH-Px activity in the liver; however, both of these enzymes decreased in activity in the brain of fish fed the highest dose (10 mg kg^{-1}), suggesting that the redox system in this tissue had been overwhelmed. In *in vitro* studies of hepatocytes, exposure to MeHg(I) at 54 or 540 µg L^{-1} increased (by 14%) and decreased (by 20%) catalase activity, respectively, and decreased SOD activity (up to 17%) at both concentrations (Neto et al., 2008). In this same study, the antioxidant GSH decreased in cells at the higher MeHg(I) concentration, there was no effect on GST activity, and glutathione disulfide reductase (GR) activity was depressed or elevated in the low and high treatments, respectively. Finally, at the highest MeHg(I) concentration, elevated production of the enzymes glucose-6-phosphate dehydrogenase (G6PDH; required for NADPH production) and δ-ALAD (δ-aminolevulinic acid dehydratase; involved in synthesizing heme groups of proteins like CAT) was observed in liver cells. The latter enzyme was inhibited in red blood cells of neotropical fish traira (*Hoplias malabaricus*) after dietborne exposures (Costa et al., 2007).

6.4.5. DIGESTION

MeHg(I) can also interfere with digestion. Intestinal tissues from two species of fish (toadfish and mummichog) exposed to aqueous MeHgCl *in vivo* showed reduced absorption of essential amino acids (leucine, isoleucine, methionine, and lysine), but this effect was not as severe as was observed after exposure to $HgCl_2$ (Farmanfarmaian and Socci, 1984). However, the macronutrient digestibility of fat, protein, and glycogen was not affected in Atlantic salmon fed $10 \, \mu g \, kg^{-1}$ MeHg(I) for 4 months (Berntssen et al., 2004).

6.4.6. METABOLISM

Dietary exposure to MeHg(I) interferes with mitochondrial respiration. Muscle tissues from zebrafish (*Danio rerio*) fed 13 mg kg^{-1} dw MeHg(I) had damaged mitochondria (i.e. disorganized cristae; see also Zaman and Pardini, 1996; Ribeiro et al., 2008), and decreased mitochondrial respiration, cytochrome *c* oxidase activity, and ATP release when these tissues were stimulated with respiratory substrates such as pyruvate (Cambier et al., 2009). The results of this study suggest that MeHg(I), like Hg(II), decouples oxidative phosphorylation processes.

6.4.7. HISTOPATHOLOGY AND CELLULAR EFFECTS

Like inorganic Hg, aqueous or dietborne MeHg(I) also causes external and internal tissue damage in fish. Some intestinal pathologies (hyperplasia) were observed after subchronic waterborne exposures of 3.2 or 5.6 µg L^{-1} MeHg(I) (Wester and Canton, 1992) and increased cell proliferation along intestinal folds after dietary exposures (Berntssen et al., 2004). The medulla of fish exposed for 4 months to 5 or 10 mg kg^{-1} had widespread vacuolation, occasional necrosis, and edema; these effects were also observed in several other regions of the brain and damage was greatest at the highest exposure level (Berntssen et al., 2003). Waterborne exposures to MeHg(I) (up to 10 µg L^{-1} for 3 months) caused hyperplasia, hypertrophy, and degeneration of the bile duct epithelium, and larger vesicular nuclei and prominent nucleoli of the liver cells, suggesting greater protein production (Wester and Canton, 1992). In addition, longer term waterborne exposures at a lower level (1 µg L^{-1} for 210 days) caused lesions, and hepatocytes with increased edema, vacuoles, and pyknotic nuclei (Liao et al., 2007).

Environmentally relevant dietary exposures of 0.075 mg kg^{-1} ww caused infiltration of leukocytes, increases in lesions, and changes in nuclear shape and the distribution of heterochromatin in liver tissues and cells (Mela et al., 2007). There was also damage in the head kidney, including necrosis, more melanomacrophages, and atypical cells and intercellular space. The kidney tissues of guppies (*Poecillia reticulata*) exposed to MeHg(I) via the water had hypertrophic or degenerated epithelial cells in the proximal tubules, and hyperplasia in the interstitial hemopoietic tissues (Wester and Canton, 1992). Red and white muscle cells in fish fed 13.5 mg kg^{-1} dw have been found with decreased space between fiber bundles and a disorganization of myofibrils in red muscle only (Ribeiro et al., 2008). Finally, damage to the pituitary, a key gland in the endocrine system, was evidenced by smaller and fewer gonadotrophs [luteinizing hormone (LH)- and follicle-stimulating hormone (FSH)-producing cells] in fish exposed to MeHg(I) (Joy and Kirubagaran, 1989).

6.4.8. BIOCHEMISTRY AND PHYSIOLOGY

Several neurotransmitters, their metabolizing enzymes, and stress hormones are affected by aqueous exposures to MeHg(I). Brain serotonin (5-HT) decreased in air-breathing catfish exposed to 40 μg L^{-1} for 90 and 180 days, likely due to decreases in its synthesis (Kirubagaran and Joy, 1990). This study also showed increases in dopamine and norepinephrine (noradrenaline) in the same tissue. These neurotransmitters (dopamine, norepinephrine, and 5-HT) are metabolized by monoamine oxidase (MAO) and the activity of this enzyme is decreased by dietborne exposure to MeHg (I) in Atlantic salmon (Berntssen et al., 2003). Equivocal effects of MeHg(I) on cholinesterase (ChE) activity have been seen. Webber and Haines (2003) saw no effects of diet exposures on brain ChE in adult golden shiners (*Notemigonus crysoleucas*), but in another study muscle ChE levels in diet-exposed fish were only 63% of controls (Costa et al., 2007). Finally, plasma cortisol was suppressed by 40% and 75% in air-breathing catfish exposed to MeHg(I) (40 μg L^{-1}) for 90 and 180 days, respectively (Kirubagaran and Joy, 1991).

6.4.9. IMMUNE RESPONSES

There is a paucity of studies that have examined how MeHg(I) affects the immune response in fish. One study by Wester and Canton (1992) showed that granulomas, a mass of immune cells, increased after 3 month exposures to waterborne MeHg(I).

7. ESSENTIALITY OR NON-ESSENTIALITY OF MERCURY

Mercury, in either its inorganic or its organic form, is not known to have any positive and essential role in growth, reproduction, or survival of fish.

8. BIOCONCENTRATION AND BIOMAGNIFICATION OF MERCURY

8.1. Bioconcentration

Bioconcentration, or the accumulation of an element into the fish at concentrations that exceed those found in its surrounding water, has been well established for both Hg(II) and MeHg(I). Bioconcentration factors [BCFs; calculated after fish tissues reach steady state as the ratio of Hg(II) or MeHg(I) in fish versus Hg(II) or MeHg(I) in water] for aquatic organisms average 4955 and 8955 for Hg(II) and MeHg(I), respectively, and BCFs are typically inversely related to concentrations in water (McGeer et al., 2003). In wild fish, exposure from both water and food occurs and bioaccumulation is a more appropriate term; bioaccumulation factors (BAFs) are also calculated as the ratio of THg or MeHg(I) in fish versus the respective concentration in water. BAFs in wild fish are up to 10^6 and tend to be highest in the large-bodied, piscivorous species (with long lifespans) and lowest in the small-bodied, primary or secondary consumers (with short lifespans).

8.2. Biomagnification

Methylmercury is one of the few metals known to biomagnify through food webs and it is this process that leads to elevated and sometimes toxic concentrations in fish, especially those that are piscivorous. It is well known that MeHg(I) biomagnifies through freshwater (e.g. Cabana and Rasmussen, 1994) and marine (e.g. Campbell et al., 2005) food webs at tropical, temperate, and Arctic latitudes, and biomagnification factors (BMFs; equilibrium ratios of Hg in predator over Hg in prey) range from 5 to 15 between a piscivorous fish and its prey (Wiener et al., 2003). Studies of Hg biomagnification in aquatic food webs are increasingly using stable isotopes of nitrogen (δ^{15}N) to assess the relative trophic positioning of organisms and quantify the average trophic transfer of this metal (e.g. Wyn et al., 2009). Log Hg versus δ^{15}N are significantly related within a food web (Fig. 5.1),

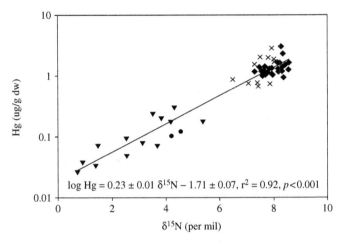

Fig. 5.1. Log-transformed Hg [μg g^{-1} dw; MeHg(I) in invertebrates, THg in fish] versus δ^{15}N (per mil) for biota from acidic Beaverskin Lake, Kejimkujik National Park, Nova Scotia (data from Wyn et al., 2009). Symbols: diamonds for yellow perch, X for forage fish (banded killifish, golden shiner), circles for zooplankton, and inverted triangles for littoral invertebrates (dragonfly larvae, amphipods, littoral chironomids, mayfly larvae, caddisfly larvae).

and this slope can then be compared across systems that differ in their physical, chemical, and biological characteristics to understand the factors that affect trophic transfer of Hg.

9. CHARACTERIZATION OF UPTAKE ROUTES

9.1. Inorganic Mercury

9.1.1. GILLS

The gills appear to be the primary uptake route of Hg(II) in non-feeding fish (Olson and Fromm, 1973). Though it varies with concentration, uptake via this route can make a substantial contribution to total Hg body burden (Simon and Boudou, 2001).

Uptake mechanisms and rates across biological membranes are dependent on the speciation of inorganic Hg (Gutknecht, 1981). This process is therefore potentially sensitive to the same factors that affect aqueous Hg(II) speciation, such as pH, salinity, and redox, and it is likely that differences in Hg speciation between freshwater and seawater affect uptake mechanisms.

9.1.1.1. Influences of Water Chemistry on Uptake. The neutral Hg complex, HgCl$_2$, rapidly crosses biological membranes through passive diffusion

(Gutknecht, 1981). Therefore, water chemistry conditions that favor this speciation, or other uncharged Hg compounds, likely increase the rate of uptake. For instance, addition of $CaCl_2$ or KCl almost doubles gill accumulation of Hg(II) compared to ion-poor water (Klinck et al., 2005). In contrast, the presence of DOC can reduce Hg(II) uptake, thereby providing protection to the fish (Playle, 1998; Pickhardt et al., 2006). Uptake at the gills of rainbow trout exposed to $400 \mu g \, L^{-1}$ ($2.0 \mu M$) Hg(II) is reduced to background levels with the addition of 7–10 mg C L^{-1} of DOC to the water (Playle, 1998).

9.1.1.2. Uptake Mechanisms: Active Versus Passive. Studies of uptake mechanisms are inconclusive, but there is a strong case for passive uptake in freshwater fish. Inorganic Hg accumulation occurs evenly throughout the gill epithelia, not within a specific cell type such as chloride cells (Olson and Fromm, 1973; Wicklund-Glynn et al., 1994; Jagoe et al., 1996). Furthermore, proposed active uptake mechanisms would involve binding of Hg(II) ions to cation-exchange sites, such as Na and Ca channels. However, the inability of Na and Ca channel blockers to inhibit uptake indicates that these channels are not involved (Wicklund-Glynn et al., 1994; Klinck et al., 2005).

Hg(II) enters the gills through the lamellar surface (Olson and Fromm, 1973), and accumulates to a greater extent in the primary than secondary lamellar epithelia (Jagoe et al., 1996). The epithelial mucus layer must be considered as a potential binding site, and possibly barrier, to further uptake, owing to the strong affinity of Hg(II) for fish mucus (Part and Lock, 1983). The rate and mechanisms of Hg(II) uptake through the mucus layer are currently unknown. The highest concentrations of Hg in gills are found in the interlamellar spaces, an indication of the ability of mucus membrane to bind Hg ions (Olson and Fromm, 1973). Binding to mucus can also lead to Hg(II) loss from the gill surface when the mucus is sloughed off. In seawater ionic species dominate and uptake into the gills may involve more active mechanisms, such as binding to ion-exchange sites in ion-transport cells.

9.1.1.3. Rate of Uptake. The kinetic uptake rate at the gills is unknown; however, since the gills are the primary site of aqueous Hg(II) uptake, the whole-body uptake rates in Table 5.3 for aqueous Hg(II) exposure can provide an estimate. These rates are slower than those for MeHg(I) uptake in the same conditions (also present in Table 5.3), and have been shown to be negatively related to body size (Newman and Doubet, 1989; Suseno et al., 2010).

9.1.2. GUT

Hg(II) can enter the gut with food, or with water during drinking. The amount accumulated from food is dependent on the type of food or prey,

Table 5.3

Whole-body kinetic uptake rate constant, k_u (L g⁻¹ day⁻¹), of aqueous Hg(II) and MeHg(I) in various fish species and water types

Common name	Species	Water type	Mercury exposure concentration (μg L^{-1})		Fish weight (g)	Temperature (°C)	Uptake rate constant k_u (L g^{-1} day^{-1})		Reference
			Hg(II)	MeHg(I)			Hg(II)	MeHg(I)	
Mosquitofish	Gambusia affinis	Freshwater	0.041	0.024	0.3–0.6	17	0.052–0.078	0.185–0.338	Pickhardt et al. (2006)
Redear sunfish	Lepomis microlophus	Freshwater	0.041	0.024	0.6–1.2	17	0.038–0.051	0.454–1.28	Pickhardt et al. (2006)
Sweetlips	Plectorhinchus gibbosus	Artificial seawater	0.109–10.90	0.020–1.832	(3.0–3.5 cm)	23	0.195	4.515	Wang and Wong (2003)
Tilapia	Oreochromis mossambicus	Brackish	0.421–10.90	–	(3.2–5.8 cm)	–	0.00619–0.00873	–	Suseno et al. (2010)

but generally is low compared to MeHg(I). Drinking is limited in freshwater fish, but occurs to a greater extent in marine fish (Evans, 1981) and may contribute to aqueous Hg(II) exposure.

Uptake of Hg(II) across the intestinal epithelium to the blood is limited (Boudou and Ribeyre, 1983, 1985; Ribeiro et al., 1999; Berntssen et al., 2004). The gut mucosal membrane functions as an effective barrier to absorption because of the high affinity of mucus for Hg(II) ions ($K_d = 66$ μg L^{-1}) (Part and Lock, 1983). As a result, the mucosa can contain between 78.4 and 85.1% of the THg found in the gut (stomach to hind intestine) depending on the region (Hoyle and Handy, 2005). The posterior intestine can contain THg concentrations up to 42 times higher than whole-body [THg] burdens (Boudou and Ribeyre, 1985).

9.1.2.1. Region of Uptake. The isolated gut mucosa accumulates Hg(II) in different regions than the isolated intact intestine. In the latter of rainbow trout, accumulation occurs as: esophagus > posterior intestine > stomach > mid intestine > pyloric ceca. Alternatively, in the isolated gut mucosa it is mid intestine > posterior intestine > esophagus > pyloric ceca > stomach (Hoyle and Handy, 2005). Other findings report highest accumulation in the posterior intestine (Boudou and Ribeyre, 1985), or from highest to lowest as: upper intestine > lower intestine > stomach > rectum (Pentreath, 1976b). These variable findings are likely due to differences in experimental conditions and duration, but accumulation seems to occur mainly below the pyloric ceca.

9.1.2.2. Rate of Uptake. Uptake from the gut is limited, with approximately 0.006–1.4% of administered Hg(II) taken up through the intestine into the blood in isolated-perfused rainbow trout intestine (Handy and Penrice, 1993; Hoyle and Handy, 2005). In whole fish this is estimated to be higher, but also variable, ranging from 8.5 to 51.3% (Wang and Wong, 2003; Pickhardt et al., 2006; Suseno et al., 2010). In terms of kinetics, the uptake across the intestine occurs in a saturable and dose-dependent manner. The estimated maximum uptake rate (V_{max}) is 28.0 μg g^{-1} h^{-1} (103 nmol g^{-1} h^{-1}), and the apparent affinity, K_m, is approximately 330 μg L^{-1} (1.2 μmol L^{-1}) in rainbow trout isolated intestine perfused with 30 mg L^{-1} HgCl$_2$ (100 μmol L^{-1}) (Hoyle and Handy, 2005).

9.1.2.3. Proposed Mechanisms of Uptake. To be taken up into the blood, inorganic Hg must first pass through the mucosal (apical) membrane of epithelial cells. Hoyle and Handy (2005) have proposed an uptake model for this process (Fig. 5.2).

In the intestine, THg is evenly distributed in the epithelia and intestinal villi without accumulation in specific areas, suggesting both passive

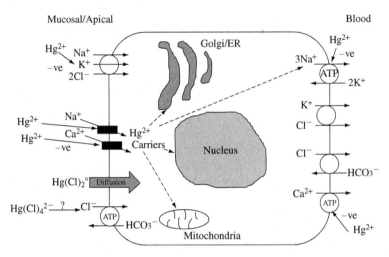

Fig. 5.2. Proposed tentative working model of inorganic Hg absorption across vertebrate intestine. Inorganic Hg enters the cell through mucosal Na^+ channels (Büsselberg, 1995), and also by diffusion of electroneutral complexes (Karniski, 1992). Hg uptake on $Na^+/K^+/2Cl^-$ cotransporter family is excluded since these are inhibited by Hg^{2+} (Kinne-Saffran and Kinne, 2001). Uptake of anionic Hg complexes on the anion exchanger cannot be excluded. Mercury is moved across the cell by chaperones (not shown for clarity) including metallothionein and Cu carriers with homology to the bacterial Hg chaperone MerP (Lu et al., 1999). Mercury exit from the cell is energy dependent and, although the precise pathway has not been identified, it could include incidental vesicular transport via Cu chaperones and the export of Hg complexes on basolateral anion exchangers (Endo et al., 1997). Mercury extrusion on either the Na^+/K^+-ATPase or Ca^{2+}-ATPase is unlikely. Reprinted from Hoyle and Handy (2005), with permission from Elsevier.

diffusion and active uptake through Na^{2+} channels (Ribeiro et al., 2002), similar to the model presented by Hoyle and Handy (2005).

9.1.3. OTHER ROUTES

9.1.3.1. Skin. Following waterborne Hg(II) exposure, the skin accumulates up to 37.7% of the Hg body burden (Weisbart, 1973; Pentreath, 1976c). This accumulation can be attributed to the affinity of Hg^{2+} for mucus (Part and Lock, 1983), but the amount absorbed through skin is unknown.

9.1.3.2. Olfactory Epithelia and Water-Exposed Sensory Nerves and Receptors. Waterborne Hg(II) is taken up by olfactory receptors, transported along olfactory nerves via axonal transport, and accumulated in the olfactory bulbs. These are essential sensory organs, whose impairment presents toxicological concerns (Borg-Neczak and Tjälve, 1996). This uptake route provides an important means of access to the brain by avoiding the otherwise impermeable blood–brain barrier (Rouleau et al., 1999). In brown trout (*Salmo trutta*) and rainbow trout the greatest accumulation from this uptake route occurs in the olfactory bulbs as well as the corpus cerebelli, rhombencephalon, mesencephalon, and eminentia granulares of the brain (Rouleau et al., 1999). Axonal transport is limited to the primary nerve pathways since Hg(II) is unable to cross nerve synapses into secondary neurons, particularly in the olfactory bulbs (Borg-Neczak and Tjälve, 1996; Rouleau et al., 1999). A similar uptake route may occur in the lateral line mechanoreceptors (Rouleau et al., 1999).

9.2. Methylmercury

9.2.1. GILLS

Aquatic concentrations of waterborne MeHg(I) are low in natural waters so although the gills present a large surface area for uptake, this route is minimal in wild fish. Uptake through the gills comprises at most 15% of total uptake, with the remainder occurring through the gut (Hall et al., 1997). Similar to Hg(II), the mechanism and rate of MeHg(I) uptake across biological membranes are influenced by Hg speciation.

9.2.1.1. Influences of Water Chemistry on Uptake. As expected, water chemistry parameters that change the speciation of MeHg(I) also influence its uptake. Conditions that favor the formation of MeHgCl (more hydrophobic) over MeHgOH (less hydrophobic) also favor accumulation of MeHg(I) in the gills; these conditions are low pH (<5) and high Cl concentrations ($> 3.5 \times 10^{-2} \text{g L}^{-1}$). In seawater, uptake is likely to be increased compared to freshwater, owing to the high Cl concentrations and dominance of MeHgCl species (Block et al., 1997). In addition, water hardness can affect uptake, with rates more than double in soft water (30 mg L^{-1} as $CaCO_3$) than in hard water (385 mg L^{-1} as $CaCO_3$) (Rodgers and Beamish, 1983), although this effect is disputed by Block et al. (1997).

9.2.1.2. Proposed Mechanisms of Uptake. Passive diffusion is the assumed mechanism of uptake for MeHg(I) (Olson and Fromm, 1973; Boudou et al., 1991; Pedersen et al., 1998), but energy-mediated mechanisms have also been implicated (Pickhardt et al., 2006). Clear evidence for either is minimal, but Block et al. (1997) suggest that the positive relationship between uptake and hydrophobicity of MeHg(I) species indicates passive mechanisms. Regardless, it remains unclear whether uptake is an active or a passive process, or a combination of these. Following uptake through the gill epithelium, MeHg(I) enters the blood and is distributed to other tissues.

9.2.1.3. Uptake Rate. Uptake of MeHg(I) through the isolated rainbow trout gill has a rate of $0.92 \text{ L g}^{-1} \text{ day}^{-1}$, when exposed to 163 μg L^{-1} (0.75 μM) Me-^{203}Hg (Pedersen et al., 1998). Whole-body uptake rate constants from waterborne exposures, shown in Table 5.3, can provide an estimate of gill uptake rates since gills are the primary uptake route in these conditions. Methylmercury uptake (as described by k_u) is greater than Hg (II) uptake by 3.6–25-fold in freshwater fish (Pickhardt et al., 2006), and by 23.2-fold in marine fish (Wang and Wong, 2003). Based on this, it seems that MeHg(I) is absorbed through the gills into the blood much more rapidly than Hg(II), and that uptake in marine fish is greater than in freshwater fish, as predicted by Block et al. (1997).

9.2.2. Gut

In wild fish, over 90% of MeHg(I) exposure is from the diet (Hall et al., 1997). Accordingly, gut uptake is a dominant process and has been the focus of many laboratory studies.

9.2.2.1. Region of Uptake. Studies regarding which gut region accumulates the greatest MeHg(I) following exposure are not in agreement. In separate studies, the following regions have been identified as the primary areas of uptake and accumulation: the posterior intestine (below the pyloric ceca) (Boudou and Ribeyre, 1985), pyloric ceca (Giblin and Massaro, 1973), stomach (Rouleau et al., 1998; Ribeiro et al., 1999), anterior intestine (above the stomach) (Pentreath, 1976b), and both the anterior and posterior intestine regions (Rouleau et al., 1998; Leaner and Mason, 2002a). Little accumulation of MeHg(I) occurs in the lumen, however, indicating rapid uptake into intestinal tissues (Rouleau et al., 1998; Ribeiro et al., 1999).

9.2.2.2. Rate of Uptake. In both marine and freshwater fish, approximately 76–92% of MeHg(I) that enters the gut also enters the blood, and is assimilated into other tissues (Boudou and Ribeyre, 1985; Wang and Wong, 2003; Leaner and Mason, 2004; Pickhardt et al., 2006). This transfer from the gut to the blood is estimated to have a rate constant (k) of $3.43–4.01 \text{ day}^{-1}$ (Leaner and Mason, 2004), and assimilation into tissues is 3.3–5 times faster than that of dietary Hg(II) (Boudou and Ribeyre, 1985). To the authors' knowledge, kinetic characteristics of this process (e.g. K_m, V_{max}) for MeHg(I) remain unspecified. In addition, accumulation in the intestine of the Japanese eel (*Anguilla japonica*) is significantly increased in seawater conditions versus freshwater. This may be a result of increased drinking rates in saline conditions, or increased uptake through the gut (Yamaguchi et al., 2004).

Solubilization of MeHg(I) within the gut controls its bioavailability for uptake. This process is primarily dependent on the relative amount of amino acids in the intestinal and gastric fluid, as well as other digestive tract factors, including pH and enzyme content (Leaner and Mason, 2002b). These factors must therefore be included when considering the rate and mechanisms of uptake, as outlined in the following section.

9.2.2.3. Proposed Mechanisms of Uptake. Uptake by gut tissues is suggested to be comprised of about 40% active and 60% passive mechanisms, and proposed active mechanisms are both specific and non-specific (Leaner and Mason, 2002a). Details of non-specific active transport mechanisms are unknown but transport is inhibited by cold temperatures (e.g. 4°C) and ouabain (an Na^+/K^+-ATPase inhibitor). Specific active mechanisms are better known and can involve uptake through neutral amino-acid carriers (L-neutral amino acid carrier), that have a high affinity for L-enantiomers. These carriers target MeHg–L-cysteine complexes, so uptake is dependent on the solubilization processes that produce MeHg–thiol complexes (such as MeHg–L-cysteine) in the gastric and intestinal fluids (Leaner and Mason, 2002b). These processes are controlled by digestive functions, as well as by the composition of the substrate being digested. For instance, solubilization is reduced in substrates with less protein (e.g. sediment) in comparison to substrates with more (e.g. worms), likely resulting in decreased MeHg(I) uptake through the gut (Leaner and Mason, 2002b). In the absence of complexing agents, such as amino acids, MeHgCl is the primary species in the gut, and its uptake is thought to occur through both passive and non-specific active mechanisms (Leaner and Mason, 2002a). These transport mechanisms are assumed to be the same for both the apical and basolateral membranes of the intestinal epithelial cells (Leaner and Mason, 2002a).

9.2.3. OTHER ROUTES

9.2.3.1. Skin, Olfactory Epithelia, and Water-Exposed Sensory Nerves and Receptors. Uptake of MeHg(I) through the skin, olfactory epithelia, or sensory nerves and receptors, has not been studied in fish. Given that concentrations of MeHg(I) species are very low in natural water bodies, the potential for uptake via these routes is minimal.

10. CHARACTERIZATION OF INTERNAL HANDLING

10.1. Inorganic Mercury

10.1.1. BIOTRANSFORMATION: *IN VIVO* METHYLATION

In fish exposed only to Hg(II), MeHg(I) can comprise up to 28% of total muscle Hg (Simon and Boudou, 2001) and can also accumulate in red blood cells (Baatrup et al., 1986). This may be a result of *in vivo* methylation. The liver has been identified as a main site of *in vivo* methylation (Imura et al., 1972). The relative contribution of this process to total MeHg(I) in wild fish is unknown, but likely to be low compared to dietary MeHg(I) uptake.

10.1.2. TRANSPORT THROUGH THE BLOODSTREAM

Following exposure, Hg(II) initially associates almost exclusively (91%) with plasma (Olson and Fromm, 1973), but over time (> 150 days), associates approximately 10 times more with red blood cells (Schultz et al., 1996). Because of its strong affinity for thiol groups, Hg(II) in plasma is bound to compounds such as cysteine, albumin, or glutathione (Hughes, 1957).

10.1.3. ACCUMULATION IN SPECIFIC ORGANS

Blood functions to transport and redistribute Hg(II) throughout various tissues over time. This distribution has been kinetically modeled by Schultz et al. (1996) and Pickhardt et al. (2006), and is shown in Fig. 5.3 and Table 5.4. Shortly after exposure, accumulation is highest in either the gills or the intestine, depending on whether the exposure is aqueous or dietary, respectively (Boudou et al., 1991). The kidney is a target organ for accumulation, with the majority of Hg(II) concentrated in the proximal tubule epithelial cells (Baatrup et al., 1986). The mechanism responsible for the uptake in kidney epithelial cells is unknown, but accumulation is similar regardless of nephron structure (complex or aglomerular) (Greif and Duvigneaud, 1959). The spleen and liver are also accumulation sites. Given that GSH is involved in the excretion of MeHg(I) in fish bile (Ballatori and Boyer, 1986), accumulation of THg in the liver may be due to inadequate GSH levels, as seen in mammalian systems (Bridges and Zalups, 2006).

10.1.4. SUBCELLULAR PARTITIONING OF MERCURY

Following Hg(II) exposure, both Hg(II) and MeHg(I) are found intracellularly in the kidney and liver (Baatrup et al., 1986); the MeHg(I) may be a product of *in vivo* methylation. In the posterior kidney, most Hg(II) is found in the lysosomes (enzyme-containing vesicles) of nephron epithelial cells. MeHg(I) in the kidney is located in the periphery of cells, and is contained

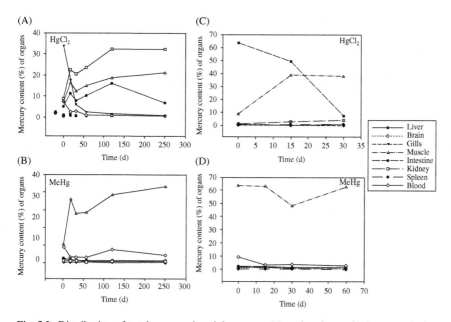

Fig. 5.3. Distribution of total mercury in rainbow trout (*Oncorhynchus mykiss*) organs during depuration, following exposure to (A) aqueous inorganic mercury, (B) aqueous methylmercury, (C) dietary inorganic mercury and (D) dietary methylmercury. Figure produced from data published by Boudou and Ribeyre (1983), and republished with permission from John Wiley & Sons.

in granules in smooth muscle cells (Baatrup et al., 1986). In the liver, more than 50% of all Hg is in the cytosol (Bose et al., 1993). Inorganic Hg is restricted to the cytosol and lysosomes, whereas MeHg(I) accumulates in moderate amounts in the mitochondria, microsomes (fragmented endoplasmic reticulum), nucleus, and nucleolus, and in small amounts in lysosomes (Baatrup et al., 1986; Bose et al., 1993). Similarly, in the gills more than 62% of THg in cells is present in the cytosol, or the soluble fraction (Cosson, 1994).

10.1.5. DETOXIFICATION AND STORAGE MECHANISMS

Detailed knowledge of the toxic mechanisms of Hg(II) is scarce. MT induction in the liver and gills is well documented following aqueous Hg(II) exposure, with increases up to 22% found in the liver (Cosson, 1994; Monteiro et al., 2010). Dietary exposure to Hg(II) caused highest MT induction in kidney, while dietary MeHg(I) exposure caused highest MT induction in the liver (Berntssen et al., 2004). The exact intracellular role of MT in detoxification within these tissues is unclear.

Table 5.4

Distribution ranking of total mercury in various fish tissues following laboratory exposures to Hg(II) and MeHg(I), and under natural field conditions

Common name	Species	Exposure route	Exposure concentration	Fish mass (g)	Temperature (°C)	Length of exposure (days)	Tissue distribution ranking	Reference
Inorganic mercury								
Black bullhead	Ictalurus melas	Waterborne	35, 140 µg L^{-1}	171.1	room	10	Kidney > gills > liver > muscle[c]	Elia et al. (2003)
Blue tilapia	Oreochromis aureus	Waterborne	100 µg L^{-1}	(9–13 cm)	25.5–28.5	0.5	Kidney > spleen > gills > intestine > brain > testes > liver	Allen (1994)
Plaice	Pleuronectes platessa	Waterborne	0.02 µg L^{-1}	40–45	10	60	Gills > spleen > blood > kidney > liver > heart > skin	Pentreath (1976c)
Plaice	Pleuronectes platessa	Dietary	–	80–100	10	5[a]	Upper intestine > stomach > liver > lower intestine > gills > kidney	Pentreath (1976b)
Goldfish	Carassius auratus	Intraperitoneal injection	1.54 µg	10	Room	28[a]	Trunk kidney > spleen > head kidney > intestine > gonad > liver > swim bladder	Weisbart (1973)
Methylmercury								
Japanese eel	Anguilla japonica	Waterborne (freshwater)	3.0 µg L^{-1}	200	20	3	Gills > spleen > liver > posterior intestine > blood > heart > kidney	Yamaguchi et al. (2004)

Japanese eel	Anguilla japonica	Waterborne (seawater)	2.5 μg L^{-1}	200	20	3	Blood>gills >spleen>liver >heart>anterior intestine>kidney	Yamaguchi et al. (2004)
Plaice	Pleuronectes platessa	Waterborne	0.015–0.030 μg L^{-1}	40–50	10	64	Muscle>skin >gills>bone >liver>blood cells>kidney	Pentreath (1976a)
Rainbow trout	Oncorhynchus mykiss	Dietary	31 μg g^{-1}	3–6	12	14	Spleen>liver >kidney >intestine >gills and whole body>muscle >brain	Baatrup and Danscher (1987)
Sheepshead minnows	Cyprinodon variegatus	Dietary	16 μg g^{-1}	0.9–1.2	23	1.5[a]	Liver>intestine >gills>blood >rest[b]	Leaner and Mason (2004)
Rainbow trout	Oncorhynchus mykiss	Dietary	0.87 μg day^{-1}	40	10	30	Posterior intestine>kidney >blood >spleen>liver >muscle>brain >gills	Boudou et al. (1985)

(Continued)

Table 5.4 (*continued*)

Common name	Species	Exposure route	Exposure concentration	Fish mass (g)	Temperature (°C)	Length of exposure (days)	Tissue distribution ranking	Reference
Field studies								
Striped bass	*Roccus saxitilis*	Natural exposure	–	–	–	Constant	Liver > muscle > heart > gonad > brain > gills > blood[b,c]	Cizdziel et al. (2003)
Largemouth bass	*Micropterus salmoides*	Natural exposure	–	–	–	Constant	Muscle > liver > gonads > blood[b,c]	Cizdziel et al. (2003)
Channel catfish	*Ictalurus punctatus*	Natural exposure	–	–	–	Constant	Liver > muscle > gonads > blood[b,c]	Cizdziel et al. (2003)

Tissues were sampled immediately following a given exposure duration (days) (except those marked with[a]).
[a]Duration (days) between end of exposure and tissue sampling, during which the fish were not exposed (i.e. depuration duration).
[b]Relative tissue distributions of the kidneys and spleen were not reported.
[c]Relative tissue distributions of the intestine and spleen were not reported.

Sequestration of Hg(II) in the lysosomes of the liver and posterior kidney (Baatrup et al., 1986) functions as a storage mechanism, removing Hg(II) from cellular processes. GSH is suggested to be involved in Hg detoxification through both sequestration and antioxidant functions (Elia et al., 2003; Monteiro et al., 2010). Hepatic GSH levels have shown both increases and decreases following Hg (II) exposure (Allen et al., 1988; Elia et al., 2003; Monteiro et al., 2010).

10.2. Methylmercury

10.2.1. BIOTRANSFORMATION: *in vivo* DEMETHYLATION

Increases over time in the proportion of inorganic to total Hg have been both measured and visualized (using autometallography) in tissues. As a result, demethylation is suspected to occur mainly in the liver, but also in the kidney and gills (Burrows and Krenkel, 1973; Olson et al., 1978; Baatrup and Danscher, 1987; Cizdziel et al., 2003; Gonzalez et al., 2005). There is a lack of conclusive evidence regarding this process, and further investigation in this area is needed.

10.2.2. TRANSPORT THROUGH THE BLOODSTREAM

In the bloodstream, approximately 90% of MeHg(I) is quickly bound by red blood cells (Olson and Fromm, 1973; Giblin and Massaro, 1975; Ribeiro et al., 1999), while about 4% in the plasma remains unbound to plasma proteins (Schultz and Newman, 1997). MeHg(I) is taken up into erythrocytes preferentially when it is free of ligands such as cysteine. Approximately 90% of MeHg(I) within erythrocytes binds to hemoglobin thiol groups within the cell (Giblin and Massaro, 1975, 1977). This binding is reversible, so erythrocytes function as an effective transport mechanism, redistributing MeHg(I) throughout the body (Giblin and Massaro, 1977). The rate of MeHg(I) loss from red blood cells to the surrounding medium is determined by the relative proportion of thiol groups inside and outside of these cells, and adequate thiol concentrations are needed in the external medium to draw MeHg(I) out. The binding affinity and carrying capacity of MeHg(I) in erythrocytes are variable, and dependent on the thiol concentration of hemoglobin (Giblin and Massaro, 1975).

10.2.3. ACCUMULATION IN SPECIFIC ORGANS

Regardless of MeHg(I) exposure route, tissue distribution is similar but highest in gills and intestine following aqueous and dietary exposures, respectively (McKim et al., 1976; Boudou and Ribeyre, 1983) (Table 5.4).

Shortly after exposure, MeHg(I) accumulates in the blood before redistributing to other tissues. This distribution has been kinetically modeled by Schultz and Newman, (1997), McCloskey et al. (1998), Rouleau et al. (1998), Ribeiro et al. (1999), and Leaner and Mason (2004) (Fig. 5.3).

Muscle is the primary site of MeHg(I) accumulation, comprising the majority of THg body burden (Olson and Fromm, 1973; Boudou and Ribeyre, 1983). This accumulation is time dependent and, after a lag period (Amlund et al., 2007), concentrations increase in the muscle as MeHg(I) redistributes from other tissues (see Fig. 5.3). In muscle, MeHg(I) is present almost exclusively as MeHg–cysteine complexes (Lemes and Wang, 2009), and likely accumulates here owing to its high affinity for thiol groups such as those on cysteine.

Accumulation in the spleen reflects that in the blood. Giblin and Massaro (1973) and Olson et al. (1978) have suggested that levels in the spleen are a result of sequestration of red blood cells in this organ, and later losses, due to the breakdown of these cells. The liver and kidney are also sites of accumulation. The liver is involved in MeHg(I) elimination, and its presence here is probably related to this process. The kidney accumulates Hg to a lesser degree following MeHg(I) exposure than after Hg(II) exposure (see Table 5.4 and Fig. 5.3), mainly in the microvilli and lumen of the proximal tubules (Baatrup and Danscher, 1987). Although the mechanism of this accumulation is unclear, MeHg–cysteine complexes are targeted for uptake (Giblin and Massaro, 1977).

10.2.4. Subcellular Partitioning of Mercury

Intracellular distribution following MeHg(I) exposure is tissue specific. Both MeHg(I) and Hg(II) are found within kidney and liver cells, following MeHg(I) exposure (Baatrup and Danscher, 1987). In hepatocytes, 42–80% of THg is in the cytosol, with smaller portions in the cytoplasmic debris, endoplasmic reticulum, ribosomes, mitochondria, nuclei, lipid droplets, and membranes (Passino and Kramer, 1982; Barghigiani et al., 1989). Both Hg(II) and MeHg(I) are present in hepatic lysosomes (Baatrup and Danscher, 1987). In the posterior kidney, MeHg(I) is associated with numerous organelles as well as dispersed throughout the cytoplasm (Olson et al., 1978; Baatrup and Danscher, 1987); however, Hg(II) is restricted to lysosomes (Baatrup and Danscher, 1987). Also in the kidney MeHg(I) accumulates in vesicles (assumed to be endosomes), in lysosomes, and on the surface of mitochondria and cell membranes. It also associates with the nuclear envelope, however accumulation of MeHg(I) in the nucleus is limited (Baatrup and Danscher, 1987). In muscle, over 92% of THg is present in the insoluble cellular fraction rather than the cytosol (Barghigiani et al., 1989).

10.2.5. DETOXIFICATION AND STORAGE MECHANISMS

In contrast to Hg(II), MeHg(I) does not induce, bind to, or undergo detoxification by MT (Olson et al., 1978; Barghigiani et al., 1989). Muscle is thought to protect other tissues by storing MeHg(I) and removing it from circulation and cellular function. Detailed mechanisms of MeHg(I) storage in muscle are not known, but MeHg(I) is incorporated into the protein fraction as a MeHg–cysteine complex (Amlund et al., 2007; Lemes and Wang, 2009). Within liver and posterior kidney cells, MeHg(I) is sequestered in lysosomes (Baatrup and Danscher, 1987); this is undoubtedly a storage mechanism to remove MeHg(I) from cellular function. The response of GSH following MeHg(I) exposure in both wild and laboratory fish is variable, and the role that this compound plays in detoxification (i.e., whether through conjugation or antioxidant function) remains unclear (Yamaguchi et al., 2004; Larose et al., 2008).

11. CHARACTERIZATION OF EXCRETION ROUTES

Although variable, on average more than 70% of THg in whole wild fish is present as MeHg(I) (e.g. Hammerschmidt et al., 1999; Wyn et al., 2009); therefore, MeHg(I) excretion is of greater concern than Hg(II). Whole-body excretion of MeHg(I) occurs in two or more phases, with a slow phase associated with elimination from the muscle, and a fast phase associated with other organs (Burrows and Krenkel, 1973; Giblin and Massaro, 1973; Trudel and Rasmussen, 1997). Excretion from the fast phase is addressed below. Elimination mechanisms from muscle are unknown, but may be linked to protein synthesis and breakdown (Niimi, 1983). Hemoglobin is able to compete for MeHg(I) in MeHg(I)–cysteine complexes, so it is possible that MeHg(I) is recycled in the blood during protein breakdown and regeneration processes, prolonging the excretion process of this metal (Giblin and Massaro, 1975).

From the half-lives listed in Table 5.5, it can be seen that the elimination of MeHg(I) is slower than that of Hg(II); Trudel and Rasmussen (1997) estimate excretion of MeHg(I) to be 2.8 times slower and rates are influenced by exposure duration, body size (negative relationships), and water temperature (positive relationship) (Trudel and Rasmussen, 1997).

11.1. Inorganic Mercury

11.1.1. GILLS

High loss rates are seen in the gills following aqueous and dietborne exposures to Hg(II), and may be indicative of excretion (Weisbart, 1973;

Table 5.5

Estimated half-lives (days) of Hg(II) and MeHg(I) excretion in fish following differing administration routes, temperatures and durations

Common name	Species name	Administration route	Fish weight (g)	Temperature (°C)	Exposure duration (days)	Depuration duration (days)	Half-life (days)	Reference
Inorganic mercury								
Mosquitofish	Gambusia affinis	Waterborne	1.8–1.9[a]	20	6	–	1.4	Newman and Doubet (1989)
Plaice	Pleuronectes platessa	Waterborne	42	10	60	81	135	Pentreath (1976c)
Sweetlips	Plectorhinchus gibbosus	Waterborne	–	23	7 (4 h day^{-1})	7	9.67	Wang and Wong (2003)
Sweetlips	Plectorhinchus gibbosus	Waterborne	–	23	7 (4 h day^{-1})	28	25.4	Wang and Wong (2003)
Sweetlips	Plectorhinchus gibbosus	Dietary	–	23	7 (4 h day^{-1})	7	7.51	Wang and Wong (2003)
Sweetlips	Plectorhinchus gibbosus	Dietary	–	23	7 (4 h day^{-1})	28	13.1	Wang and Wong (2003)
Goldfish	Carassius auratus	Intraperitoneal injection	10	Room	Instant	28	23.7	Weisbart (1973)
Channel catfish	Ictalurus punctatus	Dorsal aortic cannula	250	20–22	Instant	156	722	Schultz et al. (1996)
Methylmercury								
Bluegill	Lepomis macrochirus	Waterborne	3–5	24	4	95	130	Burrows and Krenkel (1973)
Sweetlips	Plectorhinchus gibbosus	Waterborne	–	23	7 (4 h day^{-1})	7	40.2	Wang and Wong (2003)
Sweetlips	Plectorhinchus gibbosus	Waterborne	–	23	7 (4 h day^{-1})	28	55.7	Wang and Wong (2003)

Common name	Species	Exposure type						Reference
Atlantic cod	*Gadus morhua* L.	Dietary	325	11.1	86	88	377	Amlund et al. (2007)
Thornback ray	*Raja clavata*	Dietary	107	10	7	33	323	Pentreath (1976d)
Plaice	*Pleuronectes platessa*	Dietary	61	10	85	61	275	Pentreath (1976b)
Sweetlips	*Plectorhinchus gibbosus*	Dietary	–	23	7 (4 h day^{-1})	7	72.8	Wang and Wong (2003)
Sweetlips	*Plectorhinchus gibbosus*	Dietary	–	23	7 (4 h day^{-1})	28	69.8	Wang and Wong (2003)
Channel catfish	*Ictalurus punctatus*	Forced dietary	840	21.2	Instant	>62.5	29.2	McCloskey et al. (1998)
Yellow perch	*Perca flavescens*	Dietary and waterborne	1.8–6.8	–	Unknown	440	489	Van Walleghem et al. (2007)
Channel catfish	*Ictalurus punctatus*	Dorsal aortic cannula	840	21.2	Instant	>125	41.7	McCloskey et al. (1998)

[a]Weight calculated from dry weight assuming 80% moisture content.

Handy and Penrice, 1993). This has not been thoroughly investigated, however, and could result from turnover of the gill mucus layer, which can accumulate Hg(II) (Olson and Fromm, 1973).

11.1.2. KIDNEY

Total Hg is excreted in the urine of the aglomerular toadfish following administration of a mercurial diuretic (Greif and Duvigneaud, 1959), and the bladder can accumulate levels of Hg similar to the kidney (Pentreath, 1976c); both indicate renal excretion. The relative importance of this excretion route is unknown, but is likely minimal in the environment. It is important to note that most references cited herein do not distinguish between the anterior, mid, or posterior kidney segments. Those that are not distinguished are referred to here simply as the kidney. The various segments accumulate different amounts of Hg (Greif and Duvigneaud, 1959), so distinction would be beneficial.

11.1.3. LIVER/BILE

Accumulation of Hg(II) in the bile following exposure is low compared to all other tissues (Greif and Duvigneaud, 1959; Baatrup et al., 1986; Allen, 1994; Schultz et al., 1996). It is probable that THg enters the bile following conjugation with GSH in the liver, and that biliary excretion is dependent on GSH levels, as is the case with MeHg(I).

Following aqueous Hg(II) exposure, accumulation in the intestine of freshwater teleosts is not expected, but has been found. This may indicate the excretion of Hg–complexes into the intestine via the bile (Allen, 1994; Ribeiro et al., 2000; Simon and Boudou, 2001). From the intestine these may be reabsorbed, and transported via the blood to the kidney or other tissues. Allen et al. (1988) and Allen (1994) have indicated that the liver plays an important role in excretion and interorgan distribution of Hg(II). The relative rate and efficiency of this excretion are not known, but are expected to be considerable.

11.1.4. GUT

Excretion through the gut differs depending on whether the uptake route is dietary or aqueous. Dietary Hg(II) is poorly absorbed through the intestine (Boudou and Ribeyre, 1983, 1985), and excretion is rapid, with 50–80% excreted in less than 15 days (Boudou et al., 1991; Ribeiro et al., 2000). The retained Hg is partitioned mostly in the gut mucosa, and excreted eventually through the replacement of mucosa cells (Boudou and Ribeyre, 1985). The excretion rate following aqueous exposure is slower in comparison by a factor of 2 (Wang and Wong, 2003). This is logical, given that following aqueous exposure, assimilation into the blood and tissues is greater (Boudou et al., 1991; Pickhardt et al., 2006). As outlined previously, the liver and bile play an important role in the excretion and redistribution of Hg. However, the contribution of the bile to Hg concentrations of the intestine, and feces, is unknown.

11.2. Methylmercury

11.2.1. GILLS

Excretion of MeHg(I) from the gills is also largely unknown. Transfer of Hg to the gills from the blood has been estimated to have a rate constant of $k = 0.72-1.05\,day^{-1}$ following dietary exposure. Assuming it is then excreted from the gills, this is a limited excretion route in comparison to the transfer rates between other tissue compartments, which are approximately two to 12 times higher (Leaner and Mason, 2004).

11.2.2. KIDNEY

It appears that renal excretion of MeHg(I) is unstudied in fish. The kidney accumulates less THg following MeHg(I) exposure than Hg(II) exposure; therefore, renal excretion is likely a negligible route.

11.2.3. LIVER/BILE

Total Hg accumulates to low levels in the bile compared to other tissues, in both freshwater and marine fish exposed to MeHg(I) (Giblin and Massaro, 1973; Ballatori and Boyer, 1986; Yamaguchi et al., 2004). Excretion of THg in bile depends on the rate of GSH excretion, indicating that conjugation with GSH occurs in the liver (Ballatori and Boyer, 1986).

Accumulation of Hg in the intestine of freshwater fish is not expected, but does occur following aqueous MeHg(I) exposures. This indicates the transfer of THg from the bile to the intestine, where it can be excreted in the feces, or reabsorbed (Ribeiro et al., 1999; Simon and Boudou, 2001).

11.2.4. GUT

The majority of absorbed MeHg(I) is excreted in the feces, but elimination is slow, with an estimated average daily excretion of 0.05% of a dose ($0.5\,mg\,kg^{-1}$ dose) (Giblin and Massaro, 1973). THg can enter the gut transferred from the blood (Leaner and Mason, 2004) or through the proposed route of biliary excretion (Giblin and Massaro, 1973; Ballatori and Boyer, 1986; Yamaguchi et al., 2004). Transfer from the blood to the gut is estimated to be one of the fastest transfer rates between tissues, with a rate constant of $k=7.11-7.34\,day^{-1}$ (Leaner and Mason, 2004). The rate of transfer from the bile to the intestine, however, is estimated to be low (Ballatori and Boyer, 1986).

11.2.5. EGGS

Female fish can accumulate MeHg(I) in their eggs, which are deposited into the surrounding environment during spawning (maternal transfer), representing a type of excretion (Hammerschmidt et al., 1999; Latif et al., 2001; Drevnick et al., 2006a). The MeHg(I) content of eggs is related to the size and age of female fish, as well as the MeHg(I) content of the maternal diet during oogenesis, and comprises less than 85% of the THg content of eggs (Hammerschmidt et al., 1999; Hammerschmidt and Sandheinrich, 2005). In freshwater iteroparous teleosts, transfer to eggs is less than 2.3% of the body burden, and not a substantial excretion route (Hammerschmidt et al., 1999). In contrast, the eggs of semelparous fish accumulate about 20% of the maternal body burden (Drevnick et al., 2006a); this may be due to the differing resource allocation between these reproductive strategies.

12. BEHAVIORAL EFFECTS OF MERCURY

12.1. Inorganic Mercury

There is growing evidence that a number of key behaviors – predator avoidance and reproductive, feeding, and social behaviors – are impaired at environmentally relevant exposures to contaminants such as Hg(II) (see review by Scott and Sloman, 2004). In addition, behavioral effects can be found at lower concentrations than those affecting growth or tissues, and changes in behavior are undoubtedly linked to the effects of Hg(II) on the endocrine and nervous systems. As one example, Hg(II) reduced survival of fish in a simple predator–prey environment by increasing their susceptibility to predation (Kania and O'Hara, 1974). In another study, significant reductions in feeding rates of mummichog were seen after exposures for 1 week to 20 μg L^{-1} or after 2 weeks to 10 μg L^{-1} (Weis and Khan, 1990). In this study, Hg(II) was more potent at reducing feeding than was MeHg(I).

12.2. Methylmercury

Fish behaviors are also affected by MeHg(I) exposure, including those associated with feeding, reproduction, and predator avoidance. As with other endpoints, earlier life stages are more sensitive than adults; tissue residues of 0.27 and 0.52 μg MeHg(I) g^{-1} ww have been linked to changes in behaviors in fry and adults, respectively (Beckvar et al., 2005).

The ability of fish to compete with other fish for food or to capture prey decreases when they are exposed to MeHg(I). For example, aqueous exposures of grayling (*Thymallus thymallus*) embryos to concentrations ranging from 0.16 to 20 µg L^{-1} for only 10 days resulted in reduced abilities of adults to catch *Daphnia magna* (by 16–24%). The pre-exposed fish were two- to six-fold less successful at competing with unexposed fish for food (Fjeld et al., 1998). MeHg(I) also increases the time it takes fish to capture prey and this effect, as well as reductions in prey capture success, is greater when exposure occurs at the embryonic than larval stages (Zhou et al., 2001). If exposure occurs throughout early development rather than just at the embryo stage, lower concentrations of MeHg(I) will affect predation success (Zhou et al., 2001).

Overall activity and swimming performance of fish are also reduced as a result of MeHg(I) exposure. Dietborne exposures decreased swimming activity in Atlantic salmon parr (Berntssen et al., 2003) and fathead minnow (Sandheinrich and Miller, 2006). Similarly, for waterborne exposures, studies have found a loss in posture in rainbow trout fry and fingerlings during acute toxicity experiments (Wobeser, 1975a), and more collisions between larval mummichog exposed as embryos to 5 or 10 µg L^{-1} (Ososkov and Weis, 1996).

Predator avoidance is also impaired by MeHg(I) exposure, a response that appears to be due, at least in part, to the effects of Hg on visual acuity (Hawryshyn et al., 1982). Exposures of MeHg(I) (>2 µg L^{-1}) during the first 24 h after fertilization resulted in decreased responses to visual cues (Weber et al., 2008). Predator avoidance through schooling is a commonly studied behavior for small-bodied fish, and golden shiners fed 0.959 mg MeHg(I) kg^{-1} ww for 3 months showed reduced schooling behavior when exposed to a model predator (Webber and Haines, 2003).

13. MOLECULAR CHARACTERIZATION OF MERCURY TRANSPORTERS, STORAGE PROTEINS, AND CHAPERONES

Forms of MT that are inducible by Hg(II) have been molecularly characterized in the zebrafish, common carp, and tilapia (*Oreochromis mossambicus* and *Oreochromis aureus*) (Chan et al., 2004; Yan and Chan, 2004; Chan and Chan, 2008). The MT gene is highly conserved in vertebrates, including teleosts, with the exception of the variable 5′ gene promoter region. The distal metal regulatory elements on the promoter region in particular play an important role in MT induction in exposures to metals such as Hg(II) (Yan and Chan, 2004; Chan and Chan, 2008).

14. GENOMIC AND PROTEOMIC STUDIES

A handful of studies have examined gene expression in MeHg(I)-exposed wild or laboratory fish and most used dietary exposures at environmentally relevant concentrations. In skeletal muscle, changes in gene expression associated with mitochondrial metabolism are the most notable, specifically decreased expression of genes linked to cytochrome c oxidase (Gonzalez et al., 2005; Cambier et al., 2009, 2010). Genes associated with protein synthesis, increased oxidative stress and apoptosis, regulation of mitosis, and cellular metabolism were also affected (Gonzalez et al., 2005; Cambier et al., 2010).

Gene expression in the liver has been investigated in both laboratory MeHg (I)-exposed and wild fish. Expression is sex specific, with the majority of changes seen in male fish. These include: upregulation of genes associated with mitochondrial metabolism, oxidative stress, detoxification processes, DNA repair and apoptosis; and downregulation of genes associated with immune response (Gonzalez et al., 2005; Klaper et al., 2006, 2008; Moran et al., 2007). Following acute exposure, genes associated with protein synthesis in the liver are most affected (downregulated) (Klaper et al., 2008).

Endocrine disruption by MeHg is implicated in both wild and laboratory fish, owing to altered expression of genes in the liver involved in gonadotropin pathways as well as hypothalamus–pituitary–gonadal axis function (Klaper et al., 2006; Moran et al., 2007). Differential expression of vitellogenin mRNA (a phosphoglycolipoprotein involved in egg production) is also seen between males and females following exposure to MeHg, with increases and decreases occurring in males and females, respectively (Klaper et al., 2006).

Although gene expression changes are not detected in the brain following MeHg(I) exposure (Gonzalez et al., 2005), changes in the proteome have been shown to occur in the brain of Beluga sturgeon (*Huso huso*) (Keyvanshokooh et al., 2009). The most notable of these is decreased levels of β-tubulin, a protein related to structure and function of the cellular cytoskeleton. Alterations in levels of proteins involved in protein folding, metabolism, cell division, and signal transduction also occur, likely affecting important brain functions (Keyvanshokooh et al., 2009).

15. KNOWLEDGE GAPS AND FUTURE DIRECTIONS

The effects and fate of Hg(II) and MeHg(I) in fish remain a very active area of research, and we now understand that effects of earlier exposures can manifest at later stages of development (Weis, 2009). An increasing number of studies is examining reproductive effects of MeHg(I) on fish to better understand whether it is affecting the health of wild populations (see reviews by Crump and Trudeau, 2009; Sandheinrich and Wiener, 2010). It has been well established that fish health in highly contaminated ecosystems (e.g. Minamata Bay) was affected, but considerably less is known about fish in Hg-sensitive systems receiving atmospheric inputs (Wiener et al., 2003). Fish in these lakes, streams, and rivers can accumulate levels of Hg in their tissues that are above those believed to be without effect (Beckvar et al., 2005).

Despite more than five decades of research on Hg in fish, some knowledge gaps remain. These include the following:

- It is not understood how chronic Hg exposure affects the reproductive success of wild fish.
- Limited information exists about Hg excretion and the relative involvement of various organs in these processes. These details, combined with uptake mechanisms, need to be examined to properly understand Hg accumulation in fish.
- Studies relating evidence of *in vivo* biotransformation of Hg (II) and MeHg(II) (methylation and demethylation respectively) have mostly been conducted before the 1990s, with more recent studies showing contradictory results to earlier studies. There is a need for comprehensive studies in this area, especially regarding *in vivo* demethylation. This biotransformation may potentially play an important role in the detoxification and internal handling of MeHg(I), and requires further investigation.
- There are currently only a handful of genomic and proteomic-based studies in this area, but this number is likely to increase quickly as this is a rapidly growing research field.

REFERENCES

Allen, P. (1994). Distribution of mercury in the soft-tissues of the blue tilapia *Oreochromis aureus* (Steindachner) after acute exposure to mercury-(II) chloride. *Bull. Environ. Contam. Toxicol.* **53**, 675–683.

Allen, P., Min, S. Y., and Keong, W. M. (1988). Acute effects of mercuric-chloride on intracellular GSH levels and mercury distribution in the fish *Oreochromis aureus. Bull. Environ. Contam. Toxicol.* **40**, 178–184.

Amlund, H., Lundebye, A., and Berntssen, M. H. G. (2007). Accumulation and elimination of methylmercury in Atlantic cod (*Gadus morhua* L.) following dietary exposure. *Aquat. Toxicol.* **83**, 323–330.

ANZECC and ARMCANZ (2000). Aquatic ecosystems. In: *Australian and New Zealand Guidelines for Fresh and Marine Water Quality* (*National Water Quality Strategy, no. 4*), Vol. 1. Canberra: Australian and New Zealand Environment and Conservation Council and Agriculture and Resource Management Council of Australia and New Zealand.

Baatrup, E., and Danscher, G. (1987). Cytochemical demonstration of mercury deposits in trout liver and kidney following methyl mercury intoxication – differentiation of 2 mercury pools by selenium. *Ecotoxicol. Environ. Saf.* **14**, 129–141.

Baatrup, E., Nielsen, M. G., and Danscher, G. (1986). Histochemical demonstration of 2 mercury pools in trout tissues – mercury in kidney and liver after mercuric-chloride exposure. *Ecotoxicol. Environ. Saf.* **12**, 267–282.

Babiarz, C. L., Hurley, J. P., Hoffmann, S. R., Andren, A. W., Shafer, M. M., and Armstrong, D. E. (2001). Partitioning of total mercury and methylmercury to the colloidal phase in freshwaters. *Environ. Sci. Technol.* **35**, 4773–4782.

Ballatori, N., and Boyer, J. L. (1986). Slow biliary elimination of methyl mercury in the marine elasmobranchs, *Raja erinacea* and *Squalus acanthias*. *Toxicol. Appl. Pharmacol.* **85**, 407–415.

Banerjee, S., and Bhattacharya, S. (1995). Histopathological changes induced by chronic nonlethal levels of elsan, mercury, and ammonia in the small-intestine of *Channa punctatus* (Bloch). *Ecotoxicol. Environ. Saf.* **31**, 62–68.

Bano, Y., and Hasan, M. (1989). Mercury induced time-dependent alterations in lipid profiles and lipid-peroxidation in different body organs of catfish *Heteropneustes fossilis*. *J. Environ. Sci. Health B* **24**, 145–166.

Bano, Y., and Hasan, M. (1990). Histopathological lesions in the body organs of catfish (*Heteropneustes fossilis*) following mercury intoxication. *J. Environ. Sci. Health B* **25**, 67–85.

Barghigiani, C., Pellegrini, D., and Carpene, E. (1989). Mercury binding-proteins in liver and muscle of flat fish from the northern Tyrrhenian Sea. *Comp. Biochem. Physiol. C* **94**, 309–312.

Beckvar, N., Dillon, T. M., and Read, L. B. (2005). Approaches for linking whole-body fish tissue residues of mercury or DDT to biological effects threshold. *Environ. Toxicol. Chem.* **24**, 2094–2105.

Berntssen, M. H. G., Aatland, A., and Handy, R. D. (2003). Chronic dietary mercury exposure causes oxidative stress, brain lesions, and altered behaviour in Atlantic salmon (*Salmo salar*) parr. *Aquat. Toxicol.* **65**, 55–72.

Berntssen, M. H. G., Hylland, K., Julshamn, K., Lundebye, A. K., and Waagbo, R. (2004). Maximum limits of organic and inorganic mercury in fish feed. *Aquacult. Nutr.* **10**, 83–97.

Birge, W. J., Black, J. A., Westerman, A. G., and Hudson, J. E. (1979). The effects of mercury on reproduction of fish and amphibians. In: The Biogeochemistry of Mercury in the Environment (Topics in Environmental Health Series) (J.O Nriagu, ed.), Vol. 3, pp. 629–655. Elsevier, Amsterdam.

Block, M., Pärt, P., and Wicklund-Glynn, A. (1997). Influence of water quality on the accumulation of methyl (203) mercury in gill tissue of minnow (*Phoxinus phoxinus*). *Comp. Biochem. Physiol. C* **118**, 191–197.

Bols, N. C., Brubacher, J. L., Ganassin, R. C., and Lee, L. E. J. (2001). Ecotoxicology and innate immunity in fish. *Dev. Comp. Immunol.* **25**, 853–873.

Borg-Neczak, K., and Tjälve, H. (1996). Uptake of Hg-203($^{2+}$) in the olfactory system in pike. *Toxicol. Lett.* **84**, 107–112.

Bose, S., Ghosh, P., Ghosh, S., and Bhattacharya, S. (1993). Distribution kinetics of inorganic mercury in the subcellular fractions of fish liver. *Sci. Total Environ.* (*Suppl.* **1**), 533–538.

Boudou, A., and Ribeyre, F. (1983). Contamination of aquatic biocenoses by mercury compounds: an experimental ecotoxicological approach. In: *Aquatic Toxicology (Advances in Environmental Science and Technology)* (J.O Nriagu, ed.), Vol. 13, pp. 73–116. John Wiley and Sons, New York.

Boudou, A., and Ribeyre, F. (1985). Experimental study of trophic contamination of *Salmo gairdneri* by 2 mercury-compounds – HgCl$_2$ and CH$_3$HgCl – analysis at the organism and organ levels. *Water Air Soil Pollut.* **26**, 137–148.

Boudou, A., Delnomdedieu, M., Georgescauld, D., Ribeyre, F., and Saouter, E. (1991). Fundamental roles of biological barriers in mercury accumulation and transfer in freshwater ecosystems – (analysis at organism, organ, cell and molecular-levels). *Water Air Soil Pollut.* **56**, 807–822.

Bridges, C. C., and Zalups, R. K. (2006). Molecular mimicry as a mechanism for the uptake of cysteine S-conjugates of methylmercury and inorganic mercury. *Chem. Res. Toxicol.* **19**, 1117–1118.

Burrows, W. D., and Krenkel, P. A. (1973). Studies on uptake and loss of methylmercury-203 by bluegills (*Lepomis macrochirus* Raf). *Environ. Sci. Technol.* **7**, 1127–1130.

Büsselberg, D. (1995). Calcium channels as target sites of heavy metals. *Toxicol. Lett.* **82–83**, 255–261.

Cabana, G., and Rasmussen, J. B. (1994). Modeling food-chain structure and contaminant bioaccumulation using stable nitrogen isotopes. *Nature* **372**, 255–257.

Cambier, S., Bénard, G., Mesmer-Dudons, N., Gonzalez, P., Rossignol, R., Bréthes, D., and Bourdineaud, J. P. (2009). At environmental doses, dietary methylmercury inhibits mitochondrial energy metabolism in skeletal muscles of the zebrafish (*Danio rerio*). *Int. J. Biochem. Cell Biol.* **41**, 791–799.

Cambier, S., Gonzalez, P., Durrieu, G., Maury-Brachet, R., Boudou, A., and Bourdineaud, J. P. (2010). Serial analysis of gene expression in the skeletal muscles of zebrafish fed with a methylmercury-contaminated diet. *Environ. Sci. Technol.* **44**, 469–475.

Campbell, L. M., Norstrom, R. J., Hobson, K. A., Muir, D. C. G., Backus, S., and Fisk, A. T. (2005). Mercury and other trace elements in a pelagic Arctic marine food web (Northwater Polynya, Baffin Bay). *Sci. Total Environ.* **351–352**, 247–263.

CCME (1991). Interim Canadian Environmental Quality Criteria for Contaminated Sites. CCME Report EPC-CS34. Canadian Council of Ministers of the Environment, Winnipeg.

CCME (2007). Canadian water quality guidelines for the protection of aquatic life: summary table. Updated December, 2007. In: Canadian Environmental Quality Guidelines (1999). Publication No. 1299. Canadian Council of Ministers of the Environment, Winnipeg.

Chan, P. C., Shiu, C. K. M., Wong, F. W. Y., Wong, J. K. Y., Lam, K. L., and Chan, K. M. (2004). Common carp metallothionein-1 gene: cDNA cloning, gene structure and expression studies. *Biochim. Biophys. Acta* **1676**, 162–171.

Chan, W. W. L., and Chan, K. M. (2008). Cloning and characterization of a tilapia (*Oreochromis aureus*) metallothionein gene promoter in Hepa-T1 cells following the administration of various heavy metal ions. *Aquat. Toxicol.* **86**, 59–75.

Cizdziel, J., Hinners, T., Cross, C., and Pollard, J. (2003). Distribution of mercury in the tissues of five species of freshwater fish from Lake Mead, USA. *J. Environ. Monit.* **5**, 802–807.

Cosson, R. P. (1994). Heavy-metal intracellular balance and relationship with metallothionein induction in the gills of Carp – after contamination by Ag, Cd, and Hg following pretreatment with Zn or not. *Biol. Trace Elem. Res.* **46**, 229–245.

Costa, J. R. M. A., Mela, M., de Assis, H. C. D., Pelletier, E., Randi, M. A. F., and Ribeiro, C. A. D. (2007). Enzymatic inhibition and morphological changes in *Hoplias malabaricus* from dietary exposure to lead(II) or methylmercury. *Ecotoxicol. Environ. Saf.* **67**, 82–88.

Crump, K. L., and Trudeau, V. L. (2009). Mercury-induced reproductive impairment in fish. *Environ. Toxicol. Chem.* **28**, 895–907.

CSIR (1995). South African Water Quality Guidelines for Coastal Marine Waters, *Vol. 1*. Natural Environment. Department of Water Affairs and Forestry, Council for Scientific and Industrial Research, Pretoria.

CSIR (1996). South African Water Quality Guidelines, *Vol. 7*. Aquatic Ecosystems. Department of Water Affairs and Forestry, Council for Scientific and Industrial Research, Pretoria.

Devlin, E. W. (2006). Acute toxicity, uptake and histopathology of aqueous methyl mercury to fathead minnow embryos. *Ecotoxicology* **15**, 97–110.

Doudoroff, P., and Katz, M. (1953). Critical review of literature on the toxicity of industrial wastes and their components to fish: II The Metals, as Salts. *Sewage Ind. Wastes* **25**, 802–839.

Drevnick, P. E., and Sandheinrich, M. B. (2003). Effects of dietary methylmercury on reproductive endocrinology of fathead minnows. *Environ. Sci. Technol.* **3–7**, 4390–4396.

Drevnick, P. E., Horgan, M. J., Oris, J. T., and Kynard, B. E. (2006a). Ontogenetic dynamics of mercury accumulation in northwest Atlantic Sea lamprey (*Petromyzon marinus*). *Can. J. Fish. Aquat. Sci.* **63**, 1058–1066.

Drevnick, P. E., Sandheinrich, M. B., and Oris, J. T. (2006b). Increased ovarian follicular apoptosis in fathead minnows (*Pimephales promelas*) exposed to dietary methylmercury. *Aquat. Toxicol.* **79**, 49–54.

Duncan, D. A., and Klaverkamp, J. F. (1983). Tolerance and resistance to cadmium in white suckers (*Catostomus commersoni*) previously exposed to cadmium, mercury, zinc, or selenium. *Can. J. Fish. Aquat. Sci.* **40**, 128–138.

Elia, A. C., Galarini, R., Taticchi, M. I., Dörr, A. J. M., and Mantilacci, L. (2003). Antioxidant responses and bioaccumulation in *Ictalurus melas* under mercury exposure. *Ecotoxicol. Environ. Saf.* **55**, 162–167.

Endo, T., Kimura, O., Sakata, M., and Shaikh, Z. A. (1997). Mercury uptake by LLC-PK1 cells: dependence on temperature and membrane potential. *Toxicol. Appl. Pharmacol.* **146**, 294–298.

Evans, D. H. (1981). Osmotic and ionic regulation by freshwater and marine fishes. In: Environmental Physiology of Fishes. NATO Advanced Study Institutes Series, Series A: Life Sciences (M.A Ali, ed.), Vol. 35, p. 97. Plenum Press, Montreal.

Farmanfarmaian, A., and Socci, R. (1984). Inhibition of essential amino acid absorption in marine fishes by mercury. *Mar. Environ. Res.* **14**, 185–199.

Fjeld, E., Haugen, T., and Vøllestad, L. (1998). Permanent impairment in the feeding behavior of grayling (*Thymallus thymallus*) exposed to methylmercury during embryogenesis. *Sci. Total Environ.* **213**, 247–254.

Friedmann, A., Watzin, M., Brinck-Johnsen, T., and Leiter, J. (1996). Low levels of dietary methylmercury inhibit growth and gonadal development in juvenile walleye (*Stizostedion vitreum*). *Aquat. Toxicol.* **35**, 265–278.

Gardner, G. R. (1975). Chemically induced lesions in estuarine or marine teleosts. In: *The Pathology of Fishes* (W.E. Ribelin and G. Migaki, eds), pp. 657–693. University of Wisconsin Press, Madison, WI.

Giari, L., Simoni, E., Manera, M., and Dezfuli, B. (2008). Histo-cytological responses of *Dicentrarchus labrax* (L.) following mercury exposure. *Ecotoxicol. Environ. Saf.* **70**, 400–410.

Giblin, F. J., and Massaro, E. J. (1973). Pharmacodynamics of methyl mercury in rainbow-trout (*Salmo gairdneri*) – tissue uptake, distribution and excretion. *Toxicol. Appl. Pharmacol.* **24**, 81–91.

Giblin, F. J., and Massaro, E. J. (1975). Erythrocyte transport and transfer of methylmercury to tissues of rainbow trout (*Salmo gairdneri*). *Toxicol.* **5**, 243–254.

Giblin, F. J., and Massaro, E. J. (1977). The uptake of methylmercury and methylmercury cysteine by rainbow trout (*Salmo gairdneri*) kidney. *Gen. Pharmacol.* **8**, 103–107.

Gill, G. A., and Bruland, K. W. (1990). Mercury speciation in surface fresh-water systems in California and other areas. *Environ. Sci. Technol.* **24**, 1392–1400.

Gill, T. S., Tewari, H., and Pande, J. (1990). Use of the fish enzyme-system in monitoring water-quality effects of mercury on tissue enzymes. *Comp. Biochem. Physiol. C* **97**, 287–292.

Gonzalez, P., Dominique, Y., Massabuau, J. C., Boudou, A., and Bourdineaud, J. P. (2005). Comparative effects of dietary methylmercury on gene expression in liver, skeletal muscle, and brain of the zebrafish (*Danio rerio*). *Environ. Sci. Technol.* **39**, 3972–3980.

Greif, R. L., and Du Vigneaud, M. (1959). Distribution of a radiomercury-labelled diuretic (Chlormerodrin) in tissues of marine fish. Biol. Bull. (Woods Hole, MA, U.S.) **117**, 251–257.

Gutknecht, J. (1981). Inorganic mercury (Hg^{2+}) transport through lipid bilayer-membranes. J. Membr. Biol. **61**, 61–66.

Hall, B. D., Bodaly, R. A., Fudge, R. J. P., Rudd, J. W. M., and Rosenberg, D. M. (1997). Food as the dominant pathway of methylmercury uptake by fish. Water Air Soil Pollut. **100**, 13–24.

Hammerschmidt, C. R., and Sandheinrich, M. B. (2005). Maternal diet during oogenesis is the major source of methylmercury in fish embryos. Environ. Sci. Technol. **39**, 3580–3584.

Hammerschmidt, C. R., Wiener, J. G., Frazier, B. E., and Rada, R. G. (1999). Methylmercury content of eggs in yellow perch related to maternal exposure in four Wisconsin lakes. Environ. Sci. Technol. **33**, 999–1003.

Hammerschmidt, C. R., Sandheinrich, M. B., Wiener, J. G., and Rada, R. G. (2002). Effects of dietary methylmercury on reproduction of fathead minnows. Environ. Sci. Technol. **36**, 877–883.

Handy, R. D., and Penrice, W. S. (1993). The influence of high oral doses of mercuric-chloride on organ toxicant concentrations and histopathology in rainbow trout, Oncorhynchus mykiss. Comp. Biochem. Physiol. C **106**, 717–724.

Hara, T. J., Law, Y. M. C., and MacDonald, S. (1976). Effects of mercury and copper on olfactory response in rainbow-trout, Salmo gairdneri. J. Fish. Res. Board Can. **33**, 1568–1573.

Hawryshyn, C. W., Mackay, W. C., and Nilsson, T. H. (1982). Methyl mercury induced visual deficits in rainbow trout. Can. J. Zool. **60**, 3127–3133.

Hilmy, A. M., El Domiaty, N. A., Daabees, A. Y., and Moussa, F. I. (1987). Short-term effects of mercury on survival, behaviour, bioaccumulation and ionic pattern in the catfish (Clarias lazera). Comp. Biochem. Physiol. C **87**, 303–308.

Hoyle, I., and Handy, R. D. (2005). Dose-dependent inorganic mercury absorption by isolated perfused intestine of rainbow trout, Oncorhynchus mykiss, involves both amiloride-sensitive and energy-dependent pathways. Aquat. Toxicol. **72**, 147–159.

Hughes, W. L. (1957). A physicochemical rationale for the biological activity of mercury and its compounds. Ann. N. Y. Acad. Sci. **65**, 454–460.

Imura, N., Pan, S. K., and Ukita, T. (1972). Methylation of inorganic mercury with liver homogenate of tuna fish. Chemosphere **1**, 197–201.

Jagoe, C. H., Faivre, A., and Newman, M. C. (1996). Morphological and morphometric changes in the gills of mosquitofish (Gambusia holbrooki) after exposure to mercury (II). Aquat. Toxicol. **34**, 163–183.

Joy, K. P., and Kirubagaran, R. (1989). An immunocytochemical study on the pituitary gonadotropic and thyrotropic cells in the catfish, Clarias batrachus after mercury treatment. Biol. Struct. Morphog. **2**, 67–70.

Kania, H. J., and O'Hara, J. (1974). Behavioral alterations in a simple predator–prey system due to sublethal exposure to mercury. Trans. Am. Fish. Soc. **103**, 134–136.

Karniski, L. P. (1992). Hg^{2+} and Cu^+ are ionophores, mediating Cl^-/OH exchange in liposomes and rabbit brush border membranes. J. Biol. Chem. **267**, 19218–19225.

Keyvanshokooh, S., Vaziri, B., Gharaei, A., Mahboudi, F., Esmaili-Sari, A., and Shahriari-Moghadam, M. (2009). Proteome modifications of juvenile beluga (Huso huso) brain as an effect of dietary methylmercury. Comp. Biochem. Physiol. D **4**, 243–248.

Khan, A. T., and Weis, J. (1987a). Toxic effects of mercuric-chloride on sperm and egg viability of 2 populations of mummichog, Fundulus heteroclitus. Environ. Pollut. **48**, 263–273.

Khan, A. T., and Weis, J. S. (1987b). Effects of methylmercury on sperm and egg viability of two populations of killifish (*Fundulus heteroclitus*). *Arch. Environ. Contam. Toxicol.* **16**, 499–505.

Khan, A. T., and Weis, J. S. (1987c). Effect of methylmercury on egg and juvenile viability in two populations of killifish. *Fundulus heteroclitus. Environ. Res.* **44**, 272–278.

Kinne-Saffran, E., and Kinne, R. K. H. (2001). Inhibition by mercuric chloride of Na–K–2Cl cotransport activity in rectal gland plasma membrane vesicles isolated from *Squalus acanthias. Biochim. Biophys. Acta Biomembr.* **1510**, 422–451.

Kirubagaran, R., and Joy, K. P. (1988a). Toxic effects of three mercurial compounds on survival, and histology of the kidney of the catfish *Clarias batrachus* (L.). *Ecotoxicol. Environ. Saf.* **15**, 171–179.

Kirubagaran, R., and Joy, K. P. (1988b). Inhibition of testicular 3β-hydroxyy-D₅-steroid dehydrogenase (3βHSD) activity in the catfish *Clarias batrachus* (L.) by mercurials. *Ind. J. Exp. Biol.* **26**, 907–908.

Kirubagaran, R., and Joy, K. P. (1988c). Toxic effects of mercuric chloride, methylmercuric chloride, and Emisan (an organic mercurial fungicide) on ovarian recrudescence in the catfish. *Clarias batrachus* (L.). *Bull Environ Contam Toxicol.* **47**, 902–909.

Kirubagaran, R., and Joy, K. P. (1990). Changes in brain monoamine levels and monoamine-oxidase activity in the catfish, *Clarias batrachus*, during chronic treatments with mercurials. *Bull. Environ. Contam. Toxicol.* **45**, 88–93.

Kirubagaran, R., and Joy, K. P. (1991). Changes in adrenocortical pituitary activity in the catfish, *Clarias batrachus* (L.), after mercury treatment. *Ecotoxicol. Environ. Saf.* **22**, 36–44.

Kirubagaran, R., and Joy, K. P. (1992).). Toxic effects of mercury on testicular activity in the fresh-water teleost, *Clarias batrachus* (L.). *J. Fish Biol.* **41**, 305–315.

Kirubagaran, R., and Joy, K. P. (1994). Effects of short-term exposure to methylmercury chloride and its withdrawal on serum levels of thyroid-hormones in the catfish *Clarias batrachus. Bull. Environ. Contam. Toxicol.* **53**, 166–170.

Klaper, R., Rees, C. B., Drevnick, P., Weber, D., Sandheinrich, M., and Carvan, M. J. (2006). Gene expression changes related to endocrine function and decline in reproduction in fathead minnow (*Pimephales promelas*) after dietary methylmercury exposure. *Environ. Health Perspect.* **114**, 1337–1343.

Klaper, R., Carter, B. J., Richter, C. A., Drevnick, P. E., Sandheinrich, M. B., and Tillitt, D. E. (2008). Use of a 15 k gene microarray to determine gene expression changes in response to acute and chronic methylmercury exposure in the fathead minnow *Pimephales promelas* rafinesque. *J. Fish Biol.* **72**, 2207–2280.

Klinck, J., Dunbar, M., Brown, S., Nichols, J., Winter, A., Hughes, C., and Playle, R. C. (2005). Influence of water chemistry and natural organic matter on active and passive uptake of inorganic mercury by gills of rainbow trout (*Oncorhynchus mykiss*). *Aquat. Toxicol.* **72**, 161–175.

Krishnani, K. K., Azad, I. S., Kailasam, M., Thirunavukkarasu, A. R., Gupta, B. P., Joseph, K. O., Muralidhar, M., and Abraham, M. (2003). Acute toxicity of some heavy metals to *Lates calcarifer* fry with a note on its histopathological manifestations. *J. Environ. Sci. Health A* **38**, 645–655.

Kumar, A., and Gupta, A. K. (2006). Acute toxicity of mercury to the fingerlings of Indian major carps (catla, rohu and mrigal) in relation to water hardness and temperature. *J. Environ. Biol.* **27**, 89–92.

Larose, C., Canuel, R., Lucotte, M., and Di Giulio, R. T. (2008). Toxicological effects of methylmercury on walleye (*Sander vitreus*) and perch (*Perca flavescens*) from lakes of the boreal forest. *Comp. Biochem. Physiol. C* **147**, 139–149.

Latif, M. A., Bodaly, R. A., Johnston, T. A., and Fudge, R. J. P. (2001). Effects of environmental and maternally derived methylmercury on the embryonic and larval stages of walleye (*Stizostedion vitreum*). *Environ. Pollut.* **111**, 139–148.

Leaner, J. J., and Mason, R. P. (2002a). Methylmercury accumulation and fluxes across the intestine of channel catfish *Ictalurus punctatus*. *Comp. Biochem. Physiol. C* **132**, 247–259.

Leaner, J. J., and Mason, R. P. (2002b). Factors controlling the bioavailability of ingested methylmercury to channel catfish and Atlantic sturgeon. *Environ. Sci. Technol.* **36**, 5124–5129.

Leaner, J. J., and Mason, R. P. (2004). Methylmercury uptake and distribution kinetics in sheepshead minnows, *Cyprinodon variegatus*, after exposure to CH₃Hg-spiked food. *Environ. Toxicol. Chem.* **23**, 2138–2146.

Lemes, M., and Wang, F. (2009). Methylmercury speciation in fish muscle by HPLC-ICP-MS following enzymatic hydrolysis. *J. Anal. At. Spectrom.* **24**, 663–668.

Liao, C. Y., Zhou, O. F., Fu, J. J., Shi, J. B., Yuan, C. G., and Jiang, G. B. (2007). Interaction of methylmercury and selenium on the bioaccumulation and histopathology in medaka (*Oryzias latipes*). *Environ. Toxicol.* **22**, 69–77.

Lindberg, S., Bullock, R., Ebinghaus, R., Engstrom, D., Feng, X. B., Fitzgerald, W., Pirrone, N., Prestbo, E., and Seigneur, C. (2007). A synthesis of progress and uncertainties in attributing the sources of mercury in deposition. *Ambio* **36**, 19–32.

Lu, Z. H., Cobine, P., Dameron, C. T., and Solioz, M. (1999). How cells handle copper: a view from microbes. *J. Trace Elem. Exp. Med.* **12**, 347–360.

Mason, R. P., and Gill, G. A. (2005). Mercury in the marine environment. In: *Mercury Sources, Measurements, Cycles and Effects. Mineralogical Association of Canada Short Course Series* (M.B Parsons and J.W Percival, eds), Vol. 34, pp. 179–216. Mineralogical Association of Canada, Halifax.

Matta, M. B., Linse, J., Cairncross, C., Francendese, L., and Kocan, R. M. (2001). Reproductive and transgenerational effects of methylmercury or aroclor 1268 on *Fundulus heteroclitus*. *Environ. Toxicol. Chem.* **20**, 327–335.

Mayer, F. L. (1987). *Acute Toxicity Handbook of Chemicals to Estuarine Organisms*. United States Environmental Protection Agency, Washington, DC, EPA Report 600/8–87/017.

McCloskey, J. T., Schultz, I. R., and Newman, M. C. (1998). Estimating the oral bioavailability of methylmercury to channel catfish (*Ictalurus punctatus*). *Environ. Toxicol. Chem.* **17**, 1524–1529.

McCrary, J. E., and Heagler, M. G. (1997). The use of a simultaneous multiple species acute toxicity test to compare the relative sensitivities of aquatic organisms to mercury. *J. Environ. Sci. Health A* **32**, 73–81.

McGeer, J. C., Brix, K. V., Skeaff, J. M., DeForest, D. K., Brigham, S. I., Adams, W. J., and Green, A. (2003). Inverse relationship between bioconcentration factor and exposure concentration for metals: implications for hazard assessment of metals in the aquatic environment. *Environ. Toxicol. Chem.* **22**, 1017–1037.

McKim, J. M., Olson, G. F., Holcombe, G. W., and Hunt, E. P. (1976). Long-term effects of methylmercuric chloride on three generations of brook trout (*Salvelinus fontinalis*): toxicity, accumulation, distribution, and elimination. *J. Fish. Res. Bd Can.* **33**, 2726–2739.

Mela, M., Randi, M., Ventura, D., Carvalho, C., Pelletier, E., and Ribeiro, C. (2007). Effects of dietary methylmercury on liver and kidney histology in the neotropical fish *Hoplias malabaricus*. *Ecotoxicol. Environ. Saf.* **68**, 426–435.

Monteiro, D. A., Rantin, F. T., and Kalinin, A. L. (2010). Inorganic mercury exposure: toxicological effects, oxidative stress biomarkers and bioaccumulation in the tropical freshwater fish matrinxa, *Brycon amazonicus* (Spix and Agassiz, 1829). *Ecotoxicology* **19**, 105–123.

Moran, P. W., Aluru, N., Black, R. W., and Vijayan, M. M. (2007). Tissue contaminants and associated transcriptional response in trout liver from high elevation lakes of Washington. *Environ. Sci. Technol.* **41**, 6591–6597.

Neto, F. F., Zanata, S. M., de Assis, H. C. S., Nakao, L. S., Randi, M. A. F., and Ribeiro, C. A. O. (2008). Toxic effects of DDT and methyl mercury on the hepatocytes from *Hoplias malabaricus*. *Toxicol. In Vitro* **22**, 1705–1713.

Newman, M. C., and Doubet, D. K. (1989). Size-dependence of mercury (II) accumulation kinetics in the mosquitofish, *Gambusia affinis* (Baird and Girard). *Arch. Environ. Contam. Toxicol.* **18**, 819–825.

Niimi, A. J. (1983). Physiological effects of contaminant dynamics on fish. In: *Aquatic Toxicology. Advances in Environmental Science and Technology* (J.O Nriagu, ed.), Vol. 13, pp. 207–246. John Wiley and Sons, New York.

Olson, K. R., and Fromm, P. O. (1973). Mercury uptake and ion distribution in gills of rainbow trout (*Salmo gairdneri*): tissue scans with an electron microscope. *J. Fish. Res. Board Can.* **30**, 1575–1578.

Olson, K. R., Squibb, K. S., and Cousins, R. J. (1978). Tissue uptake, subcellular-distribution, and metabolism of (CH_3HgCl)-C14 and (CH_3HgCl)-Hg_2O_3 by rainbow trout, *Salmo gairdneri*. *J. Fish. Res. Board Can.* **35**, 381–390.

Ososkov, I., and Weis, J. S. (1996). Development of social behavior in larval mummichogs after embryonic exposure to methylmercury. *Trans. Am. Fish. Soc.* **125**, 983–987.

Pacyna, E. G., Pacyna, J. M., Sundseth, K., Munthe, J., Kindbom, K., Wilson, S., Steenhuisen, F., and Maxson, P. (2010). Global emission of mercury to the atmosphere from anthropogenic sources in 2005 and projections to 2020. *Atmos. Environ.* **44**, 2487–2499.

Parsons, M. B., and Percival, J. B. (2005). A brief history of mercury and its environmental impact. In: *Mercury Sources, Measurements, Cycles and Effects. Mineralogical Association of Canada Short Course Series* (M.B Parsons and J.W Percival, eds), Vol. 34, pp. 1–16. Mineralogical Association of Canada, Halifax.

Pärt, P., and Lock, R. A. C. (1983). Diffusion of calcium, cadmium and mercury in a mucous solution from rainbow trout. *Comp. Biochem. Physiol. C* **76**, 259–263.

Passino, D. R. M., and Kramer, J. M. (1982). Sub-cellular distribution of mercury in liver of lake trout (*Salvelinus namaycush*). *Experientia* **38**, 689–690.

Pedersen, T. V., Block, M., and Pärt, P. (1998). Effect of selenium on the uptake of methyl mercury across perfused gills of rainbow trout *Oncorhynchus mykiss*. *Aquat. Toxicol.* **40**, 361–373.

Pentreath, R. J. (1976a). Accumulation of organic mercury from sea-water by plaice, *Pleuronectes platessa* L. *J. Exp. Mar. Biol. Ecol.* **24**, 121–132.

Pentreath, R. J. (1976b). Accumulation of mercury from food by plaice, *Pleuronectes platessa* L. *J. Exp. Mar. Biol. Ecol.* **25**, 51–65.

Pentreath, R. J. (1976c). Accumulation of inorganic mercury from sea-water by plaice, *Pleuronectes platessa* L. *J. Exp. Mar. Biol. Ecol.* **24**, 103–119.

Pentreath, R. J. (1976d). Accumulation of mercury by thornback ray, *Raja clavata* L. *J. Exp. Mar. Biol. Ecol.* **25**, 131–140.

Pevzner, R. A., Hernadi, L., and Salanki, J. (1986). Effect of mercury on the fish (*Aiburnus aiburnus*) chemoreceptor taste buds. A scanning electron microscopic study. *Acta Biol. Hung.* **37**, 159–167.

Pickhardt, P. C., Stepanova, M., and Fisher, N. S. (2006). Contrasting uptake routes and tissue distributions of inorganic and methylmercury in mosquitofish (*Gambusia affinis*) and redear sunfish (*Lepomis microlophus*). *Environ. Toxicol. Chem.* **25**, 2132–2142.

Playle, R. C. (1998). Modelling metal interactions at fish gills. *Sci. Total Environ.* **219**, 147–163.

Radhakrishnaiah, K., Suresh, A., and Sivaramakrishna, B. (1993). Effect of sublethal concentration of mercury and zinc on the energetics of a fresh-water fish *Cyprinus carpio* (Linnaeus). *Acta Biol. Hung.* **44**, 375–385.

Ram, R. N., and Joy, K. P. (1988). Mercurial induced changes in the hypothalamo-neurohypophysical complex in relation to reproduction in the teleostean fish, *Channa punctatus* (Bloch). *Bull. Environ. Contam. Toxicol.* **41**, 329–336.

Ram, R. N., and Sathyanesan, A. G. (1983). Effect of mercuric chloride on the reproductive cycle of the teleostean fish *Channa punctatus. Bull. Environ. Contam. Toxicol.* **30**, 24–27.

Rana, S. V. S., Singh, R., and Verma, S. (1995). Mercury-induced lipid peroxidation in the liver, kidney, brain and gills of a fresh water fish. Channa punctatus. *Jpn. J. Ichthyol.* **42**, 255–259.

Rehnberg, B. C., and Schreck, C. B. (1986). Acute metal toxicology of olfaction in coho salmon: behavior, receptors, and odor-metal complexation. *Bull. Environ. Contam. Toxicol.* **36**, 579–586.

Rehwoldt, R., Menapace, L. W., Nerrie, B., and Alessandrello, D. (1972). The effect of increased temperature upon the acute toxicity of some heavy metal ions. *Bull. Environ. Contam. Toxicol.* **8**, 91–96.

Ribeiro, C. A. O., Rouleau, C., Pelletier, E., Audet, C., and Tjälve, H. (1999). Distribution kinetics of dietary methylmercury in the arctic charr (*Salvelinus alpinus*). *Environ. Sci. Technol.* **33**, 902–907.

Ribeiro, C. A. O., Pelletier, E., Pfeiffer, W. C., and Rouleau, C. (2000). Comparative uptake, bioaccumulation, and gill damages of inorganic mercury in tropical and Nordic freshwater fish. *Environ Res.* **83**, 286–292.

Ribeiro, C. A. O., Belger, L., Pelletier, É., and Rouleau, C. (2002). Histopathological evidence of inorganic mercury and methyl mercury toxicity in the arctic charr (*Salvelinus alpinus*). *Environ. Res.* **90**, 217–225.

Ribeiro, C. A. O., Nathalie, M. D., Gonzalez, P., Yannick, D., Jean-Paul, B., Boudou, A., and Massabuau, J. C. (2008). Effects of dietary methylmercury on zebrafish skeletal muscle fibres. *Environ. Toxicol. Pharmacol.* **25**, 304–309.

Roales, R. R., and Perlmutter, A. (1974). Toxicity of methylmercury and copper, applied singly and jointly, to the blue gourami, *Trichogaster trichopterus. Bull. Environ. Contam. Toxicol.* **12**, 633–639.

Rodgers, D. W., and Beamish, F. W. H. (1983). Water quality modifies uptake of waterborne methylmercury by rainbow trout, *Salmo gairdneri. Can. J. Fish. Aquat. Sci.* **40**, 824–828.

Rouleau, C., Gobeil, C., and Tjälve, H. (1998). Pharmacokinetics and distribution of dietary tributyltin compared to those of methylmercury in the American plaice *Hippoglossoides platessoides. Mar. Ecol. Prog. Ser.* **171**, 275–284.

Rouleau, C., Borg-Neczak, K., Gottofrey, J., and Tjälve, H. (1999). Accumulation of waterborne mercury(II) in specific areas of fish brain. *Environ. Sci. Technol.* **33**, 3384–3389.

Rytuba, J. (2003). Mercury from mineral deposits and potential environmental impact. *Environ. Geol.* **43**, 326–338.

Rytuba, J. J. (2005). Geogenic and mining sources of mercury to the environment. In: *Mercury Sources, Measurements, Cycles and Effects. Mineralogical Association of Canada Short Course Series* (M.B Parsons and J.W Percival, eds), Vol. 34, pp. 21–41. Mineralogical Association of Canada, Halifax.

Sanchez-Dardon, J., Voccia, I., Hontela, A., Chilmonczyk, S., Dunier, M., and Boermans, H. (1999). Immunomodulation by heavy metals tested individually or in mixtures in rainbow trout (*Oncorhynchus mykiss*) exposed *in vivo. Environ. Toxicol. Chem.* **18**, 1492–1497.

Sandheinrich, M. B., and Miller, K. M. (2006). Effects of dietary methylmercury on reproductive behavior of fathead minnows (*Pimephales promelas*). *Environ. Toxicol. Chem.* **25**, 3053–3057.

Sandheinrich, M. B., and Wiener, J. G. (2010). Methylmercury in freshwater fish: recent advances in assessing toxicity of environmentally relevant exposures. In: *Environmental Contaminants in Biota Interpreting Tissue Concentrations* (W.N. Beyer and J.P. Meador, eds). 2nd edn Taylor and Francis, Boca Raton, FL.

Schultz, I. R., and Newman, M. C. (1997). Methyl mercury toxicokinetics in channel catfish (*Ictalurus punctatus*) and largemouth bass (*Micropterus salmoides*) after intravascular administration. *Environ. Toxicol. Chem* **16**, 990–996.

Schultz, I. R., Peters, E. L., and Newman, M. C. (1996). Toxicokinetics and disposition of inorganic mercury and cadmium in channel catfish after intravascular administration. *Toxicol. Appl. Pharmacol.* **140**, 39–50.

Scott, G., and Sloman, K. (2004). The effects of environmental pollutants on complex fish behaviour: integrating behavioural and physiological indicators of toxicity. *Aquat. Toxicol.* **68**, 369–392.

Sharp, J. R., and Neff, J. M. (1980). Effects of the duration of exposure to mercuric chloride on the embryogenesis of the estuarine teleost, *Fundulus heteroclitus*. *Mar. Environ. Res.* **3**, 195–213.

Sharp, J. F., and Neff, J. M. (1982). The toxicity of mercuric chloride and methylmercuric chloride to *Fundulus heteroclitus* embryos in relation to exposure conditions. *Environ. Biol. Fish.* **7**, 277–284.

Simon, O., and Boudou, A. (2001). Direct and trophic contamination of the herbivorous carp *Ctenophatyngodon idella* by inorganic mercury and methylmercury. *Ecotoxicol. Environ. Saf.* **50**, 48–59.

Snarski, V. M., and Olson, G. F. (1982). Chronic toxicity and bioaccumulation of mercuric chloride in the fathead minnow (*Pimephales promelas*). *Aquat. Toxicol.* **2**, 143–156.

Stinson, C. M., and Mallatt, J. (1989). Branchial ion fluxes and toxicant extraction efficiency in lamprey (*Petromyzon marinus*) exposed to methylmercury. *Aquat. Toxicol.* **15**, 237–251.

Suresh, A., Sivaramakrishna, B., Victoriamma, P. C., and Radhakrishnaiah, K. (1992). Comparative study on the inhibition of acetylcholinesterase activity in the fresh-water fish *Cyprinus carpio* by mercury and zinc. *Biochem. Int.* **26**, 367–375.

Suseno, H., Pws, S. H., Budiawan, and Wisnubroto, D. S. (2010). Effects of concentration, body size, and food type on the bioaccumulation of Hg in farmed tilapia *Oreochromis mossambicus*. *Aust. J. Basic Appl. Sci.* **4**, 792–799.

Tierney, K. B., Baldwin, D. H., Hara, T. J., Ross, P. S., Scholz, N. L., and Kennedy, C. J. (2010). Olfactory toxicity in fishes. *Aquat. Toxicol.* **96**, 2–26.

Trudel, M., and Rasmussen, J. B. (1997). Modeling the elimination of mercury by fish. *Environ. Sci. Technol.* **31**, 1716–1722.

Tsai, C. L., Jang, T. H., and Wang, L. H. (1995). Effects of mercury on serotonin concentration in the brain of tilapia. Oreochromis mossambicus. *Neurosci. Lett.* **184**, 208–211.

USEPA. (2009). *National Recommended Water Quality Criteria*. United States Environmental Protection Agency, Washington, DC.

Vallee, B. L., and Ulmer, D. D. (1972). Biochemical effects of mercury, cadmium, and lead. *Annu. Rev. Biochem.* **41**, 91–92.

Van Walleghem, J., Blanchfield, P., and Hintelmann, H. (2007). Elimination of mercury by yellow perch in the wild. *Environ. Sci. Technol.* **41**, 5895–5901.

Verma, S., and Tonk, I. (1983). Effect of sublethal concentrations of mercury on the composition of liver, muscles and ovary of *Notopterus notopterus*. *Water Air Soil Pollut.* **20**, 287–292.

Verma, S., Jain, M., and Tonk, I. (1983). *In vivo* effect of mercuric-chloride on tissue ATPases of *Notopterus notopterus*. *Toxicol. Lett.* **16**, 305–309.

Voccia, I., Krzystyniak, K., Dunier, M., Flipo, D., and Fournier, M. (1994). *In vitro* mercury-related cytotoxicity and functional impairment of the immune cells of rainbow trout (*Oncorhynchus mykiss*). *Aquat. Toxicol.* **29**, 37–48.

Wang, W. X., and Wong, R. S. K. (2003). Bioaccumulation kinetics and exposure pathways of inorganic mercury and methylmercury in a marine fish, the sweetlips *Plectorhinchus gibbosus*. *Mar. Ecol. Prog. Ser.* **261**, 257–268.

Webber, H. M., and Haines, T. A. (2003). Mercury effects on predator avoidance behavior of a forage fish, golden shiner (*Notemigonus crysoleucas*). *Environ. Toxicol. Chem.* **22**, 1556–1561.

Weber, D. N., Connaughton, V. P., Dellinger, J. A., Klemer, D., Udvadia, A., and Carvan, M. J. (2008). Selenomethionine reduces visual deficits due to developmental methylmercury exposures. *Physiol. Behav.* **93**, 250–260.

Weis, J. S. (2009). Reproductive, developmental, and neurobehavioral effects of methylmercury in fishes. *J. Environ. Sci. Health C* **27**, 212–225.

Weis, J., and Khan, A. (1990). Effects of mercury on the feeding behavior of the mummichog, *Fundulus heteroclitus* from a polluted habitat. *Mar. Environ. Res.* **30**, 243–249.

Weis, P., and Weis, J. (1977). Methylmercury teratogenesis in killifish, *Fundulus heteroclitus*. *Teratology* **16**, 317–325.

Weisbart, M. (1973). Distribution and tissue retention of mercury-203 in goldfish (*Carassius auratus*). *Can. J. Zool.* **51**, 143–150.

Wester, P. W., and Canton, H. H. (1992). Histopathological effects in *Poecillia reticulata* (Guppy) exposed to methylmercury chloride. *Toxicol. Pathol.* **20**, 81–92.

Wicklund-Glynn, A. W., Norrgren, L., and Mussener, A. (1994). Differences in uptake of inorganic mercury and cadmium in the gills of the zebrafish, *Brachydanio rerio*. *Aquat. Toxicol.* **30**, 13–26.

Wiener, J. G., Krabbenhoft, D. P., Heinz, G. H., and Scheuhammer, A. M. (2003). Ecotoxicology of mercury. In: *Handbook of Ecotoxicology* (D.J. Hoffman, B.A. Rattner, G.A. Burton, Jr. and J. Cairns, eds), 2nd edn, pp. 409–464. CRC Press, Boca Raton, FL.

Wobeser, G. (1975a). Acute toxicity of methylmercury chloride and mercuric chloride for rainbow trout (*Salmo gairdneri*) fry and fingerlings. *J. Fish. Res. Board Can.* **32**, 2005–2013.

Wobeser, G. (1975b). Prolonged oral administration of methyl mercury chloride to rainbow trout (*Salmo gairdneri*) fingerlings. *J. Fish. Res. Board Can.* **32**, 2015–2023.

Wyn, B., Kidd, K. A., Burgess, N. M., and Curry, R. A. (2009). Mercury biomagnification in the food webs of acidic lakes in Kejimkujik National Park and National Historic Site, Nova Scotia. *Can. J. Fish. Aquat. Sci.* **66**, 1532–1545.

Yamaguchi, M., Yasutake, A., Nagano, M., and Yasuda, Y. (2004). Accumulation and distribution of methylmercury in freshwater- and seawater-adapted eels. *Bull. Environ. Contam. Toxicol.* **73**, 257–263.

Yan, C. H. M., and Chan, K. M. (2004). Cloning of zebrafish metallothionein gene and characterization of its gene promoter region in HepG2 cell line. *Biochim. Biophys. Acta* **1679**, 47–58.

Zaman, K., and Pardini, R. S. (1996). An overview of the relationship between oxidative stress and mercury and arsenic. *Toxic Subst. Mech.* **15**, 151–181.

Zelson, P. R., and Cagan, R. H. (1979). Biochemical studies of taste sensation. 8. Partial characterization of alanine-binding taste receptor-sites of catfish *Ictalurus punctatus* using mercurials, sulfhydryl reagents, trypsin and phospholipase *c*. *Comp. Biochem. Physiol. B* **64**, 141–147.

Zhou, T., Scali, R., and Weis, J. S. (2001). Effects of methylmercury on ontogeny of prey capture ability and growth in three populations of larval *Fundulus heteroclitus*. *Arch. Environ. Contam. Toxicol.* **41**, 47–54.

6

ARSENIC

DENNIS O. McINTYRE
TYLER K. LINTON

Arsenic (As) is a moderately toxic, naturally abundant element with no known nutritional or metabolic roles. The chemical form of As in surface waters is dependent on redox potential, pH, and biological processes; however, the thermodynamically stable arsenate predominates in both

Homeostasis and Toxicology of Non-Essential Metals: Volume 31B
FISH PHYSIOLOGY

freshwaters and saltwaters. Arsenic concentrations are much more variable in freshwaters than in estuaries and oceans. Natural and anthropogenic sources can cause high concentrations of As in freshwaters without apparent effects on fish. Arsenite is generally more toxic than arsenate. Chronic effects on fish are primarily linked to increased maintenance due to inhibited energy-linked functions. Effects appear to occur when fish tissue concentrations reach 2–5 mg kg^{-1} wet weight. Marine fish accumulate much more As than freshwater fish, presumably because of the higher levels of arsenobetaine in their prey. Biomagnification does not occur; rather, As concentrations decrease as trophic level increases. Approximately 90% of As in fish is organic As, with arsenobetaine being the dominant species in marine fish; speciation in freshwater fish is much more variable. Few environmental situations invoke homeostatic control mechanisms by fish, except those involving uncommonly high inorganic waterborne and dietary exposure.

1. CHEMICAL SPECIATION IN FRESHWATER AND SALTWATER

1.1. Freshwater

Several As compounds occur in the environment (Table 6.1). The concentration of total As (inorganic plus organic As) in freshwaters can vary considerably depending on the local geochemical environment and anthropogenic influences. Typical natural background levels, as reported

Table 6.1
Arsenic compounds commonly found in the aquatic environment

Name	Abbreviation	Chemical formula
Arsenite (arsenous acid)	As(III)	$As(OH)_3$
Arsenate (arsenic acid)	As(V)	$AsO(OH)_3$
Monomethylarsonic acid	MMA(V)	$CH_3AsO(OH)_2$
Monomethylarsonous acid	MMA(III)	$CH_3As(OH)_2$
Dimethylarsinic acid	DMA(V)	$(CH_3)_2AsO(OH)$
Dimethylarsinous acid	DMA(III)	$(CH_3)_2As(OH)$
Dimethylarsinoyl ethanol	DMAE	$(CH_3)_2AsOCH_2CH_2OH$
Trimethylarsine oxide	TMAO	$(CH_3)_3AsO$
Trimethylarsonium ion	Me_4As^+	$(CH_3)_4As^+$
Arsenobetaine	AB	$(CH_3)_3As^+CH_2COO$
Arsenocholine	AC	$(CH_3)_3As^+CH_2CH_2OH$
Trimethylarsine	TMA(III)	$(CH_3)_3As$
Arsenic containing ribosides	Arsenosugars	

by Smedley and Kinniburgh (2002), can range from 0.13 to 2.1 μg L^{-1} in
rivers (Andreae et al., 1983; Froelich et al., 1985; Seyler and Martin, 1991)
and from 0.06 to 1.9 μg L^{-1} in lakes (Baur and Onishi, 1969; Reuther, 1992;
Azcue et al., 1994; Azcue and Nriagu, 1995) (Table 6.2). Natural As
concentrations can reach very high levels in areas influenced by high As

Table 6.2
Total arsenic concentrations in fresh surface waters

Water body and location	Average total As concentration or range (μg L^{-1})	References
Baseline: river water		
Various	0.83 (0.13–2.1)	Andreae et al. (1983); Froelich et al. (1985); Seyler and Martin (1991)
Norway	0.25 (< 0.02–1.1)	Lenvik et al. (1978)
Southeast USA	0.15–0.45	Waslenchuk (1978)
USA	2.1	Sonderegger and Ohguchi (1988)
Dordogne, France	0.7	Seyler and Martin (1990)
Po River, Italy	1.3	Pettine et al. (1992)
Polluted European rivers	4.5–45	Seyler and Martin (1990)
River Danube, Bavaria	3 (1–8)	Quentin and Winkler (1974)
Schelde catchment, Belgium	0.75–3.8 (up to 30)	Andreae and Andreae (1989)
Baseline: lake water		
British Columbia, Canada	0.28 (< 0.2–0.42)	Azcue et al. (1994, 1995)
Ontario, Canada	0.7	Azcue and Nriagu (1995)
France	0.73–9.2 (high Fe)	Seyler and Martin (1989)
Japan	0.38–1.9	Baur and Onishi (1969)
Sweden	0.06–1.2	Reuther (1992)
Groundwater and geothermal influenced		
Rio Loa basin, Chile	120–27,000	Romero et al. (2003)
Northern Chile	190–21,800	Cáceres et al. (1992)
Northern Chile	400–450	Sancha (1999)
Cordoba, Argentina	7–114	Lerda and Prosperi (1996)
Sierra Nevada, USA	0.20–264	Benson and Spencer (1983)
Waikato, New Zealand	32 (28–36)	McLaren and Kim (1995)
Madison and Missouri Rivers, USA	44 (19–67)	Robinson et al. (1995)
Mining influenced		
Ron Phibun, Thailand	218 (4.8–583)	Williams et al. (1996)
Ashanti, Ghana	284 (< 2–7900)	Smedley et al. (1996)
British Columbia, Canada	17.5 (< 0.2–556)	Azcue et al. (1994)
Presa River, Corsica	2.13–3200	Culioli et al. (2009)
Moira lake, Ontario, Canada	43 (4–94)	Diamond (1995)

(Continued)

Table 6.2 (*continued*)

Water body and location	Average total As concentration or range (μg L^{-1})	References
Lakes, Northwest Territories, Canada	556–5500	Wagemann et al. (1978); Azcue et al. (1994)
Subarctic lakes, Northwest Territories, Canada	270 (64–530)	Bright et al. (1996)
Mutare River, Zimbabwe	13–96	Jonnalagadda and Nenzou (1996a, b, 1997)
Ontario, Canada	35–100	Azcue and Nriagu (1995)
Arsenic-based pesticides influenced		
Finfeather Lake, Texas, USA	7900 (6000–8600)	Crearley (1973)
Municipal lake, Texas, USA	3200 (1700–4400)	Crearley (1973)
Maurice river, NJ, USA	2222 (1320–4160)	Faust et al. (1987)
Union lake, NJ, USA	86.1 (27.1–267)	Faust et al. (1987)

Adapted from Smedley and Kinniburgh (2002) and Ng et al. (2001).

levels in groundwater. Arsenic in surface waters fed mainly from As-rich groundwater in the Rio Loa basin in Chile average 1400 μg L^{-1}, with concentrations reaching 10,000 μg L^{-1} in the Rio Salado (Romero et al., 2003). Anthropogenic sources can also cause elevated concentrations of As. Past mining activities in Corsica have led to 9450 mg As kg^{-1} in the sediment, 3200 μg As L^{-1} in the Presa River, and 314 μg As L^{-1} in the larger Bravona River (Culioli et al., 2009). Arsenic can also reach high levels (100–5000 μg L^{-1}) in areas of sulfide mineralization and mining (Mandal and Suzuki, 2002).

Arsenic can occur in several oxidation states, but in natural freshwaters mostly in the inorganic form as oxyanions of trivalent arsenite [As(III)] or pentavalent arsenate [As(V)], with monomethylarsonous acid (MMA) and dimethylarsinic acid (DMA) being minor components (Watt and Le, 2003; Baeyens et al., 2007). Some As(III) and As(V) species can interchange oxidation state depending on redox potential (Eh), pH, and biological processes. In well-oxygenated water, the major chemical form in which As appears to be thermodynamically stable is the As(V) ion. Redox affected As speciation in the stratified water column of Lake Pavin in France. Inorganic As(V) was the dominant species in the oxygen-rich surface waters, whereas inorganic As(III) was dominant below the redoxcline where oxygen was depleted and pH was low (Seyler and Martin, 1989). The proportion of inorganic As(III) was considerably higher than expected from thermodynamic calculations, suggesting that other processes such as sorption, adsorption, precipitation, and biological mediation are affecting As speciation. Arsenite can also constitute relatively higher proportions of the total As near inputs of arsenite-dominated industrial effluent and in waters with a component of geothermal water (Andreae and Andreae, 1989).

Owing to its chemical similarity to phosphate, As(V) is actively taken up by algae. Algae can reduce As(V) to As(III) or methylate it to form MMA or DMA, which is subsequently excreted (Cullen and Reimer, 1989). Algae only produce As(III) during the log growth phase, which results in peak As (III) concentrations at the beginning of algal blooms (Hasegawa et al., 2010).

Eutrophic lakes were found to contain a higher percentage of organic arsenicals (30–60%) as mainly methylated and ultraviolet-labile fractions (range 0.2–0.6 μg L^{-1}) during the summer (Hasegawa et al., 2010). During the winter months, however, the inorganic forms of arsenic [As(V) and As (III)] were the dominant species (60–85%). The ultraviolet-labile fraction (more complex organoarsenicals) was higher in eutrophic lakes than in mesotrophic lakes during both summer and winter, suggesting that the conversion of inorganic As to organoarsenicals is higher in eutrophic lakes than in mesotrophic or oligotrophic lakes (Hasegawa et al., 2010).

1.2. Saltwater

Arsenic levels in saltwater are much less variable than in freshwater, with a typical range in open water of 0.5–2.0 μg L^{-1} (Franseconi and Kuehnelt, 2002) (Table 6.3). Neff (1997) reported an average concentration of 1.7 μg L^{-1} (range 1–3 μg L^{-1}) in clean coastal and ocean waters. Concentrations are more varied and somewhat higher in estuaries owing to inputs from rivers and salinity or redox gradients, but typically remain below 4 μg L^{-1} (Smedley and Kinniburgh, 2002).

As in freshwater, As can occur in estuarine and seawater in four valency states (+5, +3, 0, and −3) (Neff, 1997). Elemental As is not soluble in water and therefore very rare. Also similar to freshwater, As(V) and As(III) are the dominant forms of inorganic As in seawater. In well-oxygenated seawater, As(V) is the primary species which appears to be thermodynamically stable (Mandal and Suzuki, 2002). The ratio of inorganic As(V) to inorganic As(III) based on thermodynamic calculation should be 1026:1 for oxygenated seawater at pH 8.1, whereas measured ratios range from only 0.1:1 to 10:1. The reduction of As(V) to As(III) by biological activity is in part the reason for this large disparity, leading to the presence of thermodynamically unstable dissolved forms, principally MMA, DMA, and As(III) (Tanaka and Santosa, 1995). The abundance of these As species is dependent on algal

Table 6.3
Total arsenic concentrations in saltwater

Water body and location	Average total As concentration or range ($\mu g\ L^{-1}$)	Reference
Baseline: estuarine water		
Oslofjord, Norway	0.7–2.0	Abdullah et al. (1995)
Saanich Inlet, British Columbia, Canada	1.2–2.5	Peterson and Carpenter (1983)
Rhône Estuary, France	2.2 (1.1–3.8)	Seyler and Martin (1990)
Krka Estuary, Yugoslavia	0.13–1.8	Seyler and Martin (1991)
Southampton water, UK	0.76–1.0	Howard and Comber (1989)
Tagus Estuary, Portugal	24.3	de Bettencourt and Andreae (1991)
Beaulieu Estuary, UK	~1	Howard et al. (1999)
Mining and industry influenced		
Loire Estuary, France	Up to 16	Seyler and Martin (1990)
Tamar Estuary, UK	2.7–8.8	Howard et al. (1988)
Schelde Estuary, Belgium	1.8–4.9	Andreae and Andreae (1989)
Baseline: seawater		
Deep Pacific and Atlantic	1.0–1.8	Cullen and Reimer (1989)
Coastal Malaysia	1.0 (0.7–1.8)	Yusof et al. (1994)
Coastal Spain	1.5 (0.5–3.7)	Navarro et al. (1993)
North-west coast of Spain	1.0–12.5	Soto et al. (1996)
Coastal Australia	1.3 (1.1–1.6)	Maher (1985)
Causeway, Tampa Bay, Florida, USA	1.77	Braman and Foreback (1973)
Tidal flat, Tampa Bay, Florida, USA	2.28	Braman and Foreback (1973)
McKay Bay, Florida, USA	1.48	Braman and Foreback (1973)
Southern California Bight, USA	~1–2	Andreae (1978)
Northeast Pacific and California Shelf	~1–2	Andreae (1979)
East Indian Ocean surface	0.87	Santosa et al. (1994)
Antarctic Ocean surface	1.1	Santosa et al. (1994)
North Indian Ocean surface	0.85	Santosa et al. (1994)
China Sea surface	0.64	Santosa et al. (1994)

Adapted from Smedley and Kinniburgh (2002) and Franseconi and Kuehnelt (2002).

densities, and therefore, the influence of algae on As speciation is seasonal. Although As(V) was in general the dominant inorganic species, a study of the coastal waters of Japan and the Persian Gulf as well as open ocean waters (e.g. Indian, Antarctic, Pacific) found that organic As occupies as much as 50% of the total As in nearshore waters (Tanaka and Santosa, 1995). Organic As was usually less than 20% of total As in the open ocean.

2. SOURCES (NATURAL AND ANTHROPOGENIC) OF ARSENIC AND ECONOMIC IMPORTANCE

2.1. Sources

Arsenic is widely distributed in the Earth's crust, ranking 20th in abundance and occurring at 2–5 mg kg^{-1} in soil (Mandal and Suzuki, 2002; Garelick et al., 2008). Arsenic is present in more than 200 mineral species; the most abundant is arsenopyrite (FeAsS), dominant in mineral veins (Smedley and Kinniburgh, 2002). Arsenic also occurs to a large extent in two other sulfidic ores, orpiment (As_2S_3) and realgar (As_4S_4) (Baeyens et al., 2007). Arsenic is primarily obtained through recovery during mining operations of copper, lead, cobalt, and gold ores (Azcue et al., 1994). China produces about 50% of the world's As through gold mining operations, followed by Chile, Peru, and Morocco (Brooks, 2007). The USA does not currently produce As.

About two-thirds of the total atmospheric flux of As (about 28,000 tonnes per year) is of anthropogenic origin, the rest coming mainly from volcanoes and low-temperature volatilization (Chilvers and Peterson, 1987). Arsenic is released as arsenic trioxide (As_2O_3) mainly adsorbed on particles, which are dispersed by the wind and returned to the Earth by wet or dry deposition. The major sources of As to the oceans are from rivers and upwelling of As-rich deep ocean water (Waslenchuk, 1978; Cutter and Cutter, 1995), and wet/dry deposition contributes only a very minor fraction (Cutter, 1993).

The concentration of As in natural freshwaters is likely controlled by solid–solution interactions for soil, interstitial waters, and groundwaters (Smedley and Kinniburgh, 2002). Arsenic is released in soil as a result of weathering of arsenopyrite or other primary sulfide minerals. Environmental factors controlling the weathering of arsenopyrite are moisture (hydrolysis), pH, temperature, solubility, redox characteristics of the species, and reactivity of the species with CO_2/H_2O (Ahuja, 2008). Arsenic-rich geothermal water can result in very high natural contamination of surface waters (Garelick et al., 2008).

Most As entering the aquatic environment is of natural origin, but anthropogenic inputs can be locally significant from mining and smelting of non-ferrous metals, herbicide manufacturing, and burning of fossil fuels. Concentrations over 5000 μg L^{-1} have been reported (Table 6.2). Because gold- and As-bearing minerals coexist, As contamination of surface waters is frequently associated with gold mining activities (Garelick et al., 2008).

2.2. Economic Importance

Historically, As compounds have been used for pharmaceutical and medicinal purposes, while in modern times As compounds have been primarily used in agriculture (as pesticides), wood preservation, livestock (feed additives), electronics (solar cells, semiconductor applications), industries (in glassware and ceramics), and metallurgy (as alloys and battery plates) (Madhavan and Subramanian, 2006; Rosen and Liu, 2009).

The USA is the world's leading consumer of As, which is mainly used for the production of the wood preservative copper chrome arsenate (CCA) (Brooks, 2007). Approximately 70% of the As produced throughout the world is used in the pressure treatment of timber by CCA, 22% in agricultural chemicals, and the remainder in glass, pharmaceuticals, and non-ferrous alloys (Gomez-Caminero et al., 2001). Human health concerns have led to a voluntary reduction in the manufacture of CCA and its use in most residential settings, including decks and playsets.

3. ENVIRONMENTAL SITUATIONS OF CONCERN

The most widespread environmental problem is exposure of humans to naturally high concentrations of As in groundwater. Poisoning of humans from drinking As-rich groundwater has been documented in Taiwan, China, Vietnam, Hungary, Spain, Poland, India, Bangladesh, Sri Lanka, Chile, Argentina, Mexico, and several places in North America (Mandal and Suzuki, 2002). Ecological effects due to As exposure are uncommon compared to the well-documented impacts on human health, and are typically related to mining activities (Wagemann et al., 1978; de Rosemond et al., 2008) or the release of arsenical herbicides (Sandhu, 1977; Sorensen et al., 1980).

4. A SURVEY OF ACUTE AND CHRONIC AMBIENT WATER QUALITY CRITERIA IN VARIOUS JURISDICTIONS IN FRESHWATER AND SALTWATER

The United States Environmental Protection Agency's (EPA's) acute criterion is typically calculated using the four lowest genus mean acute values (GMAVs) composed of acute median lethal effect concentrations (LC50s) from short-term (48–96 h) laboratory tests (Stephan et al., 1985). The 5th percentile value is determined and considered protective of 95% of the species in a natural assemblage. For As, the four lowest GMAVs (874–2400 µg L^{-1}) were all for invertebrates (*Gammarus, Simocephalus, Ceriodaphnia, Daphnia*), which, when calculated according to the standard algorithm approach, resulted in a 5th percentile value of 670 μg L^{-1}, termed a final acute value (FAV). The FAV was then divided by 2 to obtain the acute criterion of 340 μg L^{-1} (Table 6.4). The FAV was also divided by an empirically derived acute-to-chronic ratio (4.6) to obtain the chronic criterion of 150 μg L^{-1} (Table 6.4). A much lower chronic criterion of 5 μg L^{-1} was derived by the Canadian Council of Ministers of the Environment (CCME, 2001) by dividing the lowest estimate of toxicity for As, the 14 day EC50 (growth) of 50 μg L^{-1} for the alga *Scenedesmus obliquus*, by a safety factor of 10.

The UK derived its freshwater chronic criterion of 0.5 μg L^{-1} by first dividing the lowest chronic effect concentration, 10 μg L^{-1} for reduced reproduction in *Daphnia pulex*, by 2 and then by an assessment factor of 10 (Table 6.4). Since the minimum database requirements were met for the South African As freshwater criteria, no safety factors were applied (Table 6.4). A target value of 12 μg L^{-1} was set for coastal waters for primary producers, primary consumers, and secondary consumers. Australia and New Zealand derived separate chronic criteria for As(III) and As(V),

Table 6.4
Aquatic life criteria for various jurisdictions

Jurisdiction	Reference	Medium	Acute (μg L^{-1})	Chronic (μg L^{-1})	Notes
USA	USEPA (1985)	FW	340	150	SSD
		SW	69	36	SSD
Canada	CCME (2001)	FW		5	AF
		SW		12.5 (interim)	AF
UK	Water Framework Directive	FW	8	0.5	AF
		SW	1.1	0.6	AF
South Africa	CSIR (1996)	FW	130	20	
		SW		12	Target value
Australia/New Zealand	ANZECC and ARMCANZ (2000)	FW As (III)		24	SSD
		SW As(III)		2.3	AF
		FW As(V)		13	SSD
		SW As(V)		4.5	AF
The Netherlands	RIVM (1997)	FW		25	MPC using statistical extrapolation
		SW		9.5	AF of 10 applied to lowest NOEC of 95 μg L^{-1}

FW: freshwater; SW: saltwater; SSD: species sensitivity distribution; AF: assessment factor; MPC: maximum permissible concentrations; NOEC: no observed effect concentration.

using datasets for seven and five taxonomic groups, respectively, allowing the use of their recommended statistical distribution method and high reliability criteria. The datasets were less robust for the marine species, and therefore, assessment factors [100 for As(III) and 200 for As(V)] were applied to the lowest available chronic effect values (see Table 6.4). The freshwater dataset used by the Netherlands contained 15 no observed effect concentrations (NOECs) for species from six taxonomic groups, allowing for statistical extrapolation using a log-logistic frequency distribution, which resulted in a chronic criterion of 25 µg L^{-1} (see Table 6.4). The saltwater criterion (9.5 µg L^{-1}) was derived by applying an assessment factor of 10 to the lowest NOEC of 95 µg L^{-1}.

5. MECHANISMS OF TOXICITY

5.1. Mode of Action

Our current knowledge of the mode of action of As stems almost exclusively from studies with mammals. In mammalian cells, trivalent arsenicals exert toxicity by binding to thiols or vicinal sulfhydryl groups of enzymes, receptors, or coenzymes, and, upon accumulation in mitochondria, inhibit energy-linked functions (Hughes, 2002). Arsenic(III) in particular has a higher affinity for dithiols than monothiols (Delnomdedieu et al., 1993), and thus, is known to inhibit pyruvate dehydrogenase (Szinicz and Forth, 1988). Pyruvate dehydrogenase, which requires the cofactor lipoic acid (a dithiol) for enzymatic activity, oxidizes pyruvate to acetyl-CoA, thereby fueling the citric acid cycle for ATP production in mitochondria. MMA(III) is a more potent inhibitor of pyruvate dehydrogenase than inorganic As(III) (Petrick et al., 2001). Methylated trivalent arsenicals such as MMA(III) are also potent inhibitors of glutathione (GSH) reductase (Styblo et al., 1997) and thioredoxin reductase (Lin et al., 1999), which could alter cellular redox status and eventually lead to cytotoxicity (Hughes, 2002).

Arsenic(V), by comparison, does not react as readily with sulfhydryl groups, but can replace inorganic phosphate in the enzymatic reaction of glyceraldehyde 3-phosphate dehydrogenase. The production of an intermediate arsenophosphoglycerate has been shown to inhibit ATP production *in vitro* (Dixon, 1997).

5.2. Acute Toxicity

Arsenic is far less acutely toxic to fish than are most metals (Spehar and Fiandt, 1986; Buhl and Hamilton, 1990, 1991), with juvenile life stages more sensitive than endogenously feeding life stages to inorganic As (Spotila and Paladino, 1979; Buhl and Hamilton, 1990, 1991). Acute waterborne inorganic As toxicity is also relatively slow acting. The incipient median lethal concentration (LC50) values for adult brook trout and juvenile goldfish were not reached until after approximately 11 and 14 days, respectively (Cardwell et al., 1976). The incipient LC50 values for both fish species were approximately 30% lower than their respective 96 h LC50.

Among freshwater and saltwater fish species, the acute toxicity of As(III) is remarkably similar (Table 6.5). The study of Shaw et al. (2007a) on euryhaline killifish (*Fundulus heteroclitus*) supports this observation

Table 6.5
Acute toxicity of inorganic As(III) to freshwater and saltwater fish

Common name	Species name	Range of 96 h LC50 (μg L^{-1})[a]	References
Freshwater fish			
Muskellunge	*Esox masquinongy*	1100–16,000	Spotila and Paladino (1979)
Giant gourami	*Colisa fasciata*	6090	Pandey and Shukla (1982)
Zebrafish	*Danio rerio*	7050	Seok et al. (2007)
Fathead minnow	*Pimephales promelas*	9900–27,000	Cardwell et al. (1976); Lima et al. (1984); Spehar and Fiandt (1986); Broderius et al. (1995); Dyer et al. (1993)
Snake-head	*Channa punctata*	10,900–76,000	Shukla et al. (1987); Roy and Bhattacharya (2006)
Golden shiner	*Notemigonus crysoleucas*	12,500	Hartwell et al. (1989)
Sailfin molly	*Poecilia latipinna*	12,500	Abdelghani et al. (1980)
Bloater	*Oregonus hoyi*	13,000–20,000	Passino and Kramer (1980)
Arctic grayling	*Thymallus arcticus*	13,700	Buhl and Hamilton (1991)
Flagfish	*Jordanella floridae*	14,400–55,500	Cardwell et al. (1976); Lima et al. (1984)
Smallmouth bass	*Micropterus dolomieui*	> 15,000	Hiltibran (1967)
Bluegill	*Lepomis macrochirus*	15,300–72,000	Inglis and Davis (1972); Cardwell et al. (1976)
Rainbow trout	*Oncorhynchus mykiss*	16,000–23,000	Mayer and Ellersieck (1986); Buhl and Hamilton (1991)
Killifish	*Fundulus heteroclitus*	16,450–17,260	Shaw et al. (2007a, b)
Striped catfish	*Mystus vittatus*	22,000	Gupta and Chakrabarti (1993)
Barb	*Barbus javanicus*	24,170	Gupta and Chakrabarti (1993)
Chinook salmon	*Oncorhynchus tshawytscha*	25,100	Hamilton and Buhl (1990)
Channel catfish	*Ictalurus punctatus*	25,900–41,600	Clemens and Sneed (1959)

(*Continued*)

Table 6.5 (*continued*)

Common name	Species name	Range of 96 h LC50 (μg L^{-1})[a]	References
Brook trout	*Salvelinus fontinalis*	25,800	Cardwell et al. (1976)
Tilapia	*Oreochromis mossambicus*	28,680–71,700	Hwang and Tsai (1993); Liao et al. (2004)
Asiatic knifefish	*Notopterus notopterus*	30,930	Ghosh and Chakrabarti (1990)
Western mosquitofish	*Gambusia affinis*	34,020	Johnson (1978)
Goldfish	*Carassius auratus*	44,900	Cardwell et al. (1976)
Indian catfish	*Clarias batrachus*	63,500	Ghosh et al. (2006)
Indian major carp	*Labeo rohita*	94,300	Palaniappan and Vijayasundaram (2008)
Saltwater fish			
Tigerfish	*Terapon jarbua*	3380	Krishnakumari et al. (1983)
Sheepshead minnow	*Cyprinodon variegatus*	12,700	Cardin (1985)
Four-spine stickleback	*Apeltes quadracus*	14,950	Cardin (1985)
Atlantic silverside	*Menidia menidia*	16,030	Cardin (1985)
Killifish	*Fundulus heteroclitus*	16,730	Shaw et al. (2007a)
Tilapia	*Oreochromis mossambicus*	26,500	Hwang and Tsai (1993)
Thick-lipped gray mullet	*Chelon labrosus*	27,300	Taylor et al. (1985)
Mud dab	*Limanda limanda*	28,500	Taylor et al. (1985)

[a]All values expressed on a dissolved metal basis assuming a conversion factor of 1.0 for converting total to dissolved As (per USEPA, 2009).

(Table 6.5). In contrast, Hwang and Tsai (1993) found waterborne inorganic As(III) to be more lethal to seawater-acclimated tilapia (*Oreochromis mossambicus*) (96 h LC50 of 26,500 mg As L^{-1}) than to freshwater-acclimated tilapia (71,700 mg As L^{-1}).

In direct comparisons, As(III) is up to seven-fold more acutely toxic to fish than As(V) (Abdelghani et al., 1980; McGeachy and Dixon, 1989; Suhendrayatna et al., 2002b). Waterborne inorganic As also appears considerably more acutely toxic to fish than organic (methylated) As compounds. In the adult sailfin molly (*Poecilia latipinna*), monosodium methanearsenate (MSMA) was 100 times less acutely toxic than As(III) (Abdelghani et al., 1980).

Unlike several metals, acute As does not appear to be affected by other abiotic water quality characteristics such as total water hardness, pH, or dissolved organic carbon (DOC) (USEPA, 1985). The 96 h LC50 values of juvenile bluegill exposed to sodium arsenite at 52, 209, and 365 mg L^{-1} (as $CaCO_3$) total water hardness (Ca:Mg ratios of 1.15) were virtually identical: 15,300, 16,400, and 15,400 μg As L^{-1}, respectively (Inglis and Davis, 1972). Similar findings were reported for juvenile striped bass (*Morone saxitilis*) exposed to waterborne inorganic As(V) in soft (40 mg L^{-1} as $CaCO_3$) and hard (285 mg L^{-1} as $CaCO_3$) water (Palawski et al., 1985).

The influence of water temperature on acute As toxicity to fish appears variable. The 144 h LC50 of juvenile trout exposed to As(V) at 5°C (114,100 μg As L^{-1}) was nearly twice as high as the 144 h LC50 of juvenile trout exposed at 15°C (58,000 μg As L^{-1}) (McGeachy and Dixon, 1989). Temperature had no influence on acute toxicity of As(III) to juvenile trout: 144 h LC50s of 17,700 and 20,700 μg As L^{-1}, respectively.

5.3. Chronic Toxicity

Chronic sensitivity of fish to As in general is not substantially different than their acute sensitivity. For example, chronic values based on growth in early life-stage tests are not substantially lower than typical acute effect concentrations. Acute to chronic toxicity ratios (ACRs) for waterborne inorganic As(III) for two freshwater fish species (fathead minnow and flagfish) are remarkably similar at approximately 4.7 and 4.9 (Call et al., 1983; Lima et al., 1984). However, an ACR for waterborne inorganic As(V) to fathead minnows was 28 owing in part to the much greater acute value for this species (DeFoe, 1982). Growth was more sensitive than survival in all three of these early life-stage toxicity tests.

5.3.1. ARSENIC EFFECTS ON GROWTH FROM WATERBORNE EXPOSURE

Chronic effect levels of waterborne inorganic As(III) based on growth are consistent across species, with values between approximately 3000 μg L^{-1} for fathead minnow and flagfish fry (Call et al., 1983; Lima et al., 1984) and 4900 μg L^{-1} for juvenile rainbow trout (Rankin and Dixon, 1994). Speyer (1974) found that 1000 μg L^{-1} of waterborne inorganic As(III) decreased growth of juvenile rainbow trout after only 21 days of exposure. Likewise, Gilderhus (1966) found that both growth and survival of immature bluegill sunfish exposed in outdoor ponds to a single application of a commercial formulation of herbicide containing 2320 μg L^{-1} as As(III) were reduced after 16 weeks.

Depressed food intake is a common correlate of the chronic effect of As(III) on fish growth. The major portion of growth reduction in fingerling rainbow trout exposed to 9640 μg As(III) L^{-1} was explained by depressed appetite, noticeable after only the first day of exposure (McGeachy and Dixon, 1990). Coincident with appetite suppression, many of the fish that perished from long-term exposure also exhibited severe erosion of the mandibular and olfactory region, suggesting that disruption of chemoreception was involved (see also Pedlar and Klaverkamp, 2002). Growth inhibition at 18,000 μg L^{-1} As(V), however, was accompanied by severely depleted liver glycogen, reflecting an overall depressed energy status (McGeachy and Dixon, 1990).

Tsai and Liao (2006) used a combination of laboratory bioassay and biodynamic modeling in tilapia (*O. mossambicus*) to distinguish between three possible mechanisms of As(III)-related growth inhibition at >1000 μg L^{-1} of total waterborne As: (1) increased cost of growth, (2) increased cost of maintenance, and (3) decreased feeding. The results support the supposition that decreased *ad libitum* feeding accounted for the observed reduction in growth. Subsequent modeling illustrated that the growth trajectories during an entire lifespan would be decreased significantly at 1770 (61%) and 3560 (68%) μg L^{-1} As(III).

5.3.2. ARSENIC EFFECTS ON GROWTH FROM DIETARY EXPOSURE

Chronic toxicity of As to fish via the diet is similarly manifested as a reduction in growth. Oladimeji et al. (1984a) exposed juvenile rainbow trout for up to 8 weeks to As [added as sodium arsenite, As(III)] in diets ranging from 10 to 30 mg kg^{-1} wet weight (ww) food. Growth of trout exposed to the highest dietary As(III) treatment was significantly reduced, by approximately 25%.

Cockell et al. (1991) exposed juvenile rainbow trout for 12–24 weeks to a series of diets containing disodium arsenate heptahydrate [inorganic As (V)]. This compound was employed because the authors had earlier determined that it was more toxic to juvenile rainbow trout than dietary As trioxide, dimethylarsinic acid, or arsanilic acid (Cockell and Hilton, 1988). Signs of dietary As toxicity (an approximate 13% reduction in growth) occurred at 33 mg kg^{-1} ww food, and reduction in food consumption (affected as early as the second day of exposure) occurred consistently at 44 mg kg^{-1} ww food and above (42% reduction in growth). A similar study with adult lake whitefish (*Coregonus clupeaformis*) fed diets containing disodium arsenate heptahydrate for 64 days generally corroborated the above results, although the effect on growth rate was more variable over time and not statistically significant (Pedlar and Klaverkamp, 2002; Pedlar et al., 2002).

The dietary effect concentrations for growth inhibition observed in the above laboratory-based exposures are directly in line with the total As concentration in live diets of *Lumbriculus variegatus* cultured in metal-contaminated sediments from the Clark Fork River Basin, Montana, USA (approximately 25 mg As kg^{-1} ww food) and subsequently fed to juvenile rainbow trout (Hansen et al., 2004).

5.3.3. ARSENIC EFFECTS ON REPRODUCTION

The compensatory mechanisms invoked to deal with the increased maintenance costs of As exposure may impinge on reproduction. For example, Liao et al. (2006) predicted that fertility was one of the population growth parameters most sensitive to the effect of inorganic waterborne As (III) at concentrations exceeding 1000 μg L^{-1} As(III).

Full life-cycle reproduction studies with fish have yet to be conducted. A few very recent studies, however, are informative. Boyle et al. (2008) fed pairs of sexually mature zebrafish with a natural diet of polychaete worms (*Nereis diversicolor*) collected from an unimpacted area, and a metal-impacted estuary. A 47% decline in cumulative egg production and a 36% decline in number of spawning pairs were observed in fish fed the contaminated diet (approximately 24 mg As kg^{-1} ww food). Inorganic arsenicals accounted for the majority (59%) of the As in the metal-contaminated diet, whereas the majority of As in the diet from the unimpacted area was organoarsenicals (31–39% as arsenobetaine). Tissue analysis of the zebrafish revealed that As was the only contaminant significantly accumulated in fish fed *N. diversicolor* from the impacted site. This study provides the first plausible mechanistic link between a naturally occurring As-enriched diet and reproductive impairment in fish. Furthermore, hepatic vitellogenin (Vtg) transcript levels (mRNA) were reduced 1.5-fold in these females compared to those fed the uncontaminated metal diet. The observation is consistent with very recent findings that inorganic As interferes with several steroid receptor-mediated pathways, including the hypothalamic–pituitary–gonadal (HPG) axis that involves estrogen (Davey et al., 2007).

The possibility of a direct effect of As on male fertility (not directly measured in the Boyle et al., 2008, study), however, cannot be ruled out. Yamaguchi et al. (2007) tested the direct effects of inorganic As(V) on spermatogenesis of sexually mature Japanese eel (*Anguilla japonica*) testes *in vitro*. Exposure of testicular tissues to 750 μg L^{-1} of As(V) in combination with 11-ketotestosterone (11-KT) decreased the proliferation of germ cells. In a follow-up study, Celino et al. (2008) showed that exposure of Japanese eel testes to as little as 7.5 μg L^{-1} inorganic As(V) for 6–15 days in culture inhibited germ-cell proliferation *in vitro*. The effect of As(V) on spermatogenesis was linked to inhibition of 11-KT synthesis, possibly through inhibiting the expression and activity of 3β-hydroxysteriod dehydrogenase, an enzyme essential for the biosynthesis of 11-KT.

5.3.4. OTHER CHRONIC EFFECTS OF ARSENIC ON FISH

Recent literature has focused largely on the molecular effects of As on fish. The majority of these studies examined effects on the immune system or those associated with oxidative stress. A limited number of studies also focused on alterations in gene expression. Table 6.6 provides a summary of some of the effects reported in these studies as well as the total As concentrations in specific organs eliciting the response.

Notable among the studies focusing on the immune system is that of Nayak et al. (2007). Effects on the ability of newly hatched zebrafish larvae to clear a viral load and a diminished induction of essential antiviral and antibacterial cytokines were observed at waterborne inorganic As(III) concentrations as low as $2 \mu g L^{-1}$ (with apparent developmental effects in hatched fish). Datta et al. (2009) observed extensive immunotoxicity of the head kidney in the Indian catfish (*Clarias batrachus*) from 30 day exposure to waterborne inorganic As(III) of approximately $75 \mu g L^{-1}$, apparently due to accumulation of As in that organ (Table 6.6).

Potentially complicit with the effects of waterborne inorganic As(III) on immunotoxicity in fish is a direct effect on the expression of genes associated with the immune response system at equally low concentrations (Mattingly et al., 2009). Gonzalez et al. (2006) show that differential effects on gene expression in killifish larvae can stem from maternal transfer from parents exposed to waterborne inorganic As(III) at a moderately high concentration of $230 \mu g L^{-1}$ (total As in whole-body tissue of parental fish equal to 0.02 mg kg^{-1} ww), but number of eggs laid, percentage viability, and percentage hatch of larval killifish are unaffected. A potential outcome of As exposure via maternal transfer was observed in juvenile rainbow trout grown for 6 months after a single microinjection of an unspecified form of As into the yolk sac (0.2 mg kg^{-1}) (see Kotsanis et al., 2000). In this study, juvenile fish elicited both immunosuppression and histopathological changes in liver and kidney. The extent of maternal transfer of As to eggs of fish is an area in need of research.

Much research within the last decade has also been devoted to observations of biomarkers of As-induced oxidative stress (and other related stress-mediated effects) in fish, where the GSH system is of particular importance (Table 6.6). Within this system, Allen et al. (2004) and Allen and Rana (2004) show that modulation of reduced GSH and activities of related enzymes is time dependent in snakehead (*Channa punctata*) exposed to $1000 \mu g L^{-1}$ As(III). Worth noting here is that modulation of some of these same biomarkers of oxidative stress in fish has been shown to be induced at much lower waterborne inorganic As(V) concentrations (e.g. $10–100 \mu g L^{-1}$), and over a much shorter duration (Ventura-Lima et al., 2009a, b).

Table 6.6

Effects of arsenic in freshwater and saltwater fish from chronic and sublethal exposure

Species	Exposure type	Form of As	Medium	Exposure duration	Effects	Effect concentration (μg As L^{-1})	Total As in tissue[a] (mg kg^{-1} ww)	Reference	Notes
Immunotoxicity and immunosuppression-related effects									
Zebrafish (embryos, one-cell stage), *Danio rerio*	Waterborne (R, U)	Inorganic As(III)	FW	Up to 4 days postfertilization through hatching	Inhibited ability to clear viral load; decreased host resistance measured by respiratory burst response; diminished induction of essential antiviral and antibacterial cytokines	2	NM	Nayak et al. (2007)	No developmental effects observed in hatched fish
Indian catfish (50–70 g), *Clarias batrachus*	Waterborne (R, U)	Inorganic As(III)	FW	30 days	Significant reduction in HK macrophage number and HK somatic index with histological and ultrastructural alterations in HK and decreased immune response (e.g. phagocytic potential) to bacterial infection	75	0.06–0.10 (HK)	Datta et al. (2009)	
Indian catfish (60–70 g), *Clarias batrachus*	Waterborne (R, U)	Inorganic As(III)	FW	21 days	Time-dependent changes in organo-somatic index, leukocyte count, and lymphocyte proliferation in liver and spleen; decrease in phagocytic activity of macrophages and increase in pathogen susceptibility and disease development	6400	NM	Ghosh et al. (2006)	

(Continued)

Table 6.6 (*continued*)

Species	Exposure type	Form of As	Medium	Effects	Exposure duration	Effect concentration (µg As L^{-1})	Total As in tissue[a] (mg kg^{-1} ww)	Reference	Notes
Indian catfish (60–70 g), *Clarias batrachus*	Waterborne (R, U)	Inorganic As(III)	FW	Significant reduction in HK macrophage number and HK somatic index with histological and ultrastructural alterations in HK and decreased serum immunoglobulin and HK-B cells; increase in pathogen susceptibility and disease leading to 65% mortality after 7 days	150 days	6400	NM	Ghosh et al. (2007)	
Rainbow trout (juvenile), *Oncorhynchus mykiss*	Single microinjection into yolk sac	Not specified	FW	Significant decrease in total number of lymphocytes (white cells), and hence leukocytes; atypical lymphocytes; histopathological changes in kidney and liver	6 months postinjection	NA	0.2 (embryo)	Kotsanis et al. (2000)	Authors note that dose applied is far higher than embryos would experience in nature

Oxidative stress and stress-mediated gene expressive effects

Species	Exposure type	Form of As	Medium	Effects	Exposure duration	Effect concentration (µg As L^{-1})	Total As in tissue[a] (mg kg^{-1} ww)	Reference	Notes
Snakehead (adult), *Channa punctata*	Waterborne (R, U)	Inorganic As(III)	FW	Time-dependent modulation of GSH system in liver and kidney: glutathione-S-transferases, glutathione peroxidase, glutathione reductase, catalase	90 days	1000	0.12–0.39 (liver), 0.07–0.36 (kidney)	Allen and Rana (2004)	

Species	Exposure	Arsenic form	Water	Duration	Effect	Concentration (µg L⁻¹)	Reference	Notes
Snakehead (adult), *Channa punctata*	Waterborne (R, U)	Inorganic As(III)	FW	90 days	Time-dependent modulation of lipid peroxidation and reduced GSH	1000	0.06–0.47 (liver), 0.12–0.37 (kidney), 0.17–0.47 (gill), 0.21–0.46 (muscle)	Allen et al. (2004)	Increased resistance concluded because of the decrease in As in tissues at 90 days
Common carp (mature), *Cyprinus carpio*	Waterborne (R, U)	Inorganic As(III)	FW	2 days	Significant increase in reduced GSH, decreased GR activity, and total antioxidant capacity against peroxyl radicals in liver only; decreased ROS concentrations (at 1000 µg L⁻¹ only) in gills	1000	0.094 (liver), 0.12 (gill)	Ventura-Lima et al. (2009a)	No increase in total As in liver and gill tissue after 2 days compared to control levels (at both 100 and 1000 µg As (III) L⁻¹)
Zebrafish, *Danio rerio*	Waterborne (R, M)	Inorganic As(V)	FW	2 days	Increase in GSH levels and GCL activity and decrease of oxygen consumption [100 µg As(V) L⁻¹ only] in gills	10	<0.02–0.99 at 10 and 100 µg As (V) L⁻¹, respectively (gill)	Ventura-Lima et al. (2009b)	No variations in antioxidant system GR, GST, and CAT activities in fish exposed to various As concentration or ROS and TBARS content
Common carp (mature), *Cyprinus carpio*	Waterborne (R, U)	Inorganic As(V)	FW	2 days	Decreased GR activity, increased total antioxidant capacity against peroxyl radicals, and G6PDH and reduced GSH [at 1000 µg As(V) L⁻¹ only] in liver	100	0.038 (liver); 0.059–2.2 at 100 and 1000 µg As (V) L⁻¹, respectively (gill)	Ventura-Lima et al. (2009a)	No increase in total As in liver and gill tissue after 2 days compared to control levels at 100 µg As (V) L⁻¹; significant increase in gills only at 1000 µg As(V) L⁻¹ (predominantly as MMA and DMA)
Killfish, *Fundulus heteroclitus*	Waterborne	Inorganic As(V)	SW (20 ppt)	14 days	Inhibited LDH-B mRNA levels and enzyme activity	787	9.6 (liver)	Bears et al. (2006)	No effect on stress-responsive gene, PEPCK; no weight loss observed in As-exposed fish or controls

(Continued)

Table 6.6 (*continued*)

Species	Exposure type	Form of As	Medium	Exposure duration	Effects	Effect concentration (μg As L^{-1})	Total As in tissue[a] (mg kg^{-1} ww)	Reference	Notes
Other cellular, developmental, and gene expressive effects									
Zebrafish (0.25–1.5 h postfertilization), *Danio rerio*	Waterborne (R, U)	Inorganic As(III)	FW	24–48 h	99 differentially expressed genes; 79 associated with immune response (53%), cancer, and gastrointestinal disease in humans	7	NM	Mattingly et al. (2009)	
Killifish (juvenile), *Fundulus heteroclitus*	Waterborne (S, M) and maternal transfer	Inorganic As(III)	SW (20 ppt)	10 days prespawning F0 with maternal transfer and grow out of F1 for 6 weeks	No effect on number of eggs laid, % viability or % hatch; significant increase in % deformed larvae; upregulated genes included parvalbumin, myosin light chain 2, type II keratin and tropomyosin	230	0.02 (whole-body parents)	Gonzalez et al. (2006)	
Indian catfish (60–70 g), *Clarias batrachus*	Waterborne (R, U)	Inorganic As(III)	FW	30 days	Increased hepatosomatic index and apoptotic hepatocytes with time-dependent changes in liver enzymes and histological and ultrastructural alterations	75	NM	Datta et al. (2007)	No mortality noted in fish; assuming 0.5 μM dose given as As$_2$O$_3$, and not As
Snakehead (adult), *Channa punctata*	Waterborne (S, U)	Inorganic As(III)	FW	14 days	Increased lesions in hepatic cells: karyolysis, apoptosis, and peliosis hepatitis; transient increase in liver enzyme activity; degeneration of	2900	NM	Roy and Bhattacharya (2006)	

Organism	Exposure	Form	Water	Duration	Effect	Concentration		Reference	Notes
					kidney tissue; histopathological recovery by a simultaneous rise in the MT level and expression of hsp70				
Zebrafish (transgenic larvae), *Danio rerio*	Waterborne (S, U)	Inorganic As(III)	FW	4 days	Increase in incidence of edema, uninflated swim bladder, keratosis, and trunk abnormalities (e.g. spinal lordosis, scoliosis, kyphosis); expression of hsp70 promoter–EGFP reporter–gene construct	750	NM	Seok et al. (2007)	
Zebrafish (larvae), *Danio rerio*	Waterborne (R, U)	Inorganic As(III)	FW	Between 4 and 120 h postfertilization	Reduced survival and hatching rate and morphological abnormalities (pericardial edema, dorsal curvature, flat head, and red blood cell accumulation); neural defects associated with abnormal cell proliferation and apoptosis	≥ 37,500	NM	Li et al. (2009)	Neural defects only at very high, environmentally unrealistic concentrations (150,000 µg As L^{-1})
Snakehead (adult), *Channa punctata*	Waterborne (S, U)	Inorganic As(III)	FW	14 days	Time-dependent and transient effect on structure of optic tectum and increased ACHE activity with recovery	2900	NM	Roy et al. (2006)	
Zebrafish (juvenile), *Danio rerio*	Waterborne (R, U)	Inorganic As(V)	FW	180 days	Increase in micronucleus frequency in gill cells	4040–5056	4.05 (gill)	Ramirez and Garcia (2005)	

(Continued)

Table 6.6 (*continued*)

Species	Exposure type	Form of As	Medium	Exposure duration	Effects	Effect concentration (μg As L^{-1})	Total As in tissue[a] (mg kg^{-1} ww)	Reference	Notes
Zebrafish (embryo; 120 hpf), *Danio rerio*	Waterborne (S, U)	Inorganic As(III)	FW	1 day	Increased microarray hybridizations	5992	NM	Yang et al. (2007)	
Zebrafish (transgenic; 72 hpf), *Danio rerio*	Waterborne (S, U)	Inorganic As(III)	FW	5 days	Approximate 20% decrease in survival relative to controls; 86% incidence of skin lesions (karotosis) linked to epithelial cell overproliferation (malignancy)	375	NM	Wang et al. (2006)	
Effects on biochemical composition									
Indian carp spp. (juvenile, 10–15 cm), *Labeo rohita, Cirrhinus mrigala, Catla catla*	Waterborne (S, U)	Not specified	FW	45 days	Reductions of 10–20% in lipid and carbohydrate content of muscles and gills	60	NM	Garg et al. (2009)	
Indian carp spp. (fry, 6 cm), *Labeo rohita*	Waterborne (R, U)	Inorganic As(III)	FW	14 days	Significant alteration on the major biochemical compositions: proteins, lipids, and nucleic acids of the muscle tissues	31,400	NM	Palaniappan and Vijayasundaram (2008)	LC50 reported as 124,500 μg L^{-1}
Indian carp sp. (fry, 6 cm), *Labeo rohita*	Waterborne (R, U)	Inorganic As(III)	FW	60 days	Significant alteration of the protein profile and lipid levels in gill tissue; a change in the protein amide hydrogen bonding of gill tissues	12,450	NM	Palaniappan and Vijayasundaram (2009)	

Osmoregulatory effects

Tilapia (9.5 cm), *Oreochromis mossambicus*	Waterborne (S, U)	Inorganic As(III)	FW	4 days	No effect on ultrastructure of gills or any osmoregulatory parameters	70,000		Hwang and Tsai (1993)	
Tilapia (9.5 cm), *Oreochromis mossambicus*	Waterborne (S, U)	Inorganic As(III)	SW (full strength)	4 days	Histopathological changes in gills; increased proliferation of mitochondria in chloride cells; increased plasma osmolarity and Na^+/K^+ ratio; increased Na^+/K^+-ATPase activity	15,000		Hwang and Tsai (1993)	
Killifish, *Fundulus heteroclitus*	Waterborne (S, U)	Inorganic As(III)	Soft FW to SW (full strength)	2 days in FW followed by 1 day in SW [at same As(III) concentration]	Significantly increased plasma Cl concentration and decreased CFTR protein abundance	7940	1.12 (gill)	Shaw et al (2007b)	Waterborne inorganic As (III) had no effect on actin and HSP70 protein abundance as determined by Western blot analysis

[a]Tissue concentrations reported in dry weight recalculated and expressed on a wet weight basis assuming 75% moisture content of whole body and muscle, and 80% moisture content of organs.

R: renewal test; S: static test; U: unmeasured concentrations; M: measured concentrations; FW: freshwater; SW: saltwater; HK: head kidney; GSH: glutathione; GR: glutathione reductase; ROS: reactive oxygen species; GCL: glutamate cysteine ligase; G6PDH: glucose-6-phosphate dehydrogenase; LDH-B: actate dehydrogenase B; MT: metallothionien; EGFP: enhanced green fluorescent protein; ACHE: acetylcholinesterase; CFTR: cystic fibrosis transmembrane conductance regulator; NA: not applicable; NM: not measured; CAT: catalase; TBARS: thiobarbituric acid reactive substances; MMA: monomethylarsonous acid; DMA: dimethylarsinic acid; PEPCK: phosphoenolpyruvate carboxykinase; HSP: heat shock protein.

Developmental effects, including a high incidence (86%) of skin lesions (kerotosis), edema, lordosis, and expression of hsp70 promoter–EGFP reporter-gene construct, prevail in newly hatched fish exposed to 350–750 µg L^{-1} As(III) in shorter, sublethal waterborne exposures (see Wang et al., 2006, and Seok et al., 2007, in Table 6.6). This is in contrast to several other histopathological and ultrastructural changes in tissues induced by inorganic As exposure for older life-stage fish, including chromosomal aberrations (e.g. micronuclei), which generally occur at much higher concentrations, i.e. 2900 to >37,500 µg As L^{-1} (Table 6.6). One study (Garg et al., 2009), however, has reported effects of waterborne inorganic As on biochemical composition (10–20% decrease in lipid and carbohydrate content of muscles and gills) of juvenile Indian carp at environmentally relevant concentrations as low as 60 µg As L^{-1} (Table 6.6).

Noteworthy to this discussion of other As effects on fish is the apparent effect of inorganic As on NaCl homeostasis in euryhaline fish. Studies by Hwang and Tsai (1993) and Shaw et al. (2007b) show that inorganic As(III) at approximately 8000–15,000 µg L^{-1} in water affects osmoregulation in tilapia (*O. mossambicus*) and killifish (*F. heteroclitus*). Moreover, Shaw et al. (2007b) showed that waterborne inorganic As(III) exposure for 96 h at subacute levels (< 18,000 µg L^{-1}) inhibits acclimation of freshwater-adapted killifish to increases in salinity via a freshwater-to-seawater challenge without As. Additional studies are needed, however, to determine what, if any, long-term, sublethal effects incur via exposure to As in varying levels of salinity.

5.3.5. Tissue Concentrations Associated with Chronic Response

From the above, regardless of exposure route or form of As, the tissue concentrations associated with chronic effects are remarkably similar among fish. For example, the critical body residue level (whole-fish basis) associated with the effect of waterborne inorganic As(III) on growth and survival of immature bluegill sunfish exposed to 2320 µg L^{-1} as As(III) for 16 weeks was 2.24 mg kg^{-1} ww (Gilderhus, 1966). Similarly, Tsai and Liao (2006) predicted the critical body residue causing a 10% reduction in the modeled growth trajectory of tilapia exposed to 1000 µg L^{-1} as As(III) during a lifespan to be 1.93 mg kg^{-1} ww. These critical body residues are roughly consistent with those estimated for fingerling rainbow trout exposed to waterborne inorganic As(III) and As(V) of approximately 2–4 and 4–6 mg kg^{-1} ww, respectively (McGeachy and Dixon, 1990; Rankin and Dixon, 1994).

The evaluation of critical residue levels estimated from dietary As exposure is best made at the organ level (liver, kidney) where the accumulation of total As has been linked to the concentration of inorganic

As in the diet (Gilderhuss, 1966; Cockell et al., 1991; Pedlar and Klaverkamp, 2002). The critical tissue residue concentrations in liver associated with reduced growth of juvenile rainbow trout exposed to dietary As(III) treatments of 20 and 30 mg kg^{-1} ww food, respectively, were approximately 0.7 and 1.0 mg kg^{-1} ww tissue (Oladimeji et al., 1984a). Similarly, Cockell et al. (1991) found that total As residues in livers of juvenile rainbow trout after exposure to 44 mg kg^{-1} ww food for 12 weeks were approximately 3.8 mg kg^{-1} ww tissue.

6. ESSENTIALITY OR NON-ESSENTIALITY OF ARSENIC

Arsenic is generally considered a "toxic element with no known nutritional or metabolic roles" (Rosen and Liu, 2009). The very limited data available on the beneficial, protective, and essential properties of As do not include fish (Eisler, 1988). Select organic arsenicals such as Roxarsone (4-hydroxy-3-nitrobenzenearsonic acid) are, however, common As-based additives used in chicken feed, where they are used to promote growth, kill parasites, and improve pigmentation of chicken meat (Hileman, 2007). To the best of these authors' knowledge, no studies have been initiated to investigate the possibility of such benefits from arsenicals in fish.

7. POTENTIAL FOR BIOACCUMULATION AND/OR BIOMAGNIFICATION (OR BIODIMINUTION) OF ARSENIC

In nature, the accumulation of As in fish and other aquatic organisms is a dynamic, complex, and diverse process. The degree to which As accumulates in an aquatic organism will vary with the concentration and chemical form of As in water, trophic level, species, and diet. As a consequence, the concentrations of As in the tissues of freshwater and saltwater fish and invertebrates vary considerably between species and among different sites.

Even though freshwater organisms, on average, are exposed to higher As concentrations in the water than saltwater organisms, saltwater organisms accumulate higher levels in their tissues. It has been known for some time that marine organisms have much higher As concentrations than freshwater organisms (Amlund and Berntssen, 2004; Schaeffer et al., 2006; Ciardullo et al., 2010). One hypothesis for this is the possible role of arsenobetaine (AB) as a sporadically acquired osmolyte (Schaeffer et al., 2006). Marine animals, and mollusks in particular, accumulate glycine betaine, and other similar molecules to protect against osmotic stress. It is thought AB is taken up as one of these osmolytes. Support for this hypothesis has been shown with *Mytilus edulis* that have accumulated AB proportionately to salinity (Clowes and Francesconi, 2004).

Biomagnification, the event of higher concentrations of a chemical at higher trophic levels, does not occur with As. Conversely, As diminishes through increasing trophic levels. The biodiminution of As has been reported by a number of researchers (Chen and Folt, 2000; Mason et al., 2000; Farag et al., 2003; Baeyens et al., 2007; Ikemoto et al., 2008; Culioli et al., 2009). Chen and Folt (2000), for example, found As elevated in particulate organic matter and small and large zooplankton but not in fish in an As-contaminated watershed. Although the As levels in the fish were not elevated, planktivores (alewife and killifish) had higher As concentrations (approximately 0.07–0.075 mg kg^{-1} ww) than omnivores (black crappie, bluegill sunfish and yellow perch; approximately 0.03–0.04 mg kg^{-1} ww) and a piscivore (largemouth bass; approximately 0.04 mg kg^{-1} ww), presumably due to the differences in their feeding strategies. Mason et al. (2000) evaluated several factors (e.g. fish weight, water quality) affecting As bioaccumulation in periphyton, bryophytes, invertebrates, and fish in two small Maryland streams. The concentration of As in the primary producers and invertebrates was generally above 1 mg kg^{-1} ww, whereas the concentration in sculpin, dace, and trout was below 0.05 mg kg^{-1} ww. The authors concluded that As concentrations in tissues overall tended to decrease with increasing trophic level. The same pattern of As dimunition in the food chain was found at a highly contaminated mining site in Corsica where total As in tissue decreased from macroinvertebrates to bryophytes to fish (Table 6.7) (Culioli et al., 2009). Worthy of particular note is the relatively low level of As accumulation in brown trout at a site (0.48 mg kg^{-1} ww whole body) with very high concentrations of As in the water (mean of $2,331$ µg L^{-1}) (Table 6.7). The authors state that this dramatic difference in As accumulation underscores the importance of the dietary route of exposure for As.

The bioaccumulation factor (BAF) is a measure of the capacity of a substance to accumulate in an organism or a specific tissue of an organism. The BAF is the ratio of a concentration of the chemical in the tissue of aquatic biota to the chemical's concentration in the surrounding water (USEPA, 2000). Bioconcentration factors (BCFs) are distinguished from BAFs by restricting the route of entry for the chemical to water (no dietary route). As such, BCFs for fish are only determined in the laboratory.

A compilation of BAFs for As plotted against the ambient water concentration shows an inverse relationship (Fig. 6.1). The inverse relationship between BCFs and BAFs and exposure concentrations was first discussed for metals by McGeer et al. (2003). BCFs and BAFs for Zn,

Table 6.7
Mean concentrations of arsenic and bioaccumulation factors (BAFs) in benthic organisms and fish from the Presa River

Site	C_w (μg L^{-1})	Taxon, species	C_t (mg kg^{-1}) ww	BAF
B1	2.13	Bryophytes	0.208	97
		Salmo trutta, whole body	0.0025	1.2
		Salmo trutta, muscle	0.0025	1.2
B3	108	Bryophytes	3.06	28
		Salmo trutta, whole body	0.132	1.2
		Salmo trutta, muscle	0.115	1.1
B4	43.1	*Salmo trutta*, whole body	0.04	0.93
		Salmo trutta, muscle	0.0325	0.75
P1	18.2	Bryophytes	0.845	47
P2	2331	Bryophytes	88.4	38
		Shredders: *Leuctra budtzi, Leuctra geniculata, Protonemura bucolica*	260 (114–386)	
		Scrapers: *Silonella aurata, Baetis cyrneus, Helichus substriatus, Electrogena fallax, Ancylus fluviatilis, Baetis ingridae, Silo rufesens, Esolus brevis, Limnius intermedius*	52.3 (1.26–115)	
		Collector–gatherers: *Caenis martae, Psychomyia pusilla*	42.0 (5.43–78.6)	
		Collector–filterers: *Hydropsyche cyrnotica, H. fumata*	32.5 (31.3–31.6)	
		Predators: *Rhyacophila pubescens, Isoperla insularis, Rhyacophila tarda*	5.94 (4.7–8.38)	
		Salmo trutta, whole body	0.48	0.21
		Salmo trutta, muscle	0.362	0.16

Modified from Culioli et al. (2009).
C_w: mean concentration of arsenic in water.
[a] For taxa with multiple species, C_t is the mean of all species and the range is in parentheses.

Cd, Cu, Pb, Ni, and Ag all showed a clear inverse relationship with exposure concentration. The authors speculated that active regulation and homestatic control of metal bioaccumulation by the organism may cause the metal levels in organisms to decrease with exposure concentration. Bervoets et al. (2005) found the same inverse relationship for both BAFs and aqueous exposure and biota sediment accumulation factors (BSAFs) and sediment exposure for metals in mussels. Williams et al. (2006) used laboratory and field literature data to show that As follows the same clear trend of decreasing BCFs ($r^2 = 0.79$) and BAFs ($r^2 = 0.82$) with exposure concentrations. As pointed out by McGeer et al. (2003), Bervoets et al. (2005), and Williams et al. (2006), the use of linear models to predict bioaccumulation for a chemical that has this inverse relationship with exposure may

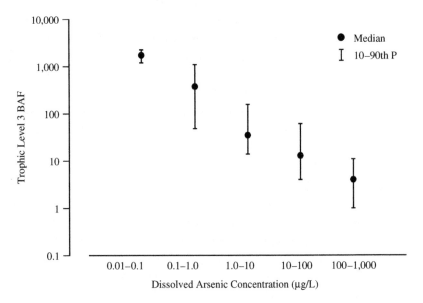

Fig. 6.1. Relationship between dissolved As in surface water and As bioaccumulation factors (BAFs) for trophic level 3 organisms (primarily fish), according to specific water As concentration ranges. Modified from EPRI (2008).

overestimate or underestimate the chemical's potential for bioaccumulation. A more precise prediction of bioaccumulation is clearly needed for ongoing risk assessment and regulatory purposes.

7.1. Arsenic Speciation in Tissues of Aquatic Animals

The vast majority of As found in edible portions of marine fish and shellfish is in an organic form (e.g. AB, arsenocholine, dimethylarsinic acid); the inorganic As species [As(III), As(V)] are a minor fraction (Table 6.8). Inorganic species may be dominant at locations with high aqueous exposure, such as in Cienfuegos Bay where acute contamination occurred owing to the accidental release of 3.7 tonnes of As (as arsenate oxides) from a local nitrofertilizer factory situated in Cienfuegos city (Fattorini et al., 2004). Concentrations of total As in these fish species, *Carranx latus* and *Lutjanus synagris*, were very high at 125 mg kg^{-1} ww.

Arsenobetaine is the predominant species of As found in marine animals (Amlund and Berntssen, 2004; Schaeffer et al., 2006), but its dominance is much less consistent in freshwater fish. One reason for this difference is the higher percentage of AB in the food of marine fish compared to freshwater fish. In a freshwater study by Soeroes et al. (2005), catfish (*Clarias gariepinus*)

Table 6.8

Arsenic speciation in saltwater and freshwater fish

Location	Common name	% Inorganic As	% Organic As	Reference
Saltwater field	Sardine, catfish	6.6	93	Rattanachongkiat et al. (2004)
Saltwater field	Squirrelfish, surgeonfish, brassy trevelly, torpedo scad, gray mullet	2.1	98	Peshut et al. (2008)
Saltwater field	Scyliorhinid shark	5.2	97	Storelli et al. (2005)
Saltwater field	Fish	99	1	Fattorini et al. (2004)
Saltwater field	Mullet	1.6	98	Liu et al. (2006)
Saltwater Aquaculture	Herring cale	0.6	75.7	Kirby et al. (2005)
Saltwater field	Snapper, flathead	0.5	99	Fabris et al. (2006)
Saltwater field	Summer flounder, Atlantic croaker, striped bass	1	99	Greene and Crecelius (2006)
Freshwater field	Northern pike, largemouth bass, yellow perch, pumpkinseed	37	54	Zheng and Hintelmann (2004)
Freshwater field	Chum and coho salmon	0.4	99.6	Foran et al. (2004)
Freshwater field	Coho and chinook salmon, white sturgeon, largescale sucker, carp, steelhead trout	9.3	91	Tetra Tech (1996)
Freshwater field	Largescale sucker, carp, smallmouth bass, northern pikeminnow	4.5	96	EVS (2000)
Freshwater field	Catfish, carp,	0.8	56	Soeroes et al. (2005)
Freshwater field	Silver carp, white bream	ND	12	Schaeffer et al. (2006)
Freshwater lab	Guppy	78	22	Maeda et al. (1990)
Freshwater field	Carp, catfish	0.6	99	Ciardullo et al. (2010)
Freshwater field	Burbot, barbell, roach, nase, marbe trout, rainbow trout, brown trout	5.8	47	Slejkovec et al. (2004)
Freshwater field	Tilapia	8.3	82	Huang et al. (2003)
Freshwater field	Tilapia	0.7	99	Moreau et al. (2007)
Freshwater field	Trout	28	53	Pizarro et al., (2003)

ND: not determined.

raised in a pond (15.1 µg L^{-1} As) and fed an artificial fishmeal diet of 2.88 mg kg^{-1} As that was mostly AB, accumulated 3–5 mg kg^{-1} dry weight (dw) As containing >90% AB. Conversely, carp (*Cyprinus carpio*) collected from a lake with no elevated As which fed on a natural diet accumulated much less As (0.06–0.4 mg kg^{-1} dw) and AB was not detected.

Several authors have reported AB to be the major As species in freshwater fish as well. Ciardullo et al. (2010) found AB to be the dominant form in eel, flathead gray mullet, chub, and carp collected from the Tiber River in Latium, Italy. Total As concentrations in these fish were high, ranging from 88.5 to 451 mg kg^{-1} ww, allowing detection of several other arsenicals including arsenous acid, methylarsonic acid, dimethylarsinic acid, trimethylarsine oxide, arsenocholine ion, tetramethylarsonium ion, oxo-arsenosugar-glycerol, oxo-arsenosugar-phosphate, oxo-arsenosugar-sulfate, thio-arsenosugar-phosphate, and three unknown As compounds. The content of AB was 88.9% in eel, 95.8% in mullet, 95.4% in chub, and 58.4% in carp. The authors state that these results are indicative of a freshwater ecosystem not polluted by inorganic As. The occurrence of AB is variable in freshwater fish and appears to be primarily linked to its uptake via the diet.

8. CHARACTERIZATION OF UPTAKE, INTERNAL HANDLING, AND EXCRETION

8.1. Uptake and Accumulation via the Gills

Laboratory studies clearly indicate a propensity for freshwater fish to accumulate As directly from water, though generally only at environmentally elevated levels (Table 6.9). These studies, limited primarily to waterborne inorganic As [as As(III) and As(V)] exposure, indicate water concentration-dependent accumulation in whole body and select organs reaching concentrations as high as 9.6, 0.37, 4.1, and 0.48 mg kg^{-1} ww in liver, kidney, gill, and muscle, respectively (Table 6.9).

Direct uptake of waterborne inorganic As [as As(III) and As(V)] and organic As [MMA(V) and DMA(V)] by adult Japanese medaka (*Oryzias latipes*) and juvenile tilapia over a 7 day exposure at multiple concentrations was thoroughly examined in a collection of studies by Suhendrayatna et al. (2002a, b) and Kuroiwa et al. (1994). Combined, these studies showed that: (1) accumulation of total As in whole-body tissue increases with increasing concentration of inorganic and organic As in water; (2) direct uptake of inorganic arsenic as As(III) from waterborne exposure is greater than As(V) – by as much 10 times for adult Japanese medaka and three times for

Table 6.9

Summary of uptake and accumulation of arsenic in fish directly from water and from food

Species	Exposure conc.: (water µg As L⁻¹, food mg kg⁻¹ ww)	Form of As	Medium	Duration	Tissue	Total As tissue[a] (mg kg⁻¹ ww)	Reference	Notes
Waterborne uptake								
Snakehead (adult), *Channa punctata*	1000	Inorganic As(III)	FW	90 days	Liver Kidney	0.12–0.39 0.07–0.36	Allen and Rana (2004)	Max. conc. attained after 60 days
Snakehead (adult), *Channa punctata*	1000	Inorganic As(III)	FW	90 days	Liver Kidney Gill Muscle	0.06–0.47 0.12–0.37 0.17–0.47 0.21–0.46	Allen et al. (2004)	Max. conc. attained after 60 days; increased resistance concluded because of the decrease in As in tissues at 90 days
Killifish, *Fundulus heteroclitus*	787	Inorganic As(V)	SW (20 ppt)	14 days	Liver	9.6	Bears et al. (2006)	
Indian catfish (50–70 g), *Clarias batrachus*	(R, U) 75	Inorganic As(III)	FW	30 days	Head Kidney	282–504	Datta et al. (2009)	
Zebrafish (juvenile), *Danio rerio*	(R, U) 4040–5056	Inorganic As(V)	FW	180 days	Gill	4.05	Ramirez and Garcia (2005)	
Killifish, *Fundulus heteroclitus*	(S, U) 7940	Inorganic As(III)	Soft FW to SW	2 days FW/1 day SW [same As (III) conc.]	Gill	1.12	Shaw et al. (2007a)	
Common carp (mature), *Cyprinus carpio*	(R, U) 1000	Inorganic As(III)	FW	2 days	Liver Gill	0.094 0.12	Ventura-Lima et al. (2009a)	No increase in total As in liver and gill tissue after 2 days compared to control levels [at both 100 and 1000 µg As(III) L⁻¹]
Common carp (mature), *Cyprinus carpio*	(R, U) 100	Inorganic As(V)	FW	2 days	Liver Gill Gill	0.038 0.059 2.2	Ventura-Lima et al. (2009a)	No increase in total As in liver and gill tissue after 2 days compared to control levels at 100 µg As(V) L⁻¹ No increase in total As in gill tissue after 2 days compared to control levels Significant increase in gills only at 1000 µg As(V) L⁻¹ (predominantly as MMA and DMA)
Zebrafish, *Danio rerio*	(R, M) 10	Inorganic As(V)	FW	2 days	Gill	< 0.02 0.99	Ventura-Lima et al. (2009b)	
Zebrafish (larvae), *Danio rerio*	(R, U) 3260	Inorganic As(III)	FW	4 days (from 2 to 6 dpf)	Whole body	0.45	Hamdi et al. (2009)	

(Continued)

Table 6.9 (*continued*)

Species	Exposure conc.: (water μg As L⁻¹, food mg kg⁻¹ ww)	Form of As	Medium	Duration	Tissue	Total As tissue[a] (mg kg⁻¹ ww)	Reference	Notes
Zebrafish (adult), *Danio rerio*	(R, U) 3260	Inorganic As(III)	FW	4 days	Brain Eye Gill Heart Intestine Liver Muscle Skin	0.075 0.325 0.575 0.425 0.4 0.9 0.475 0.2	Hamdi et al. (2009)	
Tilapia (5–8 cm), *Oreochromis mossambicus*	(S, U) 1000	Inorganic As(III)	FW	7 days	Whole body	2	Suhendrayatna et al. (2002a)	Whole-body concentration of total As increased to approx. 5.2 mg kg⁻¹ from exposure to 5 mg As(III) L⁻¹ for 7 days; As accumulation declined in fish above 10 mg As(III) L⁻¹ in water due to toxicity
Tilapia (5–8 cm), *Oreochromis mossambicus*	(S, U) 5000	Inorganic As(V)	FW	7 days	Whole body	2	Suhendrayatna et al. (2002a)	Whole-body concentration of total As increased to approx. 2.8 mg kg⁻¹ from exposure to 10 mg As(V) L⁻¹ for 7 days
Tilapia (5–8 cm), *Oreochromis mossambicus*	(S, U) 1000	Organic MMA	FW	7 days	Whole body	0.63	Suhendrayatna et al. (2002a)	Whole-body concentration of total As increased to approx. 3 mg kg⁻¹ from exposure to 10 mg MMA L⁻¹ for 7 days
Tilapia (5–8 cm), *Oreochromis mossambicus*	(S, U) 1000	Organic DMAA	FW	7 days	Whole body	0.48	Suhendrayatna et al. (2002a)	Whole-body concentration of total As increased to approx. 1.8 mg kg⁻¹ from exposure to 10 mg DMAA L⁻¹ for 7 days
Japanese medaka (adult), *Oryzias latipes*	(S, U) 1000	Inorganic As(III)	FW	7 days	Whole body	5.3	Suhendrayatna et al. (2002b)	Whole-body concentration of total As increased to approx. 53 mg kg⁻¹ from exposure to 5 mg As(III) L⁻¹ for 7 days; As accumulation declined in fish above 10 mg As(III) L⁻¹ in water due to toxicity
Japanese medaka (adult), *Oryzias latipes*	(S, U) 1000	Inorganic As(V)	FW	7 days	Whole body	1	Kuroiwa et al. (1994)	Whole-body concentration of total As increased to approx. 7.1 mg kg⁻¹ from exposure to 5 mg As(V) L⁻¹ for 7 days; As accumulation continues to increase in fish approx. 27 mg kg⁻¹ from exposure to 15 mg As(V) L⁻¹ for 7 days
Dietary uptake Rainbow trout (juvenile, 29 g), *Oncorhynchus mykiss*	10 (once daily)	Inorganic As(III), artificial diet	FW	8 weeks	Liver	0.33–0.44	Oladimeji et al. (1984a)	Maximum conc. attained at week 6 of 8; range in conc. at 30 mg kg⁻¹ ww diet:

Species	Dose	Water	Duration	Tissue	Conc.	Form/diet	Reference	Comments
				Muscle	0.12–0.35			1.2–1.6 mg kg^{-1} ww where growth and hematological effects were observed
				Gills	0.17–0.70			Maximum conc. attained at week 4 of 8; range in conc. at 30 mg kg^{-1} ww diet: 0.37–0.60 mg kg^{-1} ww where growth and hematological effects were observed
				Skin	0.24–0.30			Maximum conc. attained at week 2 of 8; range in conc. at 30 mg kg^{-1} ww diet: 0.36–0.48 mg kg^{-1} ww where growth and hematological effects were observed
Rainbow trout (fry, 3.6 g), *Oncorhynchus mykiss*	137 (4 × day)	FW	8 weeks	Whole body	1.40	Inorganic As(V), artificial diet	Cockell and Hilton (1988)	Maximum conc. attained at week 8 of 8; range in conc. at 30 mg kg^{-1} ww diet: 0.39–0.62 mg kg^{-1} ww where growth and hematological effects were observed. Significant reduction in growth compared to controls was noted at this dietary concentration
Rainbow trout (fry, 3.6 g), *Oncorhynchus mykiss*	180 (4 × day)	FW	8 weeks	Whole body	0.60	Inorganic As(III), artificial diet	Cockell and Hilton (1988)	Significant reduction in growth compared to controls was noted at this dietary concentration
Rainbow trout (fry, 4.2 g), *Oncorhynchus mykiss*	1497 (4 × day)	FW	8 weeks	Whole body	2.30	Organic DMA, artificial diet	Cockell and Hilton (1988)	Significant reduction in growth compared to controls was noted at this dietary concentration
Rainbow trout (juvenile, 7.6 g), *Oncorhynchus mykiss*	8 (3–4 × day)	FW	16 weeks	Kidney	0.21	Inorganic As(V), artificial diet	Cockell et al. (1991)	Conc. attained at 44 mg kg^{-1} ww diet was 1.5 mg kg^{-1} ww where growth and histological effects were observed
				Liver	0.13			Conc. attained at 44 mg kg^{-1} ww diet was 0.76 mg kg^{-1} ww where growth and histological effects were observed
				Carcass	0.25			Conc. attained at 44 mg kg^{-1} ww diet was 0.28 mg kg^{-1} ww where growth and histological effects were observed
Lake whitefish (adult), *Coregonus clupeaformis*	12 (3 × week)	FW	64 days	Gall bladder	0.60	Inorganic As(V), artificial diet	Pedlar and Klaverkamp (2002)	Not significantly different from control; conc. range at 120 mg kg^{-1} ww diet: 0.50–9.0 mg kg^{-1} ww where growth effects observed, peak at 10 days
				Liver	0.25			Significant increase compared to control; conc. range at 120 mg kg^{-1} ww diet: 0.88–4.3 mg kg^{-1} ww where growth effects observed, peak at 30 days
				Pyloric caeca	0.75			Not significantly different from control; conc. range at 120 mg kg^{-1} ww diet: 1.0–5.0 mg kg^{-1} ww where growth effects observed, peak at 30 days
				Stomach	0.06			Not significantly different from control; conc. range at 120 mg kg^{-1} ww diet:

(Continued)

Table 6.9 (*continued*)

Species	Exposure conc.: (water µg As L⁻¹, food mg kg⁻¹ ww)	Form of As	Medium	Duration	Tissue	Total As tissue[a] (mg kg⁻¹ ww)	Reference	Notes
					Intestine	0.14		0.14–0.72 mg kg⁻¹ ww where growth effects observed, peak at 30 days
					Kidney	0.19		Not significantly different from control; conc. range at 120 mg kg⁻¹ ww diet: 0.71–3.9 mg kg⁻¹ ww where growth effects observed, peak at 30 days
					Skin	0.16		Not significantly different from control; conc. range at 120 mg kg⁻¹ ww diet: 0.52–0.75 mg kg⁻¹ ww where growth effects observed, peak at 30 days
					Scales	0.11		Not significantly different from control; conc. range at 120 mg kg⁻¹ ww diet: 0.22–0.29 mg kg⁻¹ ww where growth effects observed, peak at 30 days
								Not significantly different from control; conc. range at 120 mg kg⁻¹ ww diet: 0.27–0.48 mg kg⁻¹ ww where growth effects observed, peak at 30 days
Tilapia (5–8 cm), *Oreochromis mossambicus*	6.4 (lab food chain)	Algae exposed to 50,000 µg L⁻¹ As (III) in water for 10 days	FW	7 days	Whole body	6.8	Suhendrayatna et al. (2002a)	
Japanese medaka (adult), *Oryzias latipes*	42 (lab food chain)	Algae exposed to 50,000 µg L⁻¹ As (III) in water for 10 days	FW	7 days	Whole body	12.5	Suhendrayatna et al. (2002b)	

[a]Tissue concentrations reported in dry weight recalculated and expressed on wet weight basis assuming 75% moisture content of whole body and muscle, and 80% moisture content of organs.

conc.: concentration; R: renewal test; S: static test; U: unmeasured concentrations; M: measured concentrations; DMA: dimethylarsinic acid; MMA: monomethylarsonous acid; FW: freshwater; SW: saltwater.

juvenile tilapia; and (3) a substantial difference potentially exists in the ability of different fish species to take up and accumulate inorganic As directly from water. This latter observation was based on medaka accumulating up to 10 times more As(III) compared to tilapia at a given (non-toxic) waterborne concentration.

Unfortunately, laboratory data on direct uptake of As from water are limited. Chen and Liao (2004) conducted a 15 day uptake/depuration bioassay to examine the accumulation kinetics of As in tilapia exposed to 1000 μg As L^{-1} and determined that the gill concentration profile was best fitted by the one-compartment model. Depuration and uptake constants were estimated to be 0.08 day^{-1} and 0.26 ml g^{-1} day^{-1}, respectively, and the depuration half-life was estimated to be 8.56 days. Importantly, the authors of this study predicted permeability and the mass transfer rate of As through tilapia gill to decrease from 1.42 to 0.82 μm day^{-1} and from 0.039 to 0.024 day^{-1}, respectively, after a period of 2 months (modeled prediction of time to steady state). The authors also estimated a limiting uptake flux ($J_{u,max}$) of 2.17 mg L^{-1} day^{-1} and a bioaffinity constant (K_M) of 3.07 mg L^{-1}, with a regression coefficient of 0.96. The high K_M value (3.07 mg L^{-1}) in general indicates a low binding affinity of As to tilapia gills, and presumably other fish as well. Moreover, the low specific affinity ($J_{u,max}/K_M = 0.71$ day^{-1}) indicates that As may share similar binding sites on the gills of teleost fish as the class B metals (i.e. sulfur- or nitrogen-seeking metals such as Cd, Cu, and Ag). A subsequent modeling exercise indicated a high bioavailability rate within the first 20 days of exposure that progressively slowed over time. Additional research is needed to further characterize and increase our understanding and significance of this route of uptake in fish.

Research suggests a possible role of aquaporin isoform, AQP3, in physiological/osmoregulatory uptake and transport processes in fish, including gills. In the teleost gills, AQP3 is present in chloride cells, where it may have a number of different functions (Cutler et al., 2007). Aquaporins are membrane channel proteins integral for mediating the bidirectional, osmotically driven flux of water and selected small amphipathic molecules across cellular membranes (Agre and Kozono, 2003). Aquaglyceroporins are a subfamily of aquaporins that translocate larger molecules such as glycerol, including trivalent metalloids such as arsenite and antimonite because in solution the trihydroxylated inorganic forms of these metalloids are mimics of glycerol (Ramirez-Solis et al., 2004). Hamdi et al. (2009) recently cloned five members of the aquaglyceroporin family (including AQP3) and investigated their expression and ability to transport As in zebrafish following short-term (4–5 day) As exposure to waterborne inorganic As(III) at approximately 3260 μg L^{-1}. The results of this study not only represent the first molecular identification of an As(III) transport system in fish, but indirectly link expression of the fish aquaglyceroporins and transport of As in a variety of zebrafish tissues, including, possibly, the gill. Continued work in this area should greatly expand our current knowledge of As uptake and accumulation in fish and aquatic animals in general.

Related to the above, Schreiber et al. (2000) showed that cystic fibrosis transmembrane conductance regulator (CFTR) activates AQP3 expressed endogenously and exogenously in oocytes of *Xenopus laevis*. Shaw et al. (2007b) recently showed that waterborne inorganic As(III) at relatively high concentrations (7940 µg L^{-1} As in water; 1.12 mg kg^{-1} ww in gill tissue) blocks the ability of killifish to adapt to increased salinity by reducing CFTR protein abundance. The apparent interplay between these functional proteins involved in water and nutrient transport, including trivalent arsenicals, and NaCl homeostasis in aquatic organisms is intriguing, and currently an active area of research focus.

8.2. Uptake and Accumulation via the Gut

Another primary route of As uptake, potentially involving some of the same transport mechanisms mentioned above, is via the gut. Table 6.9 summarizes a number of laboratory studies that assessed As uptake and accumulation in fish via the gut from dietary exposure. As with waterborne exposure, accumulation of dietary As through the stomach and intestines is concentration dependent. Inorganic As is also accumulated to greater levels in fish relative to organic forms with dietary As exposure as in waterborne exposures (see results by Cockell and Hilton, 1988, in Table 6.9). To date, however, there has been no systematic study of the molecular mechanisms associated with absorption of As in any form across intestinal epithelia of fish, or the rates at which the mechanisms might work.

Results from Oladimeji et al. (1979) are potentially revealing in this regard. The authors dosed individual young adult rainbow trout (160–211 g) with 20 µCi ^{74}As [approx. 1 mCi/µg As in the form of arsenic acid, As(V)] added to food. Arsenic distribution within tissues and clearance were followed from 0.5 to 96 h postdose. Arsenic was detected in blood within 6 h, about half of which was partitioned to erythrocytes within 12 h. At 6 h postdose, 50% of the total As present in plasma and erythrocytes was in inorganic form, decreasing to about 8% and 5%, respectively, by 96 h postdose, concomitant with a rise in the organic As fraction.

8.3. Biotransformation

The results from the Oladimeji et al. (1979) study are important for two reasons with regard to understanding biotransformation of As in fish. Firstly, the ratio of total organic to inorganic As continually increased in trout with time in muscle, kidney, and bile extracts as well as in blood, with the organic As fraction accounting for 34–64% at 6 h and over 90% after 24 h. In contrast to these other tissues, a much higher concentration of organic As (approximately 75%) was detected in liver extract after only 6 h postdose. These findings are consistent with those observed in mammals, where, once absorbed from the gastrointestinal tract, soluble trivalent and pentavalent As is widely distributed throughout the body, and where pentavalent As is reduced to the trivalent form which in turn is methylated [mostly as DMA(V) or MMA (V)] in the liver (Adams et al., 1994; Vahter, 1994).

Secondly, but in stark contrast to the situation in mammals, the majority of total As content of all the tissues (84% kidney to 98% in muscle) after 96 h postdose was composed of an organic fraction thought to be of similar form to ethyl AB (mammals do not produce AB). No equivalent studies exist for marine fish, although studies from single-dose experiments where dietary As in the form of AB was force-fed to freshwater- and saltwater-adapted Atlantic salmon (*Salmo salar*) and the marine fish Atlantic cod (*Gadus morhua* L.) show that dietary AB accumulates unchanged in muscle tissue of fish, and that seawater adaptation has no effect on either the levels of AB accumulated in muscle or the form of As (Amlund and Berntssen, 2004; Amlund et al., 2006b), which is consistent with observations of As speciation in the field (see Section 7.1).

It is important to point out that some difference in the internal handling and biotransformation of dietary As may exist in situations where the level of inorganic As exposure is high. This situation is evident in the three-step freshwater food chain studies reported by Suhendrayatna et al. (2002a, b) and Kuroiwa et al. (1994). The results of the several laboratory food chain experiments reveal that inorganic As was the dominant form after 7 day exposures in whole body where transformation of As(III) to As(V) occurs more easily than biomethylation. This latter observation may only be relevant in highly impacted sites, where a higher percentage of As body burden may be inorganic as opposed to organic As.

8.4. Excretion

From the study by Oladimeji et al. (1979), the estimated excretion rate of As (>90% of which may be organic As) in fish is low, less than 0.2% of the dose per hour over the 48 h postdose period (Oladimeji et al., 1979). The authors, based on additional study of rainbow trout following a single dose of [74]As arsenic acid, determined that the total As present in the bile and in gallbladder tissue (only 0.1–0.3% of the total As in tissues) was also very low. Thus, urinary and biliary excretion together accounted for only a small

DENNIS O. McINTYRE AND TYLER K. LINTON

portion of the total As ingested by rainbow trout, again in stark contrast to the typical paradigm in mammals (Vahter, 1994).

Despite the apparent minor role of urinary and biliary excretion in the study, the rainbow trout still lost 65% of the ingested As dose by 48 h postdose, which led the authors to suggest that the likely additional route for the excretion of inorganic As is via the gills. A subsequent follow-up study supported this hypothesis (Oladimeji et al., 1984b). The authors did note, however, large differences among individuals in their study. To the present authors' knowledge, no additional studies have been conducted since that time to corroborate these results.

In slight contrast to fish in freshwater, the major route of AB excretion in Atlantic salmon and Atlantic cod after a single oral dose is via the urine, which appeared more important to Atlantic cod (Amlund et al., 2006b). In fact, in a subsequent study (Amlund et al., 2006a), analysis revealed that elimination kinetics in muscle were distinct between the two species, with estimated elimination half-lives of 77 days in Atlantic cod compared to only 37 days in Atlantic salmon. Absorption efficiency was also estimated to be approximately two-fold higher in Atlantic cod (15%) than in Atlantic salmon (8%), and perhaps more importantly, Atlantic cod (a stenohaline species) appeared to have little capacity to regulate the resorption and secretion of AB in kidney, whereas Atlantic salmon, being a euryhaline species, appeared to show some capacity to regulate AB. As discussed above, AB behaves as an osmolyte owing to its similarity to glycine betaine, which the authors suggest euryhaline species such as Atlantic salmon could use to maintain intracellular homeostasis when experiencing changes in salinity, but Atlantic cod, because of their limited tolerance to changes in salinity, may not. Additional work is needed to confirm these results with other species, as well as other possible routes of excretion (e.g. skin, gills).

9. DETOXIFICATION AND MECHANISMS FOR TOLERANCE

While there is a number of specific genes encoded explicitly for As detoxification in bacteria (Rosen and Liu, 2009), none of them has been shown in advanced eukaryotic cells. Instead, in mammalian cells, it is presumed that As(V) is taken up by phosphate transporters (as has been shown in bacteria and yeast; see Rosenberg et al., 1977, and Persson et al., 1999, respectively), while the aquaglyceroporins play a major role in the uptake and efflux of As (III) [and MMA(III)] in cells (Agre and Kozono, 2003; Ramirez-Solis et al., 2004; Rosen and Liu, 2009). The latter represents one family of transport proteins with clear functional roles recently discovered in fish (e.g. AQP3) (Cutler et al., 2007, Hamdi et al., 2009; Section 8.1).

Another family of proteins of importance that aid in reducing cellular As concentrations in mammals, and just recently discovered in fish, involve multidrug resistance-associated protein genes (MRP1/MRP2, which encode for the efflux transporters Mrp1 and Mrp2) acting in concert with multidrug resistance gene (MDR1, which encodes for the efflux transporter P-glycoprotein). The gene encoded MRP2 is known to transport a wide range of anionic substrates, including As–glutathione conjugates. Exposure of killifish (*Fundulus heteroclitus*) to As(III) for 4–14 days increased both MRP2 expression in the apical membrane of proximal tubules and MRP2-mediated transport activity (Miller et al., 2007). Since As–glutathione efflux from cells is driven in part by MRP2, it is postulated that increased expression of this transporter might confer tolerance and protect against toxicity in fish. Shaw et al. (2007b) recently conducted a series of studies with *F. heteroclitus* to test this hypothesis and showed that increased tolerance to As in killifish following a pre-exposure regime was associated with lower concentrations in tissues (gill, liver, kidney, and whole body), and that the observation coincided with a transient increase in expression of the MPR2 gene responsible for transporting As–glutathione conjugates out of cells. Such results are the first ever outside of plants where acquired As tolerance is associated with reduced concentrations of As in tissues.

Unlike the case for other non-essential metals, there is little direct evidence of As detoxification via expression and subsequent sequestration by metallothionien (MT) or a molecular chaperone such as hsp70 (Schlenk et al., 1997; Boyle et al., 2008). Pedlar and coauthors (2002) were among the first to demonstrate hepatic MT induction in fish (lake whitefish, *Coregonus clupeaformis*) exposed to dietary As(V). MT increased significantly in fish fed diets containing 1 and 10 mg As kg^{-1} over the course of a 64 day exposure, but the effect of As on MT induction in fish fed the 100 mg As kg^{-1} diet was transient. The transient MT induction in fish receiving the high-dose diet was speculated by the authors to be due to one of three possibilities: (1) a decline in As dose, (2) overt As toxicity, and/or (3) a biphasic pattern of MT induction with continued exposure; the latter of which has been observed in climbing perch (*Anadas testudineus*) exposed to 750 µg L^{-1} As(III) for up to 30 days (Das et al., 1998).

10. BEHAVIORAL EFFECTS OF ARSENIC

Data on behavioral effects of As on fish are limited. Weir and Hine (1970) demonstrated behavioral impairment (conditional avoidance response) in goldfish (*Carassius auratus*) exposed to 100 µg AsV L^{-1} (30% impairment) and 1000 µg AsV L^{-1} (42% impairment), respectively.

No speculation of the possible mechanism for the behavioral impairment was made, but the data imply possible inhibition of the neural–sensory system.

De Castro et al. (2009) recently found that As(V) impaired long-term memory in zebrafish exposed 96 h before avoidance behavior trials at concentrations as low as 1 μg L^{-1}. DNA damage in excised brain tissue was evaluated through the K+/SDS assay, in which potassium chloride is added to sodium dodecyl sulfate (SDS), which detects covalent protein–DNA cross-links, but DNA damage in terms of protein–DNA cross-links was not verified. Measurement of elevated levels of oxidized proteins in brain tissue at 10 and 100 μg L^{-1} in the 86–131 kDa range, however, suggests that the amnesia observed in zebrafish at low, environmentally relevant levels of waterborne inorganic As(V) may be associated with oxidative stress in brain tissue of these fish.

11. MOLECULAR CHARACTERIZATION OF ARSENIC TRANSPORTERS, STORAGE PROTEINS, AND CHAPERONES

With the possible exceptions of bacteria and genes encoded for As resistance (i.e. *ars* operons), there are no known As-specific transporters, storage proteins, or molecular chaperones in biological systems. Instead, the facilitation of As transport in particular via pathways specific to other elements and micronutrients, e.g. phosphate and glucose, appears to be utilized adventitiously in prokaryotic and eukaryotic cells of all types (Rosen and Liu, 2009).

12. INTERACTIONS WITH OTHER METALS

Most of the known chemical interactions that modify As toxicity are with selenium (Se). Arsenite has been known for half a century to have a protective effect against Se poisoning (Kraus and Ganther, 1989). In a review of As–Se bonds in biology, Gailer (2007) cites numerous studies in which As antagonizes the toxicity of Se in mice, hogs, steers, dogs, chickens, and mallards. Although arsenite antagonizes inorganic Se salts, it enhances the toxicity of methylated selenium compounds.

Minimal evidence of As–Se interaction has been reported in fish, although Se was antagonistic to As toxicity in cultured cells of bluegill (*Lepomis macrochirus*) (Babich et al., 1989). The acute cytotoxicities of arsenate, and to a lesser extent arsenite, were reduced at non-toxic concentrations of selenate and selenite.

13. KNOWLEDGE GAPS AND FUTURE DIRECTIONS

Several needs exist for understanding the effects of As on fish. While there has been some research on the toxicokinetics of As in invertebrates, very little has been done on fish. Determination of uptake and elimination rate constants, as well as assimilation efficiencies of As in freshwater and saltwater fish will improve our understanding of how As is handled internally. Such information may help to explain observations that have been made for As in fish, such as biodiminution, the inverse relationship between bioaccumulation factors and the aqueous exposure concentration, and the ability for saltwater fish to accumulate greater concentrations of As than freshwater fish. More work is needed on understanding the species of As that occur in fish, particularly at different levels of inorganic waterborne As. This is largely driven by the risk of cancer to humans from consuming fish. The focus of As speciation in fish tissue to date has logically been on marine fish because of the higher concentrations that occur in them. More research is needed on what As species occur in freshwater fish (both at background and As-enriched waters) and why this will help in the risk assessment decisions for water quality and environmental health managers. Since waterborne and dietary concentrations of As are generally not good predictors of effects in fish, any effects-related research should include tissue measurements of As so that a critical tissue level could be developed. Because of possible reproductive/endocrine effects, life-cycle tests are needed.

REFERENCES

Abdelghani, A. A., Anderson, A. C., and McDonell, D. B. (1980). Toxicity of three arsenical compounds. *Can. Res.* **13**, 31–32.

Abdullah, M. I., Shiyu, Z., and Mosgren, K. (1995). Arsenic and selenium species in the oxic and anoxic waters of the Oslofjord, Norway. *Mar. Pollut. Bull.* **31**, 116–126.

Adams, M. A., Bolger, P. M., and Gunderson, E. L. (1994). Dietary intake and hazards of arsenic. In: *Arsenic Exposure and Health. Special Issue of Environmental Geochemistry and Health, Science and Technology Letters* (W.R. Chappell, C.O. Abernathy and C.R. Cothern, eds), Vol. 16, pp. 41–50. Northwood, Middlesex.

Agre, P., and Kozono, D. (2003). Aquaporin water channels: molecular mechanisms for human diseases. *FEBS Lett.* **555**, 72–78.

Ahuja, S. (2008). The problem of arsenic contamination of groundwater. In: *Arsenic Contamination of Groundwater: Mechanism, Analysis, and Remediation* (S. Ahuja, ed.), pp. 1–21. John Wiley & Sons, Hoboken.

Allen, T., and Rana, S. V. S. (2004). Effect of arsenic (AsIII) on glutathione-dependent enzymes in liver and kidney of the freshwater fish *Channa punctatus*. *Biol. Trace Elem. Res.* **100**, 39–48.

Allen, T., Singhal, R., and Rana, S. V. S. (2004). Resistance to oxidative stress in a freshwater fish *Channa punctatus* after exposure to inorganic arsenic. *Biol. Trace Elem. Res.* **98**, 63–72.

Amlund, H., and Berntssen, M. H. G. (2004). Arsenobetaine in Atlantic salmon (*Salmo salar* L.): influence of seawater adaptation. *Comp. Biochem. Physiol. C* **138**, 507–514.

Amlund, H., Francesconi, K. A., Bethune, C., Lundebye, A.-K., and Berntssen, M. H. G. (2006a). Accumulation and elimination of dietary arsenobetaine in two species of fish, Atlantic salmon (*Salmo salar* L.) and Atlantic cod (*Gadus morhua* L.). *Environ. Toxicol. Chem.* **25**, 1787–1794.

Amlund, H., Ingebrigtsen, K., Hylland, K., Ruus, A., Eriksen, D. Ø., and Berntssen, M. H. G. (2006b). Disposition of arsenobetaine in two marine fish species following administration of a single oral dose of [^{14}C]arsenobetaine. *Comp. Biochem. Physiol. C* **143**, 171–178.

Andreae, M. O. (1978). Distribution and speciation of arsenic in natural waters and some marine algae. *Deep Sea Res.* **25**, 391–402.

Andreae, M. O. (1979). Arsenic speciation in seawater and interstitial waters: the influence of biological–chemical interactions on the chemistry of a trace element. *Limnol. Oceanogr.* **24**, 440–452.

Andreae, M. O., and Andreae, T. W. (1989). Dissolved arsenic species in the Schelde Estuary and watershed, Belgium. *Estuar. Coast. Shelf Sci.* **29**, 421–433.

Andreae, M. O., Byrd, T. J., and Froelich, O. N. (1983). Arsenic, antimony, germanium and tin in the Tejo Estuary, Portugal: modelling of a polluted estuary. *Environ. Sci. Technol.* **17**, 731–737.

ANZECC and ARMCANZ (2000). *Australian and New Zealand Guidelines for Fresh and Marine Water Quality.* Australian and New Zealand Environment and Conservation Council and Agriculture and Resource Management Council of Australia and New Zealand, Canberra, Australia.

Azcue, J., and Nriagu, J. O. (1995). Impact of abandoned mine tailings on the arsenic concentrations in Moira Lake, Ontario. *J. Geochem. Explor.* **52**, 81–89.

Azcue, J. M., Murdoch, A., Rosa, F., and Hall, G. E. M. (1994). Effects of abandoned gold mine tailings on the arsenic concentrations in water and sediments of Jack of Clubs Lake, BC. *Environ. Technol.* **15**, 669–678.

Azcue, J. M., Mudroch, A., Rosa, F., Hall, G. E. M., Jackson, T. A., and Reynoldson, T. (1995). Trace elements in water, sediments, porewater, and biota polluted by tailings from an abandoned gold mine in British Columbia, Canada. *J. Geochem. Explor.* **52**, 25–34.

Babich, H., Martin-Alguacil, N., and Borenfreund, E. (1989). Arsenic–selenium interactions determined with cultured fish cells. *Toxicol. Lett.* **45**, 157–164.

Baeyens, W., de Brauwere, A., Brion, N., De Gieter, M., and Leermakers, M. (2007). Arsenic speciation in the River Zenne, Belgium. *Sci. Total Environ.* **384**, 409–419.

Baur, W. H., and Onishi, B.-M. H. (1969). Arsenic. In: *Handbook of Geochemistry* (K.H. Wedepohl, ed.), pp. 33-A-1–33-0-5. Springer, Berlin.

Bears, H., Richards, J. G., and Schulte, P. M. (2006). Arsenic exposure alters hepatic arsenic species composition and stress-mediated gene expression in the common killifish (*Fundulus heteroclitus*). *Aquat. Toxicol.* **77**, 257–266.

Benson, L. V., and Spencer, R. J. (1983). *A Hydrochemical Reconnaissance Study of the Walker River Basin, California and Nevada.* United States Geological Survey, Denver, CO, USGS Open File Report, 83–740.

Bervoets, L., Voets, J., Covaci, A., Chu, S., Qadah, D., Smolders, R., Scepens, P., and Blust, R. (2005). Use of transplanted zebra mussels (*Dreissena polymorpha*) to assess bioavailability of microcontaminants in Flemish surface waters. *Environ. Sci. Technol.* **39**, 1492–1505.

de Bettencourt, A. M. M., and Andreae, M. O. (1991). Refractory arsenic in estuarine waters. *Appl. Organomet. Chem.* **5**, 111–116.

Boyle, D., Brix, K. V., Amlund, H., Lundebye, A. K., Hogstrand, C., and Bury, N. (2008). Natural arsenic contaminated diets perturb reproduction in fish. *Environ. Sci. Technol.* **42**, 5354–5360.

Braman, R. S., and Foreback, C. C. (1973). Methylated forms of arsenic in the environment. *Science* **182**, 1247–1249.

Bright, D. A., Dodd, M., and Reimer, K. J. (1996). Arsenic in sub-Arctic lakes influenced by gold mine effluent: the occurrence of organoarsenicals and "hidden" arsenic. *Sci. Tot. Environ* **180**, 165–182.

Broderius, S. J., Kahl, M. D., and Hoglund, M. D. (1995). Use of joint toxic response to define the primary mode of toxic action for diverse industrial organic chemicals. *Environ. Toxicol. Chem.* **14**, 1591–1605.

Brooks, W. E. (2007). *2007 Minerals Yearbook, Arsenic.* United States Geological Survey, Denver, CO, September 2008.

Buhl, K. J., and Hamilton, S. J. (1990). Comparative toxicity of inorganic contaminants released by placer mining to early life stages of salmonids. *Ecotoxicol. Environ. Saf.* **20**, 325–342.

Buhl, K. J., and Hamilton, S. J. (1991). Relative sensitivity of early life stages of Arctic grayling, coho salmon, and rainbow trout to nine inorganics. *Ecotoxicol. Environ. Saf.* **22**, 184–197.

Cáceres, L., Gruttner, E., and Contreras, R. (1992). Water recycling in arid regions – Chilean case. *Ambio* **21**, 138–144.

Call, D. J., Brooke, L. T., Ahmad, N., and Richter, J. E. (1983). Toxicity and metabolism studies with EPA priority pollutants and related chemicals in freshwater organisms. In *PB83-263665.* Springfield, VA: National Technical Information Service.

Cardin, J. A. (1985). *Results of Acute Toxicity Tests Conducted with Arsenic at ERL, Narragansett.* United States Environmental Protection Agency, Narragansett, RI.

Cardwell, R. D., Foreman, D. G., Payne, T. R., and Wilbur, D. J. (1976). Acute toxicity of selected toxicants to six species of fish. In *EPA-600/3-76-008.* Duluth, MN: United States Environmental Protection Agency.

CCME (2001). Canadian Water Quality Guidelines for the Protection of Aquatic Life, Arsenic. Winnipeg: Canadian Council of Ministers of the Environment, 1999, updated 2001.

Celino, F. T., Yamaguchi, S., Miura, C., and Miura, T. (2008). Testicular toxicity of arsenic on spermatogenesis of fish. In: *Interdisciplinary Studies on Environmental Chemistry – Biological Responses to Chemical Pollutants* (Y. Murakami, K. Nakayama, S.-I. Kitamura, H. Iwata and S. Tanabe, eds), pp. 55–60. Terrapub, Tokyo.

Chen, C. Y., and Folt, C. L. (2000). Bioaccumulation and diminution of arsenic and lead in a freshwater food web. *Environ. Sci. Technol.* **34**, 3878–3884.

Chen, B.-C., and Liao, C.-M. (2004). Farmed tilapia *Oreochromis mossambicus* involved in transport and biouptake of arsenic in aquacultural ecosystems. *Aquaculture* **242**, 365–380.

Chilvers, D. C., and Peterson, P. J. (1987). Global cycling of arsenic. In: *Lead, Mercury, Cadmium and Arsenic in the Environment* (T.C. Hutchinson and K.M. Meema, eds), pp. 279–303. John Wiley & Sons, New York.

Ciardullo, S., Aureli, F., Raggi, A., and Cubadda, F. (2010). Arsenic speciation in freshwater fish: focus on extraction and mass balance. *Talanta* **81**, 213–221.

Clemens, H. P. and Sneed, K. E. (1959). Lethal doses of several commercial chemicals for fingerling channel catfish. In *Scientific Report Fisheries No. 316.* Washington, DC: United States Department of the Interior.

Clowes, L. A., and Francesconi, K. A. (2004). Uptake and elimination of arsenobetaine by the mussel *Mytilus edulis* is related to salinity. *Comp. Biochem. Physiol. C* **137**, 35–42.

Cockell, K. A., and Hilton, J. W. (1988). Preliminary investigations on the comparative chronic toxicity of four dietary arsenicals to juvenile rainbow trout (*Salmo gairdneri* R.). *Aquat. Toxicol.* **12**, 73–82.

Cockell, K. A, Hilton, J. W., and Bettger, W. J. (1991). Chronic toxicity of dietary disodium arsenate heptahydrate to juvenile rainbow trout (*Oncorhynchus mykiss*). *Arch. Environ. Contam. Toxicol.* **21**, 518–527.

Crearley, J. E. (1973). *Arsenic contamination of Finfeather and Municipal Lakes in the City of Bryan. Report.* Texas Water Quality Board, Austin, TX.

CSIR (1996). *South African Water Quality Guidelines,* Vol. 7. *Aquatic Ecosystems.* Department of Water Affairs and Forestry, Council for Scientific and Industrial Research, Pretoria.

Culioli, J.-L., Fouquoire, A., Calendini, S., Mori, C., and Orsini, A. (2009). Trophic transfer of arsenic and antimony in a freshwater ecosystem: A field study. *Aquat. Toxicol.* **94**, 286–293.

Cullen, W. R., and Reimer, K. J. (1989). Arsenic speciation in the environment. *Chem. Rev.* **89**, 713–764.

Cutler, C. P., Martinez, A.-S., and Cramb, G. (2007). Review: The role of aquaporin 3 in teleost fish. *Comp. Biochem. Physiol. A* **148**, 82–91.

Cutter, G. A (1993). Metalloids in wet precipitation on Bermuda: concentrations, sources, and fluxes. *J. Geophys. Res.* **98**, 16777–16786.

Cutter, G. A., and Cutter, L. S. (1995). Behavior of dissolved antimony, arsenic and selenium in the Atlantic Ocean. *Mar. Chem.* **49**, 295–306.

Das, D., Sarkar, D., and Bhattacharya, S. (1998). Lipid peroxidative damage by arsenic intoxication is countered by glutathione–glutathione-*S*-transferase system and metallothionein in the liver of climbing perch, *Anabas testudineus. Biomed. Environ. Sci.* **11**, 187–195.

Datta, S., Saha, D. R., Ghosh, D., Majumdar, T., Bhattacharya, S., and Mazumder, S. (2007). Sub-lethal concentration of arsenic interferes with the proliferation of hepatocytes and induces *in vivo* apoptosis in *Clarias batrachus* L. *Comp. Biochem. Physiol. C* **145**, 339–349.

Datta, S., Ghosh, D., Saha, D. R., Bhattacharaya, S., and Mazumder, S. (2009). Chronic exposure to low concentration of arsenic is immunotoxic to fish: role of head kidney macrophages as biomarkers of arsenic toxicity to *Clarias batrachus. Aquat. Toxicol.* **92**, 86–94.

Davey, J. C., Bodwell, J. E., Gosse, J. A., and Hamilton, J. W. (2007). Arsenic as an endocrine disruptor: effects of arsenic on estrogen mediated gene expression *in vivo* and in cell culture. *Toxicol. Sci.* **98**, 75–86.

de Castro, M. R., Ventura Lima, J., de Freitas, D. P. S., de Souza-Valente, R., Dummer, N. S., de Aguiar, R. B., dos Santos, L. C., Marins, L. F., Geracitano, L. A., Monserrat, J. M., and Barros, D. M. (2009). Behavioral and neurotoxic effects of arsenic exposure in zebrafish (*Danio rerio*, Teleostei: Cyprinidae). *Comp. Biochem. Physiol. C* **150**, 337–342.

de Rosemond, S., Xie, Q., and Liber, K. (2008). Arsenic concentration and speciation in five freshwater fish species from Back Bay near Yellowknife, NT, Canada. *Environ. Monit. Assess.* **147**, 199–210.

DeFoe, D. L. (1982). *Memorandum to Robert L. Spehar.* United States Environmental Protection Agency, Duluth, MN, July 9.

Delnomdedieu, M., Basti, M. M., Otvos, J. D., and Thomas, D. J. (1993). Transfer of arsenite from glutathione to dithiols: a model of interaction. *Chem. Res. Toxicol.* **6**, 598–602.

Diamond, M. L. (1995). Application of a mass balance model to assess in-place arsenic pollution. *Environ. Sci. Technol.* **29**, 29–42.

Dixon, H. B. F. (1997). The biochemical action of arsonic acids especially as phosphate analogues. *Adv. Inorg. Chem.* **44**, 191–227.

Dyer, S. D., Brooks, G. L., Dickson, K. L., Sanders, B. M., and Zimmerman, E. G. (1993). Synthesis and accumulation of stress proteins in tissues of arsenite-exposed fathead minnows (*Pimephales promelas*). *Environ. Toxicol. Chem.* **12**, 913–924.

Eisler, R. (1988). *Arsenic Hazards to Fish, Wildlife, and Invertebrates: A Synoptic Review.* United States Fish and Wildlife Service, Laurel, MD, Biological Report No. 85(1.12).

EPRI (2008). *Evaluation of US Environmental Protection Agency's Arsenic Ambient Water Quality Criteria: Speciation and Bioaccumulation Issues.* Electric Power Research Institute, Palo Alto, CA, Technical Update.

EVS (2000). *Human Health Risk Assessment of Chemical Contaminants in Four Fish Species from the Middle Willamette River, Oregon.* Oregon Department of Environmental Quality, Portland, OR.

Fabris, G., Turoczyb, N. J., and Stagnitti, F. (2006). Trace metal concentrations in edible tissue of snapper, flathead, lobster, and abalone from coastal waters of Victoria, Australia. *Ecotoxicol. Environ. Saf.* **63**, 286–292.

Fattorini, D., Alonso-Hernandez, C. M., Diaz-Asencio, M., Munoz-Caravaca, A., Pannacciulli, F. G., Tangherlini, M., and Regoli, F. (2004). Chemical speciation of arsenic in diferent marine organisms: importance in monitoring studies. *Mar. Environ. Res.* **58**, 845–850.

Farag, A. M., Skaar, D., Nimick, D. A., MacConnell, E., and Hogstrand, C. (2003). Characterizing aquatic health using salmonid mortality, physiology, and biomass estimates in streams with elevated concentrations of arsenic, cadmium, copper, lead, and zinc in the Boulder River Watershed, Montana. *Trans. Am. Fish. Soc.* **132**, 450–467.

Faust, S. D., Winka, A. J., and Belton, T. (1987). An assessment of chemical and biological significance of arsenical species in the Maurice River drainage basin (N.J.). Part I. Distribution in water and river and lake sediments. *J. Environ. Sci. Health* **A22**, 209–237.

Foran, J. A., Hites, R. A., Carpenter, D. O., Hamilton, M. C., Mathews-Amos, A., and Schwager, S. J. (2004). A survey of metals in tissues of farmed Atlantic and wild Pacific salmon. *Environ. Toxicol. Chem.* **23**, 2108–2110.

Franseconi, K. A., and Kuehnelt, D. (2002). Arsenic compounds in the environment. In: *Environmental Chemistry of Arsenic* (W.T. Frankenberger, ed.), pp. 51–94. Marcel Dekker, New York.

Froelich, P. N., Kaul, L. W., Byrd, J. T., Andreae, M. O., and Roe, K. K. (1985). Arsenic, barium, germanium, tin, dimethyl-sulfide and nutrient biogeochemistry in Charlotte Harbor, Florida, a phosphorus-enriched estuary. *Estuar. Coast. Shelf Sci.* **20**, 239–264.

Gailer, J. (2007). Arsenic–selenium and mercury–selenium bonds in biology. *Coord. Chem. Rev.* **251**, 234–254.

Garelick, H., Jones, H., Dybowska, A., and Valsami-Jones, E. (2008). Arsenic pollution sources. *Rev. Environ. Contam. Toxicol.* **197**, 17–60.

Garg, S., Gupta, R. K., and Jain, K. L. (2009). Sublethal effects of heavy metals on biochemical composition and their recovery in Indian major carps. *J. Hazard. Mater.* **163**, 1369–1384.

Ghosh, A. R., and Chakrabarti, P. (1990). Toxicity of arsenic and cadmium to a freshwater fish *Notopterus notopterus. Environ. Ecol.* **8**, 576–579.

Ghosh, D., Bhattacharya, S., and Mazumder, S. (2006). Perturbations in the catfish immune responses by arsenic: organ and cell specific effects. *Comp. Biochem. Physiol. C* **143**, 455–463.

Ghosh, D., Datta, S., Bhattachary, S., and Mazumdera, S. (2007). Long-term exposure to arsenic affects head kidney and impairs humoral immune responses of *Clarias batrachus. Aquat. Toxicol.* **81**, 79–89.

Gilderhus, P. A. (1966). Some effects of sublethal concentrations of sodium arsenite on bluegills and the aquatic environment. *Trans. Am. Fish. Soc.* **95**, 289–296.

Gomez-Caminero, A., Howe, P., Hughes, M., Kenyon, E., Lewis, D. R., Moore, M., Ng, J., Aitio, A., and Becking, G. (2001). *Arsenic and Arsenic Compounds.* World Health Organization, Geneva.

Gonzalez, H. O., Roling, J. A., Baldwin, W. S., and Bain, L. J. (2006). Physiological changes and differential gene expression in mummichogs (*Fundulus heteroclitus*) exposed to arsenic. *Aquat. Toxicol.* **77**, 43–52.

Greene, R., and Crecelius, E. (2006). Total and inorganic arsenic in mid-Atlantic marine fish and shellfish and implications for fish advisories. *Integr. Environ. Assess. Manag.* **2**, 344.

Gupta, A. K., and Chakrabarti, P. (1993). Toxicity of arsenic to freshwater fishes *Mystus vittatus* (Bloch) and *Puntius javanicus* (Blkr.). *Environ. Ecol.* **11**, 808–811.

Hamdi, M., Sanchez, M. A., Beene, L. C., Liu, Q., Landfear, S. M., Rosen, B. P., and Liu, Z. (2009). Arsenic transport by zebrafish aquaglyceroporins. *BMC Mol. Biol.* **10**, 104–115.

Hamilton, S. J., and Buhl, K. J. (1990). Safety assessment of selected inorganic elements to fry of chinook salmon (*Oncorhynchus tshawytscha*). *Ecotoxicol. Environ. Saf.* **20**, 307–324.

Hansen, J. A., Lipton, J., Welsh, P. G., Cacela, D., and Macconnell, B. (2004). Reduced growth of rainbow trout (*Oncorhynchus mykiss*) fed a live invertebrate diet pre-exposed to metal-contaminated sediments. *Environ. Toxicol. Chem.* **23**, 1902–1911.

Hartwell, S. I., Jin, J. H., Cherry, D. S., and Cairns, J., Jr. (1989). Toxicity versus avoidance response of golden shiner, *Notemigonus crysoleucas*, to five metals. *J. Fish Biol.* **35**, 447–456.

Hasegawa, H., Azizur Rahman, M., Kitahara, K., Itaya, Y., Maki, T., and Ueda, K. (2010). Seasonal changes of arsenic speciation in lake waters in relation to eutrophication. *Sci. Total Environ.* **408**, 1684–1690.

Hileman, B. (2007). Arsenic in chicken production: a common feed additive adds arsenic to human food and endangers water supplies. *Chem. Eng. News* **85**, 34–35.

Hiltibran, R. C. (1967). Effects of some herbicides on fertilized fish eggs and fry. *Trans. Am. Fish. Soc.* **96**, 414–416.

Howard, A. G., and Comber, S. D. W. (1989). The discovery of hidden arsenic species in coastal waters. *Appl. Organomet. Chem.* **3**, 509–514.

Howard, A. G., Apte, S. C., Comber, S. D. W., and Morris, R. J. (1988). Biogeochemical control of the summer distribution and speciation of arsenic in the Tamar estuary. *Estuar. Coast. Shelf Sci.* **27**, 427–443.

Howard, A. G., Hunt, L. E., and Salou, C. (1999). Evidence supporting the presence of dissolved dimethylarsinate in the marine environment. *Appl. Organomet. Chem.* **13**, 39–46.

Huang, Y.-K., Lin, K.-H, Chen, H.-W., Chang, C.-C., Liu, C.-W., Yang, M.-H., and Hsueh, Y.-M. (2003). Arsenic species contents at aquaculture farm and in farmed mouthbreeder (*Oreochromis mossambicus*) in blackfoot disease hyperendemic areas. *Food Chem. Toxicol.* **41**, 1491–1500.

Hughes, M. (2002). Arsenic toxicity and potential mechanisms of action. *Toxicol. Lett.* **133**, 1–16.

Hwang, P. P., and Tsai, Y. N. (1993). Effects of arsenic on osmoregulation in the tilapia *Oreochromis mossambicus* reared in seawater. *Mar. Biol.* **117**, 551–558.

Ikemoto, T., Phuc Cam, Tu, N., Okuda, N., Iwata, A., Omori, K., Tanabe, S., Cach Tuyen, B., and Takeuchi, I. (2008). Biomagnification of trace elements in the aquatic food web in the Mekong Delta, South Vietnam using stable carbon and nitrogen isotope analysis. *Arch. Environ. Contam. Toxicol.* **54**, 504–515.

Inglis, A., and Davis, E. L. (1972). *Effects of Water Hardness on the Toxicity of Several Organic and Inorganic Herbicides to Fish.* United States Department of the Interior, Washington, DC, Technical Paper No. 67, Bureau Sport Fisheries.

Johnson, C. R. (1978). Herbicide toxicities in the mosquito fish, *Gambusia affinis. Proc. R. Soc. Queensl* **89**, 25–27.

Jonnalagadda, S. B., and Nenzou, G. (1996a). Studies on arsenic rich mine dumps: I. Effect on the surface soil. *J. Environ. Sci. Health* **A31**, 1909–1915.

Jonnalagadda, S. B., and Nenzou, G. (1996b). Studies on arsenic rich mine dumps: III. Effect on the river water. *J. Environ. Sci. Health* A31, 2547–2555.

Jonnalagadda, S. B., and Nenzou, G. (1997). Studies on arsenic rich mine dumps: II. The heavy element uptake by vegetation. *J. Environ. Sci. Health* A32, 455–464.

Kirby, J., Maher, W., and Spooner, D. (2005). Arsenic occurrence and species in near-shore macroalgae-feeding marine animals. *Environ. Sci. Technol.* 39, 5999–6005.

Kotsanis, N., Iliopoulou-Georgudaki, J., and Kapata-Zoumbos, K. (2000). Changes in selected haematological parameters at early stages of the rainbow trout, *Oncorhynchus mykiss*, subjected to metal toxicants: arsenic, cadmium and mercury. *J. Appl. Ichthyol.* 16, 276–278.

Kraus, R. J., and Ganther, H. E. (1989). Synergistic toxicity between arsenic and methylated selenium compounds. *Biol. Trace Elem. Res.* 20, 105–113.

Krishnakumari, L., Varshney, P. K., Gajbhiye, S. N., Govindan, K., and Nair, V. R. (1983). Toxicity of some metals on the fish *Therapon jarbua* (Forsskal, 1775). *Indian J. Mar. Sci.* 12, 64–66.

Kuroiwa, T., Ohki, A., Naka, K., and Maeda, S. (1994). Biomethylation and biotransformation of arsenic in a freshwater food chain: green algae (*Chlorella vulgaris*) – shrimp (*Neocaridina denticulata*) – killifish (*Oryzias latipes*). *Appl. Organomet. Chem.* 8, 325–333.

Lenvik, K., Steinnes, E., and Pappas, A. C. (1978). Contents of some heavy metals in Norwegian rivers. *Nord. Hydrol.* 9, 197–206.

Lerda, D. E., and Prosperi, C. H. (1996). Water mutagenicity and toxicology in Rio Tercero, Cordoba, Argentina. *Water Res.* 30, 819–824.

Li, D., Lu, C., Wang, J., Hu, W., Cao, Z., Sun, D., Xi, H., and Ma, X. (2009). Developmental mechanisms of arsenite toxicity in zebrafish (*Danio rerio*) embryos. *Aquat. Toxicol.* 91, 229–237.

Liao, C. M., Tsai, J. W., Ling, M. P., Liang, H. M., Chou, Y. H., and Yang, P. T. (2004). Organ-specific toxicokinetics and dose–response of arsenic in Tilapia *Oreochromis mossambicus*. *Arch. Environ. Contam. Toxicol.* 47, 502–510.

Liao, C. M., Chiang, K.-C., and Tsai, J. W. (2006). Bioenergetics-based matrix population modeling enhances life-cycle toxicity assessment of tilapia *Oreochromis mossambicus* exposed to arsenic. *Environ. Toxicol.* 21, 154–165.

Lima, A. R., Curtis, C., Hammermeister, D. E., Markee, T. P., Northcott, C. E., and Brooke, L. T. (1984). Acute and chronic toxicities of arsenic(III) to fathead minnows, flagfish, daphnids, and an amphipod. *Arch. Environ. Contam. Toxicol.* 13, 595–601.

Lin, S., Cullen, W. R., and Thomas, D. J. (1999). Methylarsenicals and arsinothiols are potent inhibitors of mouse liver thioredoxin reductase. *Chem. Res. Toxicol.* 12, 924–930.

Liu, C.-W., Liang, C.-P., Huang, F. M., and Hsueh, Y.-M. (2006). Assessing the human health risks from exposure of arsenic through oyster (*Crassostrea gigas*) consumption in Taiwan. *Sci. Total Environ.* 361, 57–66.

Madhavan, N., and Subramanian, V. (2006). Environmental impact assessment including evolution of fluoride and arsenic contamination process in groundwater and remediation of contaminated groundwater system. In: *Sustainable Development and Management of Groundwater Resources* (M. Thangarajan, ed.), pp. 128–155. Capital Publishing Company, New Delhi.

Maeda, S., Ohki, A., Tokuda, T., and Ohmine, M. (1990). Transformation of arsenic compounds in a freshwater food chain. *Appl. Organomet. Chem.* 4, 251–254.

Maher, W. A. (1985). Arsenic in coastal waters of South Australia. *Water Res.* 19, 933–934.

Mandal, B. K., and Suzuki, K. T. (2002). Arsenic round the world: a review. *Talanta* 58, 201–235.

Mason, R. P., Laporte, J.-M., and Andres, S. (2000). Factors controlling the bioaccumulation of mercury, methylmercury, arsenic, selenium, and cadmium by freshwater invertebrates and fish. *Arch. Environ. Contam. Toxicol.* 38, 283–297.

Mattingly, C. J., Hampton, T. H., Brothers, K. M., Griffin, N. E., and Planchart, A. (2009). Perturbation of defense pathways by low-dose arsenic exposure in zebrafish embryos. *Environ. Health Perspect.* **117**, 981–987.

Mayer, F. L. J., and Ellersieck, M. R. (1986). *Manual of Acute Toxicity: Interpretation and Data Base for 410 Chemicals and 66 Species of Freshwater Animals.* United States Department of the Interior, Washington, DC, Resource Publication No. 160.

McGeachy, S. M., and Dixon, D. G. (1989). The impact of temperature on the acute toxicity of arsenate and arsenite to rainbow trout (*Salmo gairdneri*). *Ecotoxicol. Environ. Saf.* **17**, 86–93.

McGeachy, S. M., and Dixon, D. G. (1990). Effect of temperature on the chronic toxicity of arsenate to rainbow trout (*Oncorhynchus mykiss*). *Can. J. Fish. Aquat. Sci.* **47**, 2228–2234.

McGeer, J. C., Brix, K. V., Skeaff, J. M., DeForest, D. K., Brigham, S. I., Adams, W. J., and Green, A. (2003). Inverse relationship between bioconcentration factor and exposure concentration for metals: implications for hazard assessment of metals in the aquatic environment. *Environ. Toxicol. Chem.* **22**, 1017–1037.

McLaren, S. J., and Kim, N. D. (1995). Evidence for a seasonal fluctuation of arsenic in New Zealand's longest river and the effect of treatment on concentrations in drinking water. *Environ. Pollut.* **90**, 67–73.

Miller, D. S., Shaw, J. R., Stanton, C. R., Barnaby, R., Karlson, K. H., Hamilton, J. W., and Stanton, B. A. (2007). MRP2 and acquired tolerance to inorganic arsenic in the kidney of killifish (*Fundulus heteroclitus*). *Toxicol. Sci.* **97**, 103–110.

Moreau, M. F., Surico-Bennett, J., Vicario-Fisher, M., Crane, D., Gerads, R., Gersberg, R. M., and Hurlbert, S. H. (2007). Contaminants in tilapia (*Oreochromis mossambicus*) from the Salton Sea, California, in relation to human health, piscivorous birds and fish meal production. *Hydrobiologia* **576**, 127–165.

Navarro, M., Sanchez, M., Lopez, H., and Lopez, M. C. (1993). Arsenic contamination levels in waters, soils, and sludges in southeast Spain. *Bull. Environ. Contam. Toxicol.* **50**, 356–362.

Nayak, A. S., Lage, C. R., and Kim, C. H. (2007). Effects of low concentrations of arsenic on the innate immune system of the zebrafish (*Danio rerio*). *Toxicol. Sci.* **98**, 118–124.

Neff, J. M. (1997). Ecotoxicology of arsenic in the marine environment. *Environ. Toxicol. Chem.* **16**, 917–927.

Ng, J., Gomez-Caminero, A., Howe, P., Hughes, M., Kenyon, E., Lewis, D. R., Moore, M., Aitio, A., and Becking, G. (2001). *Arsenic and Arsenic Compounds* (2nd edn.). World Health Organization.

Oladimeji, A. A., Qadri, S. U., Tam, G. K. H., and deFreitas, A. S. W. (1979). Metabolism of inorganic arsenic to organoarsenicals in rainbow trout (*Salmo gairdneri*). *Ecotoxicol. Environ. Saf.* **3**, 394–400.

Oladimeji, A. A., Qadri, S. U., and deFreitas, A. S. W. (1984a). Long-term effects of arsenic accumulation in rainbow trout, *Salmo gairdneri*. *Bull. Environ. Contam. Toxicol.* **32**, 732–741.

Oladimeji, A. A., Quadri, S. U., and deFreitas, A. S. W. (1984b). Measuring the elimination of arsenic by the gills of rainbow trout (*Salmo gairdneri*) by using a two compartment respirometer. *Bull. Environ. Contam. Toxicol.* **32**, 661–668.

Palaniappan, PL. RM., and Vijayasundaram, V. (2008). Fourier transform infrared study of protein secondary structural changes in the muscle of *Labeo rohita* due to arsenic intoxication. *Food Chem. Toxicol.* **46**, 3534–3539.

Palaniappan, PL. RM., and Vijayasundaram, V. (2009). The effect of arsenic exposure and the efficacy of DMSA on the proteins and lipids of the gill tissues of *Labeo rohita*. *Food Chem. Toxicol.* **47**, 1752–1759.

Palawski, D., Hunn, J. B., and Dwyer, F. J. (1985). Sensitivity of young striped bass to organic and inorganic contaminants in fresh and saline waters. *Trans. Am. Fish. Soc.* **114**, 748–753.

Pandey, K., and Shukla, J. P. (1982). Deleterious effects of arsenic on the growth of fingerlings of a freshwater fish, *Colisa fasciatus* (Bl. & Sch.). *Acta Pharmacol. Toxicol.* **50**, 398–400.

Passino, D. R. M., and Kramer, J. M. (1980). Toxicity of arsenic and PCBs to fry of deepwater ciscoes (*Coregonus*). *Bull. Environ. Contam. Toxicol.* **24**, 527–534.

Pedlar, R. M., and Klaverkamp, J. F. (2002). Accumulation and distribution of dietary arsenic in lake whitefish (*Coregonus clupeaformis*). *Aquat. Toxicol.* **57**, 153–166.

Pedlar, R. M., Ptashynski, M. D., Evans, R., and Klaverkamp, J. F. (2002). Toxicological effects of dietary arsenic exposure in lake whitefish (*Coregonus clupeaformis*). *Aquat. Toxicol.* **57**, 167–189.

Persson, B. L., Petersson, J., Fristedt, U., Weinander, R., Berhe, A., and Pattison, J. (1999). Phosphate permeases of *Saccharomyces cerevisiae*: structure, function and regulation. *Biochim. Biophys. Acta* **1422**, 255–272.

Peshut, P. J., Morrison, R. J., and Brooks, B. A. (2008). Arsenic speciation in marine fish and shellfish from American Samoa. *Chemosphere* **71**, 484.

Peterson, M. L., and Carpenter, R. (1983). Biogeochemical processes affecting total arsenic and arsenic species distributions in an intermittently anoxic Fjord. *Mar. Chem.* **12**, 295–321.

Petrick, J. S., Jagadish, B., Mash, E. A., and Aposhian, H. V. (2001). Monomethylarsonous acid (MMAIII) and arsenite: LD50 in hamsters and *in vitro* inhibition of pyruvate dehydrogenase. *Chem. Res. Toxicol.* **14**, 651–656.

Pettine, M., Camusso, M., and Martinotti, W. (1992). Dissolved and particulate transport of arsenic and chromium in the Po River, Italy. *Sci. Total Environ.* **119**, 253–280.

Pizarro, I., Gómez, M., Cámara, C., and Palacios, M. A. (2003). Arsenic speciation in environmental and biological samples extraction and stability studies. *Anal. Chim. Acta* **495**, 85–98.

Quentin, K. E., and Winkler, H. A. (1974). Occurrence and determination of inorganic polluting agents. *Z. Bakteriol. Hyg.* **158**, 514–523.

Ramirez, O. A. B., and Garcia, F. P. (2005). Genotoxic damage in zebra fish (*Danio rerio*) by arsenic in waters from Zimapan, Hidalgo, Mexico. *Mutagenesis* **20**, 291–295.

Ramirez-Solis, A., Mukopadhyay, R., Rosen, B. P., and Stemmler, T. L. (2004). Experimental and theoretical characterization of arsenite in water: insights into the coordination environment of As-O. *Inorg. Chem.* **43**, 2954–2959.

Rankin, M. G., and Dixon, D. G. (1994). Acute and chronic toxicity of waterborne arsenite to rainbow trout (*Oncorhynchus mykiss*). *Can. J. Fish. Aquat. Sci.* **51**, 372–380.

Rattanachongkiat, S., Millward, G. E., and Foulkes, M. E. (2004). Determination of arsenic species in fish, crustacean and sediment samples from Thailand using high performance liquid chromatography (HPLC) coupled with inductively coupled plasma mass spectrometry (ICP-MS). *J. Environ. Monit.* **6**, 254–261.

Reuther, R. (1992). Geochemical mobility of arsenic in a flowthrough water-sediment system. *Environ. Technol.* **13**, 813–823.

RIVM (1997). *Maximum Permissible Concentrations and Negligible Concentrations for Metals Taking Background Concentrations into Account.* Report No. 601501 001. Bilthoven: National Institute of Public Health and the Environment.

Robinson, B., Outred, H., Brooks, R., and Kirkman, J. (1995). The distribution and fate of arsenic in the Waikato River System, North Island, New Zealand. *Chem. Spec. Bioavail.* **7**, 89–96.

Romero, L., Alonso, H., Campano, P., Fanfani, L., Cidu, R., Dadea, C., Keegan, T., Thornton, I., and Farago, M. (2003). Arsenic enrichment in waters and sediments of the Rio Loa (Second Region, Chile). *Appl. Geochem.* **18**, 1399–1416.

Rosen, B. P., and Liu, Z. (2009). Transport pathways for arsenic and selenium: a minireview. *Environ. Int.* **35**, 512–515.

Rosenberg, H., Gerdes, R. G., and Chegwidden, K. (1977). Two systems for the uptake of phosphate in *Escherichia coli*. *J. Bacteriol.* **131**, 505–511.

Roy, S., and Bhattacharya, S. (2006). Arsenic-induced histopathology and synthesis of stress proteins in liver and kidney of *Channa punctatus*. *Ecotoxicol. Environ. Saf.* **65**, 218–229.

Roy, S., Chattoraj, A., and Bhattacharya, S. (2006). Arsenic-induced changes in optic tectal histoarchitecture and acetylcholinesterase–acetylcholine profile in *Channa punctatus*: amelioration by selenium. *Comp. Biochem. Physiol. C* **144**, 16–24.

Sancha, A. M. (1999). Full-scale application of coagulation processes for arsenic removal in Chile: a successful case study. In: *Arsenic Exposure and Health Effects* (W. R. Chappell, C. O. Abernathy and R. L. Calderon, eds), pp. 373–378. Elsevier, Amsterdam.

Sandhu, S. S. (1977). Study on the post-mortem identification of pollutants in the fish killed by water pollution: detection of arsenic. *Bull. Environ. Contam. Toxicol.* **17**, 373–378.

Santosa, S. J., Wada, S., and Tanaka, S. (1994). Distribution and cycle of arsenic compounds in the ocean. *Appl. Organomet. Chem.* **8**, 273–283.

Schaeffer, R., Francesconi, K. A., Kienzl, N., Soeroes, C., Fodor, P., Varadi, L., Raml, R., Goessler, W., and Kuehnelt, D. (2006). Arsenic speciation in freshwater organisms from the river Danube in Hungary. *Talanta* **69**, 856–865.

Schlenk, D., Wolford, L., Chelius, M., Stevens, J., and Chan, K. M. (1997). Effect of arsenite, arsenate, and the herbicide monosodium methyl arsonate (MSMA) on hepatic metallothionein expression and lipid peroxidation in channel catfish. *Comp. Biochem. Physiol. C* **118**, 177–183.

Schreiber, R., Pavenstadt, H., Gregerc, R., and Kunzelmanna, K. (2000). Aquaporin 3 cloned from *Xenopus laevis* is regulated by the cystic fibrosis transmembrane conductance regulator. *FEBS Lett.* **475**, 291–295.

Seok, S. H., Baek, M. W., Lee, H. Y., Kim, D. J., Na, Y. R., Noh, K. J., Park, S. H., Lee, H. K., Lee, B. H., Ryu, D. Y., and Park, J. H. (2007). Quantitative GFP fluorescence as an indicator of arsenite developmental toxicity in mosaic heat shock protein 70 transgenic zebrafish. *Toxicol. Appl. Pharmacol.* **225**, 154–161.

Seyler, P., and Martin, J.-M. (1989). Biogeochemical processes affecting arsenic species distribution in a permanently stratified lake. *Environ. Sci. Technol.* **23**, 1258–1263.

Seyler, P., and Martin, J.-M. (1990). Distribution of arsenite and total dissolved arsenic in major French estuaries: dependence on biogeochemical processes and anthropogenic inputs. *Mar. Chem.* **29**, 277–294.

Seyler, P., and Martin, J.-M. (1991). Arsenic and selenium in a pristine river–estuarine system: the Krka, Yugoslavia. *Mar. Chem.* **34**, 137–151.

Shaw, J. R., Gabor, K., Hand, E., Lankowski, A., Durant, L., Thibodeau, R., Stanton, C. R., Barnaby, R., Coutermarsh, B., Karlson, K. H., Sato, J. D., Hamilton, J. W., and Stanton, B. A. (2007a). Role of glucocorticoid receptor in acclimation of killifish (*Fundulus heteroclitus*) to seawater and effects of arsenic. *Am. J. Physiol. Regul. Integr. Comp. Physiol* **292**, R1052–R1060.

Shaw, J. R., Jackson, B., Gabor, K., Stanton, S., Hamilton, J. W., and Stanton, B. A. (2007b). The influence of exposure history on arsenic accumulation and toxicity in the killifish, *Fundulus heteroclitus*. *Environ. Toxicol. Chem.* **26**, 2704–2709.

Shukla, J. P., Shukla, K. N., and Dwivedi, U. N. (1987). Survivality and impaired growth in arsenic treated fingerlings of *Channa punctatus*, a fresh water murrel. *Acta Hydrochim. Hydrobiol.* **15**, 307–311.

Slejkovec, Z., Bajc, Z., and Doganoc, D. Z. (2004). Arsenic speciation in freshwater fish. *Talanta* **62**, 931–936.

Smedley, P. L, and Kinniburgh, D. G. (2002). A review of the source, behavior and distribution of arsenic in natural waters. *Appl. Geochem.* **17**, 517–568.

Smedley, P. L., Edmunds, W. M., and Pelig-Ba, K. B. (1996). Mobility of arsenic in groundwater in the Obuasi area of Ghana. In: *Environmental Geochemistry and Health* (J.D. Appleton, R. Fuge and G.J.H. McCall, eds), pp. 163–181. Geological Society, London. Special Publication 113.

Soeroes, C., Goessler, W., Francesconi, K. A., Kienzl, N., Schaeffer, R., Fodor, P., and Kuehnelt, D. (2005). Arsenic speciation in farmed Hungarian freshwater fish. *J. Agric. Food Chem.* **53**, 9238–9243.

Sonderegger, J. L., and Ohguchi, T. (1988). Irrigation related arsenic contamination of a thin, alluvial aquifer, Madison River Valley, Montana, USA. *Environ. Geol. Water Sci.* **11**, 153–161.

Sorensen, E., Ramirez-Mitchel, R., Harlan, C. W., and Bell, J. S. (1980). Cytological changes in the fish liver following chronic, environmental arsenic exposure. *Bull. Environ. Contam. Toxicol.* **25**, 93–99.

Soto, E. G., Rodriguez, E. A., Rodriguez, D. P., and Fernandez, E. F. (1996). Inorganic and organic arsenic speciation in seawaters by IEC-HG-AAS. *Anal. Lett.* **29**, 2701–2712.

Spehar, R. L., and Fiandt, J. T. (1986). Acute and chronic effects of water quality criteria-based metal mixtures on three aquatic species. *Environ. Toxicol. Chem.* **5**, 917–931.

Speyer, K. R. (1974). *Some effects of combined chronic arsenic and cyanide poisoning on the physiology of rainbow trout.* MS Thesis, Concordia University, Montreal.

Spotila, J. R., and Paladino, F. V. (1979). Toxicity of arsenic to developing muskellunge fry (*Esox masquinongy*). *Comp. Biochem. Physiol. C* **62**, 67–69.

Stephan, C. E., Mount, D. I., Hansen, D. J., Gentile, J. H., Chapman, G. A., and Brungs, W. A. (1985). *Guidelines for Deriving Numerical National Water Quality Criteria for the Protection of Aquatic Organisms and their Uses.* National Technical Information Service, Springfield, VA, PB85-227049.

Storelli, M. M., Giacominelli-Stuffler, R., Storelli, A., and Marcotrigiano, G. O. (2005). Accumulation of mercury, cadmium, lead and arsenic in swordfish and bluefin tuna from the Mediterranean Sea: a comparative study. *Mar. Pollut. Bull.* **50**, 1004–1007.

Styblo, M., Serves, S. V., Cullen, W. R., and Thomas, D. J. (1997). Comparative inhibition of yeast glutathione reductase by arsenicals and arsenothiols. *Chem. Res. Toxicol.* **10**, 27–33.

Suhendrayatna, O. A., Nakajima, T., and Maeda, S. (2002a). Studies on the accumulation and transformation of arsenic in fresh organisms: II. Accumulation and transformation of arsenic compounds by *Tilapia mossambica*. *Chemosphere* **46**, 325–331.

Suhendrayatna, A. O., Nakajima, T., and Maeda, S. (2002b). Studies on the accumulation and transformation for arsenic in freshwater organisms I. Accumulation, transformation and toxicity of arsenic compounds to the Japanese medaka, *Oryzias latipes*. *Chemosphere* **46**, 319–324.

Szinicz, L., and Forth, W. (1988). Effects of As_2O_3 on gluconeogenesis. *Arch. Toxicol.* **61**, 444–449.

Tanaka, S., and Santosa, S. J. (1995). The concentration distribution and chemical form of arsenic compounds in sea water. In: *Biogeochemical Processes and Ocean Flux in Western Pacific* (H. Sakai and Y. Nozaki, eds), pp. 159–170. Terra Scientific Publishing, Tokyo.

Taylor, D., Maddock, B. G., and Mance, G. (1985). The acute toxicity of nine "grey list" metals (arsenic, boron, chromium, copper, lead, nickel, tin, vanadium and zinc) to two marine fish species. *Aquat. Toxicol.* **7**, 135–144.

Tetra Tech (1996). *Assessing Human Health Risks from Chemically Contaminated Fish in the Lower Columbia River.* Columbia River Bi-State Water Quality Program, Portland, OR.

Tsai, J. W., and Liao, C. M. (2006). A dose-based modeling approach for accumulation and toxicity of arsenic in tilapia *Oreochromis mossambicus. Environ. Toxicol.* **21**, 8–21.

USEPA (1985). *Ambient Water Quality Criteria for Arsenic – 1984.* Office of Water, United States Environmental Protection Agency, Washington, DC, EPA 440/5-84-033.

USEPA (2000). *Methodology for Deriving Ambient Water Quality Criteria for the Protection of Human Health.* Office of Water, United States Environmental Protection Agency, Washington, DC, EPA-822-B-00-004.

USEPA (2009). *National Recommended Water Quality Criteria.* Office of Water, United States Environmental Protection Agency, Washington, DC.

Vahter, M. (1994). Species differences in the metabolism of arsenic. In: *Arsenic Exposure and Health. Special Issue of Environmental Geochemistry and Health, Science and Technology Letters* (W. R. Chappell, C. O. Abernathy and C. R. Cothern, eds), Vol. 16, pp. 171–179. Northwood, Middlesex.

Ventura-Lima, J., Fattorini, D., Regoli, F., and Monserrat, J. M. (2009a). Effects of different inorganic arsenic species in *Cyprinus carpio* (Cyprinidae) tissues after short-time exposure: Bioaccumulation, biotransformation and biological responses. *Environ. Pollut.* **157**, 3479–3484.

Ventura-Lima, J., De Castro, M. R., Acosta, D., Fattorini, D., Regoli, F., De Carvalho, L. M., Bohrer, D., Geracitano, L. A., Barros, D. M., Marins, L. F. F., Da Silva, R. S., Bonan, C. D., Bogo, M. R., and Monserrat, J. M. (2009b). Effects of arsenic (As) exposure on the antioxidant status of gills of the zebrafish *Danio rerio* (Cyprinidae). *Comp. Biochem. Physiol. C* **149**, 538–543.

Wagemann, R., Snow, N. B., Rosenberg, D. M., and Lutz, A. (1978). Arsenic in sediments, water, and aquatic biota from lakes in the vicinity of Yellowknife, Northwest Territories. Canada. *Arch. Environ. Contam. Toxicol.* **7**, 169–191.

Wang, Y. H., Chen, Y. H., Wu, T. N., Lin, Y. J., and Tsai, H. J. (2006). A keratin 18 transgenic zebrafish Tg(k18(2.9):RFP) treated with inorganic arsenite reveals visible overproliferation of epithelial cells. *Toxicol. Lett.* **163**, 191–197.

Waslenchuk, D. G. (1978). The budget and geochemistry of arsenic in a continental shelf environment. *Mar. Chem.* **7**, 39–52.

Watt, C. and Le, X. C. (2003). Arsenic speciation in natural waters. In *Biogeochemistry of Environmentally Important Trace Elements.* ACS Symposium Series, Vol. 835, pp. 11–32. Washington, DC: American Chemical Society.

Weir, P. A., and Hine, C. H. (1970). Effects of various metals on behavior of conditioned goldfish. *Arch. Environ. Health* **20**, 45–51.

Williams, L., Schoof, R. A., Yeager, J. W., and Goodrich-Mahoney, J. W. (2006). Arsenic bioaccumulation in freshwater fishes. *Hum. Ecol. Risk Assess.* **12**, 904–923.

Williams, M., Fordyce, F., Paijitprapapon, A., and Charoenchaisri, P. (1996). Arsenic contamination in surface drainage and groundwater in part of the southeast Asian tin belt, Nakhon Si Thammarat Province, southern Thailand. *Environ. Geol.* **27**, 16–33.

Yamaguchi, S., Miura, C., Ito, A., Agusa, T., Iwata, H., Tanabe, S., Tuyen, B. C., and Miura, T. (2007). Effects of lead, molybdenum, rubidium, arsenic and organochlorines on spermatogenesis in fish: monitoring at Mekong Delta area and *in vitro* experiment. *Aquat. Toxicol.* **83**, 43–51.

Yang, L., Kemadjou, J. R., Zinsmeister, C., Bauer, M., Legradi, J., Muller, F., Pankratz, M., Jakel, J., and Strahle, U. (2007). Transcriptional profiling reveals barcode-like toxicogenomic responses in the zebrafish embryo. *Genome Biol.* **8**, R227.

Yusof, A. M., Ikhsan, Z. B., and Wood, A. K. H. (1994). The speciation of arsenic in seawater and marine species. *J. Radioanal. Nucl. Chem. Articles* **179**, 277–283.

Zheng, J., and Hintelmann, H. (2004). Hyphenation of high performance liquid chromatography with sector field inductively coupled plasma mass spectrometry for the determination of ultra-trace level anionic and cationic arsenic compounds in freshwater fish. *J. Anal. Atom. Spectrom.* **19**, 191–195.

7

STRONTIUM

M. JASIM CHOWDHURY
RONNY BLUST

Homeostasis and Toxicology of Non-Essential Metals: Volume 31B
FISH PHYSIOLOGY

Strontium (Sr) is one of the alkaline earth metals, with properties resembling calcium (Ca). Strontium is not classified as essential for human or other organisms, and environmental Sr is considered hazardous to organisms because of radiological hazards associated with its radionuclides. In this review, physiological and toxicological information about stable and radioactive Sr relevant to freshwater and marine fish has been summarized. After a brief description of chemistry, bioaccumulation, and toxic effects, mechanisms related to uptake, internal handling, and elimination of Sr in fish are reviewed. A comprehensive survey of old and recent literature makes it clear that Sr preferentially accumulates in the bony parts (skeleton, scales, otoliths) of fish, and its accumulation largely depends on environmental factors. Much less is known about the physiological and toxicological mechanisms of Sr in fish. Where mechanistic information on fish for an area is absent, comments on the pertinent area have been made, based on mammalian studies or those focused on Ca physiology in fish. Finally, several recommendations have been included, identifying specific areas that warrant further research.

1. CHEMICAL SPECIATION IN FRESHWATER AND SEAWATER

1.1. Partitioning and Mobility in Aquatic Environments

Strontium (Sr) is one of the alkaline earth metals, belonging to group IIB in the periodic table. It is mainly present in the dissolved pool of aquatic systems, and its concentrations vary considerably in fresh and marine environments (Table 7.1). Environmental concentrations of radiostrontium are several orders of magnitude lower than those of stable Sr, and in nature both types of Sr tend to attain equilibrium with the aquatic compartments (Smith and Beresford, 2005). Owing to its radioecological importance, ^{90}Sr has been extensively studied to understand its partitioning and mobility in rivers and lakes (Coughtrey and Thorne, 1983; Sanada et al., 2002; Xu and Marcantonio, 2004; Monte et al., 2005, 2008; IAEA, 2009; Outola et al., 2009), coastal waters (Håkanson, 2005), and marine water (Brown, 2000;

Table 7.1
Typical range of strontium concentrations in water and fish

	Units	Concentrations	References
Freshwater	μmol L^{-1}	0.02–2.6	Coughtrey and Thorne (1983); Smith and Beresford (2005)
Coastal water	μmol L^{-1}	20–86	Coughtrey and Thorne (1983)
Seawater	μmol L^{-1}	92–114	Coughtrey and Thorne (1983); Lide (1998)
Freshwater fish (muscle)	μmol kg^{-1} ww	9–42	Coughtrey and Thorne (1983); Chowdhury (2001)
Freshwater fish (bone)	μmol kg^{-1} ww	1863–3653	Coughtrey and Thorne (1983); Chowdhury (2001)
Marine fish (muscle)	μmol kg^{-1} ww	11–57	Coughtrey and Thorne (1983)
Marine fish (bone)	μmol kg^{-1} ww	765–4451	Coughtrey and Thorne (1983)

ww: wet weight.

IAEA, 2004; Benkdad et al., 2008). The major conclusion is that Sr binds to particles weakly and it is highly mobile between sediment to water, within sediment, from catchments to water systems, and via rivers to estuaries. Strontium is found mainly associated with labile geochemical phases (e.g. acid soluble, reducible, and oxidizable phases) in the sediment. The best estimates of the distribution coefficient (K_d) for Sr [as recommended by the International Atomic Energy Agency (IAEA)], expressed as the concentration ratio between the particulate phase and the dissolved phase under equilibrium conditions (L kg^{-1}), are 180, 8, and 200 for freshwater, coastal, and marine ecosystems, respectively, suggesting that Sr binding to particles is considerably less strong relative to Zn, Cd, and Pb (IAEA, 2004, 2009). Over time the sediment can behave as a reservoir slowly releasing radiostrontium into aqueous phases, where it may then be available for uptake by biota (Brown, 2000; Sanada et al., 2002). The ecological half-lives for ^{90}Sr can be longer than 10 years in freshwater bodies (Outola et al., 2009).

1.2. Speciation in Freshwater

In typical freshwater, Sr is predominantly present as free metal ions ($\sim$95%) and some complexes of sulfate and carbonate (5%). Within naturally relevant ranges of hardness (10–300 mg L^{-1}) and pH (6–8.2), the free Sr ion concentrations do not show more than 20% differences. The complexation capacity of Sr with natural organic matter is very similar to Ca, but much lower than other metals such as Al, Cu, and Pb (Nordén and

Dabek-Zlotorzynska, 1996). The complexation of Sr with fulvic acid and humic acid can only be important in soft waters (Fig. 7.1). Size fractionation techniques reveal predominant complexation of Sr (85–88%) with low molecular mass (<1–10 kDa) humic substances in natural waters (Matsunaga et al., 2004), suggesting its high potential for bioavailability.

Strontium interacts with synthetic organic ligands in aquatic systems, but less strongly than Ca (Chowdhury and Blust, 2002). Relative to Ca, the binding affinity of ethylenediaminetetraacetic acid (EDTA) and nitrilotriacetic acid (NTA) for Sr is 30–140-fold weaker, leading to disproportionately greater decrease in the free Ca ion activity in a solution. In consequence, the complexation with organic ligands can affect Sr bioavailability to fish by affecting both Sr and Ca speciation in the real-world situation where Ca is also always present (see Section 8.1.3).

1.3. Speciation in Seawater

The concentration of Sr in full-strength seawater is approximately 90 μM, and it largely remains constant across the seas and oceans, with local variations of only 2–3% (Bernat et al., 1972; De Villiers, 1999). The concentration covaries with salinity in estuarine and coastal waters (Table 7.1). Most of the Sr in the seawater is present as the free hydrated ion, only a small portion (~10%) is present as a soluble ion pair with sulfate ($SrSO_4$), and much smaller portions are paired with bicarbonate ($SrHCO_3^+$), carbonate ($SrCO_3$), fluoride (SrF^+), borate [$SrB(OH)_4^+$], and hydroxide ($SrOH^+$) (Millero, 1996). Some measurements have shown that the Sr concentration increases slightly with increasing depth owing to the action of acantharians that deposit strontium sulfate in their internal skeletons. As these organisms die, settle to the bottom, and decompose, the dissolution of strontium sulfate leads to the increase of Sr in deeper waters (De Villiers, 1999).

2. SOURCES AND ECONOMIC IMPORTANCE OF STRONTIUM

2.1. Natural Strontium

Strontium is widely found in nature as celestite ($SrSO_4$) and strontianite ($SrCO_3$), and it comprises about 0.037% of the Earth's crust (Lide, 1998). Natural Sr is a mixture of four stable isotopes (^{84}Sr, ^{86}Sr, ^{87}Sr, and ^{88}Sr), while ^{88}Sr is the most prevalent form, comprising about 83% of natural Sr (Lide, 1998). China, Mexico, Spain, Germany, and Turkey are the major producers of Sr, and the global production of celestite and strontianite in

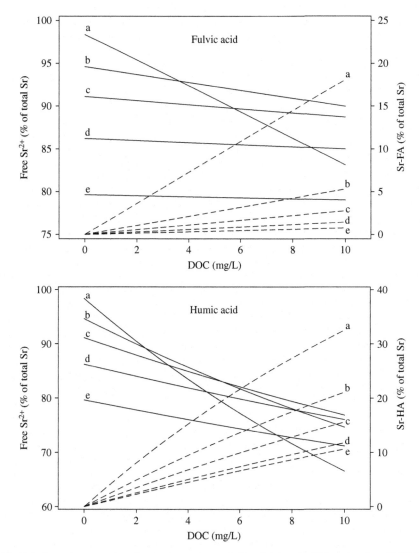

Fig. 7.1. Strontium complexation in the presence of either fulvic acid (FA) or humic acid (HA) assuming a 50% dissolved organic carbon (DOC) content in US EPA very soft to very hard waters (USEPA, 1993) and in the presence of 1 μM total Sr (temperature: $25 \pm 1°C$). Solid lines: free Sr^{2+}; broken lines: Sr–FA or Sr–HA complexes. Model calculations were performed using the WHAM VI geochemical model. a: Very soft water (hardness = 11 mg $CaCO_3$ L^{-1}, pH = 6.6); b: soft water (hardness = 42 mg L^{-1}, pH = 7.4); c: moderately hard water (hardness = 85 mg L^{-1}, pH = 7.6); d: hard water (hardness = 170 mg L^{-1}, pH = 7.8); e: very hard water (hardness = 339 mg L^{-1}, pH = 8.2).

2007 was approximately 0.6 million tonnes (USGS, 2008). Strontium compounds are used in making ceramics and glass products, pyrotechnics, paint pigments, fluorescent lights, and medicines.

2.2. Anthropogenic Strontium

The most important anthropogenic sources are nuclear industries such as nuclear power plants, nuclear fuel processing plants, nuclear waste storage facilities, and nuclear armament production facilities. ^{89}Sr and ^{90}Sr (half-lives 50.5 days and 28.1 years, respectively), produced from nuclear fission of uranium and plutonium, are by-products of these industries and can be released to the environment during controlled discharge or accidents; the Chernobyl accident in 1986 is one of the worst case examples (see Section 3). In the 1950s and 1960s, the main source was the atmospheric fallout from nuclear detonations, with a total discharge of approximately 1.0×10^{18} Bq (IPPNW, 1991). A detailed review on anthropogenic sources of radio-nuclides has been published recently by Hu et al. (2010). ^{89}Sr has medical uses, particularly in radiation therapy for bone cancer treatment. Since ^{90}Sr is a long-lived high-energy β-emitter (0.546 MeV), it is used in systems for nuclear auxiliary power (SNAP) devices for space vehicles, remote weather stations, and navigational buoys as a power source (Lide, 1998). Discharges from non-nuclear industries, such as metal refineries, wood and pulp industries, and ceramic industries are also potential sources of Sr contamination to aquatic systems.

3. ENVIRONMENTAL SITUATIONS OF CONCERN

Stable Sr in the environment is not generally considered a concern to aquatic organisms. The only known case is the Kola region of Russia, where many lakes are heavily contaminated with Sr from nearby metal mines, and the fish living in the lakes are characterized by high concentrations of tissue Sr in association with skeletal abnormalities (Moiseenko and Kudryavtseva, 2001). However, many river systems and seas are environmental repositories for radiostrontium and other radionuclides. The typical examples that have been studied for radioecological impact are the Techa River and Ural lakes in Russia (Kryshev et al., 1997, 1998; Kryshev, 2003), Pripyat River in Ukraine (Kryshev, 1995; Smith et al., 2005b), Savannah River in the USA (Carlton et al., 1999; Burger et al., 2002), and Kara, Barents, and Irish seas (Kryshev and Sazykina, 1995; Brown, 2000; Gleizon and McDonald, 2010). All of these water bodies have received various amounts of intentional or

Table 7.2
^{90}Sr bioconcentration factors (BCFs) in freshwater fish after the Chernobyl accident

Water body	Sampling year	Fish	BCF (L kg^{-1} ww)	Calcium in water (μmol L^{-1})
Cooling pond	1986	Fish (general)	100	1175
Kiev Reservoir	1987–1988	Pike and perch	99	1300
		Bream	46	
Lake Kozhanovskoe	1993–1994	Crucian carp	140	1010
		Goldfish	173	
		Perch	365	
Lake Perstok	2003	Roach and rudd	452	660
		Perch	239	
Glubokoye Lake	2003	Roach	190	738

From Smith et al. (2005b).
BCF = whole-body ^{90}Sr activity in fish (Bq kg^{-1})/^{90}Sr activity in water (Bq L^{-1}).

accidental discharges from nuclear facilities over different times since the 1940s. In addition to ^{90}Sr, these water systems have been contaminated by many other radionuclides such as ^{99}Tc, ^{129}I, ^{137}Cs, ^{237}Np, ^{241}Am, and ^{239}Pu (Hu et al., 2010). As a result, these studies were not useful for ascertaining the radiotoxicity of Sr alone to fish or other aquatic organisms.

The worst case example of a nuclear contamination event was the Chernobyl accident in 1986. The Chernobyl explosion released approximately 1.15×10^{17} Bq of ^{89}Sr and 1.0×10^{16} Bq of ^{90}Sr to the environment surrounding the reactor and over a broad geographical area in Europe and Scandinavia (Smith and Beresford, 2005). The cooling pond in the vicinity of the reactor was one of the most contaminated sites, and ^{90}Sr activities in this pond in 1986 were 0.02 ± 0.01 kBq L^{-1} in the water, 60 ± 25 kBq kg^{-1} in the sediment, and 2.0 ± 1.2 kBq kg^{-1} in fish (Kryshev, 1995). Estimates of ^{90}Sr bioconcentration factors (BCFs) for fish in some lakes in Ukraine, Russia, and Belarus are given in Table 7.2.

4. ACUTE AND CHRONIC AMBIENT WATER QUALITY CRITERIA IN VARIOUS JURISDICTIONS IN FRESHWATER AND SEAWATER

To the authors' knowledge, no country has introduced any national water quality criteria for Sr to protect freshwater or seawater organisms.

This is likely linked to the fact that environmental concentrations of stable Sr are not considered hazardous to aquatic life. The known criteria introduced by two states in the US (Michigan and Ohio) and later adopted by the province of Quebec in Canada are 40,000 μg L^{-1} (acute) and 21,000 (chronic) for the protection of freshwater life (MDEQ, 2008; Ohio EPA, 2009). The US Environmental Protection Agency (EPA) has set a limit of 4000 μg L^{-1} stable Sr for public drinking water (ATSDR, 2004). These standards are much higher than the levels reported for freshwater bodies (see Table 7.1).

Regulatory limits of radiostrontium in drinking water also commonly exist (Health Canada, 2000), but those are not relevant in this review. In one study, the maximum safe concentrations of ^{89}Sr and ^{90}Sr in surface waters for overall use are suggested to be 120 and 10.9 Bq L^{-1}, respectively (Nosov et al., 1999). As aquatic organisms remain under continuous exposure, the safe levels for surface waters should logically be much lower for aquatic life protection. The criteria value for ^{90}Sr is 10 pCi L^{-1} (0.37 Bq L^{-1}) in surface waters of the State of New Hampshire, USA (NHCAR, 2008). The aquatic life criteria for other jurisdictions are not known.

5. MECHANISMS OF TOXICITY

5.1. Acute Effects of Stable Strontium

In general, elemental Sr is not considered toxic to organisms, but limited research has been done in this area with fish or other aquatic animals. The US EPA Ecotoxicity Database (http://cfpub.epa.gov/ecotox/) has several entries for freshwater fish, but the studies, conducted decades ago, are not conclusive of the nature or extent of the toxicity because of the information gaps. There is no available report for 96 h median lethal concentration (96 h LC50) of Sr for freshwater fish, but the 28 day LC50 for rainbow trout (*Oncorhynchus mykiss*) and 7 day LC50 for goldfish (*Carassius auratus*) are 200 μg L^{-1} and 8580 μg L^{-1}, respectively, at a water hardness of 104 mg L^{-1} as CaCO$_3$ (Birge, 1978). Studies with striped bass (*Morone saxatilis*) (Dwyer et al., 1992), two seawater minnows (*Menidia beryllina* and *Cyprinodon variegatus*) (Pillard et al., 2000), and plaice (*Pleuronectes platessa*) (Holmes-Farley, 2003) indicate that sensitivity of seawater fish to Sr is even less, but depends on species and salinity, probably because of high ionic strength and particularly high Ca concentrations. The 48 h or 96 h LC50 values reported in these studies are more than 92.8 mg L^{-1}, which is approximately 11-fold greater than the Sr level in full-strength seawater. Hence, acute effects occur at high Sr levels that are environmentally unimportant. Nevertheless, the sensitivity of rainbow trout should be further verified.

The most plausible mechanism of acute toxicity is hypocalcemia, as Sr is a known Ca analogue and waterborne Sr significantly inhibits Ca uptake in whole fish and fish blood (Brungs, 1965; Suzuki et al., 1972; Chowdhury et al., 2000). However, the direct relationship between acute effects of elevated Sr in water or diet and plasma hypocalcemia in fish has not been established. High doses of Sr induce hypocalcemia in mammals as a result of increased renal excretion of Ca (Nielsen, 2004).

5.2. Acute Effects of Radioactive Strontium

It is generally assumed that no level of radiation above the natural background is completely free of the risk of random effects resulting from damage to genetic materials, particularly DNA in living cells. Lethal effects due to excessive cell death from high doses above a threshold level of radiation can occur in fish, but such effects have been less commonly studied than the long-term random effects through genetic consequences at low dose rates. Fish are generally more sensitive to radiation during early life stages (Blaylock and Griffith, 1971), which may be attributed to delicate forms of cells or tissues with high cell division rates and differentiation, and lack of cellular mechanism to repair the damage caused by ionizing radiation (Kryshev et al., 2008). Most of the earlier studies characterizing the lethal effects of radiation in freshwater and marine fish were reviewed by Polikarpov (1966) and Templeton et al. (1971). The LD50 (lethal dose resulting in 50% mortality) for adult rainbow trout ranged from 3 to 30 Gray (Gy), and for embryos, the LD50 was as low as 0.16 Gy. Carp eggs seem more resistant than eggs of salmonids, the LD50 determined from direct exposure to ^{90}Sr being greater (5 Gy) (Blaylock and Griffith, 1971). For adult marine species, comparable LD50 values (10–55 Gy) have been reported (Templeton et al., 1971), whereas the early life stages can be more sensitive to ^{90}Sr (Polikarpov, 1966). Acute mortality of fish due to high doses of radiation (~ 2 Gy day^{-1}) is not common but can occur in natural conditions after the contamination of aquatic systems with radionuclides (Kryshev and Sazykina, 1998), although radiostrontium contributes only partially to such doses.

5.3. Chronic Effects of Stable Strontium

Information on chronic effects of stable Sr is limited. In a 28 day study, Birge et al. (1981) observed the effects of waterborne Sr on the early life stages of rainbow trout during embryonic development through 4 day posthatch. The 10% effects level based on teratogenic responses and larval mortality was 49 μg L^{-1}. A significant correlation between waterborne Sr and decreased egg hatchability, but not with larval growth and mortality, has recently been demonstrated by Pyle et al. (2002) for fathead minnow (*Pimephales promelas*). However, this study is not conclusive for reproductive effects, as the early life stages were exposed to contaminated lake waters, but not to Sr alone. As Ca demand is high during early development of fish, reproductive effects of elevated Sr due to interference with Ca availability may not be ruled out.

5.4. Chronic Effects of Radioactive Strontium

Many laboratory studies were conducted in 1950–1970 to understand the chronic effects of waterborne ^{90}Sr and its sister isotope ^{90}Y on freshwater and marine species, and the results were documented in reviews by Polikarpov (1966) and Templeton et al. (1971). The effects from studies with eggs from various fish (rainbow trout, mullet, mackerel, and anchovies) include reduced and abnormal mitotic activity, increased chromosomal aberrations, reduced hatching, increased larval deformity, and early mortality. However, some of the documented results were conflicting, as for example, no significant effects were observed on early life stages of plaice (*P. platessa*) in seawater containing as high as 3.7 MBq L^{-1} of ^{90}Sr. Feeding studies, in which ^{90}Sr-containing capsules were force-fed to rainbow trout (0.185–18.5 kBq g^{-1} fish day^{-1}), have shown several effects including mortality, depression of growth, damage to the gut, and more importantly leukopenia (Templeton et al., 1971). Based on studies with laboratory-reared and natural populations of mosquitofish (*G. affinis*), it has been concluded that irradiated fish had higher fecundity, but produced a low number of healthy embryos and larvae, as the progeny carry an increased frequency of lethal genes (Blaylock and Frank, 1980). An increased level of DNA strand breaks has been observed in this irradiated mosquitofish population (Theodorakis et al., 1997).

More recently, Sazykina and Kryshev (2003) collated the effects of chronic radiation in freshwater fish, based on some landmark studies in the laboratory with ^{90}Sr and long-term observations in natural water bodies. Three main endpoints related to chronic effects at the organismal level have been categorized: morbidity resulting from damage to various physiological and metabolic systems, reproductive effects, and mortality or shortening of lifespan. The response mechanisms under these endpoints are presented in Table 7.3. Based on these endpoints, Kryshev et al. (2008) have developed a model, and predicted that a chronic radiation exposure below 10 mGy day^{-1} is not significant for a population-level effect on fish. Indeed, field

Table 7.3

Effects of chronic radiation from ^{90}Sr in fish

Endpoint	Species	Exposure[a]	Response mechanism
Morbidity	Carp (Cyprimus carpio)	Lab; 1.85–37 kBq L^{-1} in water; doses: 0.0002–5.3 Gy in kidney; 15–270 days	Dose-dependent negative changes in blood composition (decreased number of white blood cells, effects on the proportion of lymphocytes, granulocytes, and monocytes, no effects on red blood cells); weakening and delay in immune response to bacterial infections, weakening of the resistance to parasite infestation
	Carp (Cyprimus carpio)	Lab; 1–2 kBq L^{-1} in water; doses: 0.5–1 Gy in liver or muscle; 360 days	Significant increase in lipidoperoxides in liver and muscle with negative effects on cell membrane and cellular detoxification process
	Loach (Misgurnus fossilis)	Lab; 1.85–37 kBq L^{-1} in water; doses: 0.04–0.7 Gy in gonad; 90 days	Significant changes in male gonads with decrease in glycogen and increase in fatty tissues; no similar effects in female gonads
	Grass carp (Ctenopharyngodon idella), tilapia (Tilapia mossambica)	Lab; 41–150 kBq L^{-1} in water; doses: >1.5 Gy in eye; 90 days	Dysfunction of eyes with the formation of edema and degeneration of tissues
Reproduction	Tilapia (Tilapia mossambica), roach (Rutilus rutilus lacustris), goldfish (Carassius auratus), silver carp (Hypophthalmichthys molitrix)	Various life-cycle studies in the lab to long-term observations in the field; various doses	Increased number of abnormalities and mortality in developing embryos; morphological and functional abnormalities in gonads and signs of hermaphroditism; sterility of males; teratogenic effects; decrease in the production of healthy progeny by irradiated fish
Mortality or life shortening	Tilapia (Tilapia mossambica), pike (Esox lucius), goldfish (Carassius auratus)	Various life-cycle studies in the lab to long-term observations in the field; various doses	Age- and dose-dependent increase in mortality; increased mortality of offspring from irradiated parents; shorter lifespan resulting from morbidity, reproductive effects, and cytogenetic damage; decrease in population size

From Sazykina and Kryshev (2003).

[a]Radiation source in the laboratory is ^{90}Sr; radiation source in the field includes ^{90}Sr plus other radionuclides; the doses represent those used in the experiment, and all (particularly low doses) are not necessarily reflective of the responses included.

populations of fish subject to low-dose radiation thrive even with a genetic load of lethal mutations (Trabalka and Allen, 1977). Recent studies have recorded "bystander" signals from irradiated fish to non-irradiated fish, a protective mechanism expressed as proteins or other factors in the bystander fish (Smith et al., 2007; Mothersill and Seymour, 2009). However, the bystander effects specific to radiostrontium in fish are not known.

6. NON-ESSENTIALITY OF STRONTIUM

Strontium is probably present in all organisms, but its essentiality in the body has never been shown; that is, its deficiency linked with serious growth retardation and death of an animal has not been demonstrated. However, the beneficial roles of Sr in several organisms are known. Strontium can stimulate plant growth and replace Ca required by *Chlorella* (Walker, 1953). Acantharia, a free-floating marine organism, constructs its internal skeleton of strontium sulfate (da Silva and Williams, 1991). Developmental defects and abnormal swimming behavior of the marine gastropod *Aplysia californica* and several cephalopod species have been reported, when the eggs were incubated and larvae were raised in defined seawater without added Sr (Bidwell et al., 1986; Hanlon et al., 1989). Strontium given as strontium ranelate augments bone Ca by stimulating Ca^{2+}-sensing receptors in experimental animals and reduces fracture rate in osteoporotic patients, while its antioxidative properties have also been demonstrated (Nielsen, 2004; Radzki et al., 2009). There are no studies showing any important physiological role of Sr in fish.

7. POTENTIAL FOR BIOCONCENTRATION AND BIOMAGNIFICATION OF STRONTIUM

7.1. Bioconcentration

Strontium is strongly accumulated in fish, and like Ca, it primarily concentrates in bony tissues, only approximately 5% being found in the soft tissues (Chowdhury, 2001; Smith et al., 2005b; Yankovich, 2009). Organisms do not discriminate between radioactive and stable Sr for accumulation. Because of weaker competition with Ca, Sr bioaccumulates to a greater extent in organisms living in low Ca (soft) waters, relative to hard freshwaters and seawater. Because of the hazard associated with its radioisotopes, there have been numerous studies to determine the equilibrium BCF of radiostrontium for fish (Rosenthal, 1957; see reviews by Hosseini et al., 2008; IAEA, 2009; Smith et al., 2009). The majority of these studies are based on field-collected data and thus account for uptake from both water and food, so technically could be considered bioaccumulation factors (BAFs) (see Wood, Chapter 1, Vol. 31A). As commonly found for BCFs of other metals (McGeer et al., 2003), the reported BCFs for Sr in fish show strong variation ranging from less than unity to thousands depending on environmental conditions, species, target tissues, and food habit (Table 7.4).

Attempts have been made by Kryshev (2006), and Smith et al. (2009) to describe the variability by taking some of the above factors into account, but it appears that Ca concentration in the water is the major factor affecting Sr BCF in fish. Based on comprehensive datasets from freshwater bodies, they have developed several empirical relationships to describe Sr BCF in relation to Ca; the most recent ones are presented in Table 7.4. The updated Ca-normalized models account for 65–88% of the variation and can be useful to understand the degree of Sr bioconcentration ($L kg^{-1}$ wet weight) in freshwater fish. They include: $BCF_{wholebody} = 3224/[Ca]$, $BCF_{muscle} = 133/[Ca]$, and $BCF_{bone} = 9750/[Ca]$ for whole body, muscle, and bones, respectively, where [Ca] is the waterborne Ca concentration in $mg L^{-1}$ (Table 7.4). Calibrating to several seawater databases, Kryshev (2006) stated that such relationships could be applied for seawater fish as well, but not to any water bodies under non-equilibrium conditions. Whereas the influence of fish size on Sr accumulation was not significant (Smith et al., 2009), the dietary differences appear to make some contribution to the variation in Sr BCFs (Kryshev, 2006).

The conclusions that can be made based on the above studies include: (1) bioconcentration of Sr is inversely correlated with Ca concentrations in the ambient water and thus use of any default value for BCF is not recommendable (Fig. 7.2) (Kryshev, 2006); (2) waterborne Ca can account for more than 65% of variation in Sr bioconcentration in fish (Table 7.4), and the remaining variation is likely related to food habits or other factors such as water pH (Chowdhury and Blust, 2001a), but probably not fish size (Smith et al., 2009); (3) BCFs for seawater fish are smaller and less variable (Table 7.4), which may be attributed to higher but uniform Ca concentrations in the seawater; (4) the BCF for bony tissues can be 70-fold greater than that of muscle, suggesting that the BCF for fish largely reflects the bioconcentration of Sr in bony tissues (Table 7.4, Smith et al., 2009); (5) compared with Ca, the discrimination coefficient for accumulation of Sr in fish ($=BCF_{Sr}/BCF_{Ca}$) is estimated to be approximately 0.49 (Kryshev, 2006), but this again depends on Ca concentration in water.

Table 7.4
Survey of strontium bioconcentration factors (BCFs) for freshwater and marine fish

BCF (L kg^{-1})	N	Tissue	Ecosystems	Notes	Reference
3224/[Ca]	132	Whole body	Freshwater	(CI = 2923–3557), $R^2 = 82\%$	Smith et al. (2009)
133/[Ca]	19	Muscle	Freshwater	(CI = 78–231), $R^2 = 65\%$	Smith et al. (2009)
9750/[Ca]	35	Bone	Freshwater	(CI=8110–11,700), $R^2 = 88\%$	Smith et al. (2009)
3940/[Ca]	115	Whole body	Freshwater and marine	(CI = 1770–6110), $R^2 = 82.4\%$	Kryshev (2006)
4770/[Ca]	55	Whole body	Freshwater	Non-predatory fish, (CI = 3020–7520), $R^2 = $ NR	Kryshev (2006)
3420/[Ca]	33	Whole body	Freshwater	Predatory fish, (CI = 1180–5660), $R^2 = $ NR	Kryshev (2006)
17 ($\pm$23)	14	Whole body	Freshwater	Mean ($\pm$SD)	Hosseini et al. (2008)
190 (22–710)	106	Whole body	Freshwater	Mean (min–max range)	IAEA (2009)
2.8 (0.14–69)	88	Muscle	Freshwater	Mean (min–max range)	IAEA (2009)
1.1 ($\pm$0.6)	15	Gonads	Freshwater	Mean ($\pm$SD)	Yankovich (2009)
1.2 ($\pm$0.7)	15	Liver	Freshwater	Mean ($\pm$SD)	Yankovich (2009)
971 ($\pm$217)	15	Bone	Freshwater	Mean ($\pm$SD)	Yankovich (2009)
3	NR	Whole body	Marine	Generic value recommended by IAEA	Brown et al. (2004)
4–53	NR	Whole body	Marine	(min–max range), [Ca] = $\sim$270 mg L^{-1}	Kryshev (2006)
10 (3–15)	NR	Whole body	Marine	Mean (min–max range), [Ca] = $\sim$408 mg L^{-1}	Kryshev (2006)
23 ($\pm$35)	103	Whole body	Marine	Mean ($\pm$SD)	Hosseini et al. (2008)

[Ca]: Ca concentration (mg L^{-1}); see text Section 7.1 for full explanation; N: number of observations considered; CI: confidence interval; R^2: coefficient that explains the variation; NR: not reported; IAEA: International Atomic Energy Agency.

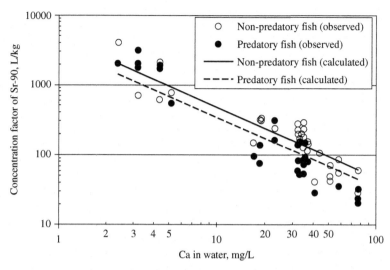

Fig. 7.2. Bioconcentration factors of ^{90}Sr for predatory and non-predatory fish in relation to Ca concentrations in water. With permission from Kryshev (2006).

7.2. Biomagnification

Several studies have been conducted to understand the trophic transfer of stable Sr or radioactive Sr through various predator–prey systems. Neither has shown any sign of biomagnification. Indeed, an opposite trend, that is, decreased concentration of Sr to higher trophic level fish, has been observed in some cases. In a comprehensive study, Wren et al. (1983) examined a series of metals including Sr at various trophic levels (clams, fish, birds, and mammals) of a Canadian Shield lake ecosystem. Mercury was the only element to exhibit biomagnification in both aquatic and terrestrial food chains. A study with fish representing six trophic levels in the Savannah River showed a decreasing trend of stable Sr levels from herbivores towards piscivores (Burger et al., 2002). Other predator–prey systems in which no biomagnification of Sr has been observed are "planktivore fish–benthic feeder fish–plankton" for stable Sr (Carraca et al., 1990) and "pike–perch" for ^{90}Sr (Beddington et al., 1989; Outola et al., 2009). The Sr BCFs for non-predatory fish appear to be greater relative to predatory fish (Kryshev, 2006). This has been explained as lower assimilation of Sr by predatory fish from bony tissues of the prey fish, compared to Sr uptake by non-predatory fish from plankton, benthos, and aquatic plants.

Concerning the food chain transfer of Sr from fish to humans, only radiostrontium is of concern, even though its biomagnification is not at all likely. Since ^{90}Sr accumulation in fish muscle tissue is low, its contribution to the radioactive dose via human consumption of fish is generally low (Outola et al., 2009). Yet, ^{90}Sr can be radiotoxic to humans because of its accumulation in bone tissues and long biological half-life.

8. CHARACTERIZATION OF UPTAKE ROUTES

8.1. Gills

8.1.1. RELATIVE CONTRIBUTION

Studies with several freshwater fish (rainbow trout, carp, goldfish) suggest that Sr uptake occurs mainly via the gills and that uptake via food is of secondary importance (Berg, 1968, Coughtrey and Thorne, 1983; Chowdhury, 2001). The contribution of water, presumably via the gills, is higher than 80% of the whole-body accumulation under natural conditions generally observed in freshwater bodies. The isotopic signature studies are somewhat contradictory; for example, the source of Sr in otoliths is 88% from freshwater in tilapia (Farrell and Campana, 1996), but 70% from food in Atlantic salmon raised in freshwater (Kennedy et al., 2000). Despite this contradiction, it is clear that the relative contribution of Sr uptake pathways appears to be affected by the status of Ca in the water and food or, more precisely, by Sr/Ca ratios in both sources (Berg, 1968; Farrell and Campana, 1996). A study with carp (Chowdhury and Blust, unpublished data) suggests that water is the predominant source in soft water (hardness <40 mg L^{-1} as $CaCO_3$) even when foodborne Sr concentration is eight-fold greater than waterborne Sr, alluding to an important role of waterborne Ca in determining the relative importance of the two Sr sources. The actual contribution of branchial uptake by freshwater fish under various Sr and Ca conditions needs further resolution.

Fish can take up Sr from seawater but bioaccumulation is relatively low because of the high Ca concentration in the water (Boroughs et al., 1956; Smith et al., 2005b). In juvenile mummichogs (*Fundulus heteroclitus*), water sources contribute 83% of Sr in otoliths (Walther and Thorrold, 2006). Given that marine fish drink water for osmotic regulation, both branchial and intestinal routes are probably involved in waterborne Sr uptake. The relative importance of the gills alone as an uptake route is not known, but is perhaps less than in freshwater fish. For Ca though, the net branchial influx in eel and tilapia in seawater is comparable to or slightly higher than in freshwater (Flik and Verbost, 1993).

8.1.2. UPTAKE MECHANISM

No molecular/cellular study has so far been done to understand the mechanism of Sr transport at fish gills. Other evidence suggests that Sr mimics Ca, entering fish using the Ca transport systems. This transcellular pathway, originally found in chloride cells (Flik et al., 1995) but more recently in other cells of the gill epithelium using molecular probes (Shahsavarani and Perry, 2006; Shahsavarani et al., 2006), involves a three-step process: passive entry through Ca^{2+}-selective channels at the apical side of the cells along an electrochemical gradient, cytosolic diffusion facilitated by a calcium-binding protein (calmodulin), and active exit to blood via a basolateral high-affinity Ca^{2+}-ATPase and less importantly via an Na^+/Ca^{2+} exchanger. Considerable evidence exists in support of Ca mimicry of Sr. Firstly, both alkaline earth metals have similar chemical properties including ionic radius (Sr = 1.12 Å, Ca = 0.99 Å; hydrated radius for both: ~ 9.6 Å), and in general, all of the three steps are implicated for Sr in different mammalian epithelia (Aidley and Stanfield, 1998; Nielsen, 2004; Shirran and Barran, 2009). Secondly, a perfused gill preparation study has detected two transport rates of Sr in rainbow trout: a passive rate of 0.5×10^{-7} cm s^{-1} and a 10-fold higher energy-dependent rate of 5.1×10^{-7} cm s^{-1}; the presence of Ca does not affect the passive rate of Sr uptake, but inhibits Sr uptake in the energy-dependent system (Schiffman, 1965). Thirdly, Sr uptake as observed in the common carp displays Michaelis–Menten-type saturation kinetics and Ca inhibits Sr uptake in a completely competitive way, suggesting a common binding site or uptake pathway on the membrane transport system (Fig. 7.3) (Chowdhury et al., 2000). Fourthly, the amino acid sequences of the calcium-sensing receptor proteins (CaSR) expressed in ion-transporting tissues (kidney, intestine, gills, and elasmobranch rectal gland) of fish have 70–91% homology with their mammalian orthologues (Loretz, 2008). It appears that Sr enters the cell via the apical Ca channels without significant interaction, but mainly competes with Ca for binding sites at the basolateral transporters. Despite the abundance of Ca in seawater, marine fish still appear to take up Ca actively at the gills using the same transport mechanisms (Flik et al., 1995). Therefore, the branchial transport system of Sr in marine fish is likely not to be different from that in freshwater fish.

8.1.3. UPTAKE KINETICS

Chowdhury, Blust and colleagues have conducted a series of studies to understand the kinetics of waterborne Sr uptake in freshwater fish and its relationship with several environmental factors (Ca, pH, temperature, and complexing agents) (Chowdhury et al., 2000; Chowdhury and Blust, 2001a,b,

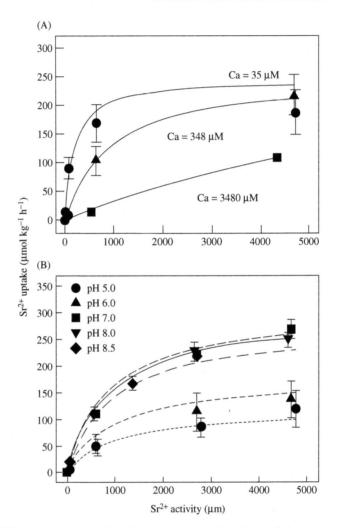

Fig. 7.3. Effects of waterborne Sr on Sr uptake in the whole body of the common carp at different waterborne Ca concentrations (A) and pH (B). Data points represent mean uptake ($\pm$ SD) in 3 h ($n = 5$–7, temperature $25 \pm 1^\circ$C, $Sr_{total} = 0.2$–10,000 μM; unless otherwise stated, $Ca_{total} = 348$ μM and pH = 8.0). The solid lines are predictions by a Michaelis–Menten type model for completely competitive inhibition of Sr uptake by Ca (A) or partially non-competitive inhibition of Sr uptake by H^+ (B). Reproduced from Chowdhury et al. (2000) and Chowdhury and Blust (2001a).

2002). These studies have extended the understanding of Sr uptake by fish from simple forms of the inverse relationship between Sr and Ca accumulations (Brungs, 1965; Suzuki et al., 1972; Smith et al., 2009) to Sr bioavailability under dynamic environmental conditions, and have shed some light on the mechanisms of Sr uptake across the epithelia. However, it is necessary to note that further studies with electrophysiological and molecular techniques are necessary to expand mechanistic understanding.

The uptake of waterborne Sr by the common carp shows saturation kinetics over a wide range of waterborne Ca (0.35–3480 μM), pH (5.0–8.5), and temperature (10–30°C), suggesting that Sr uptake is a facilitated process (Fig. 7.3), as observed for Ca (Flik et al., 1995). A Michaelis–Menten type model can be used to describe the effects of these environmental factors on Sr uptake. Calibration of the model to experimental data reveals that Sr and Ca ions inhibit each other in a completely competitive manner, competing for the same binding site, whereas H^+ inhibits the uptake of both metals in a partially non-competitive way (Chowdhury and Blust, 2001a). The non-competitive inhibition suggests that H^+ interacts with the transporter system in a region other than the binding site for Sr or Ca, so that the binding of the metal ions in the transporter is not affected. The inhibition occurs as a result of proton-induced changes in the functional characteristics of the transporter. Both types of interaction can be described by the equation (Chowdhury and Blust, 2001a):

$$j_{Sr} = J_{max\,Sr} \cdot \frac{\beta_{Sr}(H^+) + K_{iH}}{(H^+) + K_{iH}} \cdot \frac{(Sr^{2+})}{(Sr^{2+}) + K_{mSr}[1 + (Ca^{2+})/K_{iCa}]}$$

where j_{Sr} is the uptake rate of Sr (μmol kg^{-1} h^{-1}); J_{maxSr} and K_{mSr} are the maximum uptake rate (μmol kg^{-1} h^{-1}) and the half-saturation constant (μM), respectively, of Sr; K_{iCa} and K_{iH} are inhibitor constants (μM) for Ca and proton, respectively; β_{Sr} is the proportionality constant for proton inhibition of Sr uptake, and (Sr^{2+}), (Ca^{2+}), and (H^+) are free Sr, Ca, and proton activities in water (μM). Using a Monte Carlo approach, Smith et al. (2005a) derived a simplified form of the Chowdhury and Blust equation for estimating Sr uptake rates in fish.

The estimated J_{max} of Sr (293 μmol kg^{-1} h^{-1}) is twice that for Ca (159 μmol kg^{-1} h^{-1}), whereas the K_m for Sr (96 μM) and K_i for Ca (28.5 μM) suggest that binding affinity for Sr is one-third that of Ca and the system becomes saturated at a lower concentration with Ca. Such discrimination against Sr is observed in mammalian gut, but the ratio of absorbed Sr/Ca from diet is generally 0.6–0.7 (Wasserman, 1998; Nielsen, 2004). The presence of Ca affects the K_m, while pH affects J_{max} of Sr (Fig. 7.3). The pK_a (6.27) of the proton binding site, calculated from the negative logarithm of K_{iH} (0.54 μM), corresponds well to the pK_a range (6–7) of histidine imidazole in protein, suggesting the possible involvement of this ionizable group in proton-mediated effects on Sr uptake.

The temperature study reveals that the kinetic parameters (J_{max}, K_m, and K_i) for both Sr and Ca are temperature dependent, and the Q_{10} values for Sr uptake in whole fish, gills, and blood were 3.71, 2.29, and 4.05, respectively, for a temperature change from 10 to 25°C (Chowdhury and Blust, 2001b). Thermodynamic analysis of the temperature effects indicates that the activation energies (E_a) required for Sr uptake (91.9 kJ mol^{-1}) and Ca uptake (105.9 kJ mol^{-1}) in the whole body of carp were constant over the temperature range of 10–25°C and showed a break in the Arrhenius plots (Cossins and Bowler, 1987) above this temperature (Fig. 7.4). The pattern of the Arrhenius plot and the activation energy for Sr uptake (98 kJ mol^{-1}) in blood are similar to those for whole fish. In contrast, the E_{aSr} for gills was considerably smaller and constant (58.1 kJ mol^{-1}), showing no break in the Arrhenius plot over the temperature range of 10–30°C. The high energy of activation for uptake in whole body and blood compared to uptake in gills indicates that the transport of the metal across the basolateral membrane of the gill epithelium is the rate-limiting step, rather than the transport across the apical membrane. Schiffman (1965), following a gill perfusion study, revealed the same finding. Further studies are necessary to confirm the rate-liming step for Sr.

Strontium uptake by carp in the presence of organic ligands (EDTA and NTA) is not always a mere function of the free metal-ion activity, but instead suggests that certain complex species may contribute significantly to overall uptake (Chowdhury and Blust, 2002). In the presence of organic ligands, relatively more Ca is complexed because of greater binding affinity and, as a result, Ca inhibition of Sr uptake at fish gills is relatively weak. Taking into account this type of competitive interaction and free ion activities of Sr and Ca in the solution, a Michaelis–Menten-type model can describe Sr uptake in the presence of EDTA, but not NTA. In the case of NTA, the uptake rates are significantly higher than model predictions (Fig. 7.5). It seems that fish can take up SrNTA$^-$ directly from water in addition to Sr^{2+}. A similar observation has also been made for the Ca–NTA complex (Chowdhury and Blust, 2002), but the mechanism of uptake needs to be elucidated. Berg (1970) showed an increased uptake of Sr by goldfish in the presence of EDTA, which appears to be related to a decreased competition from Ca. Regardless of this, the complexation effects on Sr may have regulatory implications as chelating agents are used for decontaminating radionuclides.

Kinetic studies for Sr uptake by marine fish are not available. However, it has been shown that Sr and Ca concentrations in the seawater covary with

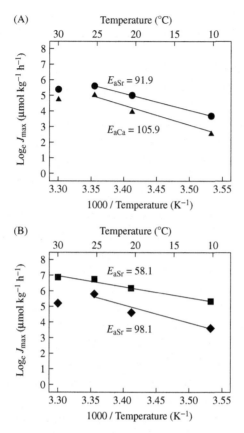

Fig. 7.4. (A) Arrhenius plot for the effect of temperature on the uptake rates of Sr (circles) and Ca (triangles) in the whole body of carp. (B) Arrhenius plot for the effect of temperature on the uptake rates of Sr in the gills (squares) and blood (diamonds) of the common carp. Data points represent the apparent maximum uptake rates (J_{max}) of Sr and Ca at pH 8.0. The Arrhenius activation energies (E_a in kJ mol^{-1}) are calculated from the slopes ($-E_a/R$) of the Arrhenius equation: $v = A.\exp[(-E_a/RT]$, where v represents the velocity of a reaction ($=J_{max}$), A the Arrhenius constant, R the universal gas constant, and T the absolute temperature. From Chowdhury and Blust (2001b).

salinity, and the Sr/Ca ratio is positively correlated with salinity and temperature in otoliths of marine fish (Martin et al., 2004; Lin et al., 2007). Unexpectedly, more Sr is incorporated into otoliths of larval spot at salinities of 25‰ than at 15‰ (Martin et al., 2004). It would be interesting to explore how Sr and Ca interact at gills or gut during uptake at various salinities, as fish encounter such situations in estuarine waters.

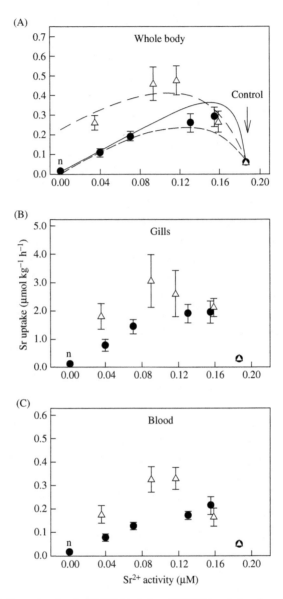

Fig. 7.5. Effect of complexation by EDTA (circles) and NTA (triangles) on Sr uptake in the whole body, gills, and blood of carp as a function of Sr^{2+} ion activity in the water containing 0.285 μM Sr and 348 μM Ca. Data points represent mean uptake rates with standard deviations ($n = 5$–6; temperature: 25°C; pH 8.0). The arrow shows the control values for whole body, gills, and blood, obtained in the absence of the ligands. A t-test indicates that the observed uptake rates, except those denoted by n for EDTA, were significantly higher than their control values ($p < 0.001$). The solid and short-dashed lines are calculations of Sr uptake by a Michaelis–Menten type competitive

8.2. Gut

The handling of Sr in fish gut is virtually unknown; this is applicable for freshwater as well as marine fish. As discussed in Section 8.1.1, dietary uptake of Sr is less important in freshwater than in marine fish, as more than 80% of the total Sr is taken up from water. Reports in favor of food as a source of Sr for freshwater fish exist but those require further verification (Ophel and Judd, 1966; Fleishman, 1973). Indeed, the actual contribution of the intestinal route appears to be dependent on Sr and Ca concentrations in the water and food. The relative contribution of gut to total Sr in seawater fish is probably higher compared to freshwater fish, mainly because of the high drinking rate of seawater fish, but less likely that it is because of dietary Sr. Intestinal Sr uptake may increase in freshwater and marine fish, rather inadvertently with Ca, at the time of gonadal maturation, when Ca demand increases significantly (Flik and Verbost, 1993).

As for the mechanism of Sr uptake, conceivably, the Ca uptake system is the major candidate. In the intestinal enterocytes of fish, similar mechanisms as described earlier for transcellular Ca transport exist. However, the transport of Ca to blood is primarily via the Na^+/Ca^{2+} exchanger and to a limited extent mediated via Ca^{2+}-ATPase (Flik and Verbost, 1993; Flik et al., 1995). Experiments *in vitro* with gut sacs have shown significant Ca uptake in both freshwater and marine fish (Flik et al., 1995; Klinck et al., 2009). Strontium uptake via Ca transport systems in mammalian gut, using both active transport and diffusional paracellular pathways, have been documented (Wesserman, 1998; ATSDR, 2004; Nielsen, 2004), although both pathways generally show preference towards the absorption of Ca by a factor of 2. Nutrient molecules such as lactose, histidine, lysine, and taurine promote or modulate the uptake of Ca, Sr, and other nutrient metals in mammals (Nielsen, 2004) and fish (Glover and Hogstrand, 2002). It appears that nutrient transporters in fish gut may also be involved in Sr uptake as nutrient complexes (Bogé et al., 2002).

8.3. Other Routes

There is no report suggesting Sr uptake from water via any extrabranchial route such as skin of freshwater fish. Integumentary uptake

---

Fig. 7.5 (Continued)

inhibition model in the presence of EDTA and NTA, respectively, considering that only the free Sr^{2+} ion is available for uptake. The long dashed line is the calculation by the model in the presence of NTA, considering that both the free Sr^{2+} and SrNTA– complex are available for uptake. From Chowdhury and Blust (2002).

of ^{89}Sr by tilapia from seawater was implicated a long time ago (Boroughs et al., 1956), but this conclusion was probably based on high ^{89}Sr burden of the integument (13–21% of the whole-body burden), which could indeed be in scales. Nevertheless, more recent information reveals the presence of ion transporting ionocytes in the skin, similar to those in the gills, and the integumentary ionocytes are implicated for active Ca uptake in *in vitro* skin preparation studies (Marshall et al., 1992; Flik et al., 1995). In general, the number of ionocytes in the fish skin is very limited relative to their abundance in the gills, and thus, the contribution of Ca uptake through this route is not probably significant. However, the number can increase in fish under stress, and some fish species (e.g. tilapia) can have a significant amount of ionocytes in opercular epithelia. Therefore, depending on fish species and environmental condition, this route may become important for Sr uptake.

9. CHARACTERIZATION OF INTERNAL HANDLING

Based on the overall information available for fish, and known mammalian studies (Nielsen, 2004), a general observation is that Sr and Ca behave similarly, but not identically regarding branchial and intestinal uptake, tissue accumulation, and renal excretion. There are some differences in Ca handling between fish and mammals. Unlike mammals, bone and intestine are not the major sites of Ca regulation in fish, and many fish (perciforms) lack bone cells, osteocytes (Lall and Lewis-McCrea, 2007). The gills play a major role for Ca homeostasis in fish. In addition, Ca metabolism is regulated by hypocalcemic and hypercalcemic hormones (stanniocalcin, cortisol, and prolactin) in fish, not by parathyroid hormones (Flik et al., 1995; Lall and Lewis-McCrea, 2007). The influence of such differences on Sr handling in fish remains to be understood.

9.1. Biotransformation

Strontium is found in ionic form or bound forms with cellular and plasma proteins such as calmodulin and albumin in the body (Nielsen, 2004; Shirran and Barran, 2009). However, like other metals Sr is not known to be metabolized into any different biochemical form with distinct characteristics.

9.2. Transport through the Bloodstream

The role of the blood in the transportation of radiostrontium was studied in tilapia decades ago (Boroughs and Reid, 1958), but the results remain in agreement with current mammalian studies (Nielsen, 2004; Höllriegl et al., 2007). In both cases, almost all of the Sr in whole blood appears to be carried by the plasma, less than 10% of the dose is found in the blood cells or on the cell walls. In another study, $7.2 \, \mu g \, L^{-1}$ of Sr was found in the erythrocyte fraction of human blood and $44 \, \mu g \, L^{-1}$ in the plasma (ATSDR, 2004). Strontium is cleared from the blood very rapidly in fish: the biological half-life of the fast component plasma clearance is less than an hour. More than 70% of the initial plasma burden of Sr in tilapia disappeared within 4 h. The early clearance of Sr and other metals is generally the consequence of rapid transfer from blood plasma to tissue compartments in fish (Boroughs and Reid, 1958; Chowdhury et al., 2004) as well as humans (Höllriegl et al., 2007), while transfer of Sr from plasma to urine is negligible within 12 h after dosing. The proportion of Sr bound to plasma protein in fish is expected to be low, as the values recorded for mammalian plasma or serum are generally 50% or lower (ATSDR, 2004).

9.3. Accumulation in Specific Organs

As a Ca analogue, Sr is mainly accumulated in calcareous tissues (skeleton, scales, fins, otoliths, opercular bones) of fish, regardless of ecosystems, environmental conditions, and uptake routes (Boroughs et al., 1956; Rosenthal, 1957; Martin and Goldberg, 1962; Brungs, 1965; Berg, 1968; Suzuki et al., 1972; Farrell and Campana, 1996; Yankovich, 2009). However, most of the earlier studies conducted with either radiotracers or stable Sr are not sufficiently detailed to allow comparison of Sr concentrations across tissues. In a study with the common carp (Chowdhury and Blust, unpublished data), the general trend for tissue concentrations is in the order: bone (including scale) $\gg$ gills $>$ muscle $\approx$ blood, but the concentrations are strongly dependent on ambient Ca concentration (Table 7.5). In the same study, the tissue concentrations for Ca showed a similar general trend, but were not reflective of the waterborne Ca concentrations at the higher level (3480 μM), suggesting regulation of tissue Ca. Overall information on relative tissue-specific distribution, based on both the laboratory experiments and field data, suggests that approximately 80–95% of Sr is found in the bony tissues. More specifically, muscle represents only 2–3%, whereas among bony tissues, scales, and skeletal bones it represents 9–20% and 70–75%, respectively (Berg, 1968; Suzuki et al., 1972; Smith et al., 2009).

Of particular interest in fisheries biology is the Sr incorporated in otoliths and scales, as its microchemistry in these aragonite tissues can be used as a biomarker of aging and environmental history of freshwater and marine fish, particularly those migrating for reproduction or other purposes

Table 7.5

Tissue and whole-body concentrations of strontium and calcium in the juvenile common carp (Cyprinus carpio) after exposure to waterborne strontium for uptake, followed by an elimination period in clean freshwater at three calcium levels.

Tissue		Day 15			Day 48		
		Ca 35 μM	Ca 348 μM	Ca 3480 μM	Ca 35 μM	Ca 348 μM	Ca 3480 μM
Blood	Sr (μmol kg^{-1} ww)	3.63±0.79	0.72±0.24	0.06±0.02	2.0±0.25	0.08±0.09	0.05±0.03
	Ca (mmol kg^{-1} ww)	0.36±0.08	1.71±0.37	3.95±2.13	0.14±0.02	1.68±0.38	1.0±0.82
Gills	Sr (μmol kg^{-1} ww)	225.0±52.47	59.2±12.1	2.87±0.72	160.2±20.8	24.4±11.0	1.76±0.70
	Ca (mmol kg^{-1} ww)	16.9±3.04	46.5±7.5	41.1±3.8	9.77±1.27	52.3±21.5	16.7±5.94
Muscle	Sr (μmol kg^{-1} ww)	3.09±0.61	0.63±0.20	0.06±0.01	2.90±0.37	0.43±0.12	0.03±0.01
	Ca (mmol kg^{-1} ww)	0.72±0.09	2.58±0.46	4.51±0.37	0.41±0.05	3.59±1.12	1.28±0.64
Bone	Sr (μmol kg^{-1} ww)	1182.3±197.3	351.2±37.2	28.9±4.5	1027.5±133.6	275.9±63.8	20.8±5.7
	Ca (mmol kg^{-1} ww)	178.8±30.1	719.3±87.6	660.2±105.2	165.5±21.5	630.9±127.9	542.1±146.0
Viscera	Sr (μmol kg^{-1} ww)	2.80±0.47	0.69±0.03	0.12±0.03	0.62±0.08	0.15±0.06	0.02±0.01
	Ca (mmol kg^{-1} ww)	0.35±0.05	1.06±0.08	1.38±0.21	0.05±0.01	0.28±0.12	0.18±0.14
Whole body	Sr (μmol kg^{-1} ww)	76.3±10.2	18.7±2.93	1.67±0.18	63.6±8.4	17.0±3.38	1.29±0.19
	Ca (mmol kg^{-1} ww)	12.9±1.66	47.6±6.50	37.0±5.16	11.2±1.46	44.9±6.0	41.1±10.9

ww: wet weight.

Data are shown as mean ± SD (n=5–7).

Carp were exposed to waterborne [Sr] of 0.37 μmol L^{-1} for 15 days for uptake, followed by an elimination period of 33 days (total 48 days) in clean freshwater at three Ca levels (35, 348, and 3480 μmol L^{-1}; temperature: 25°C, pH 8.0) (Chowdhury and Blust, unpublished data).

(Tzeng et al., 1999; Kennedy et al., 2000, 2002; Melancon et al., 2009; Collingsworth et al., 2010). In most laboratory studies, more than 80% of Sr in otoliths originates from water sources, and the remainder from diet (Farrell and Campana, 1996; Walther and Thorrold, 2006). The accreted Sr and the resultant Sr/Ca ratio reflect ambient environmental conditions including temperature, salinity, and food sources in the course of fish life stages (Martin et al., 2004; Zimmerman, 2005); the Sr/Ca ratio is positively correlated with temperature (Fig. 7.6). As a result, Sr in the annual rings and more commonly Sr/Ca ratio provides a geochemical signature to reconstruct information on fish age, stock, geographical sources, habitat use, and migration.

9.4. Subcellular Partitioning

Information on the subcellular location of Sr in fish tissues does not exist. In rats that were exposed to Sr in drinking water (1.9 mg L^{-1}) for 3 months, the Sr concentrations in the mitochondrial, lysosomal, and microsomal fractions of liver were approximately five times those in the cytosol (ATSDR, 2004). A major fraction ($\sim$50–80%) of the Sr in tissues appears to be bound to protein. In a terrestrial invertebrate (the earthworm *Aporrectodea caliginosa*), Ca and Mg were found in three major fractions of ecotoxicological importance (Vijver et al., 2006): (1) proteins and cellular debris including tissue fractions and cell membrane ($\sim$60%), (2) a granular fraction rich in Ca and phosphorus ($\sim$30%), (3) a cytosolic fraction (10%) that included microsomal, heat-denatured proteins, and heat-stable protein fractions. Indeed, all of these cross-kingdom studies grossly corroborate

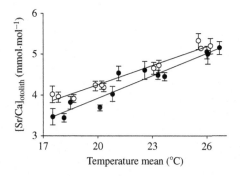

Fig. 7.6. Sr/Ca (mmol mol^{-1}) ratios in otoliths of laboratory-reared *Leiostomus xanthurus* as a function of tank temperatures (°C) at two salinity levels: 15‰ (closed circles) and 25‰ (open circles). The lines were fitted by linear least-squares regression for each of the salinity treatments. With permission from Martin et al. (2004).

each other. It is reasonable to assume that Sr would follow similar localization with Ca in fish, depending on exposure concentrations.

9.5. Detoxification and Storage Mechanisms

Detoxification of Sr by metallothionein (MT) in humans has been reported in a recent study in which human subjects were given Sr therapy against osteoporosis for 30 days (Fabrik et al., 2009). Their serum Sr concentration was as high as 10.5 mg L^{-1} and the induction of MT was up to 6.5-fold. Such a high concentration of Sr is not expected to occur in fish in the environment. No correlation between Sr and MT has been found in wild animals in the Savannah River site, which is heavily contaminated with Sr (Burger et al., 2000). However, Sr may be found in granules in any organisms, similar to Ca. Such localization may not be part of Sr detoxification, but rather an indirect effect of Ca regulation in the body. Indeed, more than 90% of Sr in fish is localized in the bone tissues including scales (Berg, 1968; Suzuki et al., 1972; Smith et al., 2009). Therefore, Sr detoxification in the soft tissues is of less significance.

Similar to Ca, bones represent storage sites for Sr in fish with likely slower exchange kinetics with the surrounding extracellular and circulatory fluids. Building blocks of bone and scales are mostly collagen and hydroxyapatite, a hydroxylated polymer of Ca and phosphate $[Ca_{10}(PO4)_6(OH)_2]$ (Lall and Lewis-McCrea, 2007). Strontium has a high affinity for bone and is incorporated in two ways: surface exchange or ionic substitution. However, Sr content is only a few percent of Ca in the bones (Dahl et al., 2001). Yet, the concentration of Sr is several orders of magnitude greater in bones than in soft tissues (Table 7.5).

9.6. Homeostatic Controls

Based on mammalian studies, it has been suggested that Sr can substitute for Ca in many physiological processes such as muscular contraction and blood clotting, and take part in active transport across biological membranes, although always less efficiently than Ca (Nielsen, 2004). However, no evidence has so far been found for homeostatic control with any feedback mechanism to regulate Sr concentration in any body compartment or fluid. The limited number of fish studies does not provide any additional evidence in support of homeostatic control of Sr. Incorporation of Sr in otoliths of fish is less precisely controlled than Ca (Farrell and Campana, 1996). A general remark is that Sr uptake is less likely to be physiologically regulated than an essential element but may be affected by factors controlling Ca concentration in the body (Thresher et al., 1994; Nielsen, 2004).

10. CHARACTERIZATION OF EXCRETION ROUTES

10.1. Gills

Circumstantial evidence suggests that Sr is excreted via branchial, intestinal, and renal routes in fish, but the relative importance of these routes has not been studied. Two early studies suggested branchial excretion of ^{90}Sr in marine fish, as a high ^{90}Sr activity in gills was recorded for a prolonged period after gastrointestinal dosing (Boroughs et al., 1956; Martin and Goldberg, 1962). Similar results have been found for another Ca-mimicking metal, Cd, in feeding studies with rainbow trout (Chowdhury et al., 2004), but the mechanism is unknown.

10.2. Kidney

No information could be found related to renal clearance of Sr in fish, but in mammals including humans, this is a major route, representing 50–60% of the total body Sr excretion (Dahl et al., 2001; ATSDR, 2004). The renal clearance of Sr is 4–10 L day^{-1} in humans and it is two to three times that of Ca. Such discrimination is thought to be related to the smaller reabsorption of Sr in renal tubules, while direct evidence based on *in vitro* studies with proximal tubules from rats revealed that Sr and Ca share a common mechanism of reabsorption, involving the active transport pathway for Ca (ATSDR, 2004). The renal clearance of Ca is approximately 0.2 L kg^{-1} day^{-1} in freshwater fish (Chowdhury and Wood, 2007). A higher renal clearance of Sr relative to Ca is likely to occur in fish, should a similar discrimination against Sr reabsorption exist. Freshwater eels and rainbow trout exhibit urine Ca concentrations of 1.5–2 mmol L^{-1} at an excretion rate of 40–70 μmol kg^{-1} fish day^{-1} (Flik and Verbost, 1993; Chowdhury and Wood, 2007). Given that the Sr level in the blood plasma is much lower than the Ca level (micromolar versus millimolar levels), the renal excretion rate of Sr in a freshwater fish would be much lower. Relatively high excretion of Sr is expected in seawater fish, as they secrete divalent ions (Ca^{2+}, Mg^{2+}) into the urine for ionic regulation.

10.3. Gut

Again, fish-specific information on the transfer of Sr from blood to the gut lumen is not available. Only one study recorded ^{90}Sr activity in the gall bladder of the Pacific mackerel after gastrointestinal dosing (Martin and Goldberg, 1962), suggesting biliary excretion. However, the intestinal excretion of Ca in cod (29 μmol kg^{-1} day^{-1}) represents 20% of the total Ca excretion and 50% of the extrarenal excretion (Flik and Verbost, 1993). The mechanisms of passive diffusion as well as Ca^{2+}-ATPase-mediated active transport have been implicated for intestinal Ca effluxes in fish (Flik et al., 1995). The gastrointestinal excretion of Sr, either from the bile or directly from the plasma, occurs in rat and human, possibly representing 40% of total excretion. In the rat, the secretion from plasma to the small intestine is four to eight times that in the large intestine. The evidence for both the passive and active transport of Sr from the serosal side to the mucosal side of the intestinal epithelium has been provided in both *in vitro* and *in situ* studies with rat (Nielsen, 2004; ATSDR, 2004).

11. BEHAVIORAL EFFECTS OF STRONTIUM

Behavioral effects including higher swimming activity, avoidance, and abnormal locomotor orientation due to X-ray or other type of irradiation (e.g. ^{60}Co, ^{137}Cs) have been reported for freshwater and marine fish (Templeton et al., 1971). However, no studies describing such effects in fish from stable or radioactive Sr were found. Reduced fitness, increased mortality, and morphological abnormalities of embryos and larvae are reported for fish that are chronically exposed to ^{90}Sr (see Section 5.4 and Table 7.3). Although not reported, underlying neurological damage in these fish due to radiation cannot be ruled out.

12. MOLECULAR CHARACTERIZATION OF STRONTIUM TRANSPORTERS, STORAGE PROTEINS, AND CHAPERONES

No Sr-specific transporters or binding proteins have so far been identified in any organism, and it is not expected that one will be found, as Sr is a non-essential metal. However, Sr is well known for its ability to substitute for Ca in physiological processes. The known Ca transporters have been tested for involvement in Sr transport in mammalian models (Wasserman, 1998; ATSDR, 2004; Hoenderop et al., 2005), and evidence exists for their presence in fish. These are the epithelial Ca channel (ECaC), plasma membrane Ca^{2+}-ATPase (PMCA), and Na$^+$/Ca^{2+} exchanger (NCX) (Flik et al., 1995; Shahsavarani et al., 2006; Hwang, 2009). ECaC is a voltage-independent apical Ca channel predominantly present in the gills, and is an orthologue of mammalian ECaC encoded for by two genes (TRPV5, TRPV6; also known as CaT1 and CaT2) belonging to the vanilloid subfamily and transient receptor potential (TRP) superfamily. Intestinal ECaC is probably an L-type voltage-dependent channel (Larsson et al., 1998). PMCA is a calmodulin-dependent, high-affinity pump, and NCX is a carrier energized by Na^+/K^+-ATPase; both are located in the basolateral side of the gill cells, although NCX is thought to be predominant in the intestinal enterocytes (Flik et al., 1995). The molecular physiology of the three transporters (ECaC, PMCA, and NCX), including their isoforms and localization in gill cells, has been provided in two recent papers (Shahsavarani et al., 2006; Hwang, 2009).

Strontium binds to calmodulin (CaM), a ubiquitous Ca-binding protein for intracellular Ca regulation and transport in fish and other organisms. CaM can bind up to four Sr ions, similar to the number of Ca ions, in the same binding sites, but the affinity for Sr is relatively low (Shirran and Barran, 2009), suggesting partially why a high concentration of Sr is needed to displace Ca from binding sites or to elicit toxicity in organisms. There are six genes encoding for a CaM molecule in zebrafish (Friedberg and Taliaferro, 2005). Another Ca-specific storage protein that is predominantly present and implicated for Sr transport in mammalian intestine is calbindin (Wasserman, 1998; ATSDR, 2004). The presence of calbindin D28K in fish gills has been reported (Zaccone et al., 1992), but its molecular physiology has not been studied. The Ca-specific chaperone, calreticulin, may be of importance for Sr binding in fish. It is located in the endoplasmic reticulum of mammalian cells and plays a key role in folding of newly synthesized proteins and glycoproteins (Michalak et al., 2002).

13. GENOMIC AND PROTEOMIC STUDIES

Strontium has not been studied in any organism to define genomic- or proteomic-level changes. One relevant study that may be of interest demonstrated proteomic changes in the gills of rainbow trout exposed to X-ray-induced bystander signals (Smith et al., 2007), and it is possible that similar changes may occur from exposure to radiostrontium. Two-dimensional gel analysis of gills from X-ray-treated trout and trout exposed to X-ray-induced bystander signals revealed expression of several proteins having properties related to the maintenance of epithelial polarity of the gills, and protection against reactive oxygen damage and lactate acidosis. Whether such changes are adaptive responses to future radiation damage or just short-term protection mechanisms is still not clear.

14. INTERACTIONS WITH OTHER METALS

From the preceding sections, it is clear that Ca is the most important element that has so far been studied for its interaction with Sr in biological systems. In general, both have been found to be competitive to each other in all mammalian and piscine studies during uptake, binding to proteins or other ligands, and elimination from the body, although biological systems favor Ca over Sr by a factor of 2–3 during transport or binding processes (see Sections 8, 9, 10, and 12). In theory, Sr should interact in a qualitatively similar way (competitive) with other Ca-mimicry metals such as Cd, Pb, Co, and Zn that may coexist in aquatic systems. However, although not studied, Sr interaction with other metals is probably negligible at its environmentally relevant concentrations. This is because the Ca concentration is generally many times higher than that of Sr in the environment. In fish, extracellular fluids have millimolar concentrations of Ca and micromolar concentrations of Sr. Because of high Ca/Sr ratios in the environment and in the body, and greater the affinity of Ca to biological systems, the effects of Sr on other metals will be obscured by Ca.

15. KNOWLEDGE GAPS AND FUTURE DIRECTIONS

Surveying old and recent studies, it seems reasonable to note that there is a wealth of radioecological information on Sr, focusing largely on ^{90}Sr, to understand its geochemical partitioning, mobility, and bioconcentration in aquatic ecosystems (see Section 1.1). There are several bioconcentration models (see Section 7.1), but all are empirically based, and bioaccumulation–effect relationships have not been characterized. Mechanistic information to describe uptake, internal handling, and elimination of stable or radioactive Sr in fish is extremely sparse, even at the organismal level. Experimental information on Sr physiology at subcellular or molecular levels virtually does not exist for fish. The caveat of knowledge gaps seems equally applicable to other aquatic animals. The current understanding of Sr physiology in fish is speculative, circumstantial, or correlational, depending on mammalian studies or those focused on Ca physiology in fish and other animal models. Toxicological information is also inadequate. While the chronic effects of ^{90}Sr on reproduction and morbidity of teleost fish are understood to a limited degree (see Section 5.2), chronic effects of stable Sr on early life stages of fish appear to be significant, but have never been experimentally verified. Some of the specific areas that warrant further research, and could actually be addressed with moderate resource allocation, are recommended below.

- Although stable Sr is thought to be benign, limited information suggests significant effects of stable Sr on egg hatchability and embryonic development of sensitive freshwater species (see Section 5.3). These effects require further verification, particularly in soft to medium-hard waters using simple exposure experiments.
- Qualitative and quantitative characterization of Sr uptake (all possible routes) and elimination in freshwater and seawater fish under various environmental conditions, need to be studied, as this information will be necessary to develop any mechanistically based bioaccumulation model for radiostrontium.
- Preliminary understanding of the mechanism of Sr transport in the gills and gut is required. For this, acclimation of fish by chronic exposure to Sr, followed by measurements of Ca and Sr fluxes using *in vivo* and *in vitro* techniques (Perry and Wood, 1985; Klinck et al., 2009), can be recommended. In addition, analyses of blood and urine samples in acclimation studies will provide further information on internal handling of Sr by fish.
- Further understanding of the mechanism of Sr transport in the gills and gut at molecular levels (mRNA and protein expression) is required. For this, acclimation studies as well as hormonal approaches (Shahsavarani and Perry, 2006) could be recommended. Adding molecular probes for the extracellular calcium-sensing receptor (CaSR) (Loretz, 2008) in this study would provide further insight into Sr effects on the modulation of Ca receptors and its possible regulatory roles in Sr uptake by fish.
- From mammalian studies, it has been postulated that radiostrontium damages blood capillaries and thereby compromises bone blood flow (Nielsen, 2004). Similar work could be recommended in fish to understand low-dose radiation effects.
- Subcellular localization of stable Sr is probably less important from toxicological perspectives, but the information will be useful for precise understanding of the internal radiation dose–effects relationship in fish for ^{90}Sr.

REFERENCES

Aidley, D. J., and Stanfield, P. R. (1998). *Ion Channels: Molecules in Action.* Cambridge University Press, Cambridge.

ATSDR (2004). *Toxicological Profile for Strontium.* Agency for Toxic Substances and Disease Registry, US Department of Health and Human Services, Atlanta, GA.

Beddington, J. R., Mills, C. A., Beards, F., Minski, M. J., and Bell, J. N. B. (1989). Long-term changes in strontium-90 concentrations within a freshwater predator–prey system. *J. Fish. Biol.* **35**, 679–686.

Benkdad, A., Laissaoui, A., El Bari, H., Benmansour, M., and IbnMajah, M. (2008). Partitioning of radiostrontium in marine aqueous suspensions: laboratory experiments and modeling studies. *J. Environ. Radioactiv.* **99**, 748–756.

Berg, A. (1968). Studies on the metabolism of calcium and strontium in freshwater fish. I. Relative contribution of direct and intestinal absorption. *Mem. Inst. Ital. Idrobid.* **23**, 161–196.

Berg, A. (1970). Studies on the metabolism of calcium and strontium in freshwater fish. III. Effect of EDTA as a chelating agent on the exchange of calcium and strontium from water. *Mem. Inst. Ital. Idrobid.* **26**, 257–267.

Bernat, M., Church, T., and Allegre, C. J. (1972). Barium and strontium concentrations in Pacific and Mediterranean seawater profiles by direct isotope dilution mass spectrometry. *Earth Planet. Sci. Lett.* **16**, 75–80.

Bidwell, J. P., Paige, J. A., and Kuzirian, A. M. (1986). Effects of strontium on the embryonic development of *Aplysia californica*. *Biol. Bull.* **170**, 75–90.

Birge, W. J. (1978). Aquatic toxicology of trace elements of coal and fly ash. In: *Energy and Environmental Stress in Aquatic Systems* (J. H. Thorp and J. W. Gibbons, eds), pp. 219–240. Department of Energy Symposium Series, Augusta, GA.

Birge, W. J., Black, J. A., and Ramey, B. A. (1981). The reproductive toxicology of aquatic contaminants. In: *Hazard Assessment of Chemicals: Current Developments* (J. Saxena and F. Fisher, eds), pp. 59–115. Academic Press, New York.

Blaylock, B. G., and Frank, M. L. (1980). Effects of chronic low-level irradiation on *Gambusia affinis*. In: *Radiation Effects on Aquatic Organisms* (N. Egami, ed.), pp. 81–91. University Park Press, Baltimore, MD.

Blaylock, B. G., and Griffith, N. A. (1971). Effects of acute beta and gamma radiation on developing embryos of carp (*Cyprinus carpio*). *Radiat. Res.* **46**, 99–104.

Bogé, G., Rocheb, H., and Baloccoc, C. (2002). Amino acid transport by intestinal brush border vesicles of a marine fish, *Boops salpa*. *Comp. Biochem. Physiol. B* **131**, 19–26.

Boroughs, H., and Reid, D. F. (1958). The role of the blood in the transportation of strontium^{90}–yittrium^{90} in teleost fish. *Biol. Bull.* **115**, 64–73.

Boroughs, H., Townsley, S. J., and Hiatt, R. W. (1956). The metabolism of radionuclides by marine organisms. I. The uptake, accumulation, and loss of strontium^{90} by marine fishes. *Biol. Bull.* **111**, 336–351.

Brown, J. (ed.). (2000). *Radionuclide Uptake and Transfer in Pelagic Food-Chains of the Barents Sea and Resulting Doses to Man and Biota*. Norwegian Radiation Protection Authority, Østerås.

Brown, J. E., Borretzen, P., Dowdall, M., Sazykina, T., and Kryshev, I. (2004). The derivation of transfer parameters in the assessment of radiological impacts on Arctic marine biota. *Arctic* **57**, 279–289.

Brungs, W. A. (1965). Experimental uptake of strontium-85 by freshwater organisms. *Health Phys.* **11**, 41–46.

Burger, J., Gaines, K. F., Boring, C. S., Stephens, W. L., Snodgrass, J., Dixon, C., McMahon, M., Shukla, S., Shukla, T., and Gochfeld, M. (2002). Metal levels in fish from the Savannah River: potential hazards to fish and other receptors. *Environ. Res. A* **89**, 85–97.

Carlton, W. H., Murphy, C. E., Jr., Jannik, G. T., and Simpkins, A. A. (1999). Radiostrontium in the Savannah river site environment. *Health Phys.* **77**, 677–685.

Carraca, S., Ferreira, A., and Coimbra, J. (1990). Sr transfer factor between different levels in the trophic chain in two dams of Douro rivers (Portugal). *Water Res.* **24**, 1497–1508.

Chowdhury, M. J. (2001). *Bioavailability of Strontium to the Common Carp, Cyprinus carpio*. PhD Thesis, University of Antwerp.

Chowdhury, M. J., and Blust, R. (2001a). A mechanistic model for the uptake of waterborne strontium in the common carp (*Cyprinus carpio* L.). *Environ. Sci. Technol.* **35**, 669–675.

Chowdhury, M. J., and Blust, R. (2001b). Effects of temperature on the uptake of waterborne strontium in the common carp (*Cyprinus carpio* L.). *Aquat. Toxicol.* **54**, 151–160.

Chowdhury, M. J., and Blust, R. (2002). Bioavailability of waterborne strontium to the common carp, *Cyprinus carpio*, in complexing environments. *Aquat. Toxicol.* **58**, 215–227.

Chowdhury, M. J., and Wood, C. M. (2007). Renal function in the freshwater rainbow trout after dietary cadmium acclimation and waterborne cadmium challenge. *Comp. Biochem. Physiol. C* **145**, 321–332.

Chowdhury, M. J., van Ginneken, L., and Blust, R. (2000). Kinetics of waterborne strontium uptake in the common carp, *Cyprinus carpio*, at different calcium levels. *Environ. Toxicol. Chem.* **19**, 622–630.

Chowdhury, M. J., McDonald, D. G., and Wood, C. M. (2004). Gastrointestinal uptake and fate of cadmium in rainbow trout acclimated to sublethal dietary cadmium. *Aquat. Toxicol.* **69**, 149–163.

Collingsworth, P. D., Van Tassell, J. J., Olesik, J. W., and Marschall, E. A. (2010). Effects of temperature and elemental concentration on the chemical composition of juvenile yellow perch (*Perca flavescens*) otoliths. *Can. J. Fish. Aquat. Sci.* **67**, 1187–1196.

Cossins, A. R., and Bowler, K. (1987). *Temperature Biology of Animals*. Chapman & Hall, London.

Coughtrey, P. J., and Thorne, M. C. (1983). *Radionuclide Distribution and Transport in Terrestrial and Aquatic Ecosystems: A Critical Review of Data*, Vol. 1. A. A. Balkema, Rotterdam.

Dahl, S. G., Allain, P., Marie, P. J., Mauras, Y., Boivin, G., Ammann, P., Tsouderos, Y., Delmas, P. D., and Christiansen, C. (2001). Incorporation and distribution of strontium in bone. *Bone* **28**, 446–453.

da Silva, J. J. R., and Williams, R. J. P. (1991). *The Biological Chemistry of the Elements*. Oxford University Press, Oxford.

De Villiers, S. (1999). Seawater strontium and Sr/Ca variability in the Atlantic and Pacific oceans. *Earth Planet. Sci. Lett.* **171**, 623–634.

Dwyer, F. J., Burch, S. A., Ingersoll, C. G., and Hunn, J. B. (1992). Toxicity of trace element and salinity mixtures to striped bass (*Morone saxatilis*) and *Daphnia magna*. *Environ. Toxicol. Chem.* **11**, 513–520.

Fabrik, I., Kukacka, J., Baloun, J., Sotornik, I., Adam, V., Prusa, R., Vajtr, D., Babula, P., and Kizek, R. (2009). Electrochemical investigation of strontium–metallothionein interactions – analysis of serum and urine of patients with osteoporosis. *Electroanalysis* **21**, 650–656.

Farrell, J., and Campana, S. E. (1996). Regulation of calcium and strontium deposition on the otoliths of juvenile tilapia, *Oreochromis niloticus*. *Comp. Biochem. Physiol. A* **115**, 103–109.

Fleishman, D. G. (1973). Accumulation of artificial radionuclides in freshwater fish. In: *Radioecology* (V. M. Klechkovskii, G. G. Polikarpov and R. M. Alesakhin, eds), pp. 347–371. John Wiley & Sons, New York.

Flik, G., and Verbost, P. M. (1993). Calcium transport in fish gills and intestine. *J. Exp. Biol.* **184**, 17–29.

Flik, G., Verbost, P. M., and Wendelaar Bonga, S. E. (1995). Calcium transport processes in fishes. In: *Fish Physiology. Cellular and Molecular Approaches to Fish Ionic Regulation* (C. M. Wood and T. J. Shuttleworth, eds), pp. 317–343. Academic Press, New York.

Friedberg, F., and Taliaferro, L. (2005). Calmodulin genes in zebrafish (revisited). *Mol. Biol. Rep.* **32**, 55–60.

Gleizon, P, and McDonald, P. (2010). Modelling radioactivity in the Irish Sea: from discharge to dose. *J. Environ. Radioactiv.* **101**, 403–413.

Glover, C. N., and Hogstrand, C. (2002). Amino acid modulation of *in vivo* intestinal zinc absorption in freshwater rainbow trout. *J. Exp. Biol.* **205**, 151–158.

Håkanson, L. (2005). A new general dynamic model predicting radionuclide concentrations and fluxes in coastal areas from readily accessible driving variables. *J. Environ. Radioactiv.* **78**, 217–245.

Hanlon, R. T., Bidwell, J. P., and Tait, R. (1989). Strontium is required for statolith development and thus normal swimming behavior of hatchling cephalopods. *J. Exp. Biol.* **141**, 187–195.

Health Canada (2000). *Canadian Guidelines for the Restriction of Radioactively Contaminated Food and Water Following a Nuclear Emergency: Guidelines and Rationale*. Health Canada, Radiation Protection Bureau, Ottawa.

Hoenderop, J. G., Nilius, B., and Bindels, R. J. (2005). Calcium absorption across epithelia. *Physiol. Rev.* **85**, 373–422.

Höllriegl, V., Li, W. B., Greiter, M., and Oeh, U. (2007). Plasma clearance and urinary excretion after intravenous injection of stable ^{84}Sr in humans. *Radiat. Protect. Dosim.* **127**, 144–147.

Holmes-Farley, R. (2003). Chemistry and the aquarium. *Advanced Aquarist's On-line Magazine*. http://www.advancedaquarist.com/issues/nov2003/chem.htm

Hosseini, A., Thørring, H., Brown, J. E., Saxén, R., and Ilus, E. (2008). Transfer of radionuclides in aquatic ecosystems – default concentration ratios for aquatic biota in the Erica Tool. *J. Environ. Radioactiv.* **99**, 1408–1429.

Hu, Q.-H., Weng, J.-Q., and Wang, J.-S. (2010). Sources of anthropogenic radionuclides in the environment: a review. *J. Environ. Radioactiv.* **101**, 426–437.

Hwang, P.-P. (2009). Ion uptake and acid secretion in zebrafish (*Danio rerio*). *J. Exp. Biol.* **212**, 1745–1752.

IAEA (2004). *Sediment Distribution Coefficients and Concentration Factors for Biota in the Marine Environment*. IAEA Technical Reports Series No. 422. Vienna: International Atomic Energy Agency. http://wwwpub.iaea.org/MTCD/publications/PDF/TRS422_web.pdf (20.12.06).

IAEA (2009). *Quantification of Radionuclide Transfer in Terrestrial and Freshwater Environments for Radiological Assessments*. IAEA-TECDOC-1616.P.616. International Atomic Energy Agency, Vienna.

IPPNW (1991). *Radioactive Heaven and Earth: The Health and Environmental Effects of Nuclear Weapons Testing in, on, above the Earth*. International Physicians for the Prevention of Nuclear War, Apex Press, New York.

Kennedy, B. P., Blum, J. D., Folt, C. L., and Nislow, K. H. (2000). Reconstructing the lives of fish using Sr isotopes in otoliths. *Can. J. Fish. Aquat. Sci.* **59**, 925–929.

Kennedy, B. P., Klaue, A., Blum, J. D., Folt, C. L., and Nislow, K. H. (2002). Using natural strontium isotopic signatures as fish markers: methodology and application. *Can. J. Fish. Aquat. Sci.* **57**, 2280–2292.

Klinck, J. S., Ng, T., and Wood, C. M. (2009). Cadmium accumulation and *in vitro* analysis of calcium and cadmium transport functions in the gastro-intestinal tract of trout following chronic dietary cadmium and calcium feeding. *Comp. Biochem. Physiol. C* **150**, 349–360.

Kryshev, A. I. (2006). ^{90}Sr in fish: a review of data and possible model approach. *Sci. Total Environ.* **370**, 182–189.

Kryshev, A. I., Sazykina, T. G., and Sanina, K. D. (2008). Modelling of effects due to chronic exposure of a fish population to ionizing radiation. *Radiat. Environ. Biophys.* **47**, 121–129.

Kryshev, I. I. (1995). Radioactive contamination of aquatic ecosystems following the Chernobyl accident. *J. Environ. Radioactiv.* **27**, 207–219.

Kryshev, I. I. (2003). Model reconstruction of ^{90}Sr concentrations in fish from 16 Ural lakes contaminated by the Kyshtym accident of 1957. *J. Environ. Radioactiv.* **64**, 67–84.

Kryshev, I. I., and Sazykina, T. G. (1995). Radiological consequences of radioactive contamination of the Kara and Barents Seas. *J. Environ. Radioactiv.* **29**, 213–223.

Kryshev, I. I., and Sazykina, T. G. (1998). Radioecological effects on aquatic organisms in the areas with high levels of radioactive contamination: environmental protection criteria. *Radiat. Prot. Dosim.* **75**, 187–191.

Kryshev, I. I., Romanov, G. N., Isaeva, L. N., and Kholina, Y. B. (1997). Radioecological state of lake in the Southern Ural impacted by radioactive release of the 1957 radiation accident. *J. Environ. Radioactiv.* **34**, 223–235.

Kryshev, I. I., Romanov, G. N., Chumichev, V. B., Sazykina, T. G., Isaeva, L. N., and Ivanitskaya, M. V. (1998). Radioecological consequences of radioactive discharges into the Techa River on the Southern Urals. *J. Environ. Radioactiv.* **38**, 195–209.

Lall, S. P., and Lewis-McCrea, L. M. (2007). Role of nutrients in skeletal metabolism and pathology in fish – an overview. *Aquaculture* **267**, 3–19.

Larsson, D., Lundgren, T., and Sundell, K. (1998). Ca^{2+} uptake through voltage-gated L-type Ca^{2+} channels by polarized enterocytes from Atlantic cod *Gadus morhua*. *J. Membr. Biol.* **164**, 229–237.

Lide, D. R. (1998). *CRC Handbook of Chemistry and Physics* (78th edn.). CRC Press, Boca Raton, FL.

Lin, S.-H., Chang, C.-W., Lizika, Y., and Tzeng, W.-N. (2007). Salinities, not diets, affect strontium/calcium ratios in otoliths of *Anguilla japonica*. *J. Exp. Mar. Biol. Ecol.* **341**, 254–263.

Loretz, C. A. (2008). Extracellular calcium-sensing receptors in fishes. *Comp. Biochem. Physiol. A* **149**, 225–245.

Marshall, W. S., Bryson, S. E., and Wood, C. M. (1992). Calcium transport by isolated skin of rainbow trout. *J. Exp. Biol.* **166**, 297–316.

Martin, D., and Goldberg, E. D. (1962). Uptake and assimilation of radiostrontium by Pacific mackerel. *Limnol. Ocean.* **7**, 76–81.

Martin, G. B., Thorrold, S. R., and Jones, C. M. (2004). Temperature and salinity effects of strontium incorporation in otoliths of larval spot (*Leiostomus xanthurus*). *Can. J. Fish. Aquat. Sci.* **61**, 34–42.

Matsunaga, T., Nagao, S., Ueno, T., Takeda, S., Amano, H., and Tkachenko, Y. (2004). Association of dissolved radionuclides released by the Chernobyl accident with colloidal materials in surface water. *Appl. Geochem.* **19**, 1581–1599.

McGeer, J. C., Brix, K. V., Skeaff, J. M., DeForest, D. K., Brigham, S. I., Adams, W. J., and Green, A. (2003). Inverse relationship between bioconcentration factor and exposure concentration for metals: implications for hazard assessment in the aquatic environment. *Environ. Toxicol. Chem.* **22**, 1017–1037.

MDEQ (2008). *Rule 57 Water Quality Values* 2008-12-10. Surface Water Quality Assessment Section, Michigan Department of Environmental Quality.

Melancon, S., Fryer, B. J., and Markham, J. L. (2009). Chemical analysis of endolymph and the growing otolith: fractionation of metals in freshwater fish species. *Environ. Toxicol. Chem.* **28**, 1279–1287.

Michalak, M., Parker, J. M. R., and Opas, M. (2002). Ca^{2+} signaling and calcium binding chaperones of the endoplasmic reticulum. *Cell Calcium* **32**, 269–278.

Millero, F. J. (1996). *Chemical Oceanography* (2nd edn.). CRC Press, Boca Raton, FL.

Moiseenko, T. I., and Kudryavtseva, L. P. (2001). Trace metal accumulation and fish pathologies in areas affected by mining and metallurgical enterprises in the Kola Region, Russia. *Environ. Pollut.* **114**, 285–297.

Monte, L., Boyer, P., Brittain, J. E., Håkanson, L., Lepicard, S., and Smith, J. T. (2005). Review and assessment of models for predicting the migration of radionuclides through rivers. *J. Environ. Radioactiv.* **79**, 273–296.

Monte, L., Boyer, P., Brittain, J. E., Goutal, N., Heling, R., Kryshev, A., Kryshev, I., Laptev, G., Luck, M., Perianez, R., Siclet, F., and Zheleznyak, M. (2008). Testing models for predicting the behaviour of radionuclides in aquatic systems. *Appl. Radiat. Isot.* **66**, 1736–1740.

Mothersill, C., and Seymour, C. (2009). Communication of ionizing radiation signals – a tale of two fish. *Int. J. Radiat. Biol.* **85**, 909–919.

NHCAR (2008). *Surface Water Quality Regulations*, Chapter ENV-Wq 1700. New Hampshire Code of Administrative Rules, State of New Hampshire, USA.

Nielsen, S. P. (2004). The biological role of strontium. *Bone* **35**, 583–588.

Nordén, M., and Dabek-Zlotorzynska, E. (1996). Study of metal–fulvic acid interactions by capillary electrophoresis. *J. Chromatogr. A* **739**, 421–429.

Nosov, A. V., Ivanov, A. B., Pechkurov, A. V., Vozzhennikov, I., and Nikonov, S. A. (1999). Setting standards for the safe level of radioactive contamination of water and bottom sediments in bodies of water. *Atom. Energy.* **86**, 398–407.

Ohio EPA (2009). *Lake Erie Basin Aquatic Life and Human Health Tier I Criteria, Tier II Values and Screening Values (SV)*. Chapter 3745-1 of the Ohio Administrative Code. Division of Surface Water, Ohio Environmental Protection Agency.

Ophel, I. L., and Judd, J. M. (1966). Experimental studies of radiostrontium accumulation by freshwater fish from food and water. In: *Radioecological Concentration Processes* (B. Aberg and F. P. Hungate, eds), pp. 859–865. Pergamon Press, Oxford.

Outola, I., Saxén, R. L., and Heinävaara, S. (2009). Transfer of ^{90}Sr into fish in Finnish lakes. *J. Environ. Radioactiv.* **100**, 657–664.

Perry, S. F., and Wood, C. M. (1985). Kinetics of branchial calcium uptake in the rainbow trout: effects of acclimation to various external calcium levels. *J. Exp. Biol.* **116**, 411–433.

Pillard, D. A., DuFresne, D. L., Caudle, D. D., Tietge, J. E., and Evans, J. M. (2000). Predicting the toxicity of major ions in seawater to mysid shrimp (*Mysidopsis bahia*), sheepshed minnow (*Cyprinodon variegates*), and inland silverside minnow (*Menidia beryllina*). *Environ. Toxicol. Chem.* **19**, 183–191.

Polikarpov, G. G. (1966). *Radioecology of Aquatic Organisms: The Accumulation, and Biological Effect of Radioactive Substances*. Reinhold, New York.

Pyle, G. G., Swanson, S. M., and Lehmkuhl, D. M. (2002). Toxicity of uranium mine receiving waters to early life stage fathead minnows (*Pimephales promelas*) in the laboratory. *Environ. Pollut.* **116**, 243–255.

Radzki, R. P., Bienko, M., Filip, M. R., Albera, E., and Kankofer, M. (2009). Effect of strontium ranelate on femur densitometry and antioxidative/oxidative status in castrated male rats. *Scand. J. Lab. Anim. Sci.* **36**, 193–201.

Rosenthal, H. L. (1957). The metabolism of strontium-90 and calcium-45 by *Lebistes*. *Biol. Bull.* **113**, 442–450.

Sanada, Y., Matsunaga, T., Yanase, N., Nagao, S., Amano, H., Takada, H., and Tkachenko, Y. (2002). Accumulation and potential dissolution of Chernobyl-derived radionuclides in river bottom sediment. *Appl. Radiat. Isot.* **56**, 751–760.

Sazykina, T. G., and Kryshev, A. I. (2003). EPIC database on the effects of chronic radiation in fish: Russian/FSU data. *J. Environ. Radioactiv.* **68**, 65–87.

Schiffman, R. H. (1965). Strontium–calcium transport across the gill of rainbow trout (*Salmo gairdnerii*). *J. Cell. Comp. Physiol.* **65**, 1–6.

Shahsavarani, A., and Perry, S. F. (2006). Hormonal and environmental regulation of epithelial calcium channel in gill of rainbow trout (*Oncorhynchus mykiss*). *Am. J. Physiol. Regul. Integr. Comp. Physiol.* **291**, R1490–R1498.

Shahsavarani, A., McNeill, B., Galvez, F., Wood, C. M., Goss, G. G., Hwang, P. P., and Perry, S. F. (2006). Characterization of a branchial epithelial calcium channel (ECaC) in freshwater rainbow trout (*Oncorhynchus mykiss*). *J. Exp. Biol.* **209**, 1928–1943.

Shirran, S. L., and Barran, P. E. (2009). The use of ESI-MS to probe the binding of divalent cations to calmodulin. *J. Am. Soc. Mass Spectrom.* **20**, 1159–1171.

Smith, J. T., and Beresford, N. A. (2005). *Chernobyl Catastrophe and Consequences*. Springer/ Praxis, Chichester.

Smith, J. T., Belova, N. V., Bulgakov, A. A., Comans, R. N. J., Konoplev, A. V., Kudelsky, A. V., Madruga, M., Voitsekhovitch, O. V., and Zibold, G. (2005a). The "AQUASCOPE" simplified model for predicting Sr-89, Sr-90, I-131, and Cs-134, Cs-137 in surface waters after a large-scale radioactive fallout. *Health Phys.* **89**, 628–644.

Smith, J. T., Voitsekhovitch, O. V., Knoplov, A. V., and Kudelsky, A. V. (2005b). Radioactivity in aquatic system. In *Chernobyl Catastrophe and Consequences*. In: (J. T. Smith and N. A. Beresford, eds), pp. 139–189. Springer/Praxis, Chichester.

Smith, J. T., Sasina, N. V., Kryshev, A. I., Belova, N. V., and Kudelsky, A. V. (2009). A review and test of predictive models for the bioaccumulation of radiostrontium in fish. *J. Environ. Radioactiv.* **100**, 950–954.

Smith, R. W., Wang, J., Bucking, C. P., Mothersill, C. E., and Seymour, C. B. (2007). Evidence for a protective response by the gill proteome of rainbow trout exposed to X-ray induced bystander signals. *Proteomics* **7**, 4171–4180.

Suzuki, Y., Nakamura, R., and Ueda, T. (1972). Accumulation of strontium and calcium in freshwater fishes of Japan. *J. Radiat. Res.* **13**, 199–207.

Templeton, W. L., Nakatani, R. E., and Heid, E. E. (1971). Radiation effects. In *Radioactivity in the Marine Environment*. National Academy of Sciences, Washington, DC, 223–239

Theodorakis, C. W., Blaylock, B. G., and Shugart, L. R. (1997). Genetic ecotoxicology I: DNA integrity and reproduction in mosquitofish exposed *in situ* to radionuclides. *Ecotoxicology* **6**, 205–218.

Thresher, R. E., Proctor, C. H., Gunn, J. S., and Harrowfield, I. R. (1994). An evaluation of electron-probe microanalysis of otoliths for delineation and identification of nursery areas in a southern temperate groundfish, *Nemadactylus macropterus* (Cheilodactylidae). *Fish. Bull.* **92**, 817–840.

Trabalka, J. R., and Allen, C. P. (1977). Aspects of fitness of a mosquitofish *Gambusia affinis* exposed to chronic low-level environmental radiation. *Radiat. Res.* **70**, 198–211.

Tzeng, W.-N., Severin, K. P., Wickström, H., and Wang, C.-H. (1999). Strontium bands in relation to age marks in otoliths of European eel *Anguilla anguilla*. *Zool. Stud.* **38**, 452–457.

USEPA (1993). *Methods for Measuring the Acute Toxicity of Effluents and Receiving Waters to Freshwater and Marine Organism* (4th edn.). Office of Research and Development, US Environmental Protection Agency, Cincinnati, OH, EPA/600/4-90/027F.

USGS (2008). *Mineral Commodity Summaries*. US Geological Survey, Reston, VA.

Vijver, M. G., Van Gestel, C. A. M., Van Straalen, N. M., Lanno, R. P., and Peijnenburg, W. J. G. M. (2006). Biological significance of metals partitioned to subcellular fractions within earthworms (*Aporrectodea caliginosa*). *Environ. Toxicol. Chem.* **25**, 807–814.

Walker, J. B. (1953). Inorganic micronutrient requirements of *Chlorella*: I. requirements for calcium (or strontium), copper, and molybdenum. *Arch. Biochem. Biophys.* **46**, 1–11.

Walther, B. D., and Thorrold, S. R. (2006). Water, not food, contributes the majority of strontium and barium deposited in the otoliths of a marine fish. *Mar. Ecol. Prog. Ser.* **311**, 125–130.

Wasserman, R. H. (1998). Strontium as a tracer for calcium in biological and clinical research. *Clin. Chem.* **44**, 437–439.

Wren, C. D., Maccrimmon, H. R., and Loescher, B. R. (1983). Examination of bioaccumulation and biomagnification of metals in a precambrian shield lake. *Water Air Soil Pollut.* **19**, 277–291.

Xu, Y., and Marcantonio, F. (2004). Speciation of strontium in particulates and sediments from the Mississippi River mixing zone. *Geochim. Cosmochim. Acta* **68**, 2649–2657.

Yankovich, T. L. (2009). Mass balance approach to estimating radionuclide loads and concentrations in edible fish tissues using stable analogues. *J. Environ Radioactiv.* **100**, 795–801.

Zaccone, G., Wendelaar Bonga, S. E., Flik, G., Fasulo, S., Licata, A., Lo Cascio, P., Mauceri, A., and Lauriano, E. R. (1992). Localization of calbindin D28K-like immunoreactivity in fish gill: a light microscopic and immunoelectron histochemical study. *Regul. Pept.* **41**, 195–208.

Zimmerman, C. E. (2005). Relationship of otolith strontium-to-calcium ratios and salinity: experimental validation for juvenile salmonids. *Can. J. Fish Aquat. Sci.* **62**, 88–97.

8

URANIUM

RICHARD R. GOULET

CLAUDE FORTIN

DOUGLAS J. SPRY

Homeostasis and Toxicology of Non-Essential Metals: Volume 31B   Copyright © 2012, Her Majesty the Queen in right of Canada
FISH PHYSIOLOGY   DOI: 10.1016/S1546-5098(11)31030-8

Release of uranium (U) to the environment is mainly through the nuclear fuel cycle. In oxic waters, U(VI) is the predominant redox state, while U(IV) is likely to be encountered in anoxic waters. The free uranyl ion (UO_2^{2+}) dominates dissolved U speciation at low pH while complexes with hydroxides and carbonates prevail in neutral and alkaline conditions. Whether the toxicity of U(VI) to fish can be predicted based on its free ion concentration remains to be demonstrated but a strong influence of pH has been shown. In the field, U accumulates in bone, liver, and kidney, but does not biomagnify. There is certainly potential for uptake of U via the gill based on laboratory studies; however, diet and/or sediment may be the major route of uptake, and may vary with feeding strategy. Uranium toxicity is low relative to many other metals, and is further reduced by increased calcium, magnesium, carbonates, phosphate, and dissolved organic matter in the water. Inside fish, U produces reactive oxygen species and causes oxidative damage at the cellular level. The radiotoxicity of enriched U has been compared with chemical toxicity and it has been postulated that both may work through a mechanism of production of reactive oxygen species. In practical terms, the potential for chemotoxicity of U outweighs the potential for radiotoxicity. The toxicokinetics and toxicodynamics of U are well understood in mammals, where bone is a stable repository and the kidney the target organ for toxic effects from high exposure concentrations. Much less is known about fish, but overall, U is one of the less toxic metals.

1. CHEMICAL SPECIATION IN FRESHWATER AND SEAWATER

Several techniques, such as time-resolved fluorescence spectroscopy, are currently available to explore analytically the speciation of uranium (U) in aqueous solutions (Moulin et al., 1998). These techniques are, however, not routinely applicable and, thus, we often have to rely on equilibrium thermodynamic calculations. As most of the thermodynamic data available on U were obtained at high concentrations, the stability constants for dimeric or polymeric species of uranyl are more adequately defined than monomeric species [such as the uranyl dihydroxo complex: $UO_2(OH)_{2(aq)}$] that only become noteworthy at very low concentrations (i.e. at environmentally significant concentrations) (Choppin and Mathur, 1991). Large sets of U thermodynamic data have been extensively reviewed in the past (e.g. Guillaumont et al., 2003 and references therein). However, the relative scarcity of data obtained in conditions that are applicable to an environmental setting increases the uncertainty of chemical speciation based on thermodynamic chemical equilibrium (Nitzsche et al., 2000; Unsworth et al., 2002; Denison and Garnier-Laplace, 2005). The speciation of U in aquatic systems has been reviewed in the past (Moulin and Moulin, 2001; Markich, 2002; Ansoborlo et al., 2006).

1.1. Solubility

In oxic surface waters, the U mineral phase most likely to precipitate is Schoepite ($UO_3 \cdot 2H_2O_{(s)}$; log $K_{sp} = 10^{-4.81}$) (Denison and Garnier-Laplace, 2005). Fig. 8.1 illustrates the strong pH-dependent solubility of Schoepite.

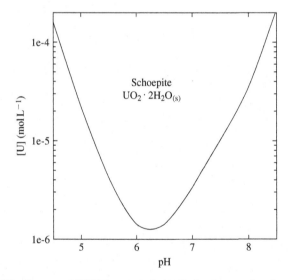

Fig. 8.1. Solubility diagram of U(VI) in a typical oxic freshwater as a function of pH. Surface water is assumed to be at equilibrium with the atmosphere ($pCO_2 = 10^{-3.51}$ atm). Total dissolved UO_2 was calculated using MINEQL+ v4.62 and stability constants from Denison and Garnier-Laplace (2005). Water composition used (in mol L^{-1}): [Ca] = 6.8×10^{-4}; [Mg] = 2.6×10^{-4}; [Na] = 4.3×10^{-4}; [K] = 3.0×10^{-4}; [Cl] = 4.8×10^{-4}; [SO$_4$] = 2.4×10^{-4}; [F] = 2.5×10^{-6}.

These calculations indicate that a minimum in solubility ($\sim 1\ \mu mol\ L^{-1}$ or $\sim 240\ \mu g\ L^{-1}$) is reached between pH 6.0 and 6.5. Experimental evidence shows that the solubility minimum can be slightly lower but can also be shifted toward a more circumneutral pH (Silva, 1992; Jang et al., 2006). Such discrepancies may be the result of poorly defined thermodynamic constants used for the calculations or differences in solution composition. At lower (< 6) and higher (>7) pH, solubility increases markedly. The presence of a strong ligand will result in an increase in uranyl solubility. In natural surface waters, natural dissolved organic matter (DOM) may play this role through metal binding. Here, DOM is referred to in the general sense and dissolved organic carbon (DOC) is the way in which measured concentrations are usually expressed, in mg L^{-1}. It can be assumed that carbon represents 50% of the DOM and that, on average, 60% of DOM is composed of humic and fulvic acids with a ratio of 1:3 (Perdue and Ritchie, 2003).

It is, however, unlikely that an inorganic ligand other than hydroxide or carbonate could contribute to notably increase U solubility. For example, a 10-fold increase in the sulfate concentration used in Fig. 8.1 would increase Schoepite solubility at pH 6.25 from $1.23\ \mu mol\ L^{-1}$ to $1.30\ \mu mol\ L^{-1}$. However, water oversaturated with carbonates (e.g. groundwater) would favor the presence of carbonato-complexes and thus facilitate U solubility to some extent.

1.2. Redox Speciation

The two dominant aqueous redox states are U(IV) and U(VI). The former will only exist in reducing conditions and is poorly soluble (Ragnarsdottir and Charlet, 2000). In the context of fish exposure to aqueous U, it can be assumed that U will be present in the U(VI) state. As for sedimentary U exposure, incidental uptake of U(IV) by fish is also plausible.

1.3. Species Distribution in Oxic Waters

As noted above, the two major inorganic ligands that significantly influence uranyl aqueous speciation are hydroxyl and carbonate ions. It follows that the key parameter that will control dissolved U speciation in surface waters is pH (Fig. 8.2). The free uranyl ion (UO_2^{2+}) is the most abundant species at pH < 5. At higher pH, hydroxo- and carbonato-complexes dominate. Other inorganic ligands usually play a minor role owing to the relatively low binding affinity for UO_2^{2+} and/or low ligand concentration. For example, uranyl complexes with sulfate or fluoride will only become significant (i.e. >1% of all species present) at low pH (Fig. 8.2) or in conditions

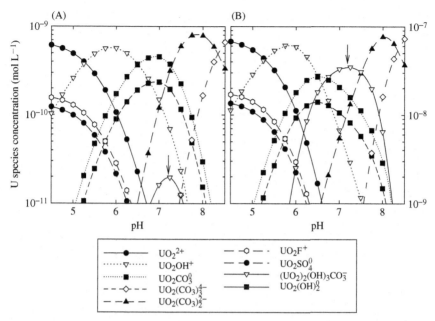

Fig. 8.2. Speciation diagram of U(VI) in a typical oxic freshwater as a function of pH. Surface water is assumed to be at equilibrium with the atmosphere ($pCO_2 = 10^{-3.51}$ atm). Total dissolved UO_2 was (a) 1 nmol L^{-1} and (b) 110 nmol L^{-1}. Species distributions were calculated using MINEQL+ v4.62 and stability constants from Denison and Garnier-Laplace (2005). Water composition used (in mol L^{-1}): [Ca] = 6.8×10^{-4}; [Mg] = 2.6×10^{-4}; [Na] = 4.3×10^{-4}; [K] = 3.0×10^{-4}; [Cl] = 4.8×10^{-4}; [SO$_4$] = 2.4 $\times 10^{-4}$; [F] = 2.5×10^{-6}. Arrows are to identify the influence of U concentration on the relative importance of the mixed species $(UO_2)_2(OH)_3CO_3^-$.

where unusually high ligand concentrations are found. Similarly, the proportion of phosphate complexes may reach significant levels but only at rather high dissolved phosphate concentrations (e.g. >0.1 µmol L^{-1} or 3 µg L^{-1}) that are not likely to be encountered in natural freshwaters. It should be noted also that U is poorly soluble in the presence of high concentrations of HPO$_4^{2-}$ (precipitates as $UO_2HPO_{4(s)}$; log $K_{sp} = 10^{-24.2}$).

At high concentrations, a uranyl dimeric ternary complex [(UO$_2$)$_2$ (OH)$_3$CO$_3^-$; see arrows in Fig. 8.2] becomes the dominant species at circumneutral pH. Such species distribution, however, is only relevant to laboratory work in the absence of any particulate or DOM such as humic and fulvic acids. The speciation of U is expected to be largely affected by the presence of DOM (Moulin et al., 1992; Unsworth et al., 2002). Metal complexation by humic and fulvic acids is difficult to predict because of their polyfunctional and heterogeneous nature. The diversity in functional groups and molecular structures results in a distribution of binding site affinities of DOM for a given metal (i.e. a continuum of log K values instead of a single one). In the last two decades some progress has been made to account for these complex reactions within equilibrium models. One of these models, the Windermere humic aqueous model (WHAM) is commonly used to estimate metal speciation in the presence of DOM and has been previously calibrated for U (Tipping, 2002). The WHAM model was used to illustrate the influence of DOM on U(VI) speciation (Fig. 8.3). Several scenarios were explored by using different DOC concentrations (1, 5, and 10 mg L^{-1}). These concentrations were selected arbitrarily but they do correspond to what would be encountered in poorly humic to highly humic waters.

Since metal complexation is often strongly influenced by the presence of Fe(III) and Al, which are ubiquitous in surface freshwaters, these elements were included in the calculations (Tipping, 2005; Lofts et al., 2008). Typically, the dissolved concentrations of Fe(III) and Al in surface waters are strongly pH dependent and are mostly controlled by dissolution reactions with a classic U-shape function (similar to Fig. 8.1). In this exercise, a concentration of 1 μmol L^{-1} for both metals was used, based on previous observations (Fortin et al., 2010). It follows that these concentrations are underestimated for the low and high ends of the pH range examined. Nevertheless, Fig. 8.3

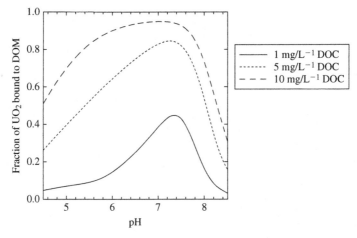

Fig. 8.3. Fraction of U(VI) bound to dissolved organic matter (DOM) in typical oxic freshwaters of varying dissolved organic carbon (DOC) concentrations as a function of pH. Surface water is assumed to be at equilibrium with the atmosphere ($p\text{CO}_2 = 10^{-3.51}$ atm). Total dissolved UO$_2$ was 1 nmol L^{-1} and species distributions were calculated using WHAM 6.0.1 and its default stability constants database. Water composition used (in mol L^{-1}): [Ca] $= 6.8 \times 10^{-4}$; [Mg] $= 2.6 \times 10^{-4}$; [Na] $= 4.3 \times 10^{-4}$; [K] $= 3.0 \times 10^{-4}$; [Cl] $= 4.8 \times 10^{-4}$; [SO$_4$] $= 2.4 \times 10^{-4}$; [F] $= 2.5 \times 10^{-6}$; [Al] $= 1 \times 10^{-6}$; [Fe(III)] $= 1 \times 10^{-6}$.

provides an overview of the important role of DOM on U speciation. Even in poorly humic water (1 mg L^{-1} DOC), the binding to DOM is predicted to be significant at circumneutral pH values. At more commonly found DOC concentrations of 5–10 mg L^{-1}, UO_2 becomes predominantly bound to DOM over a larger pH range. Thus, in natural surface waters, the speciation of UO_2 will be strongly influenced by the presence of natural DOM, even at low DOC concentrations.

At the relatively low concentrations found in seawater, dimeric species are not likely to dominate the U species distribution. The most abundant species expected are the di- and tri-carbonato complex (Djogic et al., 1986). In the open ocean with background DOC concentrations below 1 mg L^{-1}, organic complexes of U are not likely to be very important.

1.4. Kinetics of Uranium(VI) Binding

Thermodynamics can tell us about the direction a system is taking (e.g. complex formation) but does not provide any information on the rate at which these reactions are occurring. It is thus often assumed that equilibrium is attained, but this is not always the case. A key parameter in the determination of the rate of reaction of a metal is the water exchange rate constant (i.e. rate at which a water molecule leaves and enters the coordination sphere of the ion; k_{-w}) (Stumm and Morgan, 1996; Helm and Merbach, 1999). Indeed, dissolved ions are never actually free but, rather, complexed by water molecules (aquo-ions), although the term "free ion" is commonly used. Recent evidence has shown that UO_2^{2+} has four or five coordinated water molecules in its hydration sphere (Neuefeind et al., 2004). The water exchange rate constant for uranyl ions is in the range of $1.3–1.6 \times 10^6$ s^{-1} (Farkas et al., 2000), close to Co^{2+} and Fe^{2+}. Therefore, solution complexation reactions should reach equilibrium quite rapidly in surface waters. However, such an assumption remains to be validated in the presence of DOM that forms very stable complexes with UO_2^{2+}, especially in freshwaters.

The fate and transport of ions are often controlled by the reaction with suspended particles and sediments. These reactions are expected to be fast for the same reasons mentioned above. For example, the sorption of U on Goethite (iron oxyhydroxide) was shown to reach steady state within minutes to hours (Giammar and Hering, 2001). It should be noted, however, that precipitation reactions, as for many metals, can be very slow, with time scales of weeks to months (Silva, 1992; Giammar and Hering, 2001; Jang et al., 2006). Oversaturation with respect to a particular mineral phase is likely to be observed before an equilibrium can be reached. In other words, some precipitate may not form over a time scale that is expected in most laboratory experiments.

1.5. Interactions with Sediments

Sediments often act as sinks for metals. The mechanisms of U deposition seem, however, to differ between freshwater and marine sediments. In freshwater sediments, a large portion of the U is presumably associated with organic matter settling to the bottom, whereas in seawater, diffusion from the water column to the sediments is reported to be more important. In all cases, particulate U is strongly correlated to organic carbon (McManus et al., 2006; Chappaz et al., 2010). Uranium would then be immobilized by reduction from the U(VI) state to the poorly soluble U(IV) redox state (Windom et al., 2000). In Canadian Shield lakes, a small portion of U was remobilized from the sediments and linked to the dissolution of iron oxyhydroxides, resulting in an upward flux from the porewaters to the water column (Chappaz et al., 2010). It was suggested that this upward flux was due to the very low ambient dissolved concentrations (>7 ng U L^{-1}). In a lake where the dissolved U concentrations were higher, a downward flux into sediments was observed, similar to the typical profiles found in seawater (Chappaz et al., 2010).

1.6. Microbial Transformation in Sediments

The mobility and long-term stability of U in sediments are influenced by the form of the reduced product. As explained earlier, hexavalent U [U(VI)] is generally soluble and mobile, but forms sparingly soluble uraninite [U(IV) $O_{2(s)}$] upon reduction to tetravalent U (Wall and Krumholz, 2006). Lovely et al. (1991) demonstrated that bacteria can accelerate the reduction of U(VI) to U(IV) compared to abiotic reduction. Therefore, identifying and characterizing microbial U(VI) reduction products are vital for predicting U(IV) behavior when fish forage into the sediments.

2. SOURCES OF URANIUM AND ITS ECONOMIC IMPORTANCE

Nuclear reactor fuel is the main use of U. The isotope of interest for nuclear reactions is ^{235}U, because it can undergo fission and liberate energy. Uranium mills extract triuranium octooxide (U_3O_8) from crushed ore by either an acid or alkaline leaching process. Then, the U-laden solution is processed through solvent extraction which yields purified and concentrated U. Yellowcake, the end product of the milling process, is then shipped to a U refinery. The major steps involved in refining and conversion are the purification of the yellowcake to uranium trioxide (UO_3) in the UO_3 circuit and conversion of it to uranium hexafluoride (UF_6) in the UF_6 circuit. It is in the form of UF_6 that U is used for enrichment to fuel nuclear power stations. In another processing stage, the uranium dioxide (UO_2) circuit, an intermediate product from the UO_3 circuit is converted into CANDU (Canadian Deuterium Uranium) reactor-grade fuel for use in CANDU reactors.

Major uses for depleted U include armor-piercing ammunition, counterweight (e.g. in helicopter blades and airplane control surfaces), military applications (e.g. ammunition manufacturing, shielding on army tanks), and radiation shielding. The production of high-energy X-rays uses U metal as X-ray targets. Small amounts of U are used in various other industries and household products (ATSDR, 1999). Historically, U has been used in nuclear weapons. Uranium is also present as a contaminant in phosphate fertilizers (ATSDR, 1999).

In 2006, 20 countries were involved in the production of U. Canada (25%) and Australia (19%) contributed 44% of the world production, while Kazakhstan (13%), Niger (9%), Russia (8%), Namibia (8%), Uzbekistan (6%), and the USA (5%) contributed 49% (OECD, 2008).

Uranium demand is fundamentally driven by the number of operating nuclear reactors, which is ultimately driven by the demand for electricity. World demand for electricity is expected to double from 2002 through 2030 and reactor construction may increase to meet the projected increase in electricity demand and to replace aging infrastructure. Demand for nuclear energy may further increase so as to meet greenhouse gas emission targets, to desalinate seawater, to produce heat for industrial and residential purposes, and to produce hydrogen (a potential replacement for fossil fuel) by electrolysis. However, in countries where public concern with safety, security, non-proliferation, and waste disposal is not correctly addressed, the contribution of nuclear energy could be limited.

3. ENVIRONMENTAL SITUATIONS OF CONCERN

Australia, Canada, and Kazakhstan are the leading producers of U (OECD, 2008) and U concentrations in surface waters are monitored (Table 8.1). For example, at the Ranger uranium mine in northern Australia, upstream (natural) concentrations in Gulungul and Magela Creeks, sampled on a weekly basis during the annual wet season since 2002, have remained below 0.06 µg U L^{-1}. Similarly, in Ngarradj Creek upstream of the nearby Jabiluka mineral lease, U levels in upstream surface waters have remained below 0.23 µg U L^{-1}. In seawater, the total dissolved U concentration is around 14 nmol L^{-1} (3.3 µg U L^{-1}) (Meinrath et al., 2003).

Table 8.1
Aqueous concentrations of total uranium in countries involved in uranium mining

Location	n	Minimum ($\mu g\,U\,L^{-1}$)	Maximum ($\mu g\,U\,L^{-1}$)	Median ($\mu g\,U\,L^{-1}$)
Magela creek (Australia)[a]	256	0.002	0.06	0.02
Gulungul creek (Australia)[a]	192	0.02	0.23	0.07
Ngarradj creek (Australia)[a]	77	0.004	0.02	0.01
Canada (lakes)[b]	68,303	<0.05	1350	<0.05
Canada (streams)[b]	75,471	<0.05	255	0.06
Kyrgyzstan (Naryn River)[c]	5	0.36	1.24	0.85
Kyrgyzstan (Mailuu-Suu river)[c]	9	0.37	3.1	1.2
Kazakhstan (Syrdarya and Amudarya Rivers)[d]	160	0.40	41	N/A

Data from [a]Supervising Scientist Division website (http://www.environment.gov.au/ssd/monitoring/index.html#data); [b]Garrett (personal communication), summarized from National Geochemical Reconnaissance Program and Uranium Reconnaissance Program data from the Geological Survey of Canada; [c]Vasiliev et al. (2005); [d]Yuldashev et al. (2005).
N/A: not available.

The Geological Survey of Canada's National Geochemical Reconnaissance program has 36 years of data on U levels in lakes and stream water across Canada (Table 8.1) (Garrett, personal communication). Of the dissolved waterborne concentrations, 60% of the lake data and 40% of the stream data were below the detection limit of $0.05\ \mu g\ U\ L^{-1}$. Of the detectable natural background concentrations, 75% of the data were below $1\ \mu g\ U\ L^{-1}$. Higher concentrations (up to $1350\ \mu g\ U\ L^{-1}$) occurred downstream of U mining facilities some years ago and are now closer to natural background owing to improvements in treatment technology (CNSC and EC, 2009).

In Kazakhstan, a transboundary surface water quality monitoring program, developed by Kazakhstan, Kyrgyzstan, Tajikistan, Uzbekistan, and the USA, showed that natural U levels in the main rivers of these countries were below $5\ \mu g\ U\ L^{-1}$, with some more elevated levels downstream of contaminated areas. Hence, Table 8.1 indicates that the background concentrations of U in surface water can vary to a certain extent in the main countries involved in U production. Higher U concentrations ($\sim 41\ \mu g\ U\ L^{-1}$) were attributed to loading from U mining facilities discharging along these rivers.

3.1. Direct Uranium Toxicity

Currently, the various environmental effects monitoring programs undertaken by U mines and mills provide information on fish density, condition factor, liver and gonad weight, as well as benthic community structure upstream and downstream of the facilities in Canada. As further discussed in Section 4, fish are in general very tolerant to U. In addition, removal technology at operating U mines and mills site usually leads to low U levels in the water downstream of these facilities (CNSC and EC, 2009) that are below environmental quality guidelines (see Section 4). Such low levels are not likely to lead to direct U toxicity to fish.

In Australia, large monitoring studies on fish in creeks and lagoons around the Ranger Mine site indicate no effect of U mining on fish communities (Buckle et al., 2010). As in Canada, U concentrations at mine-related sites, although above background, are always still very low (e.g. $< 0.3 \, \mu g \, L^{-1}$).

3.2. Indirect Uranium Impacts

Given the low toxicity of U to fish, any effects on fish populations are likely to be indirect. Laboratory studies (Environment Canada and Health Canada, 2003; Vizon SciTec, 2004; Liber et al., 2007; CCME, 2011) indicate that invertebrates, which can be food items for fish, are more sensitive to U than fish. Field evidence indicated that concentrations of U in surface water, sediment and pore water downstream of a U mine site likely caused *in situ* toxicity to *Hyalella azteca* (Robertson and Liber, 2007). Impacts on benthic invertebrates have also been observed in Beaverlodge Lake, where U levels have continuously decreased since the early 1980s to current levels of 100 µg U L^{-1} (SENES, 2009). However, it is likely that impacts to benthic organisms are also the result of other contaminants. Nevertheless, absence of sensitive benthic invertebrates could lead to potential growth stunting in fish (Sherwood et al., 2002a, b) but it remains to be tested whether U alone could lead to potential growth stunting of fish in nature.

4. A SURVEY OF ACUTE AND CHRONIC AMBIENT WATER QUALITY CRITERIA IN VARIOUS JURISDICTIONS IN FRESHWATER AND SEAWATER

Jurisdictions develop and use environmental quality guidelines as part of regulatory programs to help manage toxic substances; this includes interpreting the significance of monitoring data and limiting direct discharges. Table 8.2 lists environmental quality guidelines that have been promulgated by jurisdictions. Acute guidelines ranged from 40 to 2300 µg U L^{-1}, the latter for water of high hardness, while chronic guidelines ranged from 0.5 to 300 µg U L^{-1}. In addition, the Institut de Radioprotection et de Sûreté Nucléaire in France has proposed the novel approach of 5 µg U L^{-1}

Table 8.2
Summary of existing uranium water quality guidelines/criteria in different jurisdictions for protection of aquatic life (unfiltered total concentrations)

Jurisdiction	Freshwater		Marine		Notes	Reference
	Acute ($\mu g\ U\ L^{-1}$)	Chronic ($\mu g\ U\ L^{-1}$)	Acute ($\mu g\ U\ L^{-1}$)	Chronic ($\mu g\ U\ L^{-1}$)		
Australia	—	6	—	—	Site-specific guideline for Magela Creek in Alligator Rivers uranium mining area: 99% protection trigger value; species sensitivity distribution; no safety factor used; invertebrates most sensitive	Hogan et al. (2003)
Australia/ New Zealand	—	0.5	—	—	Low reliability trigger value: safety factor 20	ANZECC and ARMCANZ (2000)
Canada	—	300	—	—	To protect fish and wildlife: safety factor 20	Environment Canada (1983)
Canada	33	15	—	—	Species sensitivity distribution: no safety factor used; invertebrates most sensitive	CCME (2011)
Canada (Ontario)	—	5	—	—	Interim value: safety factor 590	OMOEE (1994)
Canada (Québec)	320	14	—	—	Interim values for hardness 20–$100\ mg\ L^{-1}$	Boudreau and Guay (2002)
	2300	100			Interim values for hardness 100–$210\ mg\ L^{-1}$	
Canada (Saskatchewan)		15			Developed by the Industrial, Uranium and Hardrock Mining Unit of Saskatchewan Environment	Saskatchewan Environment (2006)
Canada (British Columbia)	—	300		100	Adopted from Environment Canada (freshwater) and NAS/ NAE Blue book (marine)	

above background (Beaugelin-Seiller et al., 2009). The following jurisdictions did not list environmental quality guidelines for U in their guideline documents: European Water Framework Directive, South Africa, or the US Environmental Protection Agency (EPA).

Reasons for differences in environmental quality guidelines between jurisdictions are due to currency of the data but equally to differing policy goals that may have different intended levels of protection, margins of safety, or toxicity modifying factors (pH, DOC, water hardness). Thus, guidelines and criteria may be based on no effects versus low effects, and different safety factors (usually a factor of 10 by which the critical toxicity value is divided; this process provides a margin of safety to account for, among others, untested species). The safety factors used in Table 8.2 range from 1 to 590. Where larger datasets are used, environmental quality guidelines are more likely to be based on invertebrates or plants, which seem to be orders of magnitude more sensitive to U than fish (Sheppard et al., 2005; CCME, 2011). Environment Canada and Health Canada (2003) recommended a predicted no effect concentration to fish of 280 µg U L^{-1}. Quebec is the only jurisdiction to establish environmental quality guidelines as a function of hardness (Boudreau and Guay, 2002), though Sheppard et al. (2005) also derived predicted no effect concentrations for fish as a function of hardness. Proposed values from Sheppard et al. (2005) were 400 µg U L^{-1} at <10 mg $CaCO_3$ L^{-1}, 2800 µg L^{-1} at 10–100 mg $CaCO_3$ L^{-1}, and 23,000 µg U L^{-1} at hardness >100 mg $CaCO_3$ L^{-1}.

For sediment, Thompson et al. (2005) published sediment quality guidelines for the protection of freshwater aquatic life using the "screening level concentration" method of the province of Ontario (Persaud et al., 1992). The guidelines are two-tiered – low and severe effect – and are based on co-occurrence of contaminants and benthic invertebrate community health. The low-effect values were 32 or 104 µg U g^{-1} sediment dry weight (dw), established by two different variations on the method. The severe-effect levels were 3410 or 5874 µg U g^{-1} dw sediment. The approach protects benthic communities which are food sources for fish, but does not address direct chemical effects on fish or the potential for bioaccumulation by fish. However, given the higher sensitivity of invertebrates relative to fish, sediment quality guidelines should protect fish as the dietary assimilation efficiency in fish is low (see Section 9.2). There are no environmental quality guidelines for tissue residues to protect fish.

5. MECHANISMS OF TOXICITY

Uranium shares properties with Ca, Sr, and Pb, is a hard Lewis acid and a class A metal (oxygen-seeking, along with the other actinides, as well as

Na, K, Mg, and Ca, but not Pb, which is borderline with some B character) (Nieboer and Richardson, 1980). Preferred ligands contain oxygen as electron donors, as opposed to sulfur or nitrogen for class B metals.

5.1. Acute Toxicity

Uranium is not highly toxic to fish. Some toxic effects are outlined in Table 8.3 (see CCME, 2011, for a detailed discussion of the toxicity of U to aquatic biota). Hamilton and Buhl (1997) ranked the acute lethality of U on the same order as lithium and arsenate.

The mechanism of toxicity under acutely lethal aqueous exposures has recently been specifically investigated. Zebrafish exposed to $100 \, \mu g \, U \, L^{-1}$ for 20 days experienced gill damage characterized by severe edema, chloride cell hyperplasia, and breakdown of structural integrity (Barillet et al., 2010), confirming that responses to U are similar to those to other metals. This inflammatory reaction in the gill causes reduced diffusion of oxygen and other gases (CO_2, ammonia), reduced blood flow, and consequently asphyxiation, as shown by Spry and Wood (1984) with rainbow trout exposed to zinc.

5.2. Chronic Toxicity

Mechanisms of chronic toxicity are largely unknown. However, the hallmark of toxicity in mammals exposed to very high concentrations of U is nephrotoxicity (ATSDR, 1999). Since much of the U transported in the blood is bound to bicarbonate (see Section 10.1), filtration in the kidney of these complexes into the urine will result in greater U deposition due to dissociation from the bicarbonate complex when the urine is acidic. Uranium concentrations that can lead to renal toxicity range from 0.1 to $3 \, \mu g \, g^{-1}$ kidney (ANSI, 1995; ATSDR, 1999; Royal Society 2001a, b; OMOE, 2008). At such levels, mammalian kidney cells can regenerate following necrosis, but histological observations suggest that function is diminished.

The only example of chronic U toxicity to fish comes from the work of Cooley et al. (2000). In lake whitefish exposed to very high dietary concentrations, Cooley et al. (2000) reported a plethora of histological changes indicative of widespread necrosis, including tubular necrosis, inflammation, hemorrhaging, and depletion of hematopoietic tissue. Despite this, there were no changes in hematocrit and only mild transient effects on serum electrolytes. The dietary concentrations ($100–10 \, 000 \, mg \, U \, kg^{-1}$) in spiked commercial trout food pellets (Cooley et al., 2000) were quite high compared to prey items collected from lakes impacted by U mining and

Table 8.3
Some illustrative effects of uranium exposure on fish (concentrations in μg U L^{-1} except where noted)

Effect	Comments
Survival	LC50 ranged from 43,500 μg U L^{-1} (Hamilton and Buhl, 1997) to 1670 μg U L^{-1} (Trapp, 1986). Older early life stages significantly more tolerant in acute, lethal exposures (Cheng et al., 2010)
Growth	LOEC from 27,860, weight and length of white sucker fry were both affected (Liber et al., 2004b), IC25 of 1300 μg U L^{-1} to fathead minnow (Vizon SciTec, 2004), reduced length and dry weight in *Mogurnda* (Cheng et al., 2010)
Reproduction	Fertilizing eggs of northern pike directly in U test solutions increased the toxicity about 10-fold (Liber et al., 2005). Times to hatch, hatch success, and survival of white sucker affected at 27,860 μg U L^{-1}
Organ	Dietary study (Cooley et al., 2000) – kidney: widespread necrosis, inflammation, hemorrhaging; liver: pathology with fatty accumulation, no change in liver somatic index. Olfactory bulb (neuropil) histological changes (Lerebours et al., 2010)
Tissue	Oxidative stress – serum lipid peroxides elevated at 10,000 μg U g^{-1} diet. Maximum increase eight-fold, declined under continued exposure (Cooley et al., 2000)
Cellular/ subcellular	Inhibition of oxidative stress response: liver catalase ↓ 32% at 20 and 58% at 500 (Buet et al., 2005), liver superoxide dismutase ↓ 20% at 20 and 46% at 500 (Buet et al., 2005). Acetylcholinesterase (brain) ↑ 30% at 100 (Barillet et al., 2007)
Genotoxicity	DNA strand breaks (comet assay) at 2025 (Lourenco et al., 2010) and 100 (Barillet et al., 2007)
Genomics	Gene regulation (upregulation and downregulation of a large number of genes in zebrafish (Lerebours et al., 2009, 2010)
No effect	Metallothionein: transient induction in liver and kidney during dietary exposure (Cooley et al., 2000) consistent with class A chemistry
Muscle RNA/DNA ratios (growth indicator) (Liber et al., 2004a, 2005)
Whole-body triglycerides (energy metabolism) (Liber et al., 2004a, 2005)
Total muscle protein (Liber et al., 2004a, 2005)
Total glutathione (liver), glutathione peroxidase (liver) (oxidant status) at 100 μg U L^{-1} (Barillet et al., 2007)
Not studied (partial list)	Behavior (incidental observations only)
Bone structure or function
ATPase activity (ion transport)

For a fuller description, especially of whole organism effects, see CCME (2011).
LOEC: lowest observed effect concentration.

milling activities. For instance, aquatic insects, a preferred food item for several benthic feeders, reached U levels up to 24 mg kg^{-1} wet weight (ww) (Swanson, 1982). In lake chub, mean U levels ranged from 0.6 and 1.5 mg kg^{-1} (Golder Associates, 2008) to 80 mg kg^{-1} whole body ww

(Swanson, 1982). Given the reported concentrations measured in prey items from these monitoring programs, the severity of necrosis in kidney cells reported by Cooley et al. (2000) is unlikely to be observed in nature, first because concentrations in prey items were lower than that required to cause toxicity in the laboratory, and second because U incorporated into actual prey organisms is likely less bioavailable than in the study of Cooley et al. (2000) where test solutions were sprayed onto commercial food.

However, some impact on kidney function cannot be ruled out, based on field data where U concentrations in fish kidney ranged from 0.08 to 4.0 μg U g^{-1} (Golder Associates, 2002). In mammals, kidney concentrations below 2.5 μg g^{-1} appeared not to cause renal damage in either humans or rats (ATSDR 1999), whereas in lake whitefish fed a contaminated diet for 100 days, proximal tubule kidney necrosis was severe in 30–40% of the test fish (Cooley et al., 2000) when U concentrations ranged from 5 to 17 μg g^{-1} ww in kidney (Cooley and Klaverkamp, 2000).

5.3. Oxidative Stress

Oxidative stress as a mechanism of toxicity has been the subject of considerable research in a variety of organisms and from a variety of insults. The production of reactive oxygen species (ROS) can cause oxidative damage at the cellular level, against which there exists a battery of cellular antioxidant defenses (DiGiulio, 1991). In the lake whitefish study (Cooley et al., 2000), serum lipid peroxides were elevated about eight-fold above controls at the highest U diet on day 30. By day 100, however, levels, though still elevated, had subsided. The effect was thus transient, but could have been involved in the tissue pathologies associated with the dietary exposure.

Several other biomarkers indicative of oxidative stress have been measured. Superoxide dismutase, catalase, and glutathione peroxidase in zebrafish responded to combined effects of radiological and chemical exposures (see discussion of radiological effects below). Catalase in liver was irreversibly inhibited in goldfish exposed to 2 mg U L^{-1} (a concentration much higher than normally encountered in the field) for 96 h followed by 96 h of depuration, but surprisingly there was no increase in lipid peroxidation (Lourenço et al., 2010). Buet et al. (2005) also found that U was inhibitory to antioxidant systems in zebrafish (superoxide dismutase and catalase). Similar responses were found for both radiological and chemical toxicity (see Barillet et al., 2007, in Section 5.4). The inhibitory effects of U on these antioxidant responses would seem maladaptive. In conclusion, the role of ROS in the mechanism of toxic action of U remains unclear.

5.4. Mixed Radiological Toxicity and Chemical Toxicity

From a theoretical standpoint, the chemical toxicity of U should outweigh radiological toxicity for the following reasons: ^{238}U is a weak alpha emitter; therefore, the low rate of decay, coupled with low penetration in tissue (about 50 μm) should result in low radiotoxicity. Mathews et al. (2009) compared a radiological benchmark thought to be protective of ecosystems (the predicted no-effect dose rate) of 10 μGy h^{-1} to a chemical benchmark of 3.2 μg U L^{-1}, derived using a species sensitivity distribution ($n = 12$, EC10 values, mostly reproductive endpoints). They found that the chemical benchmark was generally protective of any radiological effects from natural or depleted U to fish.

Differential radiological toxicity versus chemical toxicity of different isotopes was also examined directly. Uranium isotopes have the same chemical characteristics and hence the same chemical toxicity, but for any given chemical concentration, the radiation dose can be varied by adjusting the isotopic ratios. Bourrachot et al. (2008) exposed zebrafish eggs to either depleted U or ^{233}U, an artificially produced isotope with a specific activity 29,000-fold higher than ^{238}U. They found that delay in hatching was more strongly affected by ^{233}U than by ^{238}U, though the resulting mortality was the same. A unique effect of the radiation exposure seemed to be the retarded rate of tail detachment during embryonic development, an effect that was not seen at even lethal concentrations of ^{238}U.

Barillet et al. (2007) exposed male zebrafish for up to 20 days to 100 μg U L^{-1}, but with different isotopic ratios, such that the internal radiological dose was 1500-fold higher in fish exposed to ^{233}U plus depleted U compared with depleted U alone. They found similar tissue concentrations of U (no isotopic discrimination), with gills and liver having two- to three-fold higher concentrations compared to total body. Oxidative stress could theoretically be caused by both radiological and chemical effects. Indeed, biomarkers of oxidative stress responded by minor depression rather than activation (liver superoxide dismutases were depressed by about 15%; catalase activity declined, but not significantly; and there was no trend in total glutathione). Although U did accumulate in brain, there was actually a stimulation of acetylcholinesterase activity by about 30%. The only differences attributable to radioactivity were significantly depressed catalase activities halfway through the experiment, at day 10, in the higher radiological dose. The authors proposed that there were initial radiological insults, but these did not persist through the experiment. In short, radiotoxic effects have been demonstrated using an artificial isotope having a high specific activity, but the effects are subtle. The above authors have suggested a common mechanism of action through the generation of ROS. In conclusion, the present authors would agree with others (Mathews et al., 2009) that the chemical hazard of naturally occurring U outweighs the potential for radiological toxicity.

5.5. Acclimation

Beak Consultants (1989) reviewed the sensitivity of aquatic biota including fish to ionizing radiation. In their review, they indicated that resistant individuals were likely different from sensitive ones. Kolok (2001) also indicated that susceptible individuals should be genetically different from resistant ones. They reported a potentially interesting relationship between susceptible and resistant fish and allozymic variation in phosphoglucose isomerase-1 (PGI-1). While the small number of susceptible individuals makes it impractical to test this finding statistically, the importance of the PGI-1 loci to copper susceptibility in fathead minnows is consistent with the results from time-to-death studies involving the same metal and fish species (Schlueter et al., 1995, 1997). Cooley et al. (2000) raised the possibility that the transient effects seen, particularly the reductions in serum lipid peroxides and reduced kidney lesions by day 100 of the study, might be evidence of acclimation to U by lake whitefish. The phenomenon has not otherwise been studied except at the population level, where mosquitofish collected from a natural population from a site with 4.2 μg U L^{-1} in water and 6165 μg U g^{-1} had lower genetic diversity. Yet, fish from this contaminated site experienced lower mortality (25–57%) compared with naïve fish (96–98%) when exposed to 2.6 mg U L^{-1} for 7 days (Keklak et al., 1994). However, it is not clear whether this reflects phenotypic acclimation or genotypic adaptation.

6. WATER CHEMISTRY INFLUENCES ON BIOAVAILABILITY AND TOXICITY

In the absence of any evidence of passive diffusion, one can postulate that U (a non-essential element) can be accumulated by carrier-mediated transport (Köster, 2004) through a membrane carrier dedicated to an essential element. To the authors' knowledge, no studies have identified the nature of this transport system for U. Nevertheless, the literature available provides several clues on the major aqueous parameters that influence U bioavailability and toxicity. Metal accumulation and toxicity can often be predicted using the free-ion activity model (Campbell, 1995) or the biotic ligand model (BLM) (Paquin et al., 2002; see Wood, Chapter 1, Vol. 31A and Paquin et al., Chapter 9). The application of these models to U remains to be validated but the few available studies seem to support this tenet, as explored in this section.

6.1. Natural Variation in Water Chemistry

Water chemistry is important in predicting the bioavailability of U to fish and to fish prey items. Bird and Schwartz (1997) documented the average values of hardness, alkalinity, pH, and DOC in Ontario lakes on the Canadian Shield. The analysis showed that these lakes usually have low hardness (geometric mean of 4.56 mg $CaCO_3$ L^{-1}, range of 0.01–82.9 mg $CaCO_3$ L^{-1}) and low alkalinity (geometric mean of 16.6 mg $CaCO_3$ L^{-1}, range of 8.49–84.5 mg $CaCO_3$ L^{-1}), that average pH is near neutral, and that average DOC is slightly less than 5 mg L^{-1}.

In northern Australian rivers, where U mining activity occurs, pH values range from 5.8 to 8.3, alkalinity (mg $CaCO_3$ L^{-1}) from 5.3 to 480, and hardness (mg $CaCO_3$ L^{-1}) from 8.7 to 448 (Riethmuller et al., 2000).

6.2. pH

As shown in Section 1, within a pH range of 6–9, the speciation of U in surface water changes drastically, which can, in turn, have significant influence on U bioavailability and toxicity to fish. Most studies on the effect of pH on U bioavailability and toxicity were done with freshwater invertebrates (Markich et al., 2000; Fournier et al., 2004; Simon and Garnier-Laplace, 2004; Alves et al., 2008; Zeman et al., 2008) and algae (Nakajima et al., 1979; Franklin et al., 2000; Fortin et al., 2007). Freshwater bivalves appear particularly sensitive at a lower pH (5.5 vs 6.5), presumably because the resultant changes in speciation favor high relative abundance of the free uranyl ion, UO_2^{2+}. However, this increase in UO_2^{2+} concentration at lower pH values may be countered, presumably because H^+ can compete with UO_2^{2+} at the site of uptake, as observed for algae, thereby producing a protective effect (Franklin et al., 2000). In their study with freshwater bivalves, Markich et al. (2000) showed that toxicity could be empirically predicted by a combination of UO_2^{2+} and UO_2OH^+ ambient concentrations. Fortin et al. (2004, 2007), in a series of studies with a freshwater alga, proposed that uptake could be predicted by the free uranyl ion, but that pH influenced the uptake rate of the ion through two mechanisms: (1) competitive inhibition where protons compete with UO_2^{2+} ions for binding at the uptake site (as predicted by the BLM); and (2) non-competitive inhibition where protons decrease the uptake rate of UO_2^{2+} ions. The latter mechanism was further supported by François et al. (2007) for two different metals (Cd and Mn). When both pH and DOC were varied in very soft water, the toxicity to larval *Mogurnda mogurnda* after 28 days correlated best with uranyl ion concentration (modeled using HARPHRQ) (Cheng et al., 2010). In this case, the low pH (6.0) and high DOC (4.2 mg L^{-1}) medium resulted in a higher toxicity than in the neutral pH (6.7) and low DOC (2.1 mg L^{-1}) exposure medium. Most U was predicted to be bound to DOM (47–99%), but it was not sufficient to counteract the increase in toxicity that occurred at lower pH (6.0 vs 6.7). The true influence of pH on U uptake and toxicity to fish remains to be fully elucidated.

6.3. Hardness

According to the BLM, calcium (Ca^{2+}) and magnesium (Mg^{2+}) cations may reduce U uptake and potentially toxicity through competition with U for uptake sites. Relatively few studies have investigated the effect of hardness on U toxicity to fish (Tarzwell and Henderson, 1960; Parkhurst et al., 1984; Vizon SciTec, 2004). Tarzwell and Henderson (1960) reported that a 20-fold increase in hardness resulted in a 4.8-fold decrease in toxicity to fathead minnow (*Pimephales promelas*). Similarly, Parkhurst et al. (1984) indicated that a 6.5-fold increase in hardness resulted in a 4.2-fold decrease in toxicity to juvenile brook trout (*Salvelinus fontinalis*). However, in these studies, the effect of hardness was likely confounded with those of alkalinity and pH.

Vizon SciTec (2004) investigated the effect of hardness alone on U toxicity to fathead minnow and rainbow trout by adding CaSO$_4$ and keeping pH constant. In these fathead minnow and rainbow trout survival tests, higher hardness levels alleviated U toxicity. Vizon SciTec (2004) also performed an early life-stage test lasting for 31 days with rainbow trout eggs. The number of deformed alevins increased with increasing U concentration. Although the toxicity tests were conducted at only two hardness values, there was an apparent hardness effect on toxicity of U to embryo/alevins in that toxicity was greater in softer water.

Riethmuller et al. (2000) also investigated the effect of hardness alone on U toxicity to *M. mogurnda*. True water hardness was achieved by adding Ca(NO$_3$)$_2$ and Mg(NO$_3$)$_2$ to the synthetic diluent water, while other physicochemical parameters were held constant (i.e. pH 6.0±0.3 and conductivity within 10% error, over 24 h). A 50-fold increase in hardness had no effect on U toxicity to *M. mogurnda*.

6.4. Bicarbonate (Alkalinity)

There are no current data on the effect of alkalinity on the toxicity of U to fish. In an open system, carbonate concentrations and pH cannot be varied independently. It is thus impossible to test the influence of each parameter individually. The concentration of carbonates can only be manipulated at constant pH in a closed system. Only one study could be found where the pCO_2 was controlled. Tran et al. (2004) showed that, at a given total U concentration, an increase in pCO_2 reduced U accumulation in the gills and mantle of the freshwater clam *Corbicula fluminea*. The presence of carbonato-complexes in the exposure media went from 79% (low CO_2) to 95% (high CO_2), suggesting that U complexation by carbonates contributes to decreasing its bioavailability, as expected from the BLM.

6.5. Dissolved Organic Matter

DOM has the potential to bind the uranyl ion and hence reduce toxicity (see Section 1). One research group found that small changes in pH were a stronger determinant of U toxicity to larval *M. mogurnda* than low concentrations of natural DOM (Cheng et al., 2010; see Section 6.2). Nevertheless, they also showed that at constant pH, the addition of Suwannee River Fulvic Acid (SRFA) or a natural DOM could effectively reduce U toxicity to *M. mogurnda* as well as to other organisms (green hydra, *Hydra viridissima*, and the unicellular alga, *Chlorella* sp.) (Houston et al., 2009, 2010). The relative protective effect of each DOM differed among species.

6.6. Phosphorus

Finally, another variable that can also affect U speciation and potentially its bioavailability and toxicity to fish is phosphorus (P), although it has received much less attention. In surface waters, dissolved phosphorus mainly exists as phosphate although its concentrations are generally very low because phosphorus is essential to phytoplankton growth (Wetzel, 2001). In most toxicity studies such as the one completed by Vizon SciTec (2004), levels of P ranged from 50 μg P L^{-1} to 6000 μg P L^{-1} in *Lemna minor* tests. The lower concentrations are similar to levels encountered in eutrophic lakes (Wetzel, 2001), whereas the higher concentrations used in the *L. minor* tests are rarely encountered even in agricultural areas. As mentioned in Section 1, at these concentrations of P, U will precipitate as $UO_2HPO_{4(s)}$. Such conditions are unlikely to occur in surface waters near current U mines and mills in northern Saskatchewan where total P concentrations typically range from less than 10 to 60 μg P L^{-1}. This is unlikely to affect toxicity tests with fish, but the effects of the test medium used with plants/algae should be borne in mind. Mkandawire et al. (2007) indeed noted that U toxicity to *Lemna* sp. was affected by phosphate in their experimental medium, while

Fortin et al. (2004) showed that an increase in concentration of $UO_2PO_4^-$ and $UO_2HPO_4^0$ $_{(aq)}$ did not contribute to U uptake in a freshwater alga, as expected from the BLM (i.e. uptake was proportional to the free uranyl ion concentration). To the authors' knowledge, the influence of phosphate on U bioavailability to fish has not been tested directly. However, based on speciation considerations (see Section 1), it would likely decrease bioavailability through complexation and/or precipitation.

7. NON-ESSENTIALITY OF URANIUM

Uranium has no known physiological function in humans and animals (WHO, 2001). To date, there are also no reports of a metabolic function for U in fish since U is not an essential element.

8. POTENTIAL FOR BIOACCUMULATION OF URANIUM

Bioaccumulation is the process of a chemical moving from the external environment into the organism from all possible exposure routes (water, sediment, soil, air, or diet) and is expressed as a bioaccumulation factor (BAF). Bioaccumulation in aquatic organisms from water alone is measured as the bioconcentration factor (BCF) (USEPA, 1999a,b). As indicated by McGeer et al. (2003), "in general, BAF is derived from measurements in natural environments and BCF is more readily measured under laboratory conditions".

The relatively few BCFs available for U in fish ranged from 0.001 to 149 L kg^{-1} (Poston, 1982; Parkhurst et al., 1984; Labrot et al., 1999; Barillet et al., 2007; Cheng et al., 2010). This range is attributed to the different water chemistry in each of the different studies and likely also to differences in species internal biokinetics. In addition, U is usually measured only in a few tissues, which makes it difficult to accurately obtain fish whole body U concentrations (Yankovich, 2009) and can therefore increase uncertainty in BCF calculations.

Through their environmental effects monitoring programs, the U mining industry in Canada collects several fish tissue concentrations (mainly flesh, bone, and organs). From these data, BAFs were calculated using fish whole-body concentrations and U concentrations in water. Fig. 8.4 provides the range of BAFs in wild fish populations. BAFs ranged from 0.3 to 3340 L kg^{-1}, which can likely be attributed to the different water chemistry of the different study sites, where pH ranged from pH 5.1 to pH 6.9, calcium ranged from 13 to 22,000 μg L^{-1}, DOC ranged from 4 to 20 mg L^{-1}, and alkalinity ranged from 12 to 106 mg CaCO$_3$ L^{-1}.

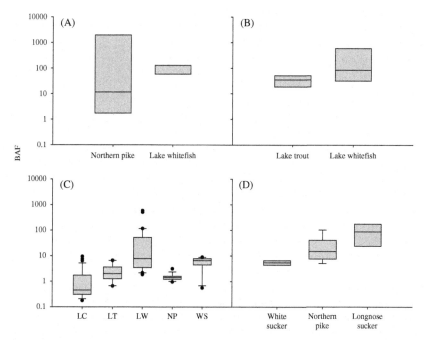

Fig. 8.4. Uranium bioaccumulation factors (BAF; L kg^{-1}) at (A) St Mary's Channel, Back Bay and Langley Bay of Lake Athabasca (CanNorth, 2006b), (B) in lakes near Elliot Lake, Ontario (Clulow et al., 1998), (C) in Lake Beaverlodge (Swanson, 1982; Golder Associates, 2002; CanNorth, 2006a), and (D) in Pow, Hidden and Collins bays of Wollaston Lake (SENES, 2007b). LC: lake chub; LT: lake trout; LW: lake whitefish; NP: northern pike; WS: white sucker.

Regarding trophic transfer and biomagnification, organisms do accumulate U, but because it has low assimilation efficiency, it does not biomagnify (Swanson, 1985; Environment Canada and Health Canada, 2003; Simon and Garnier-Laplace, 2005). Trophic transfer rates of U are low (1–13%), similar to that of Cd (Simon and Garnier-Laplace, 2005). Organisms lower on the food chain typically have higher levels of U than higher trophic level organisms (Environment Canada and Health Canada, 2003).

9. CHARACTERIZATION OF UPTAKE ROUTES

In order to determine the main route of uptake of U to fish and determine associated toxic effects, laboratory, and preferably, field experiments need to consider U uptake through the gill and the gut. In addition, the relative importance of route of uptake depends on the feeding behavior of fish.

9.1. Gills

The biokinetics of U uptake across the gill has been very little studied in fish. In studies that were designed to measure U uptake at the gill surface, there were several limitations. For instance, Poston's (1982) study with small rainbow trout (static exposure with replacement of exposure solutions) was complicated by loss of U from solution and he indicated that initial body burdens were likely due to uncleared food in the gut. This was based on the fact that during the depuration phase, fish lost 25% of the whole body load by day 6, during which time the gut would have cleared, but there was no further depuration after 25 days. There was no significant growth dilution. Likewise, Labrot et al. (1999) found that uptake in zebrafish (*Danio rerio*) appeared to reach steady state at about 0.93 μg U g^{-1} fresh weight by day 29 in fish exposed to about 150 μg U L^{-1}. Fish lost about 35% of their body burden in a single day after transfer to clean water (declined to about 0.6 μg U g^{-1} fresh weight). For the remaining 30 days there was little, if any, depuration. Finally, Lerebours et al. (2009) exposed zebrafish to 23 or 130 μg U L^{-1} for 28 days with a subsequent 8 days in clean water, to observe uptake and depuration, but primarily the study was done to determine gene expression in brain, liver, skeletal muscle, and gills (see Section 13). The uptake by gills appeared to follow single-compartment, first order kinetics, increasing asymptotically over the 28 days of exposure. Gill U concentrations did not come rapidly to steady state, in contrast to what would be predicted for a waterborne exposure. Liver U concentrations fluctuated considerably, but appeared to come to steady state by day 10. Both liver and gill tissue had order of magnitude greater concentrations (4000 μg U g^{-1} dw) than either brain or muscle (500–700 ng U g^{-1} dw, respectively). In muscle and brain tissue there were small increases until day 10 followed by a dramatic increase from day 10 to day 28. A single sample after 8 days in clean water showed significant losses from both liver and gill tissue. There was no significant depuration post-exposure in either brain or muscle (note that kidneys were not examined). Despite the preliminary nature of these studies, some body burden of U is lost quickly, and some is lost very slowly. This suggests at least two pools with very different kinetics, where bone represents the more stable compartment (OMOE, 2008).

9.2. Gut

There are currently no data on U assimilation efficiencies (AEs) in the gastrointestinal tract of fish, but they are expected to be low based on results with other metals. Simon and Garnier-Laplace (2005) reported trophic transfer efficiency on the order of 1–13% for crayfish feeding on bivalves. Reported AEs for other metals in fish range from 6% to 40% (Harrison and Stather, 1981; Wrenn et al., 1985; Hursh et al., 1969), while the International Commission on Radiological Protection (ICRP, 1979) recommends the use of a default value of 5% for U in humans. Evidence from whitefish (*Coregonus clupeaformis* and *Prosopium cylindraceum*), rainbow trout (*Oncorhynchus mykiss*), lake trout (*Salvelinus namaycush*), white sucker (*Catostomus commersoni*), and northern pike indicates that U concentrations in food passing through the gastrointestinal tract are generally higher than those in fish tissue (Swanson, 1982; Clulow et al., 1998; Waite et al., 1990), which supports the hypothesis that AE for U in fish is likely to be low.

9.3. Feeding Ecology

The feeding ecology of the different fish species can potentially affect their exposure to U in the surrounding environment. While pelagic fish will take up U from water and food, benthic fish will accumulate U from water, food, and sediments. Some studies suggest that benthic fish that prey on benthic invertebrates and spend significant time foraging in sediments accumulate the highest concentrations of U, relative to pelagic and predatory species (Emery et al., 1981; Swanson, 1982, 1983, 1985). However, Golder Associates (2002) showed that the extent of U accumulation in the different tissues depends on fish species (Fig. 8.5), with lake whitefish

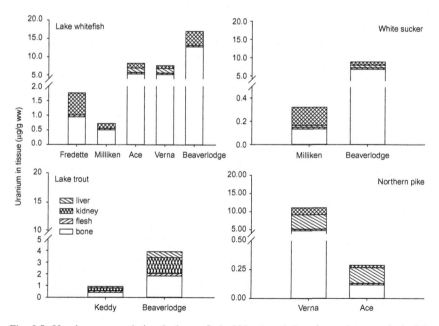

Fig. 8.5. Uranium accumulation in bone, flesh, kidney, and liver in predatory pelagic fish species (lake trout and northern pike) and benthic fish species (lake whitefish and white sucker) in Lake Beaverlodge. Data from Golder Associates (2002).

accumulating significantly more U than white sucker, both of which have similar feeding habits. Hence, how feeding ecology affects U bioaccumulation in fish remains largely unknown.

10. CHARACTERIZATION OF INTERNAL HANDLING

10.1. Transport through the Bloodstream

In mammals, once absorbed, U remains as uranyl species. Most is carried in the blood bound to small ligands such as citrate and bicarbonate. About 20% is bound by protein that consists mostly of albumin and transferrin, although Vidaud et al. (2005) suggested that U could use a wide variety of binding sites and identified additional binding proteins including ceruloplasmin, hemopexin, and two complement proteins. Uranium is rapidly cleared from blood to other tissues. Of soluble U injected intravenously into humans, only 25% remained after 5 min, 5% after 5 h and 0.5% after 20 h (ATSDR, 1999). No information is available for fish.

10.2. Accumulation in Specific Organs

In general, within fish tissues, U tends to accumulate in mineralized tissues, such as bone and scales, and to a lesser extent in the kidney and liver, with measurable accumulation in the gills, skin, and muscle, whether from field studies (Waite et al., 1990), laboratory diet (Cooley and Klaverkamp, 2000), or laboratory waterborne exposures (Barillet et al., 2007). Other tissues having measurable concentrations were gonad (Cooley and Klaverkamp, 2000; Barillet et al., 2007) and brain (Lerebours et al., 2009, 2010). Figure 8.5 provides an illustration of the compartmentalization of U from field samples of benthic feeding fish and piscivorous fish exposed to legacy contamination from U mining in the 1970s (Golder Associates, 2002). This figure supports the observation that U preferably partitions into the bones and then into the liver and kidneys.

Yankovich (2009) used a mass balance approach to demonstrate that 20% of the U mass of fish was in bone, 18% was retained in scales, 43% in edible tissue, 2.8% in gonad, 0.6% in kidney, 1.3% in liver, and the remaining 15% in other soft tissues. Hence, a mass balance approach brings a much different perspective than the ones presented in Fig. 8.5. It suggests that most of the U taken up is stored in edible tissues (43%), which could have implications for human consumption. However, this mass balance approach has not been validated in the field and should be interpreted with caution.

11. CHARACTERIZATION OF EXCRETION ROUTES

The first step towards excretion of metals is depuration of tissues. There is limited information about U depuration in fish tissue. Muscle tissue of *Carassius auratus* exposed to 450 and 2025 μg U L^{-1} was able to depurate close to control levels after 96 h of exposure (Lourenço et al., 2010). A similar ability to accumulate and depurate U was also demonstrated for *Brachydano rerio* exposed to an environmentally relevant concentration of U (151 μg U L^{-1}) (Labrot et al., 1999). However, depuration of one tissue may not necessarily lead to excretion. Lourenço et al. (2010) recorded a decrease in catalase activity even after depuration of the muscle tissue, which suggests the possible translocation of U to other organs rather than excretion out of the body.

Excretion has not specifically been measured in fish, other than the depuration rates discussed before (see Section 9.1). However, in lake whitefish fed high dietary U, Cooley et al. (2000) found the hepatocyte and biliary pathology to be possibly indicative of a biliary route of excretion. At the highest dietary loading, the concentrations in gill peaked at day 30 and thereafter subsided. It is not known whether the gill plays any role in excretion.

In mammals, most dietary U is not absorbed and is excreted via the gut. Absorbed U is rapidly excreted via the kidney (ATSDR, 1999).

12. BEHAVIORAL EFFECTS OF URANIUM

There have been no studies directed purposely at behavioral effects of U on fish. Observations of the effects of U on behavior have been incidental to other tests. Lourenço et al. (2010) conducted a feeding behavior assay in which *Carassius auratus* exposed to 100, 450, and 2 025 μg U L^{-1} for 96 h, fed on 10 adult *D. magna* for 2 min. Feeding behavior was not affected at these high U concentrations and short exposure period. At a longer exposure period of 30 days, Liber et al. (2004b) did not record unconsumed food in the bottom of test containers of *Catostomus commersoni* exposed to concentrations below 6400 μg U L^{-1}. In a dietary exposure experiment, lake whitefish (*Coregonus clupeaformis*) showed no aversion to commercial pellet food containing up to 10,000 μg U g^{-1} (Cooley and Klaverkamp, 2000). Hence, the unaffected feeding behavior of three fish species at very high dietary or waterborne U concentrations indicates that these species would likely not change their feeding behavior at currents U levels reported in the environment.

13. GENOMIC AND PROTEOMIC STUDIES

Genomic studies have become the new sensitive biomarker for the effects of toxic substances on both the detrimental effects (mechanisms of toxic action) and adaptive responses to toxic insult. The zebrafish, whose entire genome has been sequenced, has provided an especially useful model. That said, there are not many genomic or proteomic studies of U in fish. Lerebours et al. (2009, 2010) (see Table 8.3 and Section 5.3) looked at the gene regulation underlying several responses to U toxicity at the cellular level, including those coding for antioxidant response in zebrafish brain, liver, gill, and muscle exposed to 15–23 or 100–130 μg U L^{-1}. They found that there were more changes in gene expression at the lower U concentration, and suggested that the pattern of gene response depended upon the exposure concentration, with higher concentrations causing some inhibition of gene expression that overwhelmed the protective cellular capacity.

14. INTERACTIONS WITH OTHER METALS

There are very few studies on interactions of U with other metals. Mixtures of arsenate, boron, copper, molybdenum, selenate, selenite, U, vanadium, and zinc all showed strictly additive toxicity to flannelmouth sucker (*Catostomus latipinnis*) (Hamilton and Buhl, 1997). That is to say, several solutions, each comprising only a single metal, are concocted to have equal toxicity, but (usually) different metal concentration. When these were mixed together, regardless of the ratios, the resulting mixture was no more or less toxic than the individual solutions. The U toxicity curves were different from those of the other inorganics, however, in that the LC50 values remained constant from 1 to 4 days, whereas the other metals showed the more typical declines in LC50 (increased toxicity) with exposure time. In lake waters contaminated by metal mixtures due to U mining activities, toxicity was attributed to As, Mo, and Ni rather than U, which was typically below 10 μg L^{-1} (Pyle et al., 2002).

15. KNOWLEDGE GAPS AND FUTURE DIRECTIONS

In general, there is currently much less research devoted to the environmental fate, bioavailability, and indirect and direct toxicity of U to fish compared to other metals discussed in these volumes. The main reason is that the existing data show that U is not very toxic to fish and is not the primary contaminant of concern in U mine and mill effluent. Another possible reason is that extensive research has been devoted to the design and improvement of efficient wastewater treatment technology such as reverse osmosis and chemical precipitation techniques. These treatment technologies have considerably improved the quality of U mine effluents over the years such that direct U toxicity to fish is unlikely near modern U mining facilities.

Although continuous improvement in treatment technology is warranted, one also needs to accurately predict the temporal and spatial extent of both direct and indirect impacts of U to fish as important components of aquatic ecosystems, in order to inform the public and regulatory authorities. Improved predictive models are needed to accomplish this. This section includes suggestions on important research needs regarding U environmental fate, bioavailability, and toxicity to fish.

15.1. Environmental Fate

Some predictive environmental fate models already exist and have been applied to U mines and mills in Canada (SENES, 2007a) and elsewhere around the world (Monte et al., 2003, 2009a, b). The Lakeview model (SENES 2007a) considers horizontal and vertical transport of dissolved species, chemical and biochemical reactions, settling of particulate matter, and sediment exchange of contaminants with the overlying water column. Monte's model, the MOIRA-plus model, also considers chemical speciation and surface complexation to particles, but also considers transfer into the food chain. To improve these environmental fate models, there needs to be a better understanding of how U speciation at low concentrations is affected by pH, bicarbonate, phosphorus, and DOM.

Similarly, predictive environmental fate models need to better predict how U adsorbs onto suspended particles as a function of pH, calcium, magnesium, bicarbonate, phosphorus, and DOM. In turn, a better understanding is required of how the settling velocity of these particles and their rates of accumulation in sediments are affected by aggregation with other particles. Finally, further research is needed to determine how refractory the U particles in sediments are to microbial processes, and the conditions under which U will be taken up by either benthic invertebrates or fish during feeding.

15.2. Uranium Bioavailability and Biokinetics

Cooley et al. (2000) indicated that diet was the only uptake pathway of any consequence, based on field studies. While it is clear that uptake has occurred across the gill in laboratory studies there have been no studies that have quantified the relative contributions of diet versus water within the same study (e.g. Spry et al., 1988). Predictive environmental fate models will also need to be coupled to models such as the dynamic multi-pathway bioaccumulation model (DYMBAM) (Luoma and Rainbow, 2005) that consider uptake of U from water, food, and sediments. For waterborne U, a better understanding of its complexation to inorganic and organic ligands and how this will affect the availability of U to fish prey items and fish themselves is warranted. A BLM (Paquin et al., 2002) could then be coupled to the DYMBAM. However, in order to develop such a kinetic model for fish, ingestion rate of food items, assimilation efficiency, U levels in food, and efflux rate constant from food and sediments are needed (see Paquin et al., Chapter 9).

15.3. Uranium Toxicity to Fish

In this chapter, it has been demonstrated that current knowledge indicates that fish are relatively tolerant to U. Despite this, the role of toxicity modifying factors should be further examined using multivariate techniques (e.g. Cheng et al., 2010). Given chemical properties of U, interactions with calcium and magnesium should be examined as done with other ions both in the water and the diet and at the cellular level.

There is also considerable scope for investigation of mechanisms of toxicity in fish. Interference with transport of essential metals has not been examined. Biochemistry, genotoxicity, and genomics have been covered in this review. Metallothionein does not appear to be involved and this should be confirmed. Genomics studies, in particular, offer tantalizing prospects of examining homeostasis, mechanisms of toxic action, and detoxification processes, but the authors caution that these techniques can generate more heat than light unless applied in a focused manner with solid hypothesis-driven investigations.

ACKNOWLEDGMENTS

We acknowledge the protracted work of staff (mostly our students) of Environment Canada's National Guidelines and Standards Office, in developing the Canadian Water Quality Guideline for uranium (see CCME, 2011). We also acknowledge comments from the editors, Rick van Dam from the Supervising Scientist Division of Australia, and one anonymous reviewer. We would also like to thank the Canadian Nuclear Safety Commission, notably Patsy A. Thompson and Michael J. Rinker, for comments on the manuscript and for allowing time for RRG to work on this chapter. Finally, we dedicate this chapter to Steve Munger, friend and colleague at CNSC, who was instrumental in initiating this work.

REFERENCES

Alves, L. C., Borgmann, U., and Dixon, D. G. (2008). Water–sediment interactions for *Hyalella azteca* exposed to uranium-spiked sediment. *Aquat. Toxicol.* **87**, 187–199.

Ansoborlo, E., Prat, O., Moisy, P., DenAuwer, C., Guilbaud, P., Carriere, M., Gouget, B., Duffield, J., Doizi, D., Vercouter, T., Moulin, C., and Moulin, V. (2006). Actinide speciation in relation to biological processes. *Biochimie* **88**, 1605–1618.

ANSI. (1995). *Bioassay Programs for Uranium.* American National Standards Institute, Washington, DC, ANSI/HPSN13.22-1995.

ANZECC and ARMCANZ. (2000). *Australian and New Zealand Guidelines for Fresh and Marine Water Quality.* Australian and New Zealand Environment Conservation Council and Agriculture and Resource Management Council of Australia and New Zealand, Canberra.

ATSDR. (1999). *Toxicology Profile for Uranium.* Agency for Toxic Substances and Disease Registry, Atlanta, GA.

Barillet, S., Adam, C., Palluel, O., and Devaux, A. (2007). Bioaccumulation, oxidative stress, and neurotoxicity in *Danio rerio* exposed to different isotopic compositions of uranium. *Environ. Toxicol. Chem.* **26**, 497–505.

Barillet, S., Larno, V., Floriani, M., Devaux, A., and Adam-Guillermin, C. (2010). Ultrastructural effects on gill, muscle, and gonadal tissues induced in zebrafish (*Danio rerio*) by a waterborne uranium exposure. *Aquat. Toxicol.* **100**, 295–302.

Beak Consultants. (1989). *Sensitivity to Ionizing Radiation of Organisms other than Man – An Overview with Emphasis on Cellular and Ecosystem Level Effects.* Report to Environment Canada and the Atomic Energy Control Board, Ottawa.

Beaugelin-Seiller, K., Garnier-Laplace, J., and Gilbin, R. (2009). *Vers la Proposition d'une Norme de Qualité Environnementale pour l'Uranium en Eau Douce.* Institut de Radioprotection et de Surete Nucléaire, Paris.

Bird, G. A., and Schwartz, W. J. (1997). *Background Chemical and Radiological Levels in Canadian Shield Lakes and in Groundwater.* AECL, Whiteshell Laboratories, Pinawa, Manitoba, Technical Record TR-761, COG-96-524-I.

Boudreau, L., and Guay, I. (2002). *Effets de l'Uranium sur la Vie Aquatique et Détermination de Critères de Qualité de l'Eau de Surface.* Ministère de l'Environnement, Québec.

Bourrachot, S., Simon, O., and Gilbin, R. (2008). The effects of waterborne uranium on the hatching success, development, and survival of early life stages of zebrafish (*Danio rerio*). *Aquat. Toxicol.* **90**, 29–36.

Buckle, D., Humphrey, C., and Davies, C. (2010). Monitoring using fish community structure. In: *ERISS Research Summary (2008–2009)* (D. R. Jones and A. L. Webb, eds), pp. 88–92. Supervising Scientist, Australian Government, Darwin. Supervising Scientists Report 201.

Buet, A., Barillet, S., and Camilleri, V. (2005). Changes in oxidative stress parameters in fish as response to direct uranium exposure. *Radioprotection* **40** (Suppl. 1), S151–S155.

Campbell, P. G. C. (1995). Interactions between trace metals and aquatic organisms: A critique of the free-ion activity model. In: *Metal Speciation and Bioavailability in Aquatic Systems* (A. Tessier and D. R. Turner, eds), pp. 45–102. John Wiley and Sons, Chichester.

CanNorth. (2006a). *Beaverlodge Decommissioning – Spatial and Temporal Trends of Metals and Radionuclides in the Sediment of Fulton Bay.* Cameco Corporation. Canada North Environmental Services, Saskatoon.

CanNorth. (2006b). Gunnar Site Characterisation 2004 and 2005 Aquatic Assessment – Final Report. Canada North Environmental Services, Saskatoon.

CCME (2011). *Canadian Water Quality Guidelines for Uranium: Scientific Criteria Document.* Canadian Council of Ministers of the Environment, Winnipeg.

Chappaz, A., Gobeil, C., and Tessier, A. (2010). Controls on uranium distribution in lake sediments. *Geochim. Cosmochim. Acta* **74**, 203–214.

Cheng, K. L., Hogan, A. C., Parry, D. L., Markich, S. J., Harford, A. J., and van Dam, R. A. (2010). Uranium toxicity and speciation during chronic exposure to the tropical freshwater fish, *Mogurnda mogurnda. Chemosphere* **79**, 547–554.

Choppin, G. R., and Mathur, J. N. (1991). Hydrolysis of actinyl(VI) cations. *Radiochim. Acta* **52/53**, 25–28.

Clulow, F. V., Davé, N. K., Lim, T. P., and Avadhanula, R. (1998). Radionuclides (lead-210, polonium-210, thorium-230, and -232) and thorium and uranium in water, sediments, and fish from lakes near the city of Elliot Lake, Ontario, Canada. *Environ. Pollut.* **99**, 199–213.

CNSC and EC (2009). *Risk Management of Uranium Releases from Uranium Mines and Mills: 2007 Annual Report.* Canadian Nuclear Safety Commission and Environment Canada. Catalogue No. CC171-9/2007E-PDF. Ottowa: Minister of Public Works and Government Services Canada.

Cooley, H. M., and Klaverkamp, J. F. (2000). Accumulation and distribution of dietary uranium in lake whitefish (*Coregonus clupeaformis*). *Aquat. Toxicol.* **48**, 477–494.

Cooley, H. M., Evans, R. E., and Klaverkamp, J. F. (2000). Toxicology of dietary uranium in lake whitefish (*Coregonus clupeaformis*). *Aquat. Toxicol.* **48**, 495–515.

Denison, F. H., and Garnier-Laplace, J. (2005). The effects of database parameter uncertainty on uranium(VI) equilibrium calculations. *Geochim. Cosmochim. Acta* **69**, 2183–2191.

DiGiulio, R. T. (1991). Indices of oxidative stress as biomarkers for environmental contamination. In: *Aquatic Toxicology and Risk Assessment* (M. A. Mayes and M. G. Barron, eds), Vol. 14, p. 1124. American Society for Testing and Materials, Philadelphia, PA.Special Technical Publication

Djogic, R., Sipos, L., and Branica, M. (1986). Characterization of uranium(VI) in seawater. *Limnol. Oceanogr.* **31**, 1122–1131.

Emery, R. M., Kloppfer, D. C., Baker, D. A., and Soldat, J. K. (1981). Potential radiation dose from eating fish exposed to actinide contamination. *Health Phys.* **40**, 493–510.

Environment Canada. (1983). *Uranium*, Vol. 1. *Inorganic Chemical Substances* (M. C. Taylor, ed.). Environment Canada, Inland Waters Directorate, Water Quality Branch, Ottawa.

Environment Canada and Health Canada. (2003). *Priority Substances List Assessment Report: Releases of Radionuclides from Nuclear Facilities (Impact on Non-human Biota).* Environment Canada, Ottawa.

Farkas, I., Bányai, I., Szabó, Z., Wahlgren, U., and Grenthe, I. (2000). Rates and mechanisms of water exchange of UO_2^{2+}(aq) and $UO2(oxalate)F(H2O)_2^-$: A variable-temperature 17O and 19F NMR study. *Inorg. Chem.* **39**, 799–805.

Fortin, C., Dutel, L., and Garnier-Laplace, J. (2004). Uranium complexation and uptake by a green alga in relation to chemical speciation: The importance of the free uranyl ion. *Environ. Toxicol. Chem.* **23**, 974–981.

Fortin, C., Denison, F. H., and Garnier-Laplace, J. (2007). Metal–phytoplankton interactions: modeling the effect of competing ions (H^+, Ca^{2+}, and Mg^{2+}) on uranium uptake. *Environ. Toxicol. Chem.* **26**, 242–248.

Fortin, C., Couillard, Y., Vigneault, B., and Campbell, P. G. C. (2010). Determination of free Cd, Cu and Zn concentrations in lake waters by *in situ* diffusion followed by column equilibration ion-exchange. *Aquat. Geochem.* **16**, 151–172.

Fournier, E., Tran, D., Denison, F., Massabuau, J. C., and Garnier-Laplace, J. (2004). Valve closure response to uranium exposure for a freshwater bivalve (*Corbicula fluminea*): Quantification of the influence of pH. *Environ. Toxicol. Chem.* **23**, 1108–1114.

François, L., Fortin, C., and Campbell, P. G. C. (2007). pH modulates transport rates of manganese and cadmium in the green alga *Chlamydomonas reinhardtii* through non-competitive interactions: implications for an algal BLM. *Aquat. Toxicol.* **84**, 123–132.

Franklin, N. M., Stauber, J. L., Markich, S. J., and Lim, R. P. (2000). pH-dependent toxicity of copper and uranium to a tropical freshwater alga (*Chlorella sp.*). *Aquat. Toxicol.* **48**, 275–289.

Giammar, D. E., and Hering, J. G. (2001). Time scales for sorption–desorption and surface precipitation of uranyl on goethite. *Environ. Sci. Technol.* **35**, 3332–3337.

Golder Associates. (2002). *Cameco – Current Period Environmental Monitoring Program (CPEMP) for the Beaverlodge Decommissioned Site.* Golder Associates, Canada.

Golder Associates. (2008). *Technical Information Document for the Inactive Lorado Uranium Tailings Site. Appendix 3.* Golder Associates, Canada.

Guillaumont, R., Fanghanel, T., Neck, V., Fuger, J., Palmer, D. A., Grenthe, I., and Rand, M. H. (2003). *Update on the Chemical Thermodynamics of Uranium, Neptunium, Plutonium, Americium and Technetium. Chemical Thermodynamics*, Vol. 5, (Nuclear Energy Agency and Organisation for Economic Co-operation and Development), Elsevier, Oxford.

Hamilton, S. J., and Buhl, K. J. (1997). Hazard evaluation of inorganics, singly and in mixtures, to flannelmouth sucker *Catostomus latipinnis* in the San Juan River, New Mexico. *Ecotoxicol. Environ. Saf.* **38**, 296–308.

Harrison, J. D., and Stather, J. W. (1981). The gastrointestinal absorption of protactinium, uranium, and neptunium in the hamster. *Radiat. Res.* **88**, 47–55.

Helm, L., and Merbach, A. E. (1999). Water exchange on metal ions: experiments and simulations. *Coord. Chem. Rev.* **187**, 151–181.

Hogan, A. C., Van Dam, R. A., Markich, S. J., McCullough, C. and Camilleri, C. (2003). *Chronic Toxicity of Uranium to the Tropical Green Alga Chlorella sp. for the Derivation of a Site Specific Trigger Value for Magela Creek.* Internal Report 412. Darwin: Supervising Scientist, Australian Government.

Houston, M., Ng, J., Noller, B., Markich, S. and van Dam, R. (2009). Influence of dissolved organic carbon on the toxicity of uranium. In ERISS Research Summary 2007–2008 (eds. D. R. Jones and A. Webb), pp. 6–11. Supervising Scientist Report 200. Darwin: Supervising Scientist, Australian Government.

Houston, M., Ng, J., Noller, B., Markich, S. and van Dam, R. (2010). Amelioration of uranium toxicity by dissolved organic carbon. In *ERISS Research Summary 2008–2009* (eds. D. R. Jones and A. Webb), pp. 16–22. Supervising Scientist Report 201. Darwin: Supervising Scientist, Australian Government.

Hursh, J. B., Neuman, W. F., Toribara, T., Wilson, H., and Waterhouse, C. (1969). Oral ingestion of uranium by man. *Health Phys.* **17**, 619–621.

ICRP. (1979). *Limits for Intakes of Radionuclides by Workers.* International Commission on Radiological Protection. Publication 30. Pergamon Press, Oxford.

Jang, J. H., Dempsey, B. A., and Burgos, W. D. (2006). Solubility of schoepite: comparison and selection of complexation constants for U(VI). *Water Res.* **40**, 2738–2746.

Keklak, M. M., Newman, M. C., and Mulvey, M. (1994). Enhanced uranium tolerance of an exposed population of the eastern mosquitofish (*Gambusia holbrooki* Girard 1859). *Arch. Environ. Contam. Toxicol.* **27**, 20–24.

Kolok, A. S. (2001). Sublethal identification of susceptible individuals: using swim performance to identify susceptible fish while keeping them alive. *Ecotoxicology* **10**, 205–209.

Köster, W. (2004). Transport of solutes across biological membranes: prokaryotes. In: *Physicochemical Kinetics and Transport at Biointerfaces* (H. P. van Leeuwen and W. Köster, eds), pp. 271–335. John Wiley & Sons, Chichester.

Labrot, F., Narbonne, J. F., Ville, P., Saint Denis, M., and Ribera, D. (1999). Acute toxicity, toxicokinetics, and tissue target of lead and uranium in the clam *Corbicula fluminea* and the worm *Eisenia fetida*: comparison with the fish *Brachydanio rerio*. *Arch. Environ. Contam. Toxicol.* **36**, 167–178.

Lerebours, A., Gonzalez, P., Adam, C., Camilleri, V., Bourdineaud, J.-P., and Garnier-Laplace, J. (2009). Comparative analysis of gene expression in brain, liver, skeletal muscles, and gills of zebrafish (*Danio rerio*) exposed to environmentally relevant waterborne uranium concentrations. *Environ. Toxicol. Chem.* **28**, 1271–1278.

Lerebours, A., Bourdineaud, J.-P., van der Ven, K., Vandenbrouck, T., Gonzalez, P., Camilleri, V., Floriani, M., Garnier-Laplace, J., and Adam-Guillermin, C. (2010). Sublethal effects of waterborne uranium exposures on the zebrafish brain: transcriptional responses and alterations of the olfactory bulb ultrastructure. *Environ. Sci. Technol.* **44**, 1438–1443.

Liber, K., Stoughton, S. and Janz, D. (2004a). *Uranium Toxicity Testing to Early Life Stage Lake Trout* (Salvelinus namaycush). Final report to Saskatchewan Environment, Shield EcoRegion – Uranium Mining Unit, Saskatoon, Canada

Liber, K., Stoughton, S., and Rosaasen, A. (2004b). Chronic uranium toxicity to white sucker fry (*Catostomus commersoni*). *Bull. Environ. Contam. Toxicol.* **73**, 1065–1071.

Liber, K., Stoughton, S. and Janz, D. (2005). *Uranium Toxicity Testing to Early Life Stage Northern Pike* (Esox lucius). Final report to Saskatchewan Environment, Shield EcoRegion – Uranium Mining Unit, Saskatoon, Canada.

Liber, K., deRosemond, S. and Budnick, K. (2007). *Uranium Toxicity to Regionally-Representative Algae and Invertebrate species*. Report submitted to AREVA Resources and Cameco Corp.

Lofts, S., Tipping, E., and Hamilton-Taylor, J. (2008). The chemical speciation of Fe(III) in freshwaters. *Aquat. Geochem.* **14**, 337–358.

Lourenço, J., Castro, B. B., Machado, R., Nunes, B., Mendo, S., Gonçalves, F., and Pereira, R. (2010). Genetic, biochemical, and individual responses of the teleost fish *Carassius auratus* to uranium. *Arch. Environ. Contamin. Toxicol.* **58**, 1023–1031.

Lovely, D. R., Phillips, E. J. P., Gorby, Y. A., and Landa, E. R. (1991). Microbial reduction of uranium. *Nature* **350**, 413–416.

Luoma, S. N., and Rainbow, P. S. (2005). Why is metal bioaccumulation so variable? Biodynamics as a unifying concept. *Environ. Sci. Technol.* **39**, 1921–1931.

Markich, S. J. (2002). Uranium speciation and bioavailability in aquatic systems: an overview. *The Scientific World Journal* **2**, 707–729.

Markich, S. J., Brown, P. L., Jeffree, R. A., and Lim, R. P. (2000). Valve movement responses of *Velesunio angasi* (Bivalvia: Hyriidae) to manganese and uranium: an exception to the free ion activity model. *Aquat. Toxicol.* **51**, 155–175.

Mathews, M., Beaugelin-Seiller, K., Garnier-Laplace, J., Gilbin, R., Adam, C., and Della-Vedova, C. (2009). A probabilistic assessment of the chemical and radiological risks of chronic exposure to uranium in freshwater ecosystems. *Environ. Sci. Technol.* **43**, 6684–6690.

McGeer, J. C., Brix, K. V., Skeaff, J. M., Deforest, D. K., Brigham, S. I., Adams, W. J., and Green, A. (2003). Inverse relationship between bioconcentration factor and exposure concentration for metals: implications for hazard assessment of metal in the aquatic environment. *Environ. Toxicol. Chem.* **22**, 1017–1037.

McManus, J., Berelson, W. M., Severmann, S., Poulson, R. L., Hammond, D. E., Klinkhammer, G. P., and Holm, C. (2006). Molybdenum and uranium geochemistry in continental margin sediments: paleoproxy potential. *Geochim. Cosmochim. Acta* **70**, 4643–4662.

Meinrath, A., Schneider, P., and Meinrath, G. (2003). Uranium ores and depleted uranium in the environment, with a reference to uranium in the biosphere from the Erzgebirge/Sachsen, Germany. *J. Environ. Radioactiv.* **64**, 175–193.

Mkandawire, M., Vogel, K., Taubert, B., and Dudel, E. G. (2007). Phosphate regulates uranium(VI) toxicity to *Lemna gibba* L. G3. *Environ. Toxicol.* **22**, 9–16.

Monte, L., Brittain, J. E., Hakanson, L., Heling, R., Smith, J. T., and Zheleznyak, M. (2003). Review and assessment of models used to predict the fate of radionuclides in lakes. *J. Environ. Radioactiv.* **69**, 177–205.

Monte, L., Brittain, J. E., Gallego, E., Hakanson, L., Hofman, D., and Jimenez, A. (2009a). MOIRA-PLUS: a decision support system for the management of complex fresh water ecosystems contaminated by radionuclides and heavy metals. *Comp. Geosci.* **35**, 880–896.

Monte, L., Perianez, R., Boyer, P., Smith, J. T., and Brittain, J. E. (2009b). The role of physical processes controlling the behaviour of radionuclide contaminants in the aquatic environment: a review of state-of-the-art modelling approaches. *J. Environ. Radioactiv.* **100**, 779–784.

Moulin, V., Tits, J., and Ouzounian, G. (1992). Actinide speciation in the presence of humic substances in natural water conditions. *Radiochim. Acta* **58/59**, 179–190.

Moulin, C., Laszak, I., Moulin, V., and Tondre, C. (1998). Time-resolved laser-induced fluorescence as a unique tool for low-level uranium speciation. *Appl. Spectrosc.* **52**, 528–535.

Moulin, V., and Moulin, C. (2001). Radionuclide speciation in the environment: a review. *Radiochim. Acta* **89**, 773–778.

Nakajima, A., Horikoshi, T., and Sakaguchi, T. (1979). Ion effects on the uptake of uranium by *Chlorella regularis. Agric. Biol. Chem.* **43**, 625–629.

Neuefeind, J., Soderholm, L., and Skanthakumar, S. (2004). Experimental coordination environment of uranyl(VI) in aqueous solution. *J. Phys. Chem. A* **108**, 2733–2739.

Nieboer, E., and Richardson, D. H. S. (1980). The replacement of the nondescript term "heavy metals" by a biologically and chemically significant classification of metal ions. *Environ. Pollut. B* **1**, 3–26.

Nitzsche, O., Meinrath, G., and Merkel, B. (2000). Database uncertainty as a limiting factor in reactive transport prognosis. *J. Contam. Hydrol.* **44**, 223–237.

OECD. (2008). *Uranium 2007: Resources, Production and Demand.* Organization for Economical Cooperation and Development, Nuclear Energy Agency and International Atomic Energy Agency, Brussels.

OMOE. (2008). *Science Discussion Document on the Development of Air Standard for Uranium and Uranium Compounds.* Standards Development Branch, Ontario Ministry of the Environment.

OMOEE. (1994). *Water Management Policies, Guidelines, Provincial Water Quality Objectives, of the Ministry of Environment and Energy* (Reprinted 1999). Ontario Ministry of Environment and Energy, Toronto.

Paquin, P. R., Zoltay, V., Winfield, R. P., Wu, K. B., Mathew, R., Santore, R. C., and DiToro, D. M. (2002). Extension of the biotic ligand model to acute toxicity to a physiologically-based model of the survival time of rainbow trout (*Oncorhynchus mykiss*) exposed to silver. *Comp. Biochem. Physiol. C* **133**, 305–343.

Parkhurst, B. R., Elder, R. G., Meyer, J. S., Sanchez, D. A., Pennak, R. W., and Waller, W. T. (1984). An environmental hazard evaluation of uranium in a Rocky Mountain stream. *Environ. Toxicol. Chem.* **3**, 113–124.

Perdue, E. M., and Ritchie, J. D. (2003). Dissolved organic matter in freshwaters. In: *Surface and Ground Water, Weathering, and Soils* (J. I Drever, ed.), Vol. 5, pp. 273–318. Elsevier, Amsterdam.

Persaud, D., Jaagumagi, R. and Hayton, A. (1992). *Guidelines for the Protection and Management of Aquatic Sediment Quality in Ontario.* Ontario Ministry of the Environment, Toronto.

Poston, T. M. (1982). Observations on the bioaccumulation potential of thorium and uranium in rainbow trout (*Salmo gairdneri*). *Bull. Environ. Contam. Toxicol.* **28**, 682–690.

Pyle, G. G., Swanson, S. M., and Lehmkuhl, D. M. (2002). Toxicity of uranium mine receiving waters to early life stage fathead minnows (*Pimephales promelas*) in the laboratory. *Environ. Pollut.* **116**, 243–255.

Ragnarsdottir, K. V., and Charlet, L. (2000). In: *Uranium behaviour in natural environments* (J. D. Cotter-Howells, L. S. Campbell, E. Valsami-Jones and M. Batchelder, eds), pp. 333–377. Mineralogical Society of Great Britain & Ireland, London.

Riethmuller, N. Markich, S., Parry, D. and Van Dam, R. (2000). *The Effect of True Water Hardness and Alkalinity on the Toxicity of Cu and U to Two Tropical Australian Freshwater Organisms.* Supervising Scientist Report 155. Supervising Scientist, Canberra.

Robertson, E. L., and Liber, K. (2007). Bioassays with caged *Hyalella azteca* to determine *in situ* toxicity downstream of two Saskatchewan, Canada, uranium operations. *Environ. Toxicol. Chem.* **26**, 2345–2355.

Royal Society (2001a). *The Health Hazards of Depleted Uranium Munitions, Part I.* London: Royal Society. http://www.royalsociety.org/displaypagedoc.asp?id=11496.

Royal Society (2001b). *The Health Hazards of Depleted Uranium Munitions, Part II.* London: Royal Society. http://www.royalsociety.org/displaypagedoc.asp?id=11498.

Saskatchewan Environment (2006) *Surface Water Quality Objectives,* Interim edn. EPB 356. http://www.environment.gov.sk.ca/

Schlueter, M. A., Guttman, S. I., Oris, J. T., and Bailer, A. J. (1995). Survival of copper-exposed juvenile fathead minnows (*Pimephales promelas*) differs among allozyme genotypes. *Environ. Toxicol. Chem.* **14**, 1727–1734.

Schlueter, M. A., Guttman, S. I., Oris, J. T., and Bailer, A. J. (1997). Differential survival of fathead minnows, *Pimephales promelas*, as affected by copper exposure, prior population stress, and allozyme genotypes. *Environ. Toxicol. Chem.* **16**, 939–947.

SENES. (2007a). *Beaverlodge Mine Site Update of Geochemical Characterization and Modeling Assessment.* SENES Consultants, Ontario.

SENES. (2007b). *Rabbit Lake operation Integrated ERA and SOE (2000–2005).* SENES Consultants, Ontario.

SENES. (2009). *Beaverlodge Mine Site Integrated ERA and SOE 1985–2007.* SENES Consultants, Ontario.

Sheppard, S. C., Sheppard, M. I., Gallerand, M., and Sanipelli, B. (2005). Derivation of ecotoxicity thresholds for uranium. *J. Environ. Radioactiv.* **79**, 55–83.

Sherwood, G. D., Kovecses, J., Hontela, A., and Rasmussen, J. B. (2002a). Simplified food webs lead to energetic bottlenecks in polluted lakes. *Can. J. Fish. Aquat. Sci.* **59**, 1–5.

Sherwood, G. D., Pazzia, I., Moeser, A., Hontela, A., and Rasmussen, J. B. (2002b). Shifting gears: enzymatic evidence for the energetic advantage of switching diet in wild-living fish. *Can. J. Fish. Aquat. Sci.* **59**, 229–241.

Silva, R. J. (1992). Mechanisms for the retardation of uranium(VI) migration. *Mater. Res. Soc. Symp. Proc.* **257**, 323–330.

Simon, O., and Garnier-Laplace, J. (2004). Kinetic analysis of uranium accumulation in the bivalve *Corbicula fluminea*: effect of pH and direct exposure levels. *Aquat. Toxicol.* **68**, 95–108.

Simon, O., and Garnier-Laplace, J. (2005). Laboratory and field assessment of uranium trophic transfer efficiency in the crayfish *Orconectes limosus* fed the bivalve *C. fluminea*. *Aquat. Toxicol.* **74**, 372–383.

Spry, D. J., and Wood, C. M. (1984). Acid–base, plasma ion and blood gas changes in rainbow trout during short term toxic zinc exposure. *J. Comp. Physiol. B.* **154**, 149–158.

Spry, D. J., Hodson, P. V., and Wood, C. M. (1988). Relative contributions of dietary and waterborne zinc in the rainbow trout, *Salmo gairdneri*. *Can. J. Fish. Aquat. Sci.* **45**, 32–41.

Stumm, W., and Morgan, J. (1996). *Aquatic Chemistry: Chemical Equilibria and Rates in Natural Waters*. John Wiley and Sons, New York.

Swanson, S. M. (1982). *Levels and Effects of Radionuclides in Aquatic Fauna of the Beaverlodge Area (Saskatchewan)*. Saskatchewan Research Council, Saskatoon, SRC Publication No. C-806-5-E-82.

Swanson, S. M. (1983). Levels of 226Ra, 210Pb and total U in fish near a Saskatchewan uranium mine and mill. *Health Phys.* **45**, 67–80.

Swanson, S. M. (1985). Food-chain transfer of U-series radionuclides in a northern Saskatchewan aquatic system. *Health Phys.* **49**, 747–770.

Tarzwell, C. M., and Henderson, C. (1960). Toxicity of less common metals to fishes. *Ind. Wastes* **5**, 12.

Thompson, P. A., Kurias, J., and Mihok, S. (2005). Derivation and use of sediment quality guidelines for ecological risk assessment of metals and radionuclides released to the environment from uranium mining and milling activities in Canada. *Environ. Monit. Assess.* **110**, 71–85.

Tipping, E. (2002). *Cation Binding by Humic Substances*. Cambridge University Press, Cambridge.

Tipping, E. (2005). Modelling Al competition for heavy metal binding by dissolved organic matter in soil and surface waters of acid and neutral pH. *Geoderma* **127**, 293–304.

Tran, D., Massabuau, J. C., and Garnier-Laplace, J. (2004). Effect of carbon dioxide on uranium bioaccumulation in the freshwater clam *Corbicula fluminea*. *Environ. Toxicol. Chem.* **23**, 739–747.

Trapp, K. E. (1986). *Acute Toxicity of Uranium to Waterfleas* (Daphnia pulex) *and Bluegill* (Lepomis macrochirus). Report ECS-SR-30. Aiken, SC: Environmental and Chemical Sciences.

Unsworth, E. R., Jones, P., and Hill, S. J. (2002). The effect of thermodynamic data on computer model prediction of uranium speciation in natural water systems. *J. Environ. Monit.* **4**, 528–532.

USEPA (1999a). *TRI (Toxic Release Inventory) PBT Final Rule*. 64 FR 58666, October 29, 1999. http://www.epa.gov/fedrgstr/EPA-WASTE/1999/October/Day-29/f28169.htm

USEPA (1999b). *Category for Persistent, Bioaccumulative, and Toxic New Chemical Substances*. Federal Register: November 4, 1999 (Vol. 64, No. 213). Notices Page 60194–60204. From the Federal Register Online via GPO Access. wais.access.gpo.gov. DOCID:fr04no99-64.

Vasiliev, I. A., Barber, D. S., Alekhina, V. M., Mamatibtaimov, S., Betsill, D., and Passell, H. (2005). Uranium levels in the Naryn and Mailuu-Suu rivers of Kyrgyz Republic. *J. Rad. Nucl. Chem.* **263**, 107–212.

Vidaud, C., Dedieu, A., Basset, C., Plantevin, S., Dany, I., Pible, O., and Quéméneur, E. (2005). Screening of human serum proteins for uranium binding. *Chem. Res. Toxicol.* **18**, 946–953.

Vizon SciTec. (2004). *Final Report on the Toxicity Investigation of Uranium to Aquatic Organisms*. Vizon SciTec, Vancouver, CNSC Project No. R205.1.

Waite, D. T., Joshi, S. R., Sommerstad, H., Wobeser, G., and Gajadhar, A. A. (1990). A toxicological examination of whitefish (*Coregonus clupeaformis*) and northern pike (*Esox lucius*) exposed to uranium mine tailings. *Arch. Environ. Contam. Toxicol.* **19**, 578–582.

Wall, J. D., and Krumholz, L. R. (2006). Uranium reduction. *Ann. Rev. Microbiol.* **60**, 149–166.

Wetzel, R. G. (2001). *Limnology: Lake and River Ecosystems*. Academic Press, New York.

WHO. (2001). *Depleted Uranium: Sources, Exposure and Health Effects*. Department of Protection of the Human Environment, World Health Organization, Geneva, WHO/SDE/ PHE/01.1.

Windom, H., Smith, R., Niencheski, F., and Alexander, C. (2000). Uranium in rivers and estuaries of globally diverse, smaller watersheds. *Mar. Chem.* **68**, 307–321.

Wrenn, M. E., Durbin, P. W., and Howard, B. (1985). Metabolism of ingested uranium and radium. *Health Phys.* **48**, 601–633.

Yankovich, T. L. (2009). Mass balance approach to estimating radionuclide loads and concentrations in edible fish tissues using stable analogues. *J. Environ. Radioactiv.* **100**, 795–801.

Yuldashev, B. S., Salikhbaev, U. S., Kist, A. A., Radyuk, R. I., Barber, D. S., Passell, H. D., Betsill, J. D., Matthews, R., Vdovina, E. D., Zhuk, L. I., Solodukhin, V. P., Poznyak, V. L., Vasiliev, I. A., Alekhina, V. M., and Djuraev, A. A. (2005). Radioecological monitoring of transboundary rivers of the Central Asian Region. *J. Rad. Nucl. Chem.* **263**, 219–228.

Zeman, F. A., Gilbin, R., Alonzo, F., Lecomte-Pradines, C., Garnier-Laplace, J., and Aliaume, C. (2008). Effects of waterborne uranium on survival, growth, reproduction and physiological processes of the freshwater cladoceran *Daphnia magna*. *Aquat. Toxicol.* **86**, 370–378.

9

MODELING THE PHYSIOLOGY AND TOXICOLOGY OF METALS

PAUL PAQUIN

AARON REDMAN

ADAM RYAN

ROBERT SANTORE

1. Introduction
2. Model Frameworks for Evaluating Metal Accumulation
    2.1. Whole-Body and Multicompartment Bioaccumulation Models
    2.2. Physiologically Based Pharmacokinetic Models
    2.3. Food Chain and Food Web Models and Biomagnification
3. Models Relating Metal Accumulation to Effects
    3.1. Damage–Repair Models
    3.2. Biotic Ligand Models
    3.3. Physiologically Based Mechanistic Models
    3.4. Intracellular Speciation Models
4. Regulatory Applications
    4.1. Use of Models for Water Quality and Tissue-Residue Based Criteria
    4.2. Other Model Applications
5. Future Model Development Needs

Most bioaccumulation models in current use were originally developed for organic chemicals and subsequently applied to metals. As a result, they do not typically incorporate metal-specific features that affect how metals are taken up, metabolized, detoxified, and eliminated. This review describes models that are used to evaluate metal accumulation, from waterborne and dietary exposure, by aquatic organisms. It considers whole-body models (including single-compartment and multicompartment variations), physiologically based pharmacokinetic models (providing an explicit representation of individual organs), and food chain or food web models. A representative model framework and example application are described for each model. Variations are also discussed. Metal bioaccumulation models

429

Homeostasis and Toxicology of Non-Essential Metals: Volume 31B
FISH PHYSIOLOGY

have not typically been developed to predict effects on aquatic organisms. However, several that were developed for this purpose are described herein. These include damage–repair models, biotic ligand models, physiologically based mechanistic models, and intracellular speciation models. While some effects models incorporate relatively unique metal-specific features, they have primarily been applied to waterborne exposures, and received limited use for predicting effects resulting from both waterborne and dietary exposures. Studies that facilitate the refinement of both bioaccumulation and effects models, and the unification of such models, will likely benefit both the scientific and regulatory communities.

1. INTRODUCTION

Modeling of metal accumulation by aquatic organisms has been evolving for more than 40 years. Most early models were developed for organic chemicals and subsequently applied to metals. As a consequence, they tended to ignore important ways by which metals differ from organic chemicals. Recent efforts have focused on: (1) understanding how metal speciation in aquatic systems affects metal bioavailability; (2) measuring metal-specific kinetic coefficients for aquatic organisms; and (3) improving generic model formulations to include fundamental metal-specific features. Each of these efforts has contributed to the evolution of bioaccumulation models which are becoming increasingly metal and organism specific.

While many variants exist, certain basic features common to most bioaccumulation models are described first. Model types will then be compared and contrasted relative to the unique features and level of detail they consider. This will provide an initial perspective on the input requirements and degree of complexity of the models considered herein.

Metal bioaccumulation models require information pertaining to both the abiotic environment and the organism itself. For example, typical model inputs include environmental exposure concentrations and uptake and elimination rates by the organism. Details of how this is accomplished may vary from one model to another. Models have in some cases been developed solely to evaluate metal accumulation (Section 2), and in other cases to relate metal accumulation to effects (Section 3).

Metal bioaccumulation models are based upon the formulation of a mass balance about the organism (Fig. 9.1). Metal uptake may occur via waterborne and dietary exposure. As a general rule, bioaccumulation models require a characterization of the ambient water quality, including both dissolved and particulate metal concentrations. The concentration of

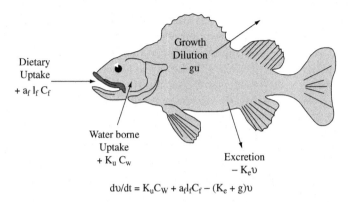

Fig. 9.1. Whole-body bioaccumulation model: schematic of mass balance model framework for a fish.

dissolved metal (or selected metal species) controls the uptake rate of waterborne metal and the particulate concentration controls dietary uptake. Upper trophic level predators require specification of dietary metal concentrations (zooplankton, small fish, etc.). Metal elimination rates are typically assumed to be proportional to tissue concentration.

Section 2 describes three types of metal accumulation model. Two are applied to individual organisms [whole-body and physiologically based pharmacokinetic (PBPK) models] and one is applied to a group of organisms (food chain or food web models). Whole-body models evaluate tissue concentrations in organisms as a whole, while PBPK models provide organ-specific concentrations. The increased specificity of PBPK models comes at the expense of increased model input requirements to characterize internal transfer processes and metal accumulation kinetics (e.g. blood distribution and flows and plasma:tissue partition coefficients for individual organs). Food web models include representative organisms comprising a food chain or food web. Each compartment corresponds to an individual organism and, though typically represented by a whole-body model, a PBPK model could be used if desired.

Section 3 describes four types of models that were developed to relate metal accumulation to effects. Although dietary exposure is not normally considered, it could be included in a damage–repair model (DRM), because the underlying equation represents a simplified whole-body accumulation model that includes uptake from both food and water. A DRM may require specification of a damage rate coefficient (analogous to a waterborne uptake rate coefficient) and a repair rate coefficient (analogous to a metal excretion rate coefficient). Cumulative damage is in turn related to a specific toxic response by the organism.

The biotic ligand model (BLM) also predicts waterborne effect levels. The BLM is a chemical equilibrium model that focuses on the influence of chemistry on metal bioavailability, accumulation at a proximate site of action of toxicity [e.g. the fish gill, the biotic ligand (BL)], and effects. The BLM evaluates a dissolved metal concentration associated with a specific degree of an effect (e.g. LC50). Although it requires a relatively detailed site-specific water chemistry characterization, it does not simulate time variable metal accumulation or effects, nor does it consider particulate (i.e. dietary) metal. Additional model parameters (e.g. BL metal binding constants) are also required. This information is frequently available for organisms and metals to which the BLM has been applied. (Note that most models do not require a detailed description of dissolved metal speciation, but could be applied in this manner. Uptake rate coefficients would need to be evaluated to be consistent with the bioavailable metal forms considered.)

The ion balance model (IBM) performs a mass balance of a particular ion about an organism as a way to predict effects. In brief, the mechanism of toxicity is inhibition of active uptake of an ion (e.g. Na^+) and the level of inhibition is related to the BL:Me concentration (computed with the BLM). The Na^+ IBM requires specification of active uptake and passive efflux rates for Na^+, permeability coefficients for Na^+ between internal fluid compartments, and a dose–response curve relating active uptake rate to BL:Me concentration (parameterized by calibration). As with the BLM, dietary uptake is not considered. The IBM simulates changes in internal Na^+ concentrations (the cumulative "damage") and Na^+ loss is associated with the response (e.g. fish death at 30% plasma Na^+ loss).

Intracellular speciation (ICS) models represent the mechanism of toxicity at the most fundamental level, predicting the metal concentration bound to a specific enzyme system, or to an enzyme-containing ligand pool. The model requires specification of total metal uptake rate by the organ (reflecting waterborne and/or dietary uptake). Additional inputs include cytosolic composition, metal binding constants, kinetic coefficients for metallothionein (MT) induction, and the like.

2. MODEL FRAMEWORKS FOR EVALUATING METAL ACCUMULATION

Most bioaccumulation models were originally developed for application to organic chemicals and subsequently applied to metals (see Paquin et al., 2003, for a review). They typically consider metal uptake (water and food), and metal elimination by the organism and are commonly grouped as: (1) whole-body bioaccumulation models, (2) PBPK models, and (3) food chain or food web models. These three types of bioaccumulation model are described in this section.

2.1. Whole-Body and Multicompartment Bioaccumulation Models

The most commonly used bioaccumulation model is the whole-body model. It is applied to individual organisms or to groups of organisms in a food chain or food web. The basic model formulation is derived from a mass balance of chemical about an organism. Because organism-specific differences are accounted for by values assigned to kinetic coefficients, the model is independent of organism type.

2.1.1. BACKGROUND AND MODEL FRAMEWORK

Bioaccumulation data are often interpreted with a kinetic model (Thomann, 1981; Wang et al., 1996; Wang and Fisher, 1997a; and reviews by Reinfelder et al., 1998, and Luoma and Rainbow, 2005). Some of the earliest applications involved studies with fish. Pentreath (1973) modeled radionuclide uptake, via the gill and from ingested water. Related experiments investigated the importance of dietary sources of trace elements. Pursuant to these results, Pentreath recognized the need for experiments to assess the potential for dietary effects during chronic exposures. Although concurrent uptake from food and water were not simulated, the test results indicated that dietary metals may be important (Pentreath, 1976, 1977) and demonstrated the utility of models for interpreting experimental results.

Concurrent with Pentreath's investigations, Thomann and co-workers were developing computer models to quantitatively evaluate Great Lakes ecosystems (Hydroscience, 1973; Thomann et al., 1974). Their original food web model was subsequently refined by adding physiologically based features (e.g. ingestion rate, growth rate) and used to evaluate organic chemical and metal accumulation by aquatic food chains (Thomann, 1981). The utility of this early model was enhanced by the growing use of radiotracer methods to measure the requisite kinetic coefficients, particularly for metals (e.g. Fisher and Reinfelder, 1995; Wang et al., 1996; Croteau et al., 2004; Croteau and Luoma, 2005). Frequently referred to as the kinetic model approach (e.g. Reinfelder et al., 1998; Wang and Fisher, 1999) and, more recently, as the dynamic multipathway bioaccumulation model (DYMBAM) (Schlekat et al., 2002; Luoma and Rainbow, 2005), it has been extensively applied to individual organisms (e.g. bivalves, as well as crustaceans and fish) (Reinfelder et al., 1998). Its widespread acceptance is likely due to its incorporation of uptake from both water and food, as well as its ease of use.

The basic model equation is based upon a mass balance of a constituent about the organism (Fig. 9.1). This yields a differential equation for the change of tissue concentration (v) with time (t):

$$\frac{dv}{dt} = k_u C_w + \alpha_f I_f C_f - (k_e + g)v \tag{1}$$

The notation and units are as follows: v = metal concentration in tissue [$\mu g\ g_{drywt}^{-1} \sim \mu g\ g_d^{-1}$], t = time [days], k_u = dissolved metal uptake rate coefficient [$L\ g_d^{-1}\ day^{-1}$], C_w = dissolved metal concentration in water [$\mu g\ L^{-1}$], α_f = metal absorption efficiency from food [%, or %/100], I_f = ingestion rate of dry food [$g\ g_d^{-1}\ day^{-1}$], C_f = metal concentration in food [$\mu g\ g^{-1}$] (dry-weight normalized), k_e = elimination or excretion rate constant [1 day^{-1}], and g = first order growth rate constant [1 day^{-1}].

Care is needed to ensure the consistent use of dry weight and wet weight units. Here, an unsubscripted g refers to grams of dry food or particulate matter concentration (e.g. g or $g\ L^{-1}$), as is routinely reported. With regard to tissue weight or tissue metal concentration, a subscript d indicates grams of dry tissue weight (g_d) or dry-weight normalized concentration (μg metal per gram dry weight of tissue $\sim \mu g\ g_d^{-1}$). A subscript w, also commonly used in the literature to denote tissue wet weight, may also be referred to herein.

The first two terms on the right side of Eq. (1) represent metal uptake from water and food, respectively. Coefficients (e.g. k_u) may be evaluated from laboratory data or by calibration to field data. Uptake rates from water and food are often related to bioenergetics (Connolly and Thomann, 1992) and the chemical flux across the gills and through the gut, respectively. The third and fourth terms of Eq. (1) represent the decrease in concentration due to excretion and growth. Additional terms may be incorporated as needed (e.g. backdiffusion of chemical).

Solving Eq. (1) yields an expression for tissue concentration as a function of time (Thomann and St. John, 1979; Thomann, 1981):

$$v(t) = v_0 e^{-(k_e+g)t} + \left[\frac{k_u C_w + \alpha_f I_f C_f}{k_e + g} \right] (1 - e^{-(k_e+g)t}) \tag{2}$$

The first term represents dissipation of the initial condition (i.e. at $t = 0$, $v = v_0$), and the second term the approach to a new steady state. Tissue concentration reaches steady state when metal uptake from food and water is balanced by metal efflux and growth dilution. The response time, which depends upon the magnitude of $k_e + g$, is independent of uptake rate. When

$(k_e+g)t > 3$, $dv/dt \sim 0$ and, pursuant to Eq. (1), the steady-state concentration (v_{ss}) is given by:

$$v_{ss} = \frac{k_u C_w + \alpha_f I_f C_f}{k_e + g} \tag{3}$$

Equation (3) is applicable to conditions where the efflux rate constants for metal accumulated from both food and water are comparable and well represented by a single k_e, as shown for marine bivalves, at least as a first approximation (Wang et al., 1996).

Variations of Eqs. (2) and (3) are available for assigning different excretion rate coefficients for metal accumulated from waterborne and dietary sources (Fisher et al., 1996), evaluating bioconcentration factors (BCFs) and bioaccumulation factors (BAFs) (Fisher et al., 1996), and evaluating trophic transfer potential (TTP), i.e. the ratio of the concentration in a predator from dietary sources to the concentration in its prey $(v_{ss,f}/C_f)$ (Fisher et al., 1996; Reinfelder et al., 1998; Wang, 2002; Luoma and Presser, 2009). The standard model is frequently used with independent measurements of key model parameter values to simulate results for complex datasets (e.g. combined waterborne and dietary exposures extending over both uptake and depuration phases). This reduces the degrees of freedom for calibration by limiting the number of adjustable coefficients. When used in this way, the model has considerable explanatory power for real world settings (e.g. Luoma and Rainbow, 2005).

Several laboratory studies have been performed to evaluate kinetic coefficients for modeling metal accumulation by fish (Reinfelder and Fisher, 1994; Baines et al., 2002; Xu and Wang, 2002; see Presser and Luoma, 2010, for a review). Alternatively, the model may be calibrated to time series data, directly. Here, the analyst is somewhat less constrained and may be able to achieve better agreement between model and data.

A fugacity-based model that is algebraically equivalent to the kinetic model has been widely used to gain additional insights into the effects of allometry, metabolism, and growth on organic chemical accumulation (Mackay and Hughes, 1984; Clark et al., 1990; Hendriks and Heikens, 2001). It has been applied to both an aquatic food web and, as a PBPK model, to mammals (Paterson and Mackay, 1987; Sharpe and Mackay, 2000). An analogous aquivalence approach is used for metals (Diamond et al., 1992; Diamond, 1999).

Several kinetic model variations have also been proposed to better explain field observations. While not developed for fish, these formulations include physiological processes that should be applicable to fish. Croteau and co-workers used stable isotopes and a two-compartment model (generic slow and fast Cu turnover pools) to evaluate kinetic coefficients for Cu accumulation by the marine bivalve *Macoma balthica* (Croteau et al., 2004; Croteau and Luoma, 2005). The analysis highlighted the increasing importance of the slow pool ($k_{e2} = 0.0038$ day^{-1}) relative to the fast pool ($k_{e1} = 0.319$ day^{-1}) as Cu accumulation progressed. Redeker and co-workers developed a two-compartment model which included biologically inactive metal (BIM) and biologically active metal (BAM) pools (Redeker and Blust, 2004; Redeker et al, 2004). While the physical attributes of these pools were not explicitly characterized, their toxicological significance (i.e. non-toxic and toxic) was recognized.

Paquin et al. (2007) formulated a one-compartment model to achieve results very similar to the two-compartment model of Croteau and co-workers. Waterborne copper uptake by *Mytilus* spp. was simulated by adding a Michaelis–Menten term to Eq. (1):

$$\frac{d\upsilon}{dt} = k_u C_w + I_{bmax}\left[\frac{C_w}{(K_{Mb} + C_w)}\right] + \alpha_f I_f C_f - E_{max}\left[\frac{\upsilon}{(K_{Me} + \upsilon)}\right] - g\upsilon$$

(4)

Acclimation was accounted for by equating the Michaelis constant ($K_{Mb} \sim \mu g\ L^{-1}$) to the dissolved Cu concentration (C_w) when Cu regulation occurs. This maintains basal uptake at one half the maximum uptake rate ($I_{bmax} \sim \mu g\ g_d^{-1}$ day^{-1}) at low C_w while passive uptake ($k_u C_w$) becomes increasingly important at high C_w. Elimination was represented as a Michaelis–Menten expression with tissue Cu (υ) a surrogate for substrate (Michaelis constant, $K_{Me} \sim \mu g\ g_d^{-1}$). This decreased the effective first order Cu excretion rate coefficient [$k_e = E_{max}/(K_{Me} + \upsilon) \sim (\mu g\ g_d^{-1}$ day$^{-1})/(\mu g\ g_d^{-1})$ ~ 1 day^{-1}] because Cu efflux approaches a maximum rate ($E_{max} \sim \mu g\ g_d^{-1}$ day^{-1}) as υ increases (K_{Me} was calibrated to an extensive dataset). The process may be similar to saturable biliary excretion, as shown to occur in mammals (Gregus and Klaassen, 1986; Houwen et al., 1990). Alternatively, the apparent decline in excretion rate with increasing υ may reflect Cu detoxification (e.g. by MT) and its subsequent sequestration in slowly excreted lysosomes. The effective first order rate coefficients ($k_e = 0.153$ day^{-1} at $\upsilon \sim 6\ \mu g\ g_d^{-1}$ and 0.0251 day^{-1} at $\upsilon \sim 150\ \mu g\ g_d^{-1}$) were comparable to measurements obtained for *M. balthica* with stable isotopes (Croteau et al., 2004).

Acclimation is likely to be an important consideration for fish as well. Pre-exposing fish to metals has been shown to influence both metal uptake kinetics and net metal accumulation for both waterborne and dietary exposures and under both laboratory and field conditions (Ausseil et al., 2002; Zhang and Wang, 2006a; Cheung et al., 2007). However, another study with juvenile rainbow trout showed that pre-exposure to 22 $\mu g\ L^{-1}$

dissolved Cu resulted in a reduced uptake rate of waterborne Cu but not dietary Cu (Kamunde et al., 2002). Organism size may also be of considerable importance and has been well studied for various aquatic organisms, including fish (Zhang and Wang, 2007).

2.1.2. EXAMPLE APPLICATION

The whole-body bioaccumulation model has been applied to the juvenile rainbow trout dataset of Harrison and Klaverkamp (1989) to illustrate its use. The fish were exposed to radiolabeled ^{109}Cd via water (C_w = 1.27 ng L^{-1}) and diet (C_f = 0.0307 μg g$_d^{-1}$). The dietary test included incidental exposure to waterborne Cd (C_w = 0.318 ng L^{-1}~25% × waterborne test concentration) due to ^{109}Cd desorption from food and feces. All ^{109}Cd measurements were adjusted to time zero and used to estimate stable Cd concentrations in excess of background Cd (which was neglected). Fish were fed 2.3% of whole-body wet weight, three times per week (average daily ingestion rate I_f = 0.00986 g g$_w^{-1}$ day^{-1}, or ~1% day^{-1}; 0.0404 g g$_d^{-1}$ day^{-1}, or ~4% day^{-1} on a whole-body dry weight basis). The initial fish weight was 46.2 g and net growth rate constants (g) for the uptake and depuration periods were estimated from observed changes in fish weights (waterborne tests: g = 0.009 day^{-1} and 0.002 day^{-1}; dietary tests: g = 0.007 day^{-1} and ~0.0 day^{-1}, respectively). The lower growth rates during depuration were not explained. Remaining kinetic parameters (k_u, α_f, and k_e) were evaluated by model calibration to experimental data.

The whole-body model calibration strategy followed the PBPK model calibration strategy of Thomann et al. (1997) for this same dataset. The model was first calibrated to the waterborne dataset to evaluate dissolved Cd uptake rate (k_u) and excretion rate (k_e) coefficients. These values were then used with the dietary dataset to evaluate Cd assimilation efficiency from food (α_f). Whole-body tissue data (waterborne and dietary test results; the latter including incidental exposure to dissolved Cd) are shown in Fig. 9.2A (unfilled and filled data points, respectively). Reasonable agreement between model and waterborne data is achieved with k_u = 0.00535 L g$_d^{-1}$ day^{-1} (5.35 mL g$_d^{-1}$ day^{-1}) and k_e = 0.0093 day^{-1} (~1% day^{-1}). Peak tissue Cd is slightly underestimated and Cd depuration is slightly overestimated. A tradeoff exists between achieving an optimum data fit during the 0–72 day uptake phase and the subsequent 56 day depuration phase.

When waterborne-derived coefficients are applied to the dietary dataset, the best fit is achieved with a dietary Cd assimilation efficiency of α_f = 4.9% (Fig. 9.2A, upper line). However, the model underestimates the rapid initial increase in tissue Cd and overestimates it during depuration. The reason is that the water-only excretion rate coefficient does not adequately describe

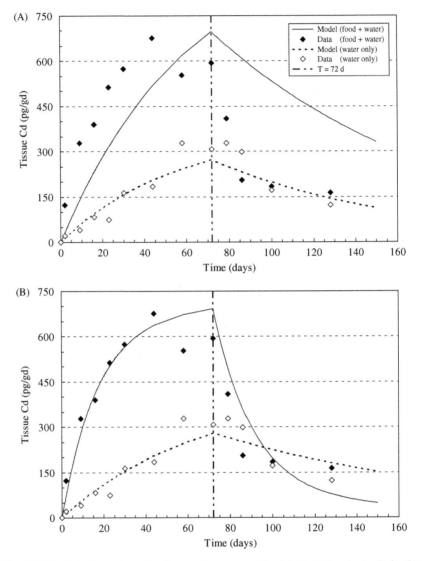

Fig. 9.2. Whole-body model calibration results compared to data for (A) a single elimination rate coefficient (0.0093 day^{-1}) and (B) independent elimination rate coefficients for Cd accumulated via waterborne and dietary uptake (of 0.00587 and 0.0531 day^{-1}, respectively). Data from Harrison and Klaverkamp (1989).

dietary accumulation kinetics. For marine bivalves and copepods at least, accumulated dietary metals may be eliminated at a faster rate than waterborne metals (Wang and Fisher, 1997b, 1998a). This may reflect a dependence of internal distribution, metabolism, and excretion on mode of uptake. Overall model–data agreement was improved when depuration rate coefficients for metal accumulated from water (k_{ew}) and food (k_{ef}) were independently evaluated (Fig. 9.2B) ($k_u = 0.020$ L g_d^{-1} day^{-1}, $\alpha_f = 0.125$, $k_{ew} = 0.00587$ day^{-1} and $k_{ef} = 0.0531$ day^{-1}).

The most notable difference in the preceding calibrations is the nearly order-of-magnitude change in depuration rate coefficients. Xu and Wang (2002) observed a qualitatively similar but less pronounced two-fold difference in Cd k_e values for the marine fish *Lutjanus argentimaculatus* (mangrove snapper; 0.025 and 0.047 day^{-1} for waterborne and dietary Cd, respectively). To the degree this response is observed for other datasets, it would be helpful to understand why. The Cd depuration rate also appears to decrease as the data approach a plateau near the end of the study. As a result, use of a constant k_e begins to underestimate whole-body Cd. As discussed in Section 2.2.2, ongoing accumulation of kidney Cd probably contributes to this whole-body Cd response (see also McGeer et al., Chapter 3).

2.2. Physiologically Based Pharmacokinetic Models

PBPK models incorporate considerably more detail than whole-body bioaccumulation models. The explicit representation of individual organs allows for the simultaneous representation of uptake and elimination kinetics for each. This increased specificity allows for more direct evaluation of target organ dose.

2.2.1. BACKGROUND AND MODEL FRAMEWORK

Torsten Teorell's studies of how diffusion affects the distribution of ions on opposite sides of charged membranes provide several examples of how mathematical models of physiological processes facilitate the interpretation of experimental results (Teorell, 1935, 1937a). His models of the distribution kinetics of substances administered by extravascular and intravascular means (Teorell, 1937b,c) led Paalzow (1995) to bestow upon him the title "the father of pharmacokinetics". The ensuing use of PBPK models was greatly facilitated by the rapidly increasing use of computers. This led to very detailed representations of individual organs and the physiological and biochemical processes they perform (e.g. see Guyton et al., 1972).

Three pharmacokinetic modeling approaches are commonly applied: generic multicompartment, clearance-based, and physiologically based approaches (see reviews by Barron et al., 1990; Landrum et al., 1992; McKim and Nichols, 1994; Nichols, 2002). Although relatively detailed information is required for model parameterization at the organ-specific level, they provide a physiologically based framework that is amenable to modeling different body sizes, species, and target organ toxicity (McKim and Nichols, 1994).

Most of the earlier PBPK model development efforts with fish focused on mechanistic investigations and ways to model organic chemical accumulation (e.g. Erickson and McKim, 1990; McKim and Erickson, 1991; McKim and Nichols, 1994). Fortunately, since many PBPK model parameter values developed for organics should also be applicable to metal accumulation by fish (e.g. organ weights, blood perfusion rates) the results remain of use for purposes herein. Other models with metals were developed for mammals and adapted for use with aquatic organisms. For example, the rainbow trout PBPK model described below was adapted from an earlier model of chromium bioaccumulation by rats (Thomann et al., 1994, 1997). Waterborne Cd uptake is represented in two steps, from water to gill and gill to blood. Dietary uptake is evaluated from food ingestion rate, exchange between dietary metal in the alimentary canal and gut tissue, and metal exchange with the blood. The blood distributes and exchanges metal with organs and tissues (gill, liver, kidney, and gut wall) and a generic storage compartment representing other tissues (e.g. muscle and bone). Excretion is assumed to occur via the kidney (first order with respect to tissue concentration) and egestion of feces.

As formulated by Thomann and co-workers, exchanges are proportional to dissolved concentration gradients between the organs and blood. For example, the mass balance for the rate of change of cadmium in the blood is given by:

$$
\frac{dV_1c_1}{dt} = E_{21}(c_{d_2} - f_dc_1) + E_{31}(c_{d_3} - f_dc_1) + E_{41}(c_{d_4} - f_dc_1) \\
+ E_{51}(c_{d_5} - f_dc_1) + E_{61}(c_{d_6} - f_dc_1)
$$

(5)

Numeric subscripts, $i = 1\text{--}7$, refer to the seven model compartments: (1) blood, (2) storage, (3) gill, (4) gut, (5) kidney, (6) liver, and (7) alimentary canal content (not directly tied to blood) (Fig. 9.3). The compartments were selected based on data availability and the need to represent organs controlling metal uptake from food and water, internal transfers, and routes of elimination by fish. Here, E_{ij} represents the bulk diffusive exchange coefficient (L day^{-1}) for dissolved concentrations, c_{di} (pg L^{-1}), between compartments i and j; c_1 is total metal concentration in compartment 1 (blood; pg L^{-1}); V_1 is the blood volume (liters, L), and f_d is the dissolved fraction of total metal in the blood. Alternative physiologically based formulations for active transport processes could readily be employed.

The relationship between dissolved (c_{di}) and total concentrations (v_i) in each compartment is expressed as a partition coefficient, π_i ([pg g$_d^{-1}$]/[pg L^{-1}]

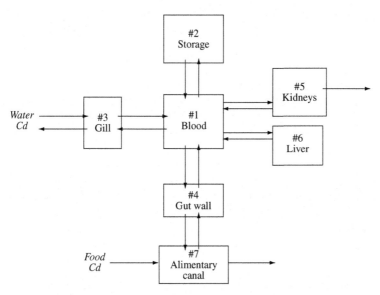

Fig. 9.3. PBPK model framework: schematic of a seven-compartment model for Cd in rainbow trout. Redrawn from Thomann et al. (1997) for Paquin et al. (2003).

~L g_d^{-1}), the ratio of total tissue metal concentration of compartment i (v_i, pg g_d^{-1}) to the dissolved concentration in compartment i (c_{di}, pg L^{-1}).

$$\pi_i = \frac{v_i}{c_{di}} \qquad (6)$$

Therefore, the dissolved concentration is given by:

$$c_{di} = \frac{v_i}{\pi_i} \qquad (7)$$

This expression for c_{di} is substituted into each diffusive exchange term (e.g. the storage compartment, $c_{d2} = v_2/\pi_2$). The model does not differentiate between arterial and venous concentrations in blood. The reason is that cardiac output is typically about 25 mL kg^{-1} min^{-1} and the blood volume is ~50 mL kg^{-1}, so circulation time is short (about 2 min) and differences in venous and arterial concentrations should be small (i.e. blood Cd is reasonably well mixed).

Mass balance equations were developed for each compartment. In contrast to the whole-body model analytical solution (Eq. 2), the system of seven differential equations for the PBPK model must be solved numerically (see Thomann et al., 1997, for equations and solution details).

Two other PBPK models that have recently been applied to metal accumulation by fish warrant mention. Liao et al. (2005), in an effort to link accumulation to effects, combined a PBPK model with an area under the curve (AUC) toxicity model for As in tilapia. More recently, Franco-Uria et al. (2010) used a generically parameterized model to predict Cd concentrations in various tissues of different fish species. The otherwise limited number of PBPK model applications to metals likely reflects the complexity of such models, including the need to evaluate organ-specific metal levels, as well as the ability of fish to regulate essential metal levels, and to detoxify and sequester excess metals. Physiological studies are needed to better understand these processes and to facilitate development of mechanistically based PBPK models for metal uptake by fish. One such collaborative effort incorporated toxicokinetics and modeling to measure the disposition of inorganic Hg and Cd in channel catfish (Schultz et al., 1996). More recently, physiologists and modelers investigated the effect of water temperature on silver uptake and depuration by fish (Nichols and Playle, 2004).

2.2.2. EXAMPLE APPLICATION

PBPK model use is illustrated by an application by Thomann et al. (1997) to the rainbow trout dataset of Harrison and Klaverkamp (1989). As previously noted, several model inputs required evaluation. When possible, they were based upon laboratory test conditions reported by the authors. These included waterborne and dietary Cd concentrations, food ingestion rate and first order growth rate coefficients (varied by phase of exposure). Others were evaluated based on the scientific literature or treated as calibration parameters. Evaluation of model inputs by Thomann et al. (1997) was in general accordance with the whole-body model analysis described previously (Section 2.1), but the greater PBPK model complexity leads to many additional model parameters. First, growth rate coefficients for each compartment were set to the whole-body growth rate coefficients. Other bioenergetic parameters were: blood volume, initial organ weights, food assimilation efficiency, and fecal egestion rate ($E_g = 0.002$ g_{feces} g_{fish}^{-1} day^{-1}). Other Cd- and compartment-specific parameters included seven diffusive exchange coefficients (E_{ij}); six organ:blood partition coefficients (π_i), dissolved fraction of Cd in blood, a gill sorption factor (a calibration variable, used to accentuate gill Cd uptake relative to loss), a biliary exchange factor, and kidney excretion rate coefficient. Clearly, the increased PBPK model complexity extends to data needs, thereby imposing increased demands on the investigators. Further details of PBPK model parameterization are provided elsewhere (Thomann et al., 1997).

Harrison and Klaverkamp (1989) reported data for each model compartment included in the PBPK model, as well as whole-body results.

Comparisons of model and data for Cd content (Cd mass, rather than Cd concentration) are shown for whole-body, gill, and kidney (Fig. 9.4A–C, respectively). The water-only exposure led to gill Cd content more than twice the accumulation of the dietary exposure (upper panel), while the reverse was true for the kidney (lower panel). An optimum overall fit of all organ data, for both exposure regimes (water only and water+food), was developed, such that the parameter values were consistent for both datasets. The authors cautioned that the set of parameter values used was not necessarily unique.

Although it was more complicated to calibrate the PBPK than the whole-body model, the calibration strategies were similar. For the PBPK model, organ-specific data were used to guide parameter evaluation for individual organs. These data also highlighted areas where refinements were needed. For example, enhanced Cd exchange between the gut wall and alimentary canal content was added to represent direct biliary transfer of Cd from the gut wall (the "bile factor", b_{47}). The observed increase in kidney Cd content over the uptake and depuration phases was attributed to sequestration of Cd by MT (see also McGeer et al., Chapter 3). This biochemical process was not explicitly represented in the PBPK model. Rather, it was simulated by

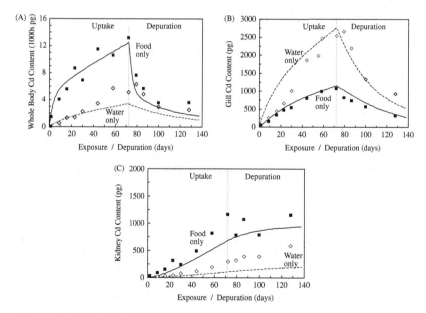

Fig. 9.4. PBPK model calibration results compared to water-only and food-only tissue Cd data for rainbow trout: (A) whole body (B) gill, and (C) kidney. Adapted from Thomann et al. (1997). **SEE COLOR PLATE SECTION.**

specifying a high kidney–blood partition coefficient in combination with a low first order excretion rate coefficient (0.00035 day^{-1}, or 0.035% day^{-1}) (Fig. 9.4C).

This progressive accumulation of kidney Cd contributed to the decrease in whole-body Cd elimination rate over time. Similar observations have been made for Cd in catfish (administered a single intravascular dose) (Schultz et al., 1996) and clams, *M. mercenaria* (a waterborne exposure) (Robinson and Ryan, 1986). These observations of kidney Cd accumulation are also consistent with data showing that rainbow trout excrete very little urinary Cd (Giles, 1988). While its relevance to fish is uncertain, the demonstrated temperature dependence of kidney Cd accumulation by *Mytilus* suggests that renal uptake may be biochemically controlled (Everaarts, 1990). Dietary Cu has also been shown to accumulate in the liver and kidney of rainbow trout, though less markedly than dietary Cd (Handy, 1992). These observations are consistent with the bi-exponential depuration pattern of whole-body Cd noted previously and attributed to the ongoing accumulation of Cd by the kidney.

The preceding results illustrate how PBPK model use inevitably leads to a more detailed consideration of data than would occur with a whole-body model. This helps to focus the direction of subsequent laboratory investigations and model development efforts.

2.3. Food Chain and Food Web Models and Biomagnification

A food chain is structured such that an organism at one trophic level serves as food (prey) for an organism (predator) at the next higher trophic level (e.g. algae → zooplankton → small fish → large fish). Food web models are slightly more complex, since a predator may feed upon more than one type of prey. Both models are used to simulate the degree of biomagnification (i.e. tissue residues increase across two or more trophic levels) resulting from uptake via both waterborne and dietary routes of exposure.

2.3.1. BACKGROUND AND MODEL FRAMEWORK

Food chain models are extensions of whole-body models. The food chain usually consists of a simplified feeding structure. For example algae and/or detritus (the base of the food chain) are eaten by zooplankton, which are eaten by small fish, which are then eaten by piscivorous fish. Each food chain compartment is represented by an equation similar to Eq. (1). Whole-body concentrations for each food chain organism are evaluated by solving the resulting system of equations. The model evaluates trophic transfer of metals as a function of physiology (e.g. uptake and bioenergetics), dietary composition, and exposure conditions (e.g. water versus diet).

As for whole-body bioaccumulation models, most food chain models were originally developed for organic chemicals (e.g. Thomann, 1978, 1979, 1981, 1989; Thomann et al., 1991; Mackay, 1989; Gobas, 1993) and subsequently applied to metals. However, one early application of a linear food chain model was an investigation of Cd bioaccumulation in Lake Erie (Thomann et al., 1974). The study demonstrated how food chain and water quality models could be linked to predict biota concentrations from ambient environmental concentrations. The model simulated trophic transfers in a pelagic food chain (e.g. water→phytoplankton→zooplankton→fish→ birds) for seven lacustrine regions. The bioaccumulation model was linked to a eutrophication model to calculate phytoplankton and zooplankton biomass. The model predicted fish tissue Cd concentrations within a factor of 3, and demonstrated that Cd uptake by periphyton reduced dissolved Cd concentrations. In addition, hydraulic residence time had an important influence on predicted biota concentrations.

More recently, a similar approach was used to evaluate trophic transfer of Se in a pelagic food web (Schlekat et al., 2004). A model previously applied to bivalves (*Potamocorbula amurensis*) (Schlekat et al., 2000) and copepods (*Temora longicornis*) (Wang and Fisher, 1998a,b) was calibrated to data for both small (73–250 μm) and large (250–500 μm) copepods feeding on marine diatoms and for mysids feeding on small or large copepods. Uptake and depuration rate coefficients and assimilation efficiencies were evaluated with radiolabeled Se in laboratory aqueous (as selenite) and dietary uptake studies (Schlekat et al., 2004). A link was demonstrated between exposure concentrations (water and diet) and Se concentrations in mysids and steady-state concentrations in both copepods and mysids. Initial model results were marginally higher than field-collected mysid data (Purkerson et al., 2003). Dietary modifications (i.e. including ingestion of sediment particulates, at a reduced Se assimilation efficiency) improved the comparison. This study concluded that predators of mysids were at less risk than predators of *P. amurensis*, a bivalve known to accumulate much higher levels of Se, and recommended that future monitoring focus on bivalve-based rather than zooplankton-based food webs. In a similar study, Stewart et al. (2004) considered two relatively simple food chains for Se: (1) phytoplankton→bivalves and (2) phytoplankton→herbivorous zooplankton→carnivorous zooplankton. Although the purpose was to predict dietary exposure levels for predators, including fish, fish accumulation was not modeled. The analysis showed that bivalves accumulated Se to much higher levels than crustaceans because they have a 10-fold lower elimination rate constant. Hence, clam-based food webs are more of a threat to predators (e.g. striped bass and white sturgeon) than are crustacean-based food webs.

Others have relied upon laboratory results to characterize trophic transfer of metals by aquatic organisms, including fish (Garnier-Laplace et al., 2000; Wang, 2002; Xu and Wang, 2002; Mathews and Fisher, 2008a,b, 2009). For example, Wang (2002) used measured ingestion rates, assimilation efficiencies, and efflux rate constants to evaluate trophic transfer by several marine food chains. The importance of metal processing strategies was highlighted. Similarly, Mathews and Fisher (2008b) modeled the transfer of seven metals (Am, Cd, Co, Cs, Mn, Se, and Zn) along a four-step food chain that included both prey and predator fish. Cs biomagnified at every trophic level, as did Se and Zn in some cases. The others did not biomagnify. More recently, Mathews and Fisher (2009) studied Am, Cd, Cs, Co, Mn, and Zn accumulation by two marine fish. They found that dietary uptake was particularly important for Mn, Cd, and Zn (>60% from diet).

For modeling purposes, the concentration in algae or detritus is commonly assigned based on measurements or calculated from a BCF times the dissolved concentration in water.

$$v_{\text{algae}} = \text{BCF} \cdot C_w \tag{8}$$

Invertebrates and fish, the next steps in the food chain, are simulated in accordance with Eq. (1). Individual model parameters (k_u, k_e, etc., as described for Eq. 1, with additional subscripts denoting organism type) are required for each upper trophic level in the food chain. Ingestion rates of food (I_f) and metal assimilation efficiencies from food (α_f) are typically assigned organism- and metal-specific values.

Food web models must consider diets that are composed of a mixture of food types. The fraction of each prey item in the predator's diet (e.g. f_1, f_2, f_3) is specified by the user.

$$\frac{dv_{\text{fish}}}{dt} = k_{u,\text{fish}} C_w + \alpha_f I_{f,\text{fish}} (f_1 C_1 + f_2 C_2 + f_3 C_3 \ldots) - (k_{e,\text{fish}} + g_i) v_{\text{fish}} \tag{9}$$

This is useful in cases where small fish consume a mixed diet of algae and small invertebrates, or larger fish consume different types of small fish and/ or invertebrates.

2.3.2. EXAMPLE APPLICATIONS

There are relatively few published examples of food chain or food web model applications to metals. This may reflect the difficulties associated with modeling metal-specific processes with a whole-body model (e.g. metal speciation and internal metabolic processes such as regulation and detoxification), difficulties compounded by consideration of several organisms.

The relationship between atmospheric loading and Hg bioaccumulation in an aquatic food chain was investigated (Knightes et al., 2009). Mercury runoff loads from a watershed model were coupled to a bioaccumulation model (Knightes, 2008) to relate atmospheric Hg loads to fish tissue Hg, and to evaluate the temporal response of tissue Hg to atmospheric loading reductions. The bioaccumulation and aquatic system simulator (BASS) model (Barber, 2001, 2003) was used to simulate fish Hg concentrations. BASS includes a bioaccumulation model similar to those discussed in Section 2.1.1. Dietary composition was varied based on species-specific preferences (e.g. large fish typically consume several prey types).

Bioenergetic and uptake parameters were compiled for more than 10 fish species across five watersheds. The diet of each fish included algae, zooplankton or benthic invertebrates and small fish, depending on fish species (Barber, 2003). Bioaccumulation parameters and dietary composition were also varied by fish size and age. Concentrations in periphyton and invertebrates were modeled using BAFs, while Hg concentrations in fish were modeled time-variably with BASS. Modeling results informed a cost–benefit analysis related to proposed Hg loading reductions.

3. MODELS RELATING METAL ACCUMULATION TO EFFECTS

Early efforts to develop bioaccumulation models for aquatic organisms focused on predicting tissue residue levels, not effects. Computed levels could then be compared to dietary standards protective of human health. Simulated tissue levels were subsequently compared to tissue benchmarks to assess effects on the organism itself. Although models vary widely in how they represent the underlying mechanisms, the objective is to relate environmental exposure conditions to effects. Four subcategories of models will be considered: damage–repair models (DRMs), biotic ligand models (BLMs), physiologically based mechanistic models (PBMMs), and intracellular speciation (ICS) models. This sequence conforms to an increased level of specificity to describe accumulation and effects. Models that address effects at the population and ecosystem level are described elsewhere (e.g. Liao et al., 2006; Mathews et al., 2009; Couture and Pyle, Chapter 9, Vol. 31A).

3.1. Damage–Repair Models

The concept of a critical body residue (CBR), most commonly applied to narcotic compounds, is based on the premise that a CBR associated with an effect is constant over a range of exposure times, compounds, and species

(McCarty, 1987; McCarty et al., 1993). Some results suggest that CBRs merit consideration for metals as well (e.g. Redeker and Blust, 2004; Salazar and Salazar, 2007a). Others posit that the CBR is not constant and that an LC50 versus time model should be used for reactive and receptor-mediated toxicants (Verhaar et al., 1999). Here, the critical area under the curve (CAUC) of LC50 versus time is constant and, even at a steady-state body burden, the LC50 continues to decrease with increasing exposure duration. Liao et al. (2005) extended the AUC approach by combining it with a PBPK model to relate arsenic accumulation in target organs to tilapia survival.

Most DRMs are extensions of whole-body accumulation models. The main difference is they provide an empirical way to quantify damage, without directly considering mechanisms, as a "black box" representation of an underlying physiological process affected by metal accumulation. With a DRM, damage can be accumulated and it can be repaired (or eliminated). The damage computation is analogous to the computation of metal accumulation in a whole-body or multicompartment model.

3.1.1. BACKGROUND AND DESCRIPTION OF MODEL FRAMEWORK

DRMs are formulated to relate a CBR to an effect, while also including the possibility of repair. They are similar to a whole-body model, without a dietary contribution, but including a damage component related to toxicity (Fig. 9.5A). Because damage is mathematically equivalent to a tissue compartment, the terms "damage" and "repair" are somewhat analogous to "uptake" and "elimination" of metal. Most DRMs are empirical models that are calibrated to toxicity data rather than whole-body metal measurements. A typical DRM formulation is a two compartment extension of Eq. (1). It includes terms for whole-body metal accumulation and accumulated damage (see below). Toxicity is a consequence of accumulated damage.

$$\frac{dv}{dt} = k_u \cdot C_w y + z k_r D - (k_e + k_d) v \tag{10}$$

and for damage:

$$\frac{dD}{dt} = k_d v - (1 - z) k_r D \tag{11}$$

where y = empirical power term [dimensionless], zD = conceptualization parameter (0 or 1) [dimensionless], k_r = repair rate coefficient [1 day^{-1}], D = accumulated damage [μg L^{-1} or μg g^{-1}], k_d = damage rate coefficient [1 day^{-1}], and remaining model parameters are consistent with Eq. (1).

(A)

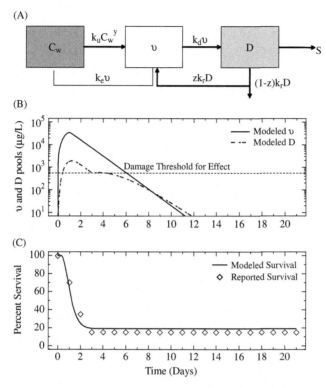

(B)

(C)

Fig. 9.5. Damage–repair model: (A) model framework (C_w: dissolved concentration; v: internal concentration; D: damage; S: survival) (adapted from Butcher et al., 2006); (B) simulated tissue metal (dashed line) and damage (dot–dash line); and (C) simulated survival (solid line) compared to *Daphnia magna* survival data. Data from Hoang et al. (2007).

This formulation, described previously by Butcher et al. (2006), is conceptually similar to the accumulation model of Redeker and Blust (2004). The tissue concentration (v) depends upon the dissolved metal concentration and uptake and elimination rate coefficients. Damage (D) is a function of internal dose and need not be associated with a well-defined physiological process. More practically, D represents metal accumulation at a generic site of action of toxicity. The value of z is either 0 or 1, depending on how D is viewed (i.e. damage or accumulation at a site of action). When $z = 0$, D represents damage, and metal in the damage pool is not allowed to contribute to the tissue metal pool.

There are various ways to relate a response to the level of damage. Butcher et al. (2006) used a survival function to combine a hazard rate with natural mortality. Survival was dependent upon the hazard associated with a damage level relative to a threshold value and to the natural mortality rate of individuals. Specifically, response over a given time interval is given by:

$$R_t = R_{t-1} e^{-\{\alpha + \beta[D_t - D_{thres}]\} \Delta t} \tag{12}$$

where R_t = predicted response (e.g. percent mortality) at time t, α = natural mortality rate coefficient, β = hazard rate coefficient, D_{thres} = threshold damage level, and Δt = duration of the time interval. For $D < D_{thres}$, hazard rate = 0, and the total response is limited to the natural response.

Mancini (1983) proposed one of the first models to relate metal accumulation to effects. Survival time was equated to the time to achieve an "equivalent mortality dose". Model variations have been used to evaluate mortality following multiple Zn pulses (Mancini, 1983; Connolly, 1985). Meyer and co-workers (1995) compared the Mancini model to one assuming that exposure concentration times duration ($c \times t$) is constant, and to the DRM of Breck (1988). The Mancini model performed well when applied to a monochloramine toxicity dataset for rainbow trout and common shiners.

Marr et al. (1996) developed relationships between Cu exposure duration, accumulation, and rainbow trout growth; growth reduction was predictable from tissue Cu for known exposure durations. A one-compartment uptake–depuration framework was used to model Co and Cu toxicity (independently and jointly) to rainbow trout, based on evaluation of an incipient lethal level (ILL). Assuming additivity over-predicted mixture toxicity (Marr et al., 1998). DRM use will be facilitated by test protocols calling for high replication and detailed records of partial mortality (Meyer et al., 1995; Butcher et al., 2006).

The applicability of CBRs to metals should be less robust than for organic chemicals because most aquatic organisms can detoxify metals (Rainbow, 2002). Application of a dynamic multicompartment accumulation model for Cd toxicity to *Tubifex tubifex* provides one demonstration of the utility of the CBR approach for metals (Redeker and Blust, 2004). Their approach was more rigorous than most other DRM applications because tissue accumulation data were used to evaluate kinetic constants. Accumulation data should be used with DRMs whenever possible.

Application of the CBR approach to nutritionally required metals is tenuous. Essential metals are well regulated at low environmental concentrations, and both essential and non-essential metals are detoxified and preferentially sequestered at elevated concentrations. Rate of uptake will further complicate CBR use because it affects metal distribution between metabolically active and inactive pools. Examples of this have been reported for copepod egg production (Hook and Fisher, 2002), zebra mussel survival (Kraak et al., 1992), and freshwater clam (*C. fluminea*) survival (Andres et al., 1999).

3.1.2. EXAMPLE APPLICATION

The DRM was applied to similar toxicity datasets for *Pimephales promelas* and *Daphnia magna*. For example, the dataset for fish included 97 toxicity tests with larval fathead minnows exposed to Cu pulses of various magnitudes ($5–125\ \mu g\ L^{-1}$), durations (3–336 h) and recovery times (0–144 h) (Butcher et al., 2006). Fish were either < 24 h old or 7 days old at the initiation of 7 day or 14 day static-renewal tests with food. Endpoints were mortality and growth. Validation tests were performed with varying pulse durations and frequencies.

The absence of tissue residue data precluded an assessment of the model's ability to predict rates of metal uptake, elimination, and net metal accumulation. Rather, parameters for Eqs. (10)–(12) were concurrently optimized to fit survival data. Caution must be used when interpreting results from this type of parameter estimation, as toxicity responses associated with variations in individual accumulation parameter values are not necessarily independent of each other.

The calibrated DRM performed reasonably well with respect to simulating *P. promelas* survival (after omitting an outlier with high natural mortality) (Butcher et al., 2006). Simulated model results for tissue Cu and damage are shown in Fig. 9.5B and model calibration results are compared to a subset of the survival data in Fig. 9.5C (*D. magna* data from Hoang et al., 2007). The adjusted r^2 after application to the complete dataset was 61.9%. The model exhibited a slight bias.

An interesting feature of this DRM is the use of a variable repair rate constant to simulate acclimation. The repair rate constant varies linearly with damage, from a naïve repair rate constant (when D is less than the damage threshold) to a fully acclimated repair rate constant (increasing when D exceeds the threshold). This is consistent with the conceptual model for acclimation of McDonald and Wood (1993), where physical damage to the fish gill is required for acclimation to occur.

Butcher and co-workers succeeded in developing a simple, empirically based DRM that predicts toxicity due to fluctuating exposure conditions. However, a more mechanistic, physiologically based approach that considers metal speciation and bioavailability and metal accumulation data would yield greater insight into processes controlling toxicity.

3.2. Biotic Ligand Models

Water chemistry can have a large impact on the concentrations of the various metal species present in a water body. Because these metal species may also differ widely in their bioavailability, the dissolved metal concentration that elicits a particular toxic response is a function of water chemistry as well. The biotic ligand model (BLM) is a computational tool that was developed to predict such effects of site-specific water chemistry on metal bioavailability and effects.

3.2.1. BACKGROUND AND MODEL FRAMEWORK

The conceptual link between metal speciation and interactions at a biotic site of action of toxicity, first described by Morel (1983), is commonly referred to as the free ion activity model (FIAM) of metal toxicity to aquatic organisms. The FIAM emanated from experiments during the 1970s and early 1980s, which highlighted the important influence of water chemistry on metal speciation, bioavailability, and toxicity (e.g. Anderson and Morel, 1978; see Campbell, 1995, for a review). They showed that even though the dissolved metal concentration associated with toxicity varied widely, the free metal ion concentration (one of several metal species that may be present) is relatively consistent. However, other cations also compete for binding at the site of action, thereby affecting metal toxicity in a manner that cannot be predicted by its free ion activity alone (Meyer et al., 1999). This is why tissue metal levels, particularly at the toxic site of action, are viewed as a better indicator of exposure and the potential for effects.

Prediction of effect levels with the BLM was a direct outgrowth of work by Playle and co-workers, who demonstrated how a chemical equilibrium model could reliably simulate metal accumulation by the gill, the target organ for acute toxicity to fish, as water chemistry varied (e.g. Playle et al., 1993a,b). The BLM uses this capability to evaluate a site-specific water-only effect level, i.e. the dissolved metal concentration associated with a critical accumulation level known to cause a specific effect on an aquatic organism (Di Toro et al., 2001; Paquin et al., 2002a). It combines information about solution equilibrium (e.g. metal complexation with inorganic or organic ligands), metal and cation binding at a presumed site of action of toxicity (the biotic ligand, BL), and the accumulated metal concentration (BL:Me) known to cause a particular effect.

Important conceptual elements of the BLM are illustrated in Fig. 9.6. Speciation reactions used by Di Toro et al. (2001) include a description of the binding of metals, protons, and cations to natural organic matter (NOM). Inorganic speciation reactions for metal complexation with carbonate, bicarbonate, hydroxides, chlorides, and other inorganic ligands are typically based on the National Institute of Standards and Technology (NIST) database of thermodynamic constants. In addition, the BLM includes reactions that represent binding of bioavailable metal and other competing cations to the BL, the proximate site of action of toxicity.

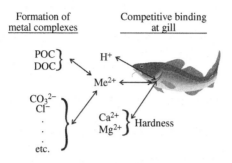

In principle, BLM computations can be performed with any chemical equilibrium model. Playle and co-workers used MINEQL with metal–organic matter interactions calibrated to gill accumulation data and, subsequently, toxicity data (Playle et al., 1993b; Janes and Playle, 1995; McGeer et al., 2000). Santore and co-workers incorporated the organic matter interactions represented in the Windermere humic aqueous model (WHAM) (Tipping, 1994) and inorganic computations based on CHESS (Santore and Driscoll, 1995). While different speciation programs should yield equivalent results, differences in reactions and binding constants used may lead to differences from one BLM implementation to another. The significance of such differences requires careful consideration.

The United States Environmental Protection Agency (USEPA) recently released a revised freshwater water quality criterion (WQC) for Cu that incorporates the Cu BLM (Santore et al., 2001; USEPA, 2007; see Section 4). It is applicable to geographically diverse waters that exhibit wide ranges of pH, ion composition, and NOM content. USEPA WQC for some metals traditionally used a hardness equation to account for bioavailability, with its applicability restricted to 25–400 mg L^{-1} as $CaCO_3$ (e.g. USEPA, 1985). This restriction was also considered for the BLM, although subsequent use has shown that it works in hard water (e.g. hardness > 1000 mg L^{-1} as $CaCO_3$) (Gensemer et al., 2002).

Although originally developed for prediction of acute toxicity, the BLM has also been used to predict the effect of water chemistry on chronic toxicity due to waterborne exposures (e.g. De Schamphelaere and Janssen, 2004a,b; Schwartz and Vigneault, 2007). The potential for dietary effects during chronic freshwater and saltwater exposures has yielded variable and inconsistent results (e.g. Hook and Fisher, 2001; De Schamphelaere et al., 2004; De Schamphelaere and Janssen, 2004c; Bielmyer et al., 2006). Because the BLM simulates interactions between waterborne metals and sensitive receptors it is not likely to be applicable to metals whose effects are primarily due to dietary exposure, such as Se (Janz, Chapter 7, Vol. 31A), Hg (Kidd and Batchelar, Chapter 5) and As (McIntyre and Linton, Chapter 6).

Borgmann and co-workers investigated how metal bioavailability influences metal accumulation and acute and chronic effects of individual metals and metal mixtures (e.g. Borgmann and Norwood, 1997; Borgmann et al., 2008, 2010). Accumulation and toxicity of metal mixtures have been difficult to simulate with consistent assumptions, because metal interactions can be competitive, anti-competitive or non-competitive (Roy and Campbell, 1995; Norwood et al., 2007; Borgmann et al., 2008, 2010). It was shown that Ca inhibition of Cd accumulation by the amphipod *Hyallela azteca* conforms to an anticompetitive interaction (Borgmann et al., 2010). The BLM could fit these data, but the fit was empirical and the underlying mechanisms were reported to be more complex than typically represented in the BLM. Playle (2004), extending his earlier work with individual metals, simulated the effects of mixtures of two to six metals. Toxicity, quantified as toxic units (TUs), was found to be more than additive for $\Sigma TU < 1.0$, strictly additive for $\Sigma TU \sim 1.0$, and less than additive for $\Sigma TU > 1.0$. These results were attributed to a non-linear BLM response, with negligible competitive interactions at low concentrations, moderate interactions at intermediate concentrations (the point of strict additivity), and marked interactions at high concentrations. Investigation of metal mixture accumulation and toxicity continues to be an area of active research.

BLM input requirements include a description of the water quality characteristics (e.g. pH, dissolved organic carbon to quantify NOM, alkalinity, and selected major ions). BLM implementations have been developed for several metals including Cd, Cu, Ni, Ag, and Zn (McGeer et al., 2000; Santore et al., 2001, 2002; Wu et al., 2003; Brooks et al., 2004; Keithly et al., 2004), while preliminary versions for Al and Pb are under development. Efforts to extend the applicability of the BLM to marine and estuarine conditions, soils, and sediments are also ongoing.

3.2.2. Example Application

Measurements have shown that gill Cu accumulation occurs to a lesser extent in hard than in soft water (Playle et al., 1992). Such observations are consistent with the known mitigating effects of hardness on metal toxicity (e.g. Zitko and Carson, 1976; Pagenkopf, 1983) and support the incorporation of competitive interactions in the BLM. However, despite its demonstrated predictive ability, there has been a systematic tendency for the copper BLM to underestimate organism sensitivity (overpredict copper effect levels) in very soft water (hardness less than $25 \, \text{mg L}^{-1}$ as $CaCO_3$) (Sciera et al., 2004; Van Genderen et al., 2005). While decreasing competition between Cu^{2+} and Ca^{2+} binding at the biotic ligand may explain the reduction of Cu effect levels as hardness decreases, the response may be further exacerbated by an increase in branchial Na^+ efflux in very soft waters. This additional mechanism is consistent with observations of reduced Na^+ content in fish in very soft waters, irrespective of metal exposure (Van Genderen et al., 2008). The BLM has traditionally incorporated the first of these effects through competitive interactions between Cu^{2+} and cation (e.g. Ca^{2+}, Mg^{2+}, Na^+, and H^+) binding to the BL, but not the latter physiological effect. It was judged that this may be the reason that dissolved Cu LC50 values have previously been overpredicted in low hardness waters (Fig. 9.7A, for fathead minnow).

An expedient way to account for this discrepancy between predicted and measured effect levels is to adjust the LA50 in low hardness waters. It is evident that calibrated LA50 values (the concentration of BL Cu that results in 50% mortality) vary systematically with hardness for $Ca^{2+} < 0.43$ mM ($\log_{10}Ca = -3.37$) (Fig. 9.7B). An adjusted LA50 can be estimated as:

$$\text{Adjusted LA50} = \text{LA50} * 10^{1.72(\text{Log(Ca)}+3.37)} \tag{13}$$

This empirical relationship can be used to adjust the LA50 and improve BLM performance in low hardness water ($< 25\ \text{mg L}^{-1}$ as $CaCO_3$), achieving considerably better LC50 predictions (Fig. 9.7C).

Alternative explanations for the apparent increase in fish sensitivity to metals in soft, low ionic strength waters have been investigated. One is that the biotic ligand binding constant changes when a fish acclimates to such waters. Evidence for this is that acclimation of fathead minnows to low ionic strength water, in the absence of Ag^+, did not increase the Ag LC50 in a post-acclimation toxicity test (Bielmyer et al., 2008). It was posited that Na^+ transport may have been upregulated in low ionic strength water, resulting in an inadvertent increase in Ag^+ uptake. Future physiological investigations should lead to an improved understanding of how to refine the BLM to improve its predictive ability in soft, low ionic strength water.

3.3. Physiologically Based Mechanistic Models

Physiologically based mechanistic models (PBMMs) are differentiated from more conventional models by their explicit representation of the mechanisms that control metal accumulation, metabolism, and effects. For instance, gill Cu or Ag accumulation by fish may be related to a gross effect (e.g. mortality, an LC50). More specifically, both metals inhibit active Na^+ uptake and, more fundamentally, the activity of one or more enzyme systems required for Na^+ uptake (see Grosell, Chapter 2, Vol. 31A; Wood,

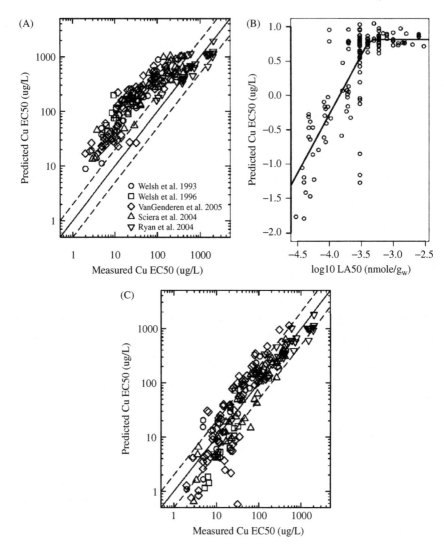

Fig. 9.7. Incorporation of the effect of soft water (hardness <25 mg L^{-1} as CaCO$_3$) on fathead minnow sensitivity to Cu. (A) Biotic ligand model (BLM)-predicted versus observed LC50 without adjusting the LA50 for test water Ca; (B) variation of LA50 with Ca; (C) BLM-predicted versus observed LC50 after adjustment of LA50 for test water Ca. For graphs A and C, the diagonal solid lines are lines of perfect agreement and predicted results between the dashed lines are within a factor of two of measured LC50 values.

Chapter 1). Incorporating such details in models elucidates how individual processes contribute to the whole organism response and helps to define future research needs. Several PBMMs will be described to illustrate the diversity of the approach. An ion-balance model (IBM) for fish will illustrate in further detail how a PBMM systematically integrates organism physiology into a useful investigatory tool.

3.3.1. BACKGROUND AND MODEL FRAMEWORK

One of the earliest and more complex examples of a PBMM was developed by Guyton and co-workers (Guyton et al., 1972). They represented each of 354 aspects of human circulatory regulation by one or more equations. This seemingly intractable system of equations was used to investigate how numerous physiological characteristics, acting in concert, elicit a systemic response (e.g. development of hypertension in a salt-loaded, renal-deficient patient). Although Guyton's early findings were controversial, subsequent independent experimental confirmation led to their widespread acceptance by medical practitioners (Hall, 2004).

Szumski and Barton (1983) proposed a PBMM to predict acute metal toxicity to fish. They reasoned that inhibition of carbonic anhydrase (CA) activity interfered with carbon dioxide (CO_2) excretion, causing acute hypoxic stress from CO_2 accumulation and reduced oxygen carrying capacity of the blood. The model considered competitive interactions at the gill and metal speciation in both the ambient water and within the gill chamber, a microenvironment having a significant influence on metal bioavailability (Playle and Wood, 1989a,b, 1990, 1991; Playle et al., 1992).

More recently, Grosell and co-workers used a whole-body IBM to characterize species sensitivity to either Cu or Ag (Grosell et al., 2002). Rather than calculate metal accumulation, they assumed a level of uptake inhibition to predict cumulative Na^+ losses and time to 30% Na^+ loss (a lethal response). They showed how Na^+ influx controls freshwater animal sensitivity to either metal (see also Grosell, Chapter 2, Vol. 31A, and Wood, Chapter 1). A particularly useful finding was that Na^+ turnover rate controls organism sensitivity to ionoregulatory inhibitors (e.g. Cu or Ag) and that it is inversely related to size (Bianchini et al., 2002; Grosell et al., 2002). This provides a rationale for modeling toxicity due to time variable metal exposures (Paquin et al., 2002b). It also suggests that differences in individual sensitivity may be explained by individual differences in Na^+ turnover rate (Kolok et al., 2002). Future investigations are needed to obtain data to test such inferences and, subsequently, refine IBMs.

Paquin et al. (2002b) developed an IBM to better understand Ag toxicity to fish. The idea is that aquatic organisms must maintain internal levels of major essential trace elements (e.g. Na^+, Ca^{2+}) within narrow limits. This ionoregulatory balance is perturbed when fish are exposed to low pH or elevated levels of an ionoregulatory stressor (e.g. Ag^+ and Cu^{2+}) (Morgan et al., 1997). Inhibition of Na^+ uptake by these metals leads to a well-characterized sequence of events, including net loss of Na^+, osmotically induced volume changes of internal fluid pools, increased blood viscosity and, finally, death via

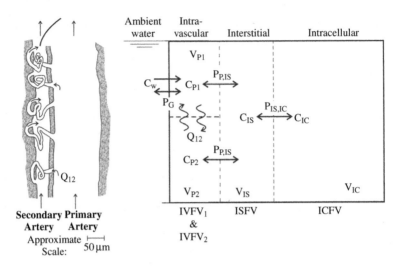

Fig. 9.8. Ion balance model (IBM): schematic of IBM fluid compartments for a freshwater fish, including details for primary and secondary vascular systems (left, after Steffensen and Lomholt, 1992) and major fluid compartments (right). **SEE COLOR PLATE SECTION.**

cardiovascular collapse (Milligan and Wood, 1982; McDonald, 1983; Wood, 1989; Wilson and Taylor, 1993; Taylor et al., 1996; Wood et al., 1996; Hogstrand and Wood, 1998). This cascading sequence of events has been associated with mortality at about 30% plasma Na^+ loss (McDonald et al., 1980; Wood, 1989; Wilson and Taylor, 1993; Wood et al., 1996; Webb and Wood, 1998; Hogstrand and Wood, 1998; Grosell et al., 2000).

A multicompartment IBM implementation incorporates sodium transfers between internal fluid compartments (see Paquin et al., 2002b, for details). In brief, mass balances of Na^+ are computed about major internal fluid compartments of a fish (Fig. 9.8). These include the intracellular fluid volume (ICFV, V_{IC}) and extracellular fluid volume (ECFV, V_{EC}). The latter includes intravascular fluid volumes (IVFV$_1$ and IVFV$_2$, V_1, V_2, respectively) and the interstitial fluid volume (ISFV, V_{IS}). The mass balance for the IVFV includes uptake of Na^+ from the water (a saturable process, formulated as a Michaelis–Menten expression), passive losses from the gill, and renal losses. Exchanges between the other fluid compartments are expressed as products of a permeability coefficient times the differences in Na^+ concentrations. This represents an expedient simplification of the actual process, which likely involves a relatively complex Donnan equilibrium condition (Krogh, 1939; Potts and Parry, 1964; Holmes and Donaldson, 1969; Potts, 1984) or differential mobilities of diffusing ions (Teorell, 1935).

The gill permeability coefficient is set to achieve a steady state balance between uninhibited active Na^+ influx and passive efflux, to account for inferred gill tissue damage caused by elevated Ag^+, and for the mitigating effects of Ca^{2+} (Paquin et al., 2002b). Other model parameters were evaluated by calibration to rainbow trout tracer data (i.e. inulin) (Nichols, 1987; Steffensen and Lomholt, 1992). Other gains and losses of Na^+ (e.g. drinking water, diet, cutaneous exchanges) are neglected.

Inhibition of active Na^+ uptake by elevated Ag^+ is rapid and reversible (Hussain et al., 1994) and is expressed via a decrease in the maximum carrier-saturated uptake rate (J_M) rather than the affinity of the transport system (e.g. Morgan et al., 1997; Bury et al., 1999a). This inhibition is represented as a sigmoidal dose–response relationship between the inhibited maximum Na^+ uptake rate (J_M*) and the chemistry-dependent BLM-calculated BL:Ag (see Wood, Chapter 1). Dose–response parameters EC50 and β are evaluated by IBM calibration to plasma Na^+ time-series data (Section 3.3.2). The controlling differential equations are numerically integrated to solve for compartmental Na^+ concentrations over time.

The essence of the multicompartment formulation is preserved when reduced to a simplified one-compartment whole-body formulation. The mass balance equation for the constituent of interest (e.g. Na^+) for the one-compartment model is:

$$\frac{V dc}{dt} = \left(\frac{J_M c_w}{c_w + K_M}\right) - P_G (c - c_w) P_r c \tag{14}$$

Here, V is the weight-normalized fluid volume (L kg_w^{-1}), c and c_w are the internal and external Na^+ concentrations (mM), K_M is the half saturation concentration for Na^+ uptake from water, P_g is the gill permeability coefficient (L kg_w^{-1} day^{-1}), P_r is the apparent permeability coefficient controlling renal losses (L kg_w^{-1} day^{-1}), and the other variables are as defined previously. The solution to Eq. (14) is:

$$c(t) = c_o e^{-Kt} + c_{ss}(1 - e^{Kt}) \tag{15}$$

The initial condition for Na^+ (c_o) is a model input and c_{ss} is the steady-state Na^+ concentration. Also:

$$K = \frac{P_T}{V} \tag{16}$$

where P_T represents the apparent total or whole-body permeability coefficient given by:

$$P_T = P_{Go} + P_r \tag{17}$$

Here,

$$P_{Go} = f_{gill} J_M [c_w/(K_M + c_w)]/[c_o - c_w] \qquad (18)$$

and

$$P_r = f_{renal} J_M [c_w/K_M + c_w]/[c_o] \qquad (19)$$

The constants f_{gill} and f_{renal} ($=1 - f_{gill}$) represent fractions of total Na^+ losses associated with passive branchial losses and urinary losses, respectively (e.g. 90% and 10%) (Wood, 1989, 1992; Curtis and Wood, 1991).

The steady-state solution, c_{ss}, is given by:

$$c_{ss} = \frac{[J_M^* c_w/(K_M + c_w) + P_{Go} c_w]}{P_T} \qquad (20)$$

3.3.2. EXAMPLE APPLICATION

The Na^+ initial condition for each internal fluid pool is set to ~150 mM to start each simulation. The active influx rate is evaluated with $J_M = 12$ mmol kg_w^{-1} day^{-1} (representative for rainbow trout in the absence of Ag^+). Water quality characteristics (including Na^+) for each test condition were input to the BLM to evaluate the associated BL:Ag concentration. This controls the inhibited Na^+ uptake rate (J_M^*). At the start ($t = 0$), all Na^+ transfers are in balance and internal Na^+ levels constant. Once Na^+ uptake is inhibited, losses via the gill and kidney exceed active uptake and plasma Na^+ declines, leading to disequilibrium and declines in Na^+ levels in the IVFV, ISFV, and ICFV.

Two datasets were used to test the IBM. First, plasma Na^+ time series data for rainbow trout (McGeer and Wood, 1998) were compared to simulated model results (Fig. 9.9A–D; see caption for details). The dataset included a control (no added Cl^- or Ag; Fig. 9.9A) and 0.03 μM (~3.3 μg L^{-1}) Ag with Cl^- concentrations of 1440, 292, and 14 μM (Fig. 9.9B–D, respectively; see caption for details). The decreasing Cl^- concentration in the Ag-amended treatments is associated with increasing Ag^+. This increases the degree of inhibition of Na^+ uptake and exacerbates the resulting decrease in both the measured and simulated plasma Na^+ levels (Fig. 9.9B–D). The second dataset monitored rainbow trout survival time (Galvez and Wood 1997), which varied in response to a competing cation (Ca^{2+}) or an inorganic complexing ligand (Cl^-). Predicted times to 30% loss of Na^+ compared well with observed times to 50% mortality (Paquin et al., 2002b). The required increase in gill permeability, inferred from otherwise inexplicable short survival times at high Ag^+ levels, was consistent with reports of enhanced

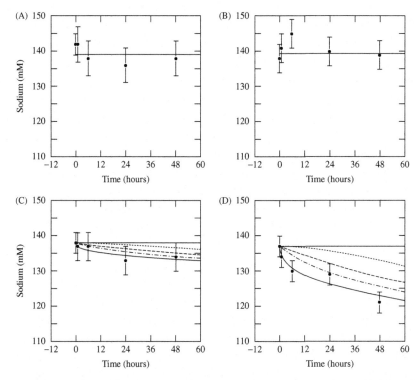

Fig. 9.9. Comparison of ion balance model (IBM)-simulated results to plasma Na⁺ data for rainbow trout in (A) the control (no added Ag) or exposed to 3.3 μg L⁻¹ Ag and KCl additions of (B) 1440 μM (C) 292 μM, and (D) 14 μM. The horizontal solid line represents a presumably constant plasma Na⁺ concentration in the absence of silver. The other four lines show the predicted sodium concentrations in the IVFV₁ (the lowest solid line, to be compared to the plasma sodium data), IVFV₂, ISFV, and ICFV (in order of lowest to highest). These four lines coincide at 1440 μM Cl⁻(graph B) because the net Na⁺ loss is negligible. In reality, the Na⁺ concentration in the ICFV compartment will be substantially lower than in the other compartments (see Paquin et al., 2002b, for ICFV modeling assumptions). Data from McGeer and Wood (1998); model from Paquin et al. (2002b).

Na⁺ efflux from rainbow trout exposed to relatively high levels of Cu (Lauren and McDonald, 1985; Wilson and Taylor, 1993).

Application of the simpler whole-body IBM (Eq. 15) to the data of Fig. 9.9A–D yields remarkably similar results to what was obtained with the more complex multicompartment model. The simulations suggest that if renal excretion is neglected ($P_r \sim 0$), then at least 30% inhibition of Na⁺ influx is required for a 30% loss of Na⁺, and hence mortality, to occur. It can be inferred from this verifiable result that Na⁺-deficient fish might survive a short-term increase in Ag^+, re-establish homeostasis with respect to Na^+, and return to normal.

3.4. Intracellular Speciation Models

Intracellular speciation (ICS) models simulate interactions at the site of action of toxicity with a high degree of specificity (Blancato and Bischoff, 1994). Though not widely used to date, continuing advances in physiological and biochemical test methods (e.g. Solioz et al., 1994; Solioz and Odermatt, 1995; Bury and Wood, 1999; Bury et al., 1999b; Lu et al., 2003; Morgan et al., 2004) should promote their increased use. An application to Cu accumulation by bivalves will illustrate ICS model use (Paquin et al., 2007). Laboratory investigations have shown that intracellular speciation has an important bearing on metal accumulation, trophic transfer, and toxicity to fish and other aquatic organisms (e.g. Hogstrand et al., 1996; Zhang and Wang, 2006b), such that the ICS approach should be applicable, generally.

3.4.1. BACKGROUND AND MODEL FRAMEWORK

Blancato and Bischoff (1994) described an ICS model that included the IBM fluid compartments described above and a cytosol with insoluble mitochondrial- and microsomal-rich fractions. Escher and Hermens (2004) recognized the consistency of dose–response relationships based on cytosolic concentrations, while external effect levels were very pH dependent (Walter, 2002, as cited by Escher and Hermens, 2004). Although these examples involved organic chemicals, the underlying concepts are also applicable to metals.

The ICS model application described here is based on the conceptual framework for metal detoxification of Mason and Jenkins (1995) (Fig. 9.10). That is, metals entering a cell are rapidly bound by one of three ligand groups. The first includes enzymes that require an optimum level of one or more essential metals (EMs). A second includes high molecular weight (HMW) proteins, other enzyme systems, and functionally important intracellular structures (e.g. nuclei, mitochondria, the endoplasmic reticulum), which may be adversely affected by non-specific binding of excess EMs or non-essential metals (NEMs). The third includes intracellular ligands that bind excess cytosolic metals, reducing their intracellular availability and interactions with target ligands, thereby mitigating toxicity. This latter pool includes glutathione (GSH, a biologically ubiquitous, high-concentration, low-affinity cytosolic ligand), metallothionein (MT, an inducible, sulfhydryl-rich high-affinity ligand), insoluble granules, and other intracellular debris. Some metal–MT complexes may be transferred into membrane-bound

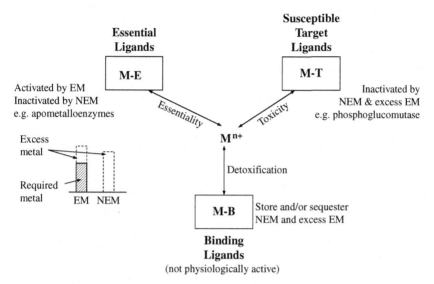

Fig. 9.10. Intracellular speciation (ICS) model framework. Metal compartmentalization is related to essentiality, toxicity, and detoxification reactions (see text for further details). Adapted from Mason and Jenkins (1995).

vesicles (i.e. lysosomes) and subsequently excreted from the organism (e.g. Viarengo et al., 1981).

The ICS model for Cu considers many of the intracellular interactions described above. In brief, reactions are considered to be at equilibrium, but not necessarily at steady state. When cellular Cu influx exceeds metabolic needs, the GSH-containing pool of low molecular weight (LMW) ligands provides a first line of defense. The GSH complexation capacity limits the initial rate of increase of intracellular free metal ion concentration (Singhal et al., 1987; Freedman et al., 1989; Zalups and Lash, 1996). This in turn limits interactions with Cu-sensitive enzymes (included in a HMW ligand pool) and other intracellular particulates. Although local equilibrium conditions may prevail, the intracellular environment is dynamic (Hogstrand and Haux, 1991). That is, cytosolic composition changes in response to changes in intracellular Cu and MT synthesis. Conversely, cytosolic concentrations of total Zn and LMW ligands in the GSH-containing pool are considered constant for modeling purposes. Excess Cu displaces bound Zn, which in turn induces MT synthesis (Palmiter, 1994). Increases in intracellular MT lead to an increase in Cu thionein (CuMT), which is transferred into lysosomes. Subsequent excretion of the Cu-containing lysosomes via a physical rather than chemical process is expected to occur at a reduced rate relative to other intracellular Cu forms, thereby reducing the effective rate of Cu excretion as tissue Cu increases. Copper binding by MT and sequestration of CuMT in lysosomes limit interaction of intracellular Cu at sites of toxic action, and thereby mitigate toxicity.

The level of specificity needed to represent a cytosolic BL or to simulate effects with an ICS model is uncertain. One approach associates the onset of adverse effects with "spillover" of Cd from the MT pool (first observed by Winge et al., 1974). Similarly, Din and Frazier (1985) injected rats with Cd and showed that inhibition of protein synthesis in rat hepatocytes was related to non-MT-bound cellular Cd in a dose-dependent manner. Other investigators have shown that effects may be associated with metal concentrations in an HMW ligand pool (e.g. Harrison et al., 1983; Brown and Parsons, 1978; Brown et al., 1990), or in a metal-sensitive fraction containing organelles and heat-sensitive proteins (described as "enzymes") (Wallace et al., 2003). Finally, it may be possible to predict effects by relating Cu accumulation to inhibition of enzyme activity for enzymes such as Na^+/K^+-ATPase (NKA) or CA. Fish gill enzyme activities of NKA and CA are inhibited by some metals and the response is directly related to the severity of ionoregulatory effects (Morgan et al., 2004). For the *Mytilus* case study described below, HMW–Cu is associated with effects (Harrison et al., 1983).

3.4.2. EXAMPLE APPLICATION

Very few ICS model applications have been reported to date and, to the authors' knowledge, none for fish. However, the ICS framework includes processes that are relevant to most aquatic organisms, including fish (e.g. Hogstrand et al., 1996; Zhang and Wang, 2006b). Thus, an ICS application to an aquatic organism, caged bivalves (*Mytilus* ssp.) deployed along a dissolved Cu gradient in San Diego Bay (~1–10 µg L^{-1}) (VanderWeele, 1996; Salazar and Salazar, 2007b), is used for illustrative purposes. Tissue Cu was well regulated over deployments of 82 days or longer at dissolved Cu concentrations <3 µg L^{-1}, but increased at higher concentrations. Bivalves at Shelter Island (SI) Yacht Basin exhibited the greatest increase in whole-body Cu; gill and digestive gland Cu increased about 34-fold (to approximately 350 µg g$_d^{-1}$), while other organs increased only about 5–10-fold. In spite of these marked increases in tissue Cu, 82 day survival was excellent (89%).

It was hypothesized that metal detoxification, the induction of MT synthesis in the gill and digestive gland in particular (Viarengo et al., 1981), might explain the preceding data. The ICS model was used with several datasets (Harrison et al., 1983; Viarengo et al., 1985; Salazar and Salazar, 2007b) to test this hypothesis. The ICS simulations treated the organ of

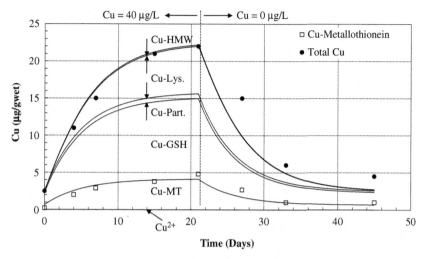

Fig. 9.11. Intracellular speciation (ICS) model simulation results compared to gill Cu data for *Mytilus edulis* exposed to 40 μg L^{-1} waterborne Cu for 21 days, then depurated for 24 days. Model results from bottom to top are the cumulative sums of concentrations of ionic Cu^{2+} (vanishingly low),+Cu–MT,+Cu–GSH,+particulate Cu,+lysosome Cu,+high molecular weight Cu. The upper line is the total gill Cu. Model results for total Cu are compared to total gill Cu data (filled dots) and to gill Cu–MT data (unfilled squares). Data from Viarengo et al. (1985).

interest (gill or digestive gland) as an independent entity. Order of magnitude estimates of cytosolic composition and equilibrium constants were assigned. Copper uptake rate (μg Cu g$_{w}^{-1}$ day^{-1}) was set to the initial slope of organ Cu accumulation data ($t \sim 0$) and coefficients related to MT kinetics, CuGSH and CuLys excretion rates, and transfer of CuMT to the CuLys compartment, were evaluated by calibration. The total gill Cu (upper curve) simulations are in reasonable agreement with both the uptake and depuration phase data (Fig. 9.11; filled dots). The model also simulates HMW Cu (the narrow interval just below total Cu), particulate Cu, free Cu (very low; coincident with the *x*-axis) and comparable concentrations of Cu associated with a GSH-containing LMW ligand pool (CuGSH), metallothionein (CuMT), and lysosomes (CuLys). The time-varying nature of gill CuMT, the only cytosolic Cu fraction measured (unfilled squares), is simulated reasonably well. Digestive gland model–data comparisons were qualitatively similar to the gill Cu results of Fig. 9.11.

Harrison et al. (1983) developed a dose–response relationship between digestive gland HMW Cu and *Mytilus* survival for Cu concentrations bracketing the test conditions of Fig. 9.11. ICS model partitioning was calibrated to their HMW Cu data to ensure that the simulated ICS HMW

Cu was reasonable. The ICS prediction of 0.39 μg HMW Cu g_w^{-1} after 21 days at 40 μg L^{-1} was comparable to dose–response test results for similar conditions (0.30 and 0.80 μg g_w^{-1} after 21 days at 25 and 50 μg L^{-1}, respectively).

Finally, the ICS model simulated HMW Cu for the 2002 caged bivalve study at SI (dissolved Cu = 7.5 μg L^{-1} for 82 days) by adjusting the digestive gland Cu uptake rate by the ratio of Cu exposure concentrations (×7.5/40). The predicted HMW Cu after 82 days (0.25 μg g_w^{-1}), when compared to the dose–response relationship of Harrison and co-workers, indicates that only about 5% mortality should be expected. This prediction is consistent with control mortality as well as the 89% survival at SI over the 82 day test duration (Salazar and Salazar, 2007b). Although preliminary in nature, these analyses illustrate an approach for performing an effects assessment based on simulations at the intracellular level.

The ICS model is capable of simulating the spillover response. That is, Cu entering the cell displaces bound Zn (e.g. from Zn–thionein) such that the concentration of Cu–thionein increases and bound Zn decreases. The increase in cytosolic Zn^{2+} induces MT synthesis such that total MT increases. If the Cu influx rate exceeds the MT synthesis rate, then MT will not effectively scavenge incoming Cu. The result is that non-MT bound intracellular Cu accumulates at an increasing rate. At this point the Cu:BL concentration (i.e. CuHMW concentration) and the expectation of adverse effects should increase.

Spillover does not necessarily occur at a fixed tissue metal level. For example, Giguere et al. (2003) examined samples of freshwater floater mussels (*Pyganodon grandis*) obtained along a gradient of metals. Rather than seeing a marked increase in the degree of Cd spillover, at some threshold level, the fractions of Cd in the HMW and LMW pools remained essentially constant in chronically exposed clams. It has been suggested that spillover may only occur when the rate of metal uptake exceeds the metal detoxification capacity of the organism. The importance of rate of uptake has been demonstrated in a number of laboratory and field investigations (Kraak et al., 1992; Andres et al., 1999; Hook and Fisher, 2002) and is consistent with the preceding ICS model structure.

While intracellular detoxification mechanisms confer a protective benefit upon fish (Hogstrand and Haux, 1991), a concern is that these same processes may lead to relatively high tissue metal levels and increased dietary exposure of predators to metals. The importance of the subcellular distribution of metals on trophic transfer and effects is therefore of considerable interest. Several studies have addressed the influence of metal form (e.g. cytosolic, granules) on dietary metal assimilation efficiency by invertebrates and fish (Nott and Nicolaidou, 1990; Wallace and Lopez, 1996; Wang and Ke, 2002;

Blackmore and Wang, 2004; Zhang and Wang, 2006b). ICS modeling should lead to an improved understanding of the implications of these processes to dietary exposure, trophic transfer potential, and effects on predators.

4. REGULATORY APPLICATIONS

Metals risk assessments require estimates of environmental concentrations (waterborne and dietary) that organisms are exposed to and the related potential for adverse effects. WQC, commonly used benchmarks in risk assessments, are derived (1) to protect aquatic organisms from exposure to elevated metal concentrations that cause adverse effects (direct toxicity), and (2) to limit metal residues in prey to protect higher level predators (secondary or indirect toxicity). An improved knowledge of fish physiology will advance our understanding of how fish accumulate and respond to metals over ranges of waterborne and dietary metal concentrations. This improved understanding should aid development of models that relate environmental exposure to fish tissue residue levels that are protective of higher level consumers (e.g. residue-based criteria for Se, Hg).

Establishing links between exposure and effects is complicated because site-specific conditions markedly affect metal bioavailability (Jarvinen and Ankley, 1999; Di Toro et al., 2001). As a result, a wide range of environmental concentrations may be associated with a given response. An advantage of tissue residue-based criteria is that they reflect the net effect of all factors influencing ambient metal bioavailability and uptake by aquatic organisms, thereby providing a direct measure of internal dose. Tissue concentration (whole-body, organ, or cytosolic fraction) alone is not sufficient, however, because it must be associated with a toxic effect. Further, because permit limits for dischargers are based on WQC, a procedure is needed to evaluate how instream concentrations (dissolved and particulate), critical tissue levels, and effects are related. Models serve as computational tools for evaluating such relationships. The adage that "all models are wrong, some are useful" is particularly relevant here, because a rationally designed and validated modeling framework may be of considerable utility within the regulatory arena.

4.1. Use of Models for Water Quality and Tissue-Residue Based Criteria

Aquatic life criteria for metals are intended to be protective of most aquatic organisms. They are frequently expressed as dissolved metal concentrations. While this approach is useful, only recently has site-specific water chemistry (other than hardness) been considered in the direct evaluation of WQC. Specifically, the BLM (Section 3.2) has been incorporated into updated WQC for Cu in the USA; and Cu, Zn and Ni in the European Union (EU) (USEPA, 2007; Comber et al., 2008; Van Sprang et al., 2009). In the USA, the Cu LA50 for a sensitive organism [i.e. the 5th percentile of the species sensitivity distribution (SSD)] is specified and an acute to chronic ratio (ACR) applied to evaluate a chronic criterion. The EU uses a chronic SSD to directly evaluate a site-specific BLM-based hazardous concentration for 5% of species, HC5 (Comber et al., 2008; Van Sprang et al., 2009). Chronic BLM implementations have been used for this purpose (e.g. De Schamphelaere and Janssen, 2004a,b,c; Schwartz and Vigneault, 2007). Although a significant advance, BLM-based WQC have not yet been universally adopted. Their use is expected to increase, but implementation will likely be gradual.

A limitation of most toxicity models (Section 3) is that, in contrast to bioaccumulation models (Section 2), they do not consider dietary exposure. Borgmann et al. (2005) proposed several ways to incorporate dietary exposure into water quality regulations for metals. The simplest is to assume that waterborne and dietary toxicity are additive and to decrement WQC based on water-only exposures to offset dietary exposure. Alternatively, WQC could be based on studies with concurrent exposure to waterborne and dietary metals that are in equilibrium. A third approach is to model metal accumulation to determine relative contributions of waterborne and dietary sources. This approach requires a relationship between exposure, accumulation, and effects (e.g. a CBR). It will be uncertain whether the CBR differs with respect to route of exposure or rate of uptake. Finally, if dietary uptake is the dominant route of exposure, then a guideline for metals in food could be adopted in place of a waterborne exposure-based guideline.

Of the preceding approaches, the latter two are amenable for use with bioaccumulation models. For example, tissue residue-based criteria have been proposed for some metals, including Hg (see Kidd and Batchelar, Chapter 5) and Se (see Janz, Chapter 7, Vol. 31A). Waterborne concentrations associated with these residue-based criteria control allowable dissolved concentrations because they are less than WQC based on water-only exposures. With respect to Hg, the USEPA has adopted a methylmercury criterion of 0.3 mg kg^{-1} wet weight of edible fish tissue (mg kg$_w^{-1}$) (USEPA, 2009). The default criterion is based upon a reference dose for humans of 0.1 µg kg^{-1} body weight day^{-1} and a lifetime average daily intake (ADI) of 17.5 g fish day^{-1} by a 70 kg adult. If a regulatory agency desires an equivalent dissolved methylmercury WQC, EPA suggests three approaches: (1) use site-specific BAFs; (2) use a scientifically defensible bioaccumulation model (e.g. QEA, 2000; EPRI, 2002; Barber, 2001); and (3) use BAFs from another site. The advantage of using mechanistic models to evaluate methylmercury bioaccumulation under different conditions (e.g. water chemistry, fish growth rates) was highlighted. One disadvantage is the potential for increased data requirements.

A whole-body residue-based criterion has been proposed for Se in fish (USEPA, 2004), with an update currently in review. Whole-body Se in fish is more readily related to Se in the ovary (the target organ), and to toxic effects, than is dissolved Se (GEI Consultants et al., 2008; Chapman et al., 2009a,b; Presser and Luoma, 2010). The modeling approach of Section 2.1 provides a framework for evaluating dissolved Se enrichment in algae and higher trophic levels (Presser and Luoma, 2010). If waterborne Se uptake is small relative to dietary uptake, and growth dilution is small relative to elimination ($g < < k_e$), then Eq. (3) reduces to an expression for the trophic transfer factor (TTF, the ratio of the steady-state tissue concentration in a predator, v_{ss}, relative to its prey, C_f):

$$TTF = \frac{v_{ss}}{C_f} = \frac{AE \cdot IR}{k_e} \tag{21}$$

Lacking kinetic coefficients for Eq. (21), TTFs may also be measured (Presser and Luoma, 2010). Regardless, the evaluation proceeds as follows. First, a partition coefficient (K_d) times a dissolved Se concentration is used to evaluate Se enrichment in algae and/or particulates (the base of the food chain). TTFs are then applied sequentially to evaluate Se concentrations at successive trophic levels, through fish. Conversely, because effluent limits are expressed as a total Se load, translation of an Se tissue concentration to dissolved concentration may also be needed (Chapman et al., 2009b; Presser and Luoma, 2010). This is accomplished by reversing the above procedure, backcalculating (using TTF and K_d) the dissolved concentration needed to meet the fish tissue Se criterion. The possibility that the Se CBR that is protective of fish (primary toxicity) may also be protective of reproductive effects on birds (another receptor of concern) that prey on fish (secondary toxicity) is under consideration.

One limitation of primary residue-based standards is that the species to be protected may not be present at a site because of elevated metal levels. A residue-based approach that avoids this difficulty is to set an allowable tissue residue level for a metal-insensitive surrogate or indicator species, to protect against adverse effects on a relatively metal-sensitive species (Luoma et al., 2009; Adams et al., 2010). As long as tissue metal concentration in the indicator species is less than the allowable threshold, the sensitive species should be protected. This approach was used for a field study on the Clark Fork River (Cain et al., 2004, 2006). More experience is needed to demonstrate the practicality of this approach from monitoring and regulatory perspectives, and the utility of modeling.

4.2. Other Model Applications

Many models have potential for use in regulatory settings. Clearly, a model that is capable of evaluating tissue metal levels resulting from waterborne and dietary exposure will be useful if a tissue residue criterion is implemented (e.g. for Hg or Se). If field observations indicate exceedences of a tissue residue standard, the model could evaluate abiotic exposure levels that will achieve compliance. If the standard is expressed on a whole-body basis, then a whole-body, PBPK, or food chain model (Section 2) might be appropriate. These models could also be used to evaluate waterborne and dietary concentrations needed to achieve a tissue residue guideline for a biomonitor species. If a tissue-based guideline is related to accumulation levels in a specific organ (e.g. Se in ovaries or eggs of fish or birds), then a PBPK model (Section 2.2) offers the specificity needed to evaluate accumulation at the site of action of toxicity. A PBPK model might also be used to design a study or interpret results of investigations directed at understanding target organ toxicity.

The BLM (Section 3.2) has already been incorporated into standard setting procedures in the USA and EU and is expected to be more widely used. Other models (e.g. damage–repair, BIM-BAM, IBM, and ICS models) will require continued development, both to improve their capabilities and to demonstrate their predictive abilities. For example, a mechanistic approach may be needed to simulate non-linear responses (e.g. essential metal regulation) and to improve current abilities to quantify how metal detoxification and compartmentalization influence long-term accumulation and effects. Incorporating these modeling capabilities into the regulatory arena will be a gradual process, one that involves ongoing laboratory testing, data generation, and model refinement. Such a process should ultimately lead to improved regulations.

5. FUTURE MODEL DEVELOPMENT NEEDS

The preceding review has considered both long-established, widely used models and recently proposed models that have received limited use to date. Several model development needs have been highlighted. These include the following.

- Continued investigation of the effect of waterborne and dietary metal bioavailability on bioaccumulation kinetics and food web dynamics is needed. Use of models will provide a

framework for quantitatively evaluating the biomagnification potential of relatively complex systems.
- The integration of metal bioaccumulation and toxicity model frameworks is needed to facilitate the concurrent consideration of how exposure to both waterborne and dietary metals is related to effects.
- Further investigation of the mechanisms of essential metal regulation and factors affecting deficiency and toxicity is warranted. Such advances will facilitate the ongoing refinement of mechanistically based predictive models.
- Further studies are required to gain an improved understanding of the:
  - speciation of metals in extracellular and intracellular fluids
  - roles of GSH, metallothionein, lysosomes, granules, etc., in metal complexation, detoxification, sequestration, and elimination
  - proximate (subcellular fractions) and biochemically specific sites of action of toxicity.
- Well-designed experiments that provide data to rigorously validate future model refinements.

The preceding research areas may have utility from both academic and practical perspectives. The point at which an academic pursuit becomes of practical utility (e.g. to satisfy the needs of environmental regulations) will undoubtedly evolve as advances in understanding are realized. Use of models in the context of such investigations, to design laboratory and field studies and for data interpretation, should enhance the ultimate utility of the results. This will, in turn, lead to the development and parameterization of more refined, mechanistically based models.

ACKNOWLEDGMENTS

Support received for completion of the model applications used as examples in Section 3.3.2 (Eastman Kodak Company, Rochester, NY) and Section 3.4.2 (International Copper Association, ICA and the Water Environment Research Foundation, WERF), as well as the contributions of Kathy Reidda and Anne Banta (research librarians at HydroQual) and Linda Jensen (manuscript preparation) are acknowledged and greatly appreciated. The early work and invaluable guidance provided by Robert Thomann, a former colleague at Hydroscience, are also recognized and much appreciated.

REFERENCES

Adams, W., Blust, R., Borgmann, U., Brix, K., DeForest, D., Green, A., McGeer, J., Meyer, J., Paquin, P., Rainbow, P., and Wood, C. M. (2010). Utility of tissue residues for predicting effects of metals on aquatic organisms. *Integr. Environ. Assess. Manag.* **7**, 75–98.

Anderson, D. M., and Morel, F. M. M. (1978). Copper sensitivity of *Gonyaulax tamarensis*. *Limnol. Oceanogr.* **23**, 283–295.

Andres, S., Baudrimont, M., Lapaquellerie, Y., Ribeyre, F., Maillet, N., Latouche, C., and Boudou, A. (1999). Field transplantation of the freshwater bivalve *Corbicula fluminea* along a polymetallic contamination gradient (River Lot, France): I. Geochemical characteristics of the sampling sites and cadmium and zinc bioaccumulation kinetics. *Environ. Toxicol. Chem.* **18**, 2462–2471.

Ausseil, O., Adam, C., Garnier-Laplace, J., Baudin, J.-P., Casellas, C., and Porcherzz, J.-M. (2002). Influence of metal (Cd and Zn) waterborne exposure on radionuclide (Cs-134, Ag-110m, and Co-57) bioaccumulation by rainbow trout (*Oncorhynchus mykiss*): a field and laboratory study. *Environ. Toxicol. Chem.* **21**, 619–625.

Baines, S. B., Fisher, N. S., and Stewart, R. (2002). Assimilation and retention of selenium and other trace elements from crustacean food by juvenile striped bass (*Morone saxatilis*). *Limnol. Oceanogr.* **43**, 646–655.

Barber, M. C. (2001). *Bioaccumulation and Aquatic System Simulator (BASS) User's Manual,* Beta Test Version 2.1.Athens, GA: US Environmental Protection Agency, Office of Research and Development, Ecosystems Research Division, EPA-600-R-01-035.

Barber, M. C. (2003). A review and comparison of models for predicting dynamic chemical bioconcentration in fish. *Environ. Toxicol. Chem.* **22**, 1963–1992.

Barron, M. G., Stehly, G. R., and Hayton, W. L. (1990). Pharmacokinetic modeling in aquatic animals. I. Models and concepts. *Aquat. Toxicol.* **18**, 61–86.

Bianchini, A., Grosell, M., Gregory, S. M., and Wood, C. M. (2002). Acute silver toxicity in aquatic animals is a function of sodium uptake rate. *Environ. Sci. Technol.* **36**, 1763–1766.

Bielmyer, G. K., Grosell, M., and Brix, K. V. (2006). Effects of silver, zinc, copper and nickel to the copepod *Acartia tonsa*, exposed via a phytoplankton diet. *Environ. Sci. Technol.* **40**, 2063–2068.

Bielmyer, G. K., Brix, K. V., and Grosell, M. (2008). Is Cl^- protection against silver toxicity due to chemical speciation? *Aquat. Toxicol.* **87**, 81–87.

Blackmore, G., and Wang, W. X. (2004). The transfer of cadmium, mercury, methylmercury, and zinc in an intertidal rocky shore food chain. *J. Exp. Mar. Biol. Ecol.* **307**, 91–110.

Blancato, J. N. and Bischoff, K. B. (1994). The application of pharmacokinetic models to predict target dose. In *Health Risk Assessment, Dermatology: Clinical and Basic Science.* EPA 600/A-94-134, 31-46. [NTIS PB94-190345]. Washington DC: US Environmental Protection Agency.

Borgmann, U., and Norwood, W. P. (1997). Identification of the toxic agent in metal-contaminated sediments from Manitouwadge Lake, Ontario, using toxicity-accumulation relationships in *Hyalella azteca*. *Can. J. Fish. Aquat. Sci.* **54**, 1055–1063.

Borgmann, U., Janssen, C. R., Blust, R. J. P., Brix, K. V., Dwyer, R. L., Erickson, R. J., Hare, L., Luoma, S. N., Paquin, P. R., Roberts, C. A., and Wang, W.-X. (2005). Incorporation of dietborne metals exposure into regulatory frameworks. In: *Toxicity of Dietborne Metals to Aquatic Biota* (J. S. Meyer, W. J. Adams, K. V. Brix, S. N. Luoma, D. R. Mount, W. A. Stubblefield and C. M. Wood, eds), pp. 153–189. SETAC Press, Pensacola, FL.

Borgmann, U., Norwood, W. P., and Dixon, D. G. (2008). Modeling bioaccumulation and toxicity of metal mixtures. *Hum. Ecol. Risk Assess.* **14**, 266–289.

Borgmann, U., Schroeder, J. A., Golding, L. A., and Dixon, D. G. (2010). Models of cadmium accumulation and toxicity to *Hyalella azteca* during 7- and 28-day exposures. *Hum. Ecol. Risk Assess.* **16**, 580–587.

Breck, J. E. (1988). Relationship among models for acute toxic effects: applications to fluctuating concentrations. *Environ. Toxicol. Chem.* **7**, 775–778.

Brooks, B. W., Stanley, J. K., White, J. C., Turner, P. K., Wu, K. B., and La Point, T. W. (2004). Laboratory and field responses to cadmium: an experimental study in effluent-dominated stream mesocosms. *Environ. Toxicol. Chem.* **23**, 1057–1064.

Brown, D. A., and Parsons, T. R. (1978). Relationship between cytoplasmic distribution of mercury and toxic effects to zooplankton and chum salmon (*Oncorhynchus keta*) exposed to mercury in a controlled ecosystem. *J. Fish. Res. Bd Can.* **35**, 880–884.

Brown, D. A., Bay, S. M., and Hershelman, G. P. (1990). Exposure of scorpionfish (*Scorpaena guttata*) to cadmium: effects of acute and chronic exposures on the cytosolic distribution of cadmium, copper and zinc. *Aquat. Toxicol.* **16**, 295–310.

Bury, N. R., and Wood, C. M. (1999). Mechanism of branchial apical silver uptake by rainbow trout is via the proton-coupled Na^+ channel. *Am. J. Physiol. Regul. Integr. C* **277**, R1385–R1391.

Bury, N. R., McGeer, J. C., and Wood, C. M. (1999a). Effects of altering freshwater chemistry on physiological responses of rainbow trout to silver exposure. *Environ. Toxicol. Chem.* **18**, 49–55.

Bury, N. R., Grosell, M., Grover, A. K., and Wood, C. M. (1999b). ATP-dependent silver transport across the basolateral membrane of rainbow trout gills. *Toxicol. Appl. Pharmacol.* **159**, 1–8.

Butcher, J., Diamond, J., Bearr, J., Latimer, H., Klaine, S. J., Hoang, T., and Bowersox, M. (2006). Toxicity models of pulsed copper exposure to *Pimephales promelas* and *Daphnia magna. Environ. Toxicol. Chem.* **25**, 2541–2550.

Cain, D. J., Luoma, S. N., and Wallace, W. G. (2004). Linking metal bioaccumulation of aquatic insects to their distribution patterns in a mining-impacted river. *Environ. Toxicol. Chem.* **23**, 1463–1473.

Cain, D. J., Buchwalter, D. B., and Luoma, S. N. (2006). Influence of metal exposure history on the bioaccumulation and subcellular distribution of aqueous cadmium in the insect *Hydropsyche californica. Environ. Toxicol. Chem.* **25**, 1042–1049.

Campbell, P. G. C. (1995). Interactions between trace metals and aquatic organisms: a critique of the free-ion activity model. In: *Metal Speciation and Bioavailability in Aquatic Systems* (A. Tessier and D. R. Turner, eds), pp. 45–102. John Wiley and Sons, New York. IUPAC.

Chapman, P. M., Adams, W. J., Brooks, M. L., Delos, C. G., Luoma, S. N., Maher, W. A., Ohlendorf, H. M., Presser, T. S., and Shaw, D. P. (2009a). *Ecological Assessment of Selenium in the Aquatic Environment: Summary of a SETAC Pellston Workshop.* SETAC Press, Pensacola FL.

Chapman, P. M., McDonald, B. G., Ohlendorf, H. M., and Jones, R. (2009b). A conceptual selenium management model. *Integr. Environ. Assess. Manag.* **5**, 461–469.

Cheung, M. S., Zhang, L., and Wang, W. X. (2007). Transfer and efflux of cadmium and silver in marine snails and fish fed pre-exposed mussel prey. *Environ. Toxicol. Chem.* **26**, 1172–1178.

Clark, K. E., Gobas, F. A. P. C., and Mackay, D. (1990). Model of organic chemical uptake and clearance by fish from food and water. *Environ. Sci. Technol.* **24**, 1203–1213.

Comber, S. D., Merrington, G., Sturdy, L., Delbeke, K., and van Assche, F. (2008). Copper and zinc water quality standards under the EU Water Framework Directive: the use of a tiered approach to estimate the levels of failure. *Sci. Total Environ.* **403**, 12–22. (Epub 2008 Jul 2)

Connolly, J. P. (1985). Predicting single-species toxicity in natural water systems. *Environ. Toxicol. Chem.* **4**, 573–582.

Connolly, J. P., and Thomann, R. V. (1992). Modeling the accumulation of organic chemicals in aquatic food chains. In: *Fate of Pesticides and Chemicals in the Environment* (J. L. Schnoor, ed.), pp. 385–406. John Wiley & Sons, New York.

Croteau, M.-N., and Luoma, S. N. (2005). Delineating copper accumulation pathways for the freshwater bivalve *Corbicula* using stable copper isotopes. *Environ. Toxicol. Chem.* **24**, 2871–2878.

Croteau, M.-N., Luoma, S. N., Topping, B. R., and Lopez, C. B. (2004). Stable metal isotopes reveal copper accumulation and loss dynamics in the freshwater bivalve *Corbicula. Environ. Sci. Technol.* **38**, 5002–5009.

Curtis, B. J., and Wood, C. M. (1991). The function of the urinary bladder *in vivo* in the freshwater rainbow trout. *J. Exp. Biol.* **155**, 567–583.

De Schamphelaere, K. A. C., and Janssen, C. R. (2004a). Development and field validation of a biotic ligand model predicting chronic copper toxicity to *Daphnia magna*. *Environ. Toxicol. Chem.* **23**, 1365–1375.

De Schamphelaere, K. A. C., and Janssen, C. R. (2004b). Bioavailability and chronic toxicity of zinc to juvenile rainbow trout (*Oncorhynchus mykiss*): comparison with other fish species and development of a biotic ligand model. *Environ. Sci. Technol.* **38**, 6201–6209.

De Schamphelaere, K. A. C., and Janssen, C. R. (2004c). Effects of chronic dietary copper exposure on growth and reproduction of *Daphnia magna*. *Environ. Toxicol. Chem.* **23**, 2038–2047.

De Schamphelaere, K. A. C., Canli, M., Van Lierde, V., Forrez, I., Vanhaecke, F., and Janssen, C. R. (2004). Reproductive toxicity of dietary zinc to *Daphnia magna*. *Aquat. Toxicol.* **70**, 233–244.

Diamond, M. L. (1999). Development of a fugacity/aquivalence model of mercury dynamics in lakes. *Water Air Soil Pollut.* **111**, 337–357.

Diamond, M. L., Mackay, D., and Welbourn, P. M. (1992). Models of multimedia partitioning of multispecies chemicals: the fugacity/aquivalence approach. *Chemosphere* **25**, 1907–1921.

Din, W. S., and Frazier, J. M. (1985). Protective effect of metallothionein on cadmium toxicity in isolated rat hepatocytes. *Biochem. J.* **230**, 395–402.

Di Toro, D. M., Allen, H. E., Bergman, H. L., Meyer, J. S., Paquin, P. R., and Santore, R. C. (2001). A biotic ligand model of the acute toxicity of metals. I. Technical basis. *Environ. Toxicol. Chem.* **20**, 2383–2396.

EPRI (2002). Dynamic mercury cycling model (D-MCM) for Windows 98/NT 4. *0/2000/XP: A Model for Mercury Cycling in Lakes*, D-MCM Version 2.0. Lafayette, CA: Electric Power Research Institute.

Erickson, R. J., and McKim, J. M. (1990). A simple flow-limited model for exchange of organic chemicals at fish gills. *Environ. Toxicol. Chem.* **9**, 159–165.

Escher, R. I., and Hermens, J. L. M. (2004). Internal exposure: linking bioavailability to effects. *Environ. Sci. Technol.* **38**, 455A–462A.

Everaarts, J. M. (1990). Uptake and release of cadmium in various organs of the common mussel, *Mytilus edulis* (L.). *Bull. Environ. Contam. Toxicol.* **45**, 560–567.

Fisher, N. S., and Reinfelder, J. R. (1995). The trophic transfer of metals in marine systems. In: *Metal Speciation and Bioavailability in Aquatic Systems* (A. Tessier and D. R. Turner, eds), pp. 363–406. John Wiley and Sons, Chichester.

Fisher, N. S., Teyssie, J.-L., Fowler, S. W., and Wang, W.-X. (1996). Accumulation and retention of metals in mussels from food and water: a comparison under field and laboratory conditions. *Environ. Sci. Technol.* **30**, 3232–3242.

Franco-Uria, A., Otero-Muras, I., Balsa-Canto, E., Alonso, A. A., and Roca, E. (2010). Generic parameterization for a pharmacokinetic model to predict Cd concentrations in several tissues of different fish species. *Chemosphere* **79**, 377–386.

Freedman, J. H., Ciriolo, M. R., and Peisach, J. (1989). The role of glutathione in copper metabolism and toxicity. *J. Biol. Chem.* **264**, 5598–5605.

Galvez, F., and Wood, C. M. (1997). The relative importance of water hardness and chloride levels in modifying the acute toxicity of silver to rainbow trout (*Oncorhynchus mykiss*). *Environ. Toxicol. Chem.* **16**, 2363–2368.

Garnier-Laplace, J., Adam, C., and Baudin, J. P. (2000). Experimental kinetic rates of food-chain and waterborne radionuclide transfer to freshwater fish: a basis for the construction of fish contamination charts. *Arch. Environ. Contam. Toxicol.* **39**, 133–144.

GEI Consultants, Golder Associates, Parametrix, University of Saskatchewan Toxicology Centre (2008). *Selenium Tissue Thresholds: Tissue Selection Criteria, Threshold Development Endpoints, and Potential to Predict Population or Community Effects in the Field.* Report submitted to North America Metals Council – Selenium Working Group, Washington DC.

Gensemer, R. W., Naddy, R. B., Stubblefield, W. A., Hockett, J. R., Santore, R. C., and Paquin, P. R. (2002). Evaluating the role of ion composition on the toxicity of copper to *Ceriodaphnia dubia* in very hard waters. *Comp. Biochem. Physiol. C* **133**, 87–97.

Giguere, A., Couillard, Y., Campbell, P. G. C., Perceval, O., Hare, L., Pinel-Alloul, B., and Pellerin, J. (2003). Steady-state distribution of metals among metallothionein and other cytosolic ligands and links to cytotoxicity in bivalves living along a polymetallic gradient. *Aquat. Toxicol.* **64**, 185–200.

Giles, M. A. (1988). Accumulation of cadmium by rainbow trout, *Salmo gairdneri*, during extended exposure. *Can. J. Fish. Aquat. Sci.* **45**, 1045–1053.

Gobas, F. A. P. C. (1993). A model for predicting the bioaccumulation of hydrophobic organic chemicals in aquatic food-webs: application to Lake Ontario. *Ecol. Model.* **69**, 1–17.

Gregus, Z., and Klaassen, C. D. (1986). Disposition of metals in rats: a comparative study of fecal, urinary, and biliary excretion and tissue distribution of eighteen metals. *Toxicol. Appl. Pharmacol.* **85**, 24–38.

Grosell, M., Hogstrand, C., Wood, C. M., and Hansen, H. J. M. (2000). A nose-to-nose comparison of the physiological effects of exposure to ionic silver versus silver chloride in the European eel (*Anguilla anguilla*) and the rainbow trout (*Oncorhynchus mykiss*). *Aquat. Toxicol.* **48**, 327–342.

Grosell, M., Nielsen, C., and Bianchini, A. (2002). Sodium turnover rate determines sensitivity to acute copper and silver exposure in freshwater animals. *Comp. Biochem. Physiol. C* **133**, 287–303.

Guyton, A. C., Coleman, T. G., and Granger, J. J. (1972). Circulation: overall regulation. *Annu. Rev. Physiol.* **34**, 13–44.

Hall, J. E. (2004). The pioneering use of systems analysis to study cardiac output regulation. *Am. J. Physiol. Regul. Integr. C* **287**, R1009–R1011.

Handy, R. D. (1992). The assessment of episodic metal pollution. II. The effects of cadmium and copper enriched diets on tissue contaminant analysis in rainbow trout (*Oncorhynchus mykiss*). *Arch. Environ. Contam. Toxicol.* **22**, 82–87.

Harrison, S. E., and Klaverkamp, J. F. (1989). Uptake and elimination and tissue distribution of dietary and aqueous cadmium by rainbow trout (*Salmo gairdneri* Richardson) and lake whitefish (*Caregonus clupeaformis* Mitchill). *Environ. Toxicol. Chem.* **8**, 87–97.

Harrison, F. L., Lam, J. R., and Berger, R. (1983). Sublethal responses of *Mytilus edulis* to increased dissolved copper. *Sci. Total Environ.* **28**, 141–158.

Hendriks, A. J., and Heikens, A. (2001). The power of size 2: Rate constants and equilibrium ratios for accumulation of inorganic substances related to species weight. *Environ. Toxicol Chem.* **20**, 1421–1437.

Hoang, T. C., Tomasso, J. R., and Klaine, S. J. (2007). An integrated model describing the toxic response of *Daphnia magna* to pulsed exposures of three metals. *Environ. Toxicol. Chem.* **26**, 132–138.

Hogstrand, D., and Haux, C. (1991). Binding and detoxification of heavy metals in lower vertebrates with reference to metallothionein. *Comp. Biochem. Physiol. C* **100**, 137–141.

Hogstrand, C., and Wood, C. M. (1998). Towards a better understanding of the bioavailability, physiology, and toxicity of silver in fish: Implications for water quality criteria. *Environ. Toxicol. Chem.* **17**, 547–561.

Hogstrand, C., Galvez, F., and Wood, C. M. (1996). Toxicity, silver accumulation and metallothionein induction in freshwater rainbow trout during exposure to different silver salts. *Environ. Toxicol. Chem.* **15**, 1102–1108.

Holmes, W. N., and Donaldson, E. M. (1969). The body compartments and the distribution of electrolytes. In: Fish Physiology, *Vol. 1.* Excretion, Ionic Regulation, and Metabolism (W. S Hoar and D. J Randall, eds), pp. 1–89. Academic Press, Orlando, FL.

Hook, S. E., and Fisher, N. S. (2001). Reproductive toxicity of metals in calanoid copepods. *Mar. Biol.* **138**, 1131–1140.

Hook, S. E., and Fisher, N. S. (2002). Relating the reproductive toxicity of five ingested metals in calanoid copepods with sulfur affinity. *Mar. Environ. Res.* **53**, 161–174.

Houwen, R., Dijkstra, M., Kuipers, F., Smit, E. P., Havinga, R., and Vonk, R. J. (1990). Two pathways for biliary copper excretion in the rat. The role of glutathione. *Biochem. Pharmacol.* **39**, 1039–1044.

Hussain, S., Meneghini, E., Moosmayer, M., Lacotte, D., and Anner, B. M. (1994). Potent and reversible interaction of silver with pure Na,K-ATPase and Na,K-ATPase-liposomes. *Biochim. Biophys. Acta.* **1190**, 402–408.

Hydroscience (1973). Limnological Systems Analysis of the Great Lakes, Phase I. Prepared for the Great Lakes Basin Commission.

Janes, N., and Playle, R. C. (1995). Modeling silver binding to gills of rainbow trout (*Oncorhynchus mykiss*). *Environ. Toxicol. Chem.* **14**, 1847–1858.

Jarvinen, A. W. and Ankley, G. T. (1999). *Linkage of Effects to Tissue Residues: Development of a Comprehensive Database for Aquatic Organisms Exposed to Inorganic and Organic Chemicals.* SETAC Technical Publication Series. Pensacola, FL: SETAC Press.

Kamunde, C., Clayton, C., and Wood, C. M. (2002). Waterborne vs. dietary copper uptake in rainbow trout and the effects of previous waterborne copper exposure. *Am.J. Physiol. Regul. Integr. C* **283**, R69–R78.

Keithly, J., Brooker, J. A., DeForest, D. K., Wu, K. B., and Brix, K. V. (2004). Acute and chronic toxicity of nickel to a cladoceran (*Ceriodaphnia dubia*) and amphipod (*Hyalella azteca*). *Environ. Toxicol. Chem.* **23**, 691–696.

Knightes, C. D. (2008). Development and test application of SERAFM: a screening-level mercury fate model and tool for evaluating wildlife exposure risk for surface water and with mercury-contaminated sediments. *Environ. Model. Soft.* **23**, 495–510.

Knightes, C. D., Sunderland, E. M., Barber, C., Johnston, J. M., and Ambrose, R. B. (2009). Application of ecosystem-scale, fate and bioaccumulation models to predict fish mercury response times to changes in atmospheric deposition. *Environ. Toxicol. Chem.* **28**, 881–893.

Kolok, A. S., Hartman, M. H., and Sershan, J. L. (2002). The physiology of copper tolerance in fathead minnows: insight from an intraspecific, correlative analysis. *Environ. Toxicol. Chem.* **21**, 1730–1735.

Kraak, M. H. S., Lavy, D., Peeters, W. H. M., and Davids, C. (1992). Chronic ecotoxicity of copper and cadmium to the zebra mussel *Dreissena polymorpha*. *Arch. Environ. Contam. Toxicol.* **23**, 363–369.

Krogh, A. (1939). *Osmotic Regulation in Aquatic Animals* (Originally published by Cambridge University Press and reprinted in English in 1965, unabridged and unaltered edition). New York: Dover Publications.

Landrum, P. F., Lee, H. II, and Lydy, M. J. (1992). Toxicokinetics in aquatic systems: model comparisons and use in hazard assessment. *Environ. Toxicol. Chem.* **11**, 1709–1725.

Lauren, D. J., and McDonald, D. G. (1985). Effects of copper on branchial ionoregulation in the rainbow trout, *Salmo gairdneri* Richardson. *J. Comp. Physiol. B* **155**, 635–644.

Liao, C.-M., Liang, H.-M., Chen, B.-C., Singh, S., Tsai, J.-W., Chou, Y.-H., and Lin, W.-T. (2005). Dynamical coupling of PBPK/PD and AUC-based toxicity models for arsenic in tilapia *Oreochromis mossambicus* from blackfoot disease area in Taiwan. *Environ. Pollut.* **135**, 221–233.

Liao, C.-M., Chiang, K. C., and Tsai, J.-W. (2006). Bioenergetics-based matrix population modeling enhances life-cycle toxicity assessment of tilapia *Oreochromis mossambicus* exposed to arsenic. *Environ. Toxicol.* **21**, 154–165.

Lu, Z. H., Dameron, C. T., and Solioz, M. (2003). The *Enterococcus hirae* paradigm of copper homeostasis: copper chaperone turnover, interactions, and transactions. *BioMetals* **16**, 137–143.

Luoma, S. N., and Presser, T. S. (2009). Emerging opportunities in management of selenium contamination. *Environ. Sci. Technol.* **43**, 8483–8487.

Luoma, S. N., and Rainbow, P. S. (2005). Why is metal bioaccumulation so variable? Biodynamics as a unifying concept, a critical review. *Environ. Sci. Technol.* **39**, 1921–1931.

Luoma, S. N., Cain, D. J., and Rainbow, P. S. (2009). Calibrating biomonitors to ecological disturbance: a new technique for explaining metal effects in natural waters. *Integr. Environ. Assess. Manag.* **6**, 199–209.

Mackay, D. (1989). Modeling the long-term behaviour of an organic contaminant in a large lake: application to PCBs in Lake Ontario. *J. Great Lakes Res.* **15**, 283–297.

Mackay, D., and Hughes, A. I. (1984). Three-parameter equation describing the uptake of organic compounds by fish. *Environ. Sci. Technol.* **18**, 439–444.

Mancini, J. L. (1983). A method for calculating effects, on aquatic organisms, of time varying concentrations. *Water Res.* **17**, 1355–1362.

Marr, J. C. A., Lipton, J., Cacela, D., Hansen, J. A., Bergman, H. L., Meyer, J. S., and Hogstrand, C. (1996). Relationship between copper exposure duration, tissue copper concentration, and rainbow trout growth. *Aquat. Toxicol.* **36**, 17–30.

Marr, J. C. A., Hansen, J. A., Meyer, J. S., Cacela, D., Podrabsky, T., Lipton, J., and Bergman, H. L. (1998). Toxicity of cobalt and copper to rainbow trout: application of a mechanistic model for predicting survival. *Aquat. Toxicol.* **43**, 225–237.

Mason, A. Z., and Jenkins, K. D. (1995). Metal detoxification in aquatic organisms. In: *Metal Speciation and Bioavailability in Aquatic Systems* (A. Tessier and D. R. Turner, eds), pp. 479–608. John Wiley and Sons, Chichester.

Mathews, T., and Fisher, N. S. (2008a). Evaluating the trophic transfer of cadmium, polonium, and methylmercury in an estuarine food chain. *Environ. Toxicol. Chem.* **27**, 1093–1101.

Mathews, T., and Fisher, N. S. (2008b). Trophic transfer of seven trace metals in a four-step marine food chain. *Mar. Ecol. Prog. Ser.* **367**, 23–33.

Mathews, T., and Fisher, N. S. (2009). Dominance of dietary intake of metals in marine elasmobranch and teleost fish. *Sci. Total Environ.* **407**, 5156–5161.

Mathews, T., Beaugelin-Seiller, K., Garnier-Laplace, J., Gilbin, R., Adam, C., and Della-Vedova, C. (2009). A probabilistic assessment of the chemical and radiological risks of chronic exposure to uranium in freshwater ecosystems. *Environ. Sci. Technol.* **43**, 6684–6690.

McCarty, L. S. (1987). Relationship between toxicity and bioconcentration for some organic chemicals. I. Examination of the relationship. In: *QSAR in Environmental Toxicology* (K. L. E. Kaiser, ed.), Vol. II, pp. 207–220. C. Reidel Publishing Company, Dordrecht.

McCarty, L. S., Mackay, D., Smith, A. D., Ozburn, G. W., and Dixon, D. G. (1993). Residue-based interpretation of toxicity and bioconcentration QSARs from aquatic bioassays: polar narcotic organics. *Ecotoxicol. Environ. Saf.* **25**, 253–270.

McDonald, D. G. (1983). The interaction of environmental calcium and low pH on the physiology of the rainbow trout, Salmo gairdneri. I. Branchial and renal net ion and H^+ fluxes. *J. Exp. Biol.* **102**, 123–140.

McDonald, D. G., and Wood, C. M. (1993). Branchial mechanisms of acclimation to metals in freshwater fish. In: *Fish Ecophysiology* (J. C. Rankin and F. B. Jensen, eds), pp. 297–321. Chapman and Hall, London.

McDonald, D. G., Hobe, H., and Wood, C. M. (1980). The influence of calcium on the physiological responses on the rainbow trout, *Salmo gairdneri*, to low environmental pH. *J. Exp. Biol.* **88**, 109–131.

McGeer, J. C., and Wood, C. M. (1998). Protective effects of water Cl^- on physiological responses to waterborne silver in rainbow trout. *Can. J. Fish. Aquat. Sci.* **55**, 2447–2454.

McGeer, J. C., Playle, R. C., Wood, C. M., and Galvez, F. (2000). A physiologically based biotic ligand model for predicting the acute toxicity of waterborne silver to rainbow trout in freshwaters. *Environ. Sci. Technol.* **34**, 4199–4207.

McKim, J. M., and Erickson, R. J. (1991). Environmental impacts on the physiological mechanisms controlling xenobiotic transfer across fish gills. *Physiol. Zool.* **64**, 39–67.

McKim, J. M., and Nichols, J. W. (1994). Use of physiologically based toxicokinetic models in a mechanistic approach to aquatic toxicology. In: Aquatic Toxicology: Molecular Biochemical, and Cellular Perspectives (D. C. Malins and G. Ostrander, eds), pp. 469–519. Lewis Publishers, Boca Raton, FL.

Meyer, J. S., Gulley, D. D., Goodrich, M. S., Szmania, D. C., and Brooks, A. S. (1995). Modeling toxicity due to intermittent exposure of rainbow trout and common shiners to monochloramine. *Environ. Toxicol. Chem.* **14**, 165–175.

Meyer, J. S., Santore, R. C., Bobbitt, J. P., DeBrey, L. D., Boese, C. J., Paquin, P. R., Allen, H. E., Bergman, H. L., and Di Toro, D. M. (1999). Binding of nickel and copper to fish gills predicts toxicity when water hardness varies, but free-ion activity does not. *Environ. Sci. Technol.* **33**, 913–916.

Milligan, C. L., and Wood, C. M. (1982). Disturbances in haematology, fluid volume distribution and circulatory function associated with low environmental pH in the rainbow trout. *Salmo gairdneri. J. Exp. Biol.* **99**, 397–415.

Morel, F. M. (1983). *Complexation: trace metals and microorganisms. Principles of Aquatic Chemistry.* Wiley Interscience, New York, 301–308.

Morgan, I. J., Henry, R. P., and Wood, C. M. (1997). The mechanism of acute silver nitrate toxicity in freshwater rainbow trout *(Oncorhynchus mykiss)* is inhibition of gill Na^+ and Cl^- transport. *Aquat. Toxicol.* **38**, 145–163.

Morgan, T. P., Grosell, M., Gilmour, K. C., Playle, R. C., and Wood, C. M. (2004). Time course analysis of the mechanism by which silver inhibits active Na^+ and Cl^- uptake in gills of rainbow trout. *Am. J. Physiol. Regul. Integr C* **287**, R234–R242.

Nichols, D. J. (1987). Fluid volumes in rainbow trout, *Salmo gairdneri*: application of compartmental analysis. *Comp. Biochem. Physiol A* **87**, 703–709.

Nichols, J. W. (2002). Modeling the uptake and disposition of hydrophobic organic chemicals in fish using a physiologically based approach. In: *The Practical Applicability of Toxicokinetic Models in the Risk Assessment of Chemicals* (J. Kruse, H. J. M. Verhaar and W. K. de Raat, eds), pp. 109–133. Kluwer Academic Publishers, Dordrecht.

Nichols, J. W., and Playle, R. C. (2004). Influence of temperature on silver accumulation and depuration in rainbow trout. *J. Fish Biol.* **64**, 1638–1654.

Norwood, W. P., Borgmann, U., and Dixon, D. G. (2007). Interactive effects of metals in mixtures on bioaccumulation in the amphipod *Hyallela azteca*. *Aquat. Toxicol.* **86**, 255–267.

Nott, J. A., and Nicolaidou, A. (1990). Transfer of metal detoxification along marine food chains. *J. Mar. Biol. Assoc. U.K* **70**, 905–912.

Paalzow, L. K. (1995). Torsten Teorell, the Father of Pharmacokinetics. *Uppsala J. Med. Sci.* **100**, 41–46.

Pagenkopf, G. K. (1983). Gill surface interaction model for trace-metal toxicity to fishes: role of complexation, pH, and water hardness. *Environ. Sci. Technol.* **17**, 342–347.

Palmiter, R. D. (1994). Regulation of metallothionein genes by heavy metals appears to be mediated by a zinc-sensitive inhibitor that interacts with a constitutively active transcription factor, MTF-1. *Proc. Natl. Acad. Sci.U.S.A.* **91**, 1219–1223.

Paquin, P. R., Gorsuch, J. W., Apte, S., Batley, G. E., Bowles, K. C., Campbell, P. G. C., Delos, C. G., Di Toro, D. M., Dwyer, R. L., Galvez, F., Gensemer, R. W., Goss, G. G., Hogstrand, C., Janssen, C. R., McGeer, J. C., Naddy, R. B., Playle, R. C., Santore, R. C., Schneider, U., Stubblefield, W. A., Wood, C. M., and Wu, K. B. (2002a). The biotic ligand model: a historical overview. Special Issue: the biotic ligand model for metals – current research, future directions, regulatory implications. *Comp. Biochem. Physiol. C* **133**, 3–35.

Paquin, P. R., Zoltay, V., Winfield, R. P., Wu, K. B., Mathew, R., Santore, R. C., and Di Toro, D. M. (2002b). Extension of the biotic ligand model of acute toxicity to a physiologically-based model of the survival time of rainbow trout (*Oncorhynchus mykiss*) exposed to silver. *Comp. Biochem. Physiol C* **133**, 305–343.

Paquin, P. R., Farley, K. J., Santore, R. C., Kavvadas, C., Mooney, K., Winfield, R. P., Wu, K. B. and Di Toro, D. M. (2003). Bioaccumulation and toxicity models. In: *Metals in Aquatic Systems: A Review of Exposure, Bioaccumulation, and Toxicity Models*, pp. 61–90. Pensacola, FL: SETAC Press.

Paquin, P. R., Mathew, R., Damiani, D., Santore, R. C. and Farley, K. J. (2007). Modeling of copper accumulation by bivalves: analysis of caged bivalve studies. In Bioavailability and Effects of Ingested Metals on Aquatic Organisms. Water Environment Research Foundation, Project 01-ECO-4T, pp. 5-1–5-14.

Paterson, S., and Mackay, D. (1987). A steady-state fugacity-based pharmacokinetic model with simultaneous multiple exposure routes. *Environ. Toxicol. Chem.* **6**, 395–408.

Pentreath, R. J. (1973). The roles of food and water in the accumulation of radionuclides by marine teleost and elasmobranch fish. In *Radioactive Contamination of the Marine Environment*, Proceedings of a Symposium, July 10–14, 1972, Seattle, WA.Vienna: International Atomic Energy Agency.

Pentreath, R. J. (1976). The accumulation of mercury from food by the plaice, *Pleuronectes platessa* L. *J. Exp. Mar. Biol. Ecol.* **25**, 51–65.

Pentreath, R. J. (1977). The accumulation of cadmium by the plaice, *Pleuronectes platessa* L. and the thornback ray, *Raja clavata* L. *J. Exp. Mar. Biol. Ecol.* **30**, 223–232.

Playle, R. C. (2004). Using multiple metal–gill binding models and the toxic unit concept to help reconcile multiple-metal toxicity results. *Aquat. Toxicol.* **67**, 359–370.

Playle, R. C., and Wood, C. M. (1989a). Water chemistry changes in the gill micro-environment of rainbow trout: experimental observations and theory. *J. Comp. Physiol. B* **159**, 527–537.

Playle, R. C., and Wood, C. M. (1989b). Water pH and aluminum chemistry in the gill micro-environment of rainbow trout during acid and aluminum exposures. *J. Comp. Physiol. B* **159**, 539–550.

Playle, R. C., and Wood, C. M. (1990). Is precipitation of aluminum fast enough to explain aluminum deposition on fish gills. *Can. J. Fish. Aquat. Sci.* **47**, 1558–1561.

Playle, R. C., and Wood, C. M. (1991). Mechanism of aluminum extraction and accumulation at the gills of rainbow trout, *Oncorhynchus mykiss* (Walbaum), in acidic soft water. *J. Fish. Biol.* **38**, 791–805.

Playle, R. C., Gensemer, R. W., and Dixon, D. G. (1992). Copper accumulation on gills of fathead minnows: influence of water hardness, complexation and pH on the gill micro-environment. *Environ. Toxicol. Chem.* **11**, 381–391.

Playle, R. C., Dixon, D. G., and Burnison, K. (1993a). Copper and cadmium binding to fish gills: modification by dissolved organic carbon and synthetic ligands. *Can. J. Fish. Aquat. Sci.* **50**, 2667–2677.

Playle, R. C., Dixon, D. G., and Burnison, K. (1993b). Copper and cadmium binding to fish gills: estimates of metal–gill stability constants and modeling of metal accumulation. *Can. J. Fish. Aquat. Sci.* **50**, 2678–2687.

Potts, W. T. W. (1984). Transepithelial potentials in fish gills. In: *Fish Physiology* (W. S Hoarand and D. J Randall, eds), Vol. 10B, pp. 105–128. Academic Press, Orlando, FL.

Potts, W. T. W., and Parry, G. (1964). *Osmotic and Ionic Regulation in Animals*. Pergamon Press, London.

Presser, T., and Luoma, S. N. (2010). A methodology for ecosystem-scale modeling of selenium. *Integr. Environ. Assess. Manag.* **6**, 685–710.

Purkerson, D. G., Doblin, M. A., Bollens, S. M., Luoma, S. N., and Cutter, G. A. (2003). Selenium in San Francisco Bay zooplankton: potential effects of hydrodynamics and food web interactions. *Estuaries* **26**, 956–969.

QEA (2000). Bioaccumulation Model QEAFDCHN. *Version 1.0.* Montvale, NJ: Quantitative Environmental Analysts, LLC.

Rainbow, P. S. (2002). Trace metal concentrations in aquatic invertebrates: why and so what? *Environ. Pollut.* **120**, 497–507.

Redeker, E. S., and Blust, R. (2004). Accumulation and toxicity of cadmium in the aquatic oligochaete *Tubifex tubifex*: a kinetic modeling approach. *Environ. Sci. Technol.* **38**, 537–543.

Redeker, E. S., Voets, L. V., and Blust, R. (2004). Dynamic model for the accumulation of cadmium and zinc from water and sediment by the aquatic oligochaete, *Tubifex tubifex*. *Environ. Sci. Technol.* **38**, 6193–6200.

Reinfelder, J. R., and Fisher, N. S. (1994). Retention of elements absorbed by juvenile fish (*Menidia menidia, Menidia beryllina*) from zooplankton prey. *Limnol. Oceanogr.* **39**, 1783–1789.

Reinfelder, J. R., Fisher, N. S., Luoma, S. N., and Wang, W.-X. (1998). Trace element trophic transfer in aquatic organisms: a critique of the kinetic model approach. In: *Paradigms of Trace Metal Bioaccumulation in Aquatic Ecosystems* (C. J. Watras, R. G. Carlton and D. B. Porcella, eds). Special Issue, *Sci. Total Environ.* **219**, 117–135.

Robinson, W. E., and Ryan, D. K. (1986). Metal interactions within the kidney, gill and digestive gland of the hard clam, *Mercenaria mercenaria*, following laboratory exposure to cadmium. *Arch. Environ. Contam. Toxicol.* **15**, 23–30.

Roy, R. R., and Campbell, P. G. C. (1995). Survival time modeling of exposure of juvenile Atlantic salmon (Salmo salar) to mixtures of aluminum and zinc in soft water at low pH. *Aquat. Toxicol.* **33**, 155–176.

Salazar, M. H., and Salazar, S. M. (2007a). Linking bioaccumulation and biological effects to chemicals in water and sediment: a conceptual framework for freshwater bivalve ecotoxicology. In: *Freshwater Bivalve Ecotoxicology* (J. L. Farris and J. H. Van Hassel, eds), pp. 215–256. SETAC Press, Pensacola, FL; CRC Press, Boca Raton, FL.

Salazar, M. H. and Salazar, S. M. (2007b). A caged marine bivalve study in San Diego Bay using *Mytilus galloprovincialis.*In Bioavailability and Effects of Ingested Metals on Aquatic Organisms. Water Environment Research Foundation, Project 01-ECO–4T.

Santore, R. C. and Driscoll, C. T. (1995). The CHESS model for calculating chemical equilibria in soils and solutions. *Chemical Equilibrium and Reaction Models*, SSSA Special Publication **42**, 357–375. Soil Science Society of America.

Santore, R. C., Di Toro, D. M., and Paquin, P. R. (2001). A biotic ligand model of the acute toxicity of metals. II. Application to acute copper toxicity in freshwater fish and *Daphnia*. *Environ. Toxicol. Chem.* **20**, 2397–2402.

Santore, R. C., Mathew, R., Paquin, P. R., and Di Toro, D. M. (2002). Application of the biotic ligand model to predicting zinc toxicity to rainbow trout, fathead minnow, and *Daphnia magna*. *Comp. Biochem. Physiol. C* **133**, 271–285.

Schlekat, C. E., Dowdle, P. R., Lee, B.-G., Luoma, S. N., and Oremland, R. S. (2000). Bioavailability of particle-associated Se to the bivalve *Potamocorbula amurensis*. *Environ. Sci. Technol.* **34**, 4504–4510.

Schlekat, C. E., Lee, B.-G., and Luoma, S. N. (2002). Dietary metals exposure and toxicity to aquatic organisms: implications for ecological risk assessment. In: *Coastal and Estuarine Risk Assessment* (M. C. Newman, M. H. Roberts and R. C. Hale, eds), pp. 151–188. Lewis Publishers, Boca Raton, FL.

Schlekat, C. E., Purkerson, D. G., and Luoma, S. (2004). Modeling selenium bioaccumulation through arthropod food webs in San Francisco Bay, California, USA. *Environ. Toxicol. Chem.* **23**, 3003–3010.

Schultz, I. R., Peters, E. L., and Newman, M. C. (1996). Toxicokinetics and disposition of inorganic mercury and cadmium in channel catfish after extravascular administration. *Toxicol. Appl. Pharmacol.* **140**, 39–50.

Schwartz, M. L., and Vigneault, B. (2007). Development and validation of a chronic copper biotic ligand model for *Ceriodaphnia dubia*. *Aquat. Toxicol.* **84**, 247–254.

Sciera, K. L., Tomasso, J. R., and Klaine, S. J. (2004). Evaluation of acute copper toxicity to larval fathead minnows (*Pimephales promelas*) in soft surface waters. *Environ. Toxicol. Chem.* **23**, 2900–2905.

Sharpe, S., and Mackay, D. (2000). A framework for evaluating bioaccumulation in food webs. *Environ. Sci. Technol.* **34**, 2373–2379.

Singhal, R. K., Anderson, M. E., and Meister, A. (1987). Glutathione, a first line of defense against cadmium toxicity. *FASEB J* **1**, 220–223.

Solioz, M., and Odermatt, A. (1995). Copper and silver transport by CopB-ATPase in membrane vesicles of *Enterococcus hirae*. *J. Biol. Chem.* **270**, 9217–9221.

Solioz, M., Odermatt, A., and Krapf, R. (1994). Copper pumping ATPases: common concepts in bacteria and man. *FEBS Lett.* **346**, 44–47.

Steffensen, J. F., and Lomholt, J. P. (1992). The secondary vascular system. In: Fish Physiology, *Vol. 12*. Part A, The Cardiovascular System (W. S Hoar, D. J Randall and A. P Farrell, eds), pp. 185–217. Academic Press, San Diego, CA.

Stewart, A. R., Luoma, S. N., Schlekat, C. E., Doblin, M. A., and Hieb, K. A. (2004). Foodweb pathway determines how selenium affects aquatic ecosystems: a San Francisco Bay case study. *Environ. Sci. Technol.* **38**, 4519–4526.

Szumski, D. S. and Barton, D. A. (1983). Development of a mechanistic model of acute heavy metal toxicity. In *Aquatic Toxicology and Hazard Assessment: Sixth Symposium*, ASTM Special Technical Publication (STP) 802 (eds. W. E. Bishop, R. D. Cardwell and B. B. Heidolph), pp. 42–72. ASTM Publication Code Number (PCN) 04-802000-16. Philadelphia, PA: American Society for Testing and Materials.

Taylor, E. W., Beaumont, M. W., Butler, P. J., Mair, J., and Mujallid, M. S. I. (1996). Lethal and sub-lethal effects of copper upon fish: a role for ammonia toxicity. In: *Toxicology of Aquatic Pollution: Physiological, Cellular and Molecular Approaches* (E. W. Taylor, ed.), pp. 85–113. Cambridge University Press, Cambridge.

Teorell, T. (1935). Studies on the "diffusion effect" upon ionic distribution I. Some theoretical considerations. *Proc. Natl. Acad. Sci. U.S.A.* **21**, 152–161.

Teorell, T. (1937a). Studies on the diffusion effect upon ionic distribution II. Experiments on ionic accumulation. *J. Gen. Physiol.* **21**, 107–122.

Teorell, T. (1937b). Kinetics of distribution of substances administered to the body. I. The extravascular modes of administration. *Arch. Int. Pharmacodyn. Ther.* **57**, 205–225.

Teorell, T. (1937c). Kinetics of distribution of substances administered to the body. II. The intravascular modes of administration. *Arch. Int. Pharmacodyn. Ther.* **57**, 226–240.

Thomann, R. V. (1978). *Size Dependent Model of Hazardous Substances in Aquatic Food Chains.* Ecological Research Series, EPA-600/3-78-036.Duluth, MN: ERL, Office of Research and Development, US Environmental Protection Agency.

Thomann, R. V. (1979). An analysis of PCB in Lake Ontario using a size-dependent food chain model. In: *Perspectives on Lake Ecosystem Modeling* (D. Scavia and A. Robertson, eds), pp. 292–320. Ann Arbor Science, Ann Arbor, MI.

Thomann, R. V. (1981). Equilibrium model of the fate of microcontaminants in diverse aquatic food chains. *Can. J. Fish. Aquat. Sci.* **38**, 280–296.

Thomann, R. V. (1989). Bioaccumulation model of organic chemical distribution in aquatic food chains. *Environ. Sci. Technol.* **23**, 699–707.

Thomann, R. V., and St. John, J. S. (1979). The fate of PCBs in the Hudson River ecosystem. *Ann. N. Y. Acad. Sci.* **320**, 610–629.

Thomann, R. V., Szumski, D. S., Di Toro, D. M., and O'Connor, D. J. (1974). A food chain model of cadmium in western Lake Erie. *Water Res.* **8**, 841–849.

Thomann, R. V., Mueller, J. A., Winfield, R. P., and Huang, C. R. (1991). Model of the fate and accumulation of PCB homologues in Hudson Estuary. *J. Environ. Eng. ASCE* **117**, 161–177.

Thomann, R. V., Snyder, C. A., and Squibb, K. S. (1994). Development of a pharmacokinetic model for chromium in the rat following subchronic exposure: 1. The importance of incorporating long-term storage compartment. *Toxicol. Appl. Pharmacol.* **128**, 189–198.

Thomann, R. V., Shkreli, F., and Harrison, S. (1997). A pharmacokinetic model of cadmium in rainbow trout. *Environ. Toxicol. Chem.* **16**, 2268–2274.

Tipping, E. (1994). WHAM–a chemical equilibrium model and computer code for waters, sediments, and soils incorporating a discrete site/electrostatic model of ion-binding by humic substances. *Comp. Geosci.* **20**, 973–1023.

USEPA (1985). *Ambient Water Quality Criteria for Copper – 1984.* Washington, DC: Office of Water Regulations and Standards, Criteria and Standards Division, US Environmental Protection Agency.

USEPA (2004). Draft Aquatic Life Water Quality Criteria for Selenium, 2004.EPA-822-D-04-001. Washington, DC: Office of Water, Office of Science and Technology, US Environmental Protection Agency.

USEPA (2007). *2007 Update of Ambient Water Quality Criteria for Copper.* EPA 822-R-03-026. Washington, DC: Office of Water, Office of Science and Technology, US Environmental Protection Agency.

USEPA (2009). *Guidance for Implementing the* January 2001 *Methylmercury Water Quality Criterion.* EPA 823-R-09-002. Washington, DC: Office of Water, US Environmental Protection Agency.

VanderWeele, D. A. (1996).*The effects of copper pollution on the bivalve* Mytilus edulis *and the amphipod* Grandidierella japonica *in Shelter Island Yacht Basin, San Diego Bay, California.* MSc thesis, San Diego State University.

Van Genderen, E. J., Tomasso, J. R., and Klaine, S. J. (2005). Influence of multiple water-quality characteristics on copper toxicity to fathead minnows (*Pimephales promelas*). *Environ. Toxicol. Chem.* **24**, 408–414.

Van Genderen, E. J., Tomasso, J. R., and Klaine, S. J. (2008). Influence of Cu exposure on whole-body sodium levels in larval fathead minnows (*Pimephales promelas*). *Environ. Toxicol. Chem.* **27**, 1442–1449.

Van Sprang, P. A., Verdonck, F. A. M., Van Assche, F., Regoli, L., and De Schamphelaere, K. A. C. (2009). Environmental risk assessment of zinc in European freshwaters: a critical appraisal. *Sci. Total Environ.* **407**, 5373–5391.

Verhaar, H. J. M., de Wolfe, W., Dyer, S., Legierse, K. C. H. M., Seinen, W., and Hermens, J. L. M. (1999). An LC50 vs time model for the aquatic toxicity of reactive and receptor-mediated compounds. Consequences for bioconcentration kinetics and risk assessment. *Environ. Sci. Technol.* **33**, 758–763.

Viarengo, A., Pertica, M., Mancinelli, G., Palmero, S., Zanicchi, G., and Orunesu, M. (1981). Synthesis of Cu-binding proteins in different tissues of mussels exposed to the metal. *Mar. Pollut. Bull.* **12**, 347–350.

Viarengo, A., Palmero, S., Zanicchi, G., Capelli, R., Vaissiere, R., and Orunesu, M. (1985). Role of metallothioneins in Cu and Cd accumulation and elimination in the gills and digestive glands of *Mytilus galloprovincialis* Lam. *Mar. Environ. Res.* **16**, 23–36.

Wallace, W. G., and Lopez, G. R. (1996). Relationship between subcellular cadmium distribution in prey and cadmium trophic transfer to a predator. *Estuaries* **19**, 923–930.

Wallace, W. G., Lee, B.-G., and Luoma, S. N. (2003). Subcellular compartmentalization of Cd and Zn in two bivalves. I. Significance of metal-sensitive fractions (MSF) and biologically detoxified metal (BDM). *Mar. Ecol.Prog. Ser.* **249**, 183–197.

Walter, H. (2002). *Kombinationswirkungen von umweltchemikalien: Zur analyse der milieuabhängigen mischungstoxizität von kontaminanten mit unbekanntem wirkmechanismus in umweltrelevanten konzentrationen.* PhD Thesis, Martin-Luther Universität, Halle Wittenberg, Germany. http://sundoc.bibliothek.uni-halle.de/diss-online/02/02A2277/t8.pdf

Wang, W.-X. (2002). Interactions of trace metals and different marine food chains. *Mar. Ecol. Prog. Ser.* **243**, 295–309.

Wang, W. X., and Fisher, N. S. (1997a). Modeling the influence of body size on trace element accumulation in the mussel *Mytilus edulis*. *Mar. Ecol. Prog. Ser.* **161**, 103–115.

Wang, W.-X., and Fisher, N. S. (1997b). Modeling metal bioavailability for marine mussels. *Rev. Environ. Contam. Toxicol.* **151**, 39–65.

Wang, W.-X., and Fisher, N. S. (1998a). Accumulation of trace elements in a marine copepod. *Limnol. Oceanogr.* **43**, 273–283.

Wang, W.-X., and Fisher, N. S. (1998b). Excretion of trace elements by marine copepods and their bioavailability to diatoms. *J. Mar. Res.* **56**, 713–729.

Wang, W.-X., and Fisher, N. S. (1999). Delineating metal accumulation pathways for marine invertebrates. *Sci. Total Environ.* **237/238**, 459–472.

Wang, W.-X., and Ke, C. (2002). Dominance of dietary intake of cadmium and zinc by two marine predatory gastropods. *Aquat. Toxicol.* **56**, 153–165.

Wang, W. X., Fisher, N. S., and Luoma, S. N. (1996). Kinetic determinations of trace element bioaccumulation in the mussel *Mytilus edulis*. *Mar. Ecol.Prog. Ser.* **140**, 91–113.

Webb, N. A., and Wood, C. M. (1998). Physiological analysis of the stress response associated with acute silver nitrate exposure in freshwater rainbow trout (Oncorhynchus mykiss). *Environ. Toxicol. Chem.* **17**, 579–588.

Wilson, R. W., and Taylor, E. W. (1993). The physiological responses of freshwater rainbow trout, *Oncorhynchus mykiss*, during acutely lethal copper exposure. *J. Comp. Physiol. B* **163**, 38–47.

Winge, D., Krasno, J., and Colucci, A. V. (1974). Cadmium accumulation in rat liver: correlation between bound metal and pathology. In: *Trace Element Metabolism in Animals–2* (W. G. Hoekstra, J. W. Suttie, H. E. Ganther and W. Mertz, eds), pp. 500–502. University Park Press, Baltimore, MD.

Wood, C. M. (1989). The physiological problems of fish in acid waters. In *Acid Toxicity and Aquatic Animals*. In: *Soc. Exp. Biol. Seminar Series* (R. Morris, E. W. Taylor, D. J. A. Brown and J. A. Brown, eds), Vol. 34, pp. 125–152. Cambridge University Press, Cambridge.

Wood, C. M. (1992). Flux measurements as indices of H$^+$ and metal effects on freshwater fish. *Aquat. Toxicol.* **22**, 239–264.

Wood, C. M., Hogstrand, C., Galvez, F., and Munger, R. S. (1996). The physiology of waterborne silver toxicity in freshwater rainbow trout (*Oncorhynchus mykiss*) 1. The effects of ionic Ag$^+$. *Aquat. Toxicol.* **35**, 93–109.

Wu, K. W., Paquin, P. R., Navab, V., Mathew, R., Santore, R. C., and Di Toro, D. M. (2003). *Development of a Biotic Ligand Model for Nickel: Phase I.*. Water Environment Research Foundation, Alexandria, VA.

Xu, Y., and Wang, W.-X. (2002). Exposure and potential food chain transfer factor of Cd, Se and Zn in marine fish *Lutjanus argentimaculatus*. *Mar. Ecol.Prog. Ser.* **238**, 173–186.

Zalups, R. K., and Lash, L. H. (1996). Interactions between glutathione and mercury in the kidney, liver, and blood. In: *Toxicology of Metals* (L. W. Chang, L. Magos and T. Suzuki, eds), pp. 145–163. CRC Press/Lewis Publishing, Boca Raton, FL.

Zhang, L., and Wang, W. X. (2006a). Alteration of dissolved cadmium and zinc uptake kinetics by metal pre-exposure in the black sea bream (*Acanthopagrus schlegeli*). *Environ. Toxicol. Chem.* **25**, 1312–1321.

Zhang, L, and Wang, W. X. (2006b). Significance of subcellular metal distribution in prey in influencing the trophic transfer of metals in a marine fish. *Limnol. Oceanogr.* **51**, 2008–2017.

Zhang, L, and Wang, W. X. (2007). Size-dependence of the potential for metal biomagnification in early life stages of marine fish. *Environ. Toxicol. Chem.* **26**, 787–794.

Zitko, V., and Carson, W. G. (1976). A mechanism of the effects of water hardness on the lethality of heavy metals to fish. *Chemosphere* **5**, 299–303.

INDEX

This index includes entries for both Homeostasis and Toxicology of Essential Metals, Volume 31A and Homeostasis and Toxicology of Non-Essential Metals, Volume 31B. The page numbers for entries from Volume 31A will be followed by an A and the page numbers for entries for Volume 31B will be followed by a B. For example, the entry for "Acute-to-chronic ratio (ACR), 17A" would be found in page 17 of Volume 31A.

485

OTHER VOLUMES IN THE
FISH PHYSIOLOGY SERIES

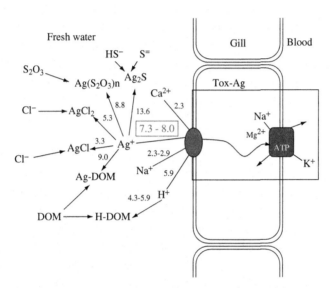

Fig. 1.8. Conceptual diagram illustrating the biotic ligand model (BLM) for Ag. The toxic mechanism (Tox-Ag) is shown as Ag^+ blocking the Mg^{2+} activation site on a basolateral Na^+/K^+-ATPase molecule (purple) in a gill ionocyte, with Ag^+ first targeting a "biotic ligand" (red) on the apical surface of the ionocyte. The numbers on the arrows represent the log K values for the various complexation and competition reactions, with the values outlined in red representing the range of log K values for Ag^+ binding to the biotic ligand used in various BLMs. Data from Janes and Playle (1995), Paquin et al. (1999), Wood et al. (1999), McGeer et al. (2000), Morgan and Wood (2004), Mann et al. (2004), and Niyogi and Wood (2004).

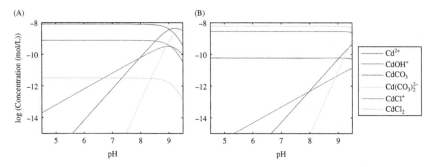

Fig. 3.1. Chemical speciation distribution diagrams for $1\ \mu g\,L^{-1}$ Cd in (A) freshwater ($1\,mM$ chloride) and (B) saltwater ($0.5\,M$ chloride). Speciation was calculated using equilibrium constants from Table 3.1 and equilibrium with atmospheric carbon dioxide.

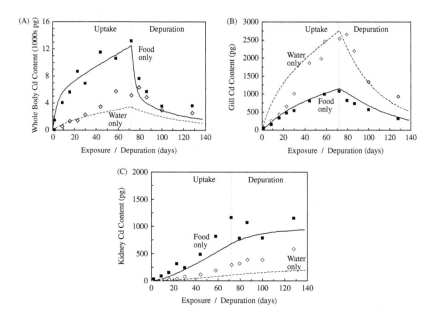

Fig. 9.4. PBPK model calibration results compared to water-only and food-only tissue Cd data for rainbow trout: (A) whole body (B) gill, and (C) kidney. Adapted from Thomann et al. (1997).

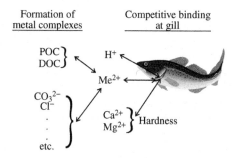

Formation of
metal complexes

Competitive binding
at gill

Fig. 9.6. Schematic of biotic ligand model (BLM) framework.

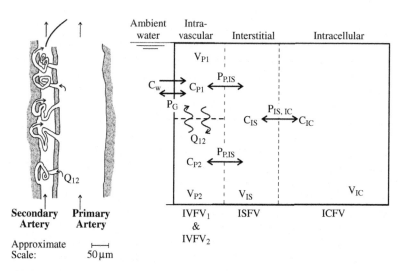

Fig. 9.8. Ion balance model (IBM): schematic of IBM fluid compartments for a freshwater fish, including details for primary and secondary vascular systems (left, after Steffensen and Lomholt, 1992) and major fluid compartments (right).

Printed in the United States
By Bookmasters